Lecture Notes in Computer Science 11316

Commenced Publication in 1973
Founding and Former Series Editors:
Gerhard Goos, Juris Hartmanis, and Jan van Leeuwen

Advanced Research in Computing and Software Science

Subline of Lecture Notes in Computer Science

More information about this series at http://www.springer.com/series/7409

George Christodoulou · Tobias Harks (Eds.)

Web and Internet Economics

14th International Conference, WINE 2018
Oxford, UK, December 15–17, 2018
Proceedings

 Springer

Editors
George Christodoulou
University of Liverpool
Liverpool, UK

Tobias Harks
Universität Augsburg
Augsburg, Germany

ISSN 0302-9743   ISSN 1611-3349 (electronic)
Lecture Notes in Computer Science
ISBN 978-3-030-04611-8   ISBN 978-3-030-04612-5 (eBook)
https://doi.org/10.1007/978-3-030-04612-5

Library of Congress Control Number: 2018961594

LNCS Sublibrary: SL3 – Information Systems and Applications, incl. Internet/Web, and HCI

This Springer imprint is published by the registered company Springer Nature Switzerland AG
The registered company address is: Gewerbestrasse 11, 6330 Cham, Switzerland

Preface

This volume contains the papers and extended abstracts presented at the 14th Conference on Web and Internet Economics (WINE 2018) held from Saturday, December 15 to Monday, December 17, 2018 in Oxford at St. Anne's College.

Over almost 20 years, researchers in theoretical computer science, artificial intelligence, and economics have joined forces to tackle problems involving incentives and computation. These problems are of particular importance in application areas like the Web and the Internet that involve large and diverse populations.

The Conference on Web and Internet Economics (WINE) is an interdisciplinary forum for the exchange of ideas and scientific progress on incentives and computation arising from these various fields. WINE 2018 built on the success of the Conference on Web and Internet Economics series (named Workshop on Internet and Network Economics until 2013), which was held annually from 2005 to 2018.

The Program Committee, consisting of 37 top researchers from the field, reviewed 119 submissions and decided to accept 36 papers. Each paper had three reviews, with additional reviews solicited as needed. We are very grateful to the Program Committee for their insightful reviews and discussions.

The review process was conducted entirely electronically via Easy Chair – we gratefully acknowledge this support. We also thank Springer for providing the proceedings and offering support for the best paper award.

The program included four invited talks by leading researchers in the field: Anna Karlin (University of Washington, USA), Paul Klemperer (University of Oxford, UK), Stefano Leonardi (Sapienza University of Rome, Italy), and Noam Nisan (Hebrew University of Jerusalem, Israel).

A special thanks go to the general chair, Paul Goldberg, to Francisco J. Marmolejo for maintaining the website, and to the organizers at St. Anne's College who provided the conference facilities. We gratefully acknowledge the sponsorship by Google and Microsoft. Last but not the least, we thank Yiannis Giannakopoulos for chairing the poster sessions.

October 2018

George Christodoulou
Tobias Harks

Organization

Program Committee

Siddharth Barman	Indian Institute of Science, India
Umang Bhaskar	Tata Institute of Fundamental Research, India
Vittorio Bilò	University of Salento, Italy
Ozan Candogan	University of Chicago, USA
Ioannis Caragiannis	University of Patras, Greece
George Christodoulou	University of Liverpool, UK
Richard Cole	New York University, USA
Jose Correa	Universidad de Chile, Chile
Shahar Dobzinski	Weizmann Institute, Israel
Paul Duetting	London School of Economics, UK
John Fearnley	University of Liverpool, UK
Aris Filos-Ratsikas	Ecole Polytechnique Fédérale de Lausanne, Switzerland
Felix Fischer	TU Berlin, Germany
Dimitris Fotakis	Yahoo Research NY and National Technical University of Athens, USA/Greece
Hu Fu	The University of British Columbia, Canada
Yiannis Giannakopoulos	Technical University of Munich, Germany
Vasilis Gkatzelis	Drexel University, USA
Yannai A. Gonczarowski	The Hebrew University of Jerusalem and Microsoft Research, Israel
Nikolai Gravin	Shanghai University of Finance and Economics, China
Tobias Harks	Augsburg University, Germany
Maria Kyropoulou	University of Essex, UK
Stefano Leonardi	Sapienza University of Rome, Italy
Pinyan Lu	Shanghai University of Finance and Economics, China
David Manlove	University of Glasgow, UK
Evangelos Markakis	Athens University of Economics and Business, Greece
Matúš Mihalák	Maastricht University, The Netherlands
Sigal Oren	Ben-Gurion University, Israel
Georgios Piliouras	Singapore University of Technology and Design, Singapore
Maria Polukarov	King's College London, UK
Qi Qi	The Hong Kong University of Science and Technology, SAR China
Aviad Rubinstein	Harvard University, USA
Daniela Saban	Stanford University, USA
Guido Schaefer	CWI Amsterdam, The Netherlands
Inbal Talgam-Cohen	Technion, Israel

Orestis Telelis	University of Piraeus, Greece
Christos Tzamos	Massachusetts Institute of Technology, USA
Marc Uetz	University of Twente, The Netherlands
Adrian Vetta	McGill University, Canada
S. Matthew Weinberg	Massachusetts Institute of Technology, USA

Additional Reviewers

Agrawal, Shubhada
Alizamir, Saed
Amanatidis, Georgios
Anagnostopoulos, Aris
Arunachaleswaran, Eshwar Ram
Bachrach, Yoram
Bailey, James
Balseiro, Santiago
Basu, Soumya
Baumeister, Dorothea
Benita, Francisco
Beyhaghi, Hedyeh
Bilò, Davide
Biswas, Arpita
Brokkelkamp, Ruben
Chen, Hongfan
Chen, Zhou
Cheung, Yun Kuen
Coey, Dominic
Cohen, Lee
de Haan, Ronald
de Jong, Jasper
de Keijzer, Bart
Deligkas, Argyrios
Dey, Palash
Dikkala, Nishanth
Essaidi, Meryem
Fanelli, Angelo
Friedler, Ophir
Gatti, Nicola
Ghosh, Abheek
Goldner, Kira
Grigoriev, Alexander
Hoefer, Martin
Hollender, Alexandros
Jin, Yaonan
Joe-Wong, Carlee

Joret, Gwenaël
Kanellopoulos, Panagiotis
Kern, Walter
Kesselheim, Thomas
Kleer, Pieter
Klimm, Max
Kominers, Scott
Koren, Moran
Kuleshov, Volodymyr
Kupfer, Ron
Lazos, Philip
Leonardos, Stefanos
Li, Minming
Li, Weian
Li, Yingkai
Lianeas, Thanasis
Lidbetter, Thomas
Lin, Yuan
Livanos, Vasileios
Loiseau, Patrick
Lundy, Taylor
Lykouris, Thodoris
Macnamara, Gregory
Mani, Ankur
Marmolejo Cossio, Francisco Javier
Meir, Reshef
Melissourgos, Themistoklis
Mnich, Matthias
Mohan, Divyarthi
Monaco, Gianpiero
Monnot, Barnabé
Moroz, Daniel
Moscardelli, Luca
Moulin, Herve
Nguyen, Thanh
Obraztsova, Svetlana
Olver, Neil

Ortega, Josue
Ovadia, Shahar
Perchet, Vianney
Plank, Benedikt
Plaut, Benjamin
Podimata, Chara
Pountourakis, Emmanouil
Procaccia, Ariel
Psomas, Alexandros
Psomas, Christos-Alexandros
Rastegari, Baharak
Rathi, Nidhi
Rios, Ignacio
Romm, Assaf
Schewior, Kevin
Schvartzman, Ariel
Shah, Nisarg
Shorrer, Ran
Simon, Sunil
Singla, Sahil
Skopalik, Alexander
Skoulakis, Stratis
Skowron, Piotr
Stamoulis, Georgios
Storms, Evan
Tang, Zhihao Gavin

Tao, Yixin
Tsikiridis, Artem
Tsipras, Dimitris
Vaish, Rohit
Velaj, Yllka
Ventre, Carmine
Venturyne, Matheus
Vinci, Cosimo
Vitercik, Ellen
Voudouris, Alexandros
Waggoner, Bo
Wang, Changjun
Wang, Hongao
Wang, Wenwei
Wang, Zihe
Wright, James
Wu, Steven
Xiao, Tao
Yan, Xiang
Zampetakis, Manolis
Zarkoob, Hedayat
Zhang, Jialin
Zhao, Alex
Zhao, Mingfei
Zhou, Kai

Contents

Regular Papers

Ordinal Approximation for Social Choice, Matching, and Facility Location Problems Given Candidate Positions

Elliot Anshelevich[✉] and Wennan Zhu[✉]

Rensselaer Polytechnic Institute, Troy, NY, USA
eanshel@cs.rpi.edu, zhuw5@rpi.edu

Abstract. In this work we consider general facility location and social choice problems, in which sets of agents $\mathcal{A}$ and facilities $\mathcal{F}$ are located in a metric space, and our goal is to assign agents to facilities (as well as choose which facilities to open) in order to optimize the social cost. We form new algorithms to do this in the presence of only *ordinal information*, i.e., when the true costs or distances from the agents to the facilities are *unknown*, and only the ordinal preferences of the agents for the facilities are available. The main difference between our work and previous work in this area is that while we assume that only ordinal information about agent preferences is known, we know the exact locations of the possible facilities $\mathcal{F}$. Due to this extra information about the facilities, we are able to form powerful algorithms which have small *distortion*, i.e., perform almost as well as omniscient algorithms but use only ordinal information about agent preferences. For example, we present natural social choice mechanisms for choosing a single facility to open with distortion of at most 3 for minimizing both the total and the median social cost; this factor is provably the best possible. We analyze many general problems including matching, k-center, and k-median, and present black-box reductions from omniscient approximation algorithms with approximation factor β to ordinal algorithms with approximation factor $1 + 2\beta$; doing this gives new ordinal algorithms for many important problems, and establishes a toolkit for analyzing such problems in the future.

1 Introduction

Many important problems involve assigning agents to facilities. For example, assigning patients to hospitals, students to universities, people to houses, etc. The target of assignment problems is usually to minimize social cost or maximize social welfare. When we consider the social cost of assignment problems, it is natural to assume the agents prefer facilities that are "closer" to them in some sense, thus the social cost of an agent is often represented by the distance between the agent and the facility it is assigned to. Besides the cost of distances, there are many other cost functions and constraints for different problems; for example, in the capacitated facility assignment problem, each facility has a maximum number of agents it can accommodate.

G. Christodoulou and T. Harks (Eds.): WINE 2018, LNCS 11316, pp. 3–20, 2018.
https://doi.org/10.1007/978-3-030-04612-5_1

In this work we consider general facility location problems, in which sets of agents $\mathcal{A}$ and facilities $\mathcal{F}$ are located in a metric space, and our goal is to assign agents to facilities (as well as choose which facilities to open) so that agents are assigned to facilities which are close to them. For example, $\mathcal{F}$ may be possible locations for opening new stores, and the goal may be that all agents have a store near them, or that the sum of agent distances to the stores they are assigned to is small, etc. This setting also captures many social choice problems, in which the facilities correspond to candidates, and the goal would be to choose a single candidate (and assign all agents to this candidate) so that the distances from the agents to the chosen candidate are small. Here the distances correspond to *spatial preferences*, i.e., the metric space represents the ideological space in which a more preferred candidate would be closer to me; see [2,17] for discussion of such spatial preferences in social choice. Our setting also captures matching and many related problems, in which we would open all facilities, but are only able to assign one agent to each facility, thus forming a matching between agents and facilities; facilities here could correspond to houses or items, for example.

If the distances between agents and facilities are known, then we can calculate the optimal solution for these assignment problems. Note that many of the facility location problems are NP-Complete, but at least it is possible to compute optimum assignments of agents to facilities (or the optimum candidates to select for social choice settings) given unlimited computational resources. For many of the settings we mentioned above, however, it is unlikely that we know the exact distances from the agents to the facilities. For social choice these distances would correspond to the cardinal preferences of voters for candidates, for example, "My cost for candidate X winning is exactly 2.35". It is far more common that only *ordinal* preferences of the agents for the candidates are known, i.e., "I prefer X to Y". Similarly, when trying to form a matching, or even in general facility location problems where we survey the agents to find out their preferences, it is much easier to elicit ordinal preferences ("I prefer to be matched with X over Y") over precise numerical preferences. These observations have recently led to a large body of work using the *utilitarian approach*, in which we assume that some latent numerical costs or utilities exist, but we only know the *ordinal* preferences of the agents, not their underlying numerical costs. See for example [2,3,10,15,20,22,28] for the social choice setting, [1,4–6] for matching and other graph problems, and [12] for facility location. These works focus on measuring the *distortion* of various algorithms: a measure of how well an algorithm behaves when using only ordinal information, as compared to the optimum algorithm which has access to the true underlying numerical information. More formally, the *distortion* [2,27] of an assignment is defined as the worst-case ratio of its social cost to the social cost of the optimal solution.

As in the work mentioned above, we assume that only ordinal information about the distances between agents and facilities is known. However, although the locations and numerical preferences of the agents are usually difficult to obtain, the locations of facilities are mostly public information. The locations of political candidates in ideological space can be reasonably well estimated based on their voting records and public statements. When forming a survey about new stores to open, we may not know exactly how much the customers would prefer one store over the other since the customer locations may be private, but the locations of the possible stores themselves are public knowledge. The main difference between our work and previous work in this area is that we assume:

While only ordinal information about agent preferences is known, we know the exact locations of the possible facilities $\mathcal{F}$.

As we discuss below, this extra information about the locations of the facilities relative to each other allows us to produce much stronger algorithms, and show much nicer bounds on distortion. In fact, in many cases, we do not even need the full information about the locations of the facilities. The main message of this paper is that having a small amount of information about the candidates in social choice settings, or the facilities in facility location, allows us to obtain solutions which are provably *close to optimal* for a large class of problems even though the only information we have about the agent preferences is ordinal, and thus it is impossible (even given unlimited computational resources) to compute the *true* optimum solution.

Our Contributions. We begin by looking at the social choice setting, in which we have a set of n agents $\mathcal{A}$ and m candidates $\mathcal{F}$ in a metric space, and we are given an ordinal ranking of each agent for the candidates. This setting was considered in e.g., [2,3,15,20,22,23,28]. In particular, for the objective of minimizing the total distance from the agents to the chosen candidate, [2] showed that Copeland and similar voting mechanisms always have distortion of at most 5, while no deterministic voting mechanism can achieve a worst-case distortion of less than 3. Finding a deterministic mechanism with distortion less than 5 has been an open problem for several years [22]. In this paper, we show that if we know the exact locations of the candidates in addition to the ordinal ranking of the agents, then there is a simple algorithm which achieves a distortion of 3, and no better bound is possible. In other words, while we do not know the true distances from agents to candidates, we can compute an outcome which is a 3-approximation *no matter what* the true distances are, as long as they are consistent with the ordinal preferences given to us. Moreover, this approximation is possible even if for each agent we are only given their favorite (i.e., top-choice) candidate: there is no need for the agents to submit a full preference ranking over all the alternatives.

We also study other objective functions in addition to minimizing the total distance from agents to the chosen alternative. We give a natural deterministic voting mechanism which has distortion at most 3 for objectives such as minimizing the median voter cost, the egalitarian objective of minimizing maximum voter cost, and many other objectives. This mechanism achieves all these approximation guarantees *simultaneously*, and moreover it does not need the exact locations of the candidates: it suffices to be given an ordinal ranking of the distances from each candidate to each other candidate. In other words, this mechanism is especially suitable for the case when candidates are a subset of voters, as our mechanism will obtain the ordinal ranking of each voter for all the candidates, and this is the only information which would be required. Note that [2] proved that *no* deterministic mechanism can achieve a distortion of better than 5 for the median objective; the reason why we are able to achieve a distortion of 3 here is precisely because we also know how each candidate ranks all the other candidates, in addition to how each voter ranks all the candidates.

We then proceed to our general facility assignment model. We are given a set of agents and a set of facilities in a metric space. The distances between facilities are given, but the distances between agents and facilities are unknown; instead we only know ordinal preferences of agents over facilities which are consistent with the true underlying distances. There could be arbitrary constraints on the assignment, such as facility capacities, or constraints enforcing that some agents cannot be (or must be) assigned to the same facility, etc. A valid assignment is to assign each agent to a facility without violating the constraints. We consider many different social cost functions to optimize. For a general class of cost functions (essentially ones which are monotone and subadditive), we give a black-box reduction which converts an algorithm for the omniscient version of this problem (i.e., the version where the true distances are known) to an ordinal algorithm with small distortion. Specifically, if we have an omniscient algorithm which always produces an assignment which is a β-approximation to the optimum, then using it we can create an ordinal algorithm which only knows the ordinal preferences of the agents instead of their true distances to the facilities, but has distortion of at most $1 + 2\beta$.

Many well-known problems fall into our facility assignment model; Table 1 summarizes some of our results. For example, classic facility location with facility costs, minimum weight bipartite matching, egalitarian bipartite matching, k-center, and k-median are all special cases. In particular our results show that if we are given unbounded computational resources, then it is always possible to form an assignment with distortion of at most 3 for these problems, and no better bound is possible simply due to the fact that we do not possess all the relevant information to compute the true optimum. This is a large improvement over previously known distortion bounds: for minimum cost ordinal matching the best-known distortion bound is n using random serial dictatorship [12]; by using the knowledge of facility locations we are able to reduce this approximation ratio to 3.

Discussion and Related Work. Ordinal approximation for the minimum social cost (or maximum social welfare) with underlying utilities/distances between agents and alternatives has been studied in many settings including social choice [2,3,10,13,15,20,22,27,28], matchings [4–6,8,12,16,21], secretary problems [25], participatory budgeting [7], general graph problems [1,4] and many other models in recent years. The general assumption of the ordinal setting is that we only have the ordinal preferences of agents over alternatives, and the goal is to form a solution that has close to optimal social cost. There are different models: social choice, matching, facility location, etc.; different objectives: minimizing social cost, maximizing social welfare, total cost objective, median

Table 1. Best known distortion of polynomial-time algorithms in different settings. "Omniscient" stands for the setting where all the distances between agents and facilities are known, and the numbers represent the best-known approximation ratios. The second column represent our setting, in which the ordinal preferences of the agents, and the numerical distances between facilities are known. The last column represents the pure ordinal setting in which only the agent ordinal preferences are known, but the distances between facilities are unknown; this setting has been previously studied, and we include the known lower bounds on the possible distortion in parentheses, including some which we prove in the full version of this paper.

	Omniscient: full distances	Agents' ordinal prefs and facility locations	Only agents' ordinal prefs (lower bounds)
Total (sum) social choice	1	3	5 (3)
Median social choice	1	3	5 (5)
Min weight bipartite matching	1	3	n (3
Egalitarian bipartite matching	1	3	- (2)
Facility location	1.488 [26]	3.976	∞ (∞)
k-center	2 [24]	5	- (-)
k-median	2.675 [11]	6.35	- ($\Omega(n)$)

objective, egalitarian objective, etc.; different assumptions on utility or cost functions: unit-sum, unit-range, metric space, etc. In this paper, we study general facility assignment problems in a metric space, and assume that the ordinal preferences of agents over alternatives are given. Unlike previous work on this topic, we also assume the locations of the alternatives are known; we show that this extra information enables us to achieve much better approximation ratios than in the pure ordinal setting for many problems.

The distortion of social choice functions was first introduced in [27], to describe the ratio between the total utility of the optimal candidate and the candidate selected by a mechanism using only ordinal preferences. [2,22,28] studied the distortion of social choice functions in a metric space; the assumption that the underlying numerical costs have this metric property allows for much better results than more general costs. In particular, for the objective of minimizing the total distance from the agents to the chosen candidate, the above papers were able to show good distortion bounds for many well-known mechanisms, in particular a bound of 5 for Copeland [2], a bound of $O(\ln m)$ for Single Transferable Vote (STV) [28], and many others. In addition, [2] proved that no deterministic mechanism can have worst-case distortion better than 3, and [28] showed that all scoring rules for m-candidates have a distortion of at least $1 + 2\sqrt{\ln m - 1}$. Goel et al. [22] showed that Ranked Pairs, and the Schulze rule have a worst-case distortion of at least 5, and the expected worst-case distortion of any (weighted)-tournament rule is at least 3. They also introduced the notion of "fairness" of social choice rules, and discussed the fairness ratio of Copeland, Randomized Dictatorship, and a general class of cost functions. Finding a deterministic mechanism with distortion less than 5 has been an open problem for several years. In this paper, we show that if we know the exact locations of the candidates in addition to the ordinal ranking of the agents, then there is a simple algorithm which achieves a distortion of 3, and no better bound is possible.

While the above work, as well as our paper, only focuses on deterministic algorithms, the distortion of randomized algorithms in social choice has also been considered, see for example [3,18,20,23]. In a slightly different flavor of result, [14,15] consider the special case where candidates are randomly and independently drawn from the set of voters. While we leave the analysis of randomized algorithms which know the location of the facilities to future work, and consider the worst-case candidate locations, it is worth pointing out that our *deterministic* algorithm achieves a distortion of 3, which is also the best known distortion bound for any *randomized* mechanism which only knows the ordinal preferences of the agents. Similarly, another common goal is to form *truthful* mechanisms with small distortion for matching and social choice, as in [5,12,20]; we focus on general mechanisms in this paper in order to understand the limitations of knowing only certain kinds of ordinal information, and leave the goal of forming truthful mechanisms for future work.

For the median objective of social choice problems, [2] showed that Copeland gives a distortion of at most 5, while *no* deterministic mechanism can achieve a distortion of better than 5. [3] also gave a randomized algorithm that has a distortion of at most 4. In this paper, we are able to improve this bound to a tight worst-case distortion of 3 by a deterministic mechanism, because we also know how each candidate ranks all the other candidates, in addition to how each voter ranks all the candidates.

The distortion of matching in a metric space has received far less attention than social choice questions. [4–6] analyzed maximum-weight metric matching; the maximization objective makes this problem far easier, and even choosing a uniformly random matching yields a distortion of a small constant. This is very different from our goal of computing a minimum-cost matching, for which no ordinal approximations better than $\Omega(n)$ are known. [12] studied facility assignment problems in a metric space; they considered the problem with or without resource augmentation, and the cases without augmentation are exactly the minimum weight bipartite matching problem. [12] showed that the approximation ratio of random serial dictatorship (RSD) is at most n, and gave a lower bound of $2^n - 1$ for the approximation ratio of serial dictatorship (SD), and a lower bound of $n^{0.26}$ for RSD. Their results are the best known ordinal approximations for this problem. In this paper, we are able to give a tight 3-approximation for the minimum weight matching problem, given the locations of facilities in addition to the agents' ordinal preferences.

2 Distortion of Social Choice Mechanisms

For the social choice problems studied in this paper, we let $\mathcal{A} = \{1, 2, \ldots, n\}$ be a set of agents, and let $\mathcal{F} = \{F_1, F_2, \ldots, F_m\}$ be a set of alternatives, which we will also refer to sometimes as *candidates* or *facilities*. We will typically use i and j to refer to agents and W, X, Y, Z to refer to alternatives. Let $\mathcal{S}$ be the set of total orders on the set of alternatives $\mathcal{F}$. Every agent $i \in \mathcal{A}$ has a preference ranking $\sigma \in \mathcal{S}$; by $X \succ_i Y$ we will mean that X is preferred over Y in ranking σ. Although we assume that each agent has a total order of preference over the alternatives and that this order is known to us, for many of our results it is only necessary that the top choice of each agent is known. We say X is i's top choice if i prefers X to every other alternative in $\mathcal{F}$. We call the vector $\sigma = (\sigma_1, \ldots, \sigma_n) \in \mathcal{S}^n$ a preference profile. We say that an alternative X pairwise defeats Y if $|\{i \in \mathcal{A} : X \succ_i Y\}| > \frac{n}{2}$. The goal is to choose a single winning alternative.

Cardinal Metric Costs. In this work we take the utilitarian view, and assume that the ordinal preferences σ are derived from underlying (latent) cardinal agent costs. Formally, we assume that there exists an arbitrary metric $d : (\mathcal{A} \cup \mathcal{F})^2 \to \mathbb{R}_{\geq 0}$ on the set of agents and alternatives. The cost incurred by agent i of alternative X being selected is represented by $d(i, X)$, which is the distance between i and X. Such spatial preferences are relatively common and well-motivated, see for example [2, 17] and the references therein. The underlying distances $d(i, X)$ are *unknown*, but unlike most previous work we *do* assume the distances between *alternatives* are given. The distance between two alternatives X and Y is denoted by $l(X, Y)$. We say that d is *consistent* with l if $\forall X, Y \in \mathcal{F}$, $d(X, Y) = l(X, Y)$.

The metric costs d naturally give rise to a preference profile. We say that d is *consistent* with σ if $\forall i \in \mathcal{A}$, $\forall X, Y \in \mathcal{F}$, if $d(i, X) < d(i, Y)$, then $X \succ_i Y$. It means that the cost of X is less than the cost of Y for agent i, so agent i prefers X over Y. As described above, we know exactly the distances l and the preferences σ, but do not know the true costs d which give rise to σ. Let $\mathcal{D}(\sigma, l)$ be the set of metrics that are consistent with σ and l; we know that one of the metrics from this possibly infinite space captures the true costs, but do not know which one.

Social Cost Distortion. We study several objective functions for social cost in this paper. First, the most common notion of social cost is the sum objective function, defined as $SC_{\Sigma}(X, \mathcal{A}) = \sum_{i \in \mathcal{A}} d(i, X)$. We also study the median objective function, $SC_{\text{med}}(X, \mathcal{A}) = \text{med}_{i \in \mathcal{A}}(d(i, X))$, as well as the egalitarian objective and many others (see Sect. 2.2). We use the notion of distortion to quantify the quality of an alternative in the worst case, similar to the notation in [10, 27]. For any alternative W, we define the distortion of W as the ratio between the social cost of W and the optimal alternative:

$$dist_\Sigma(W, \sigma, l) = \sup_{d \in \mathcal{D}(\sigma, l)} \frac{SC_\Sigma(W, \mathcal{A})}{\min_{X \in \mathcal{F}} SC_\Sigma(X, \mathcal{A})}$$

$$dist_{\mathrm{med}}(W, \sigma, l) = \sup_{d \in \mathcal{D}(\sigma, l)} \frac{SC_{\mathrm{med}}(W, \mathcal{A})}{\min_{X \in \mathcal{F}} SC_{\mathrm{med}}(X, \mathcal{A})}$$

In other words, saying that the distortion of W is at most 3 means that, no matter what the true costs d are (as long as they are consistent with the σ and l which we know), it must be that the social cost of W is within a factor of 3 of the true optimum alternative, which is impossible to compute without knowing the true costs. Because of this, a small distortion value means that there is no need to obtain the true agent costs, and the ordinal information σ (together with information l about the alternatives) is enough to form a good solution.

A social choice function f on $\mathcal{A}$ and $\mathcal{F}$ takes σ and l as input, and returns the winning alternative. We say the distortion of f is the same as the distortion of the winning alternative chosen by f on σ and l. In other words, the distortion of a social choice mechanism f on a profile σ and facility distances l is the worst-case ratio between the social cost of $W = f(\sigma, l)$, and the social cost of the true optimal alternative.

2.1 Distortion of Total Social Cost

In this section, we study the sum objective and provide a deterministic algorithm that gives a distortion of at most 3. According to [2], the lower bound on the distortion for deterministic social choice functions with only ordinal preferences (without knowing l) is 3. This occurs in the simple example with 2 alternatives which are tied with approximately half preferring each one. No matter which one is chosen, the true optimum could be the other one, and its social cost can be as much as 3 times better. Because the example in Theorem 3 from [2] only has two alternatives, knowing l does not provide any extra information, and thus that example also provides a lower bound of 3 in our setting, although we assume the distances l between facilities are known in this paper. Therefore, our mechanism achieves the best possible distortion in this setting. Note that if we only have ordinal preferences of the agents without the distances between facilities, then the best known approach so far is Copeland, which gives a distortion at most 5. Thus our results establish that by knowing the distances l between alternatives, it is possible to reduce the distortion from 5 to 3, and no better deterministic mechanism is possible.

In the following algorithm, we generate a set of projected agents as follows: Given agents $\mathcal{A}$, alternatives $\mathcal{F}$, and the preference profile σ, for each agent i denote alternative X_i as i's top choice. Then we create a new agent $\tilde{i}$ at the location of X_i in the metric space; consequently, $\forall\, Y \in \mathcal{F}$, $d(\tilde{i}, Y) = d(X_i, Y)$. Denote the set of the new agents as $\tilde{\mathcal{A}} = \{\tilde{1}, \tilde{2}, \dots, \tilde{n}\}$. For any metric d consistent with l, $d(\tilde{i}, Y) = d(X_i, Y) = l(X_i, Y)$, so the distances between agents in $\tilde{\mathcal{A}}$ and alternatives in $\mathcal{F}$ are known to us, unlike the true distances between $\mathcal{A}$ and $\mathcal{F}$.

Algorithm 1. Algorithm for the minimum total social cost.

Generate projected agent set $\tilde{\mathcal{A}}$. For each alternative $X \in \mathcal{F}$, calculate the total social cost on $\tilde{\mathcal{A}}$ by choosing X, i.e., $SC_{\sum}(X, \tilde{\mathcal{A}}) = \sum_{\tilde{i} \in \tilde{\mathcal{A}}} d(\tilde{i}, X) = \sum_{\tilde{i} \in \tilde{\mathcal{A}}} l(\tilde{i}, X)$.
Final Output: Return the alternative W that has the minimum social cost $SC_{\sum}(W, \tilde{\mathcal{A}})$.

Theorem 1. *The distortion of Algorithm 1 for minimum total social cost on $\mathcal{A}$ is at most 3.*

The proof of this is in fact a special case of our Theorem 11 for facility assignment, although it has a particularly nice clean form in this special case. We include it in the full version for completeness. Full proofs for all our results can be found in the full version of this paper at https://arxiv.org/abs/1805.03103.

2.2 Distortion of Median Social Cost

In this section, we study the median objective function, and provide a deterministic mechanism that gives a distortion of at most 3. Recall that we define the median social cost of an alternative X as $SC_{\text{med}}(X, \mathcal{A}) = \text{med}_{i \in \mathcal{A}}(d(i, X))$. We will refer to this as $\text{med}(X)$ when d and $\mathcal{A}$ are fixed. If n is even, we define median to be the $(\frac{n}{2} + 1)^{th}$ smallest value of the distances. Note that no deterministic mechanism which only knows ordinal preferences can have worst-case distortion better than 5 (Theorem 14 in [2]). With known distances between facilities, we are able to provide a natural social choice function with distortion of 3, which is also provably the best possible distortion in our setting (consider the example in Theorem 3 from [2] again). Moreover, our social choice function only uses ordinal information about the alternatives, and not the full distances l; in particular as long as we have ordinal preferences of each alternative for each other alternative (and thus a total order of the distances from each alternative to the others), then our mechanism will work properly. Such ordinal information may be easier to obtain than full distances l; for example candidates can rank all the other candidates. In particular, given agents with ordinal preferences such that the candidates are a subset of the agents, our mechanism will always form an outcome with small distortion, even if we do not know the distances l.

Note that using only agents' top choices over alternatives and the distances between alternatives, as Algorithm 1 does for the total social cost objective, is not enough to give a worst-case distortion of 3 for the median objective (see the full version). We will use the following Lemmas from [2] in the proof of our algorithm:

Lemma 2. *For any two alternatives W and Y, we have $\text{med}(W) \leq \text{med}(Y) + d(Y, W)$. [Lemma 11 in [2]]*

Lemma 3. *For any two alternatives Y and P, if P pairwise defeats (or pairwise ties) Y, then $\text{med}(Y) \geq \frac{d(Y,P)}{2}$. [Proved in Theorem 16 in [2]]*

Lemma 4. *Let W, Y be an alternatives $\in \mathcal{F}$, if W pairwise defeats (or pairwise ties) Y, then $med(W) \leq 3med(Y)$. [Proved in Theorem 8 in [2]]*

The main easy insight which we use in the formation of our algorithm comes from the following lemma.

Lemma 5. *For any three alternatives W, Y, and P, if P pairwise defeats (or pairwise ties) Y, and $d(Y, W) \leq d(Y, P)$, then $med(W) \leq 3med(Y)$.*

Proof. By Lemma 2, $med(W) \leq med(Y) + d(Y, W)$. By Lemma 3, $med(Y) \geq \frac{d(Y,P)}{2}$. And we know that $d(Y, P) \geq d(Y, W)$, thus $med(W) \leq med(Y) + d(Y, W) \leq med(Y) + d(Y, P) \leq med(Y) + 2med(Y) \leq 3med(Y)$.   □

We use a natural Condorcet-consistent algorithm to approximate the minimum median social cost with the agents' preference rankings σ and the ordinal preferences of every alternative over other alternatives. First, create the majority graph $G = (\mathcal{F}, E)$, i.e., a graph with alternatives as vertices and an edge $(Y, Z) \in E$ if Y pairwise defeats or pairwise ties Z. If a Condorcet winner (i.e. an alternative which pairwise defeats all others) exists, then we return it immediately.

Otherwise, we consider each pair of alternatives. By Lemma 4, if the edge $(W, Y) \in E$, then $med(W) \leq 3med(Y)$. When considering an alternative pair W, Y, if $(W, Y) \notin E$ and we know that there exists another alternative P which meets the conditions of Lemma 5, then we add an edge (W, Y) to G. It is not difficult to see that whenever the edge (W, Y) is in our graph, this means that $med(W) \leq 3med(Y)$. As we prove below, at the end of this process there always exists at least one alternative which has edges to all the other alternatives, and thus the distortion obtained from selecting it is at most 3, no matter which alternative is the true optimal one.

Note that from the ordinal preferences of alternatives over each other, we can get a partial order of distances between the alternatives. Denote this partial order as $\preceq$, i.e., we say that $d(W, Y) \preceq d(W, Z)$ if we know that W prefers Y to Z (we do not have information about strict preference). This is the information we have on hand: we only know the partial order of distances between pairs of alternatives which share an alternative in common. Note, however, that if there exists a cycle in this partial order, i.e., $d(Y_1, Y_2) \preceq d(Y_2, Y_3) \preceq d(Y_3, Y_4) \preceq \cdots \preceq d(Y_k, Y_1) \preceq d(Y_1, Y_2)$, then this implies that all the distances in the cycle are actually equal, and thus we can also add the relations $d(Y_1, Y_2) \succeq d(Y_2, Y_3) \succeq d(Y_3, Y_4) \succeq \cdots \succeq d(Y_k, Y_1) \succeq d(Y_1, Y_2)$. Such cycles are easy to detect (e.g., by forming a graph with a node for every alternative pair and then searching for cycles), and thus we can assume that whenever a cycle exists in our partial order, then for every pair of distances $d(W, Y)$ and $d(W, Z)$ in the cycle, we have both $d(W, Y) \preceq d(W, Z)$ and $d(W, Y) \succeq d(W, Z)$.

Lemma 6. *Consider the modified majority graph $G = (\mathcal{F}, E)$ at any point during Algorithm 2. For any edge $(W, Y) \in E$, we have that $med(W) \leq 3med(Y)$.*

Algorithm 2. Algorithm for the minimum median social cost.

If there is a Condorcet winner W, **return W as the winner**.

forall the *alternative pairs W, Y* **do**
    if $(W, Y) \notin E$ *or* $(Y, W) \notin E$ **then**
        WLOG, suppose (Y, W) exists, but (W, Y) does not exist.
        if *there exists an alternative P, such that we have $d(Y, W) \preceq d(Y, P)$ in*
        our partial order information, and P pairwise defeats (or ties) Y **then**
            Add edge (W, Y) to E;
            continue;
        end
    end
end

There must exist an alternative W such that $\forall Y \in \mathcal{F} - \{W\}$, $(W, Y) \in E$.
Return W as the winner.

Lemma 7. *At the end of Algorithm 2, there must exist an alternative W such that $\forall Y \in \mathcal{F} - \{W\}$, $(W, Y) \in E$.*

Theorem 8. *The distortion of Algorithm 2 for minimum median social cost is at most 3.*

Proof. If there is a Condorcet winner, by Lemma 4, the distortion is at most 3. Otherwise, by Lemma 7, the algorithm always returns a winner. Suppose it returns alternative W as the winner, by Lemma 6, W has a distortion at most 3 with any alternative X as the optimal solution. $\square$

Generalizing Median: Percentile Distortion

Instead of just considering the median objective, we also consider a more general objective: the α-percentile social cost. Let α-$PC(Y)$ denote the value from the set $\{d(i, Y) : i \in \mathcal{A}\}$, that α fraction of the values lie below α-$PC(Y)$. Thus median is a special case when $\alpha = \frac{1}{2}$, $\text{med}(Y) = \frac{1}{2}$-$PC(Y)$. It was shown in [2] Theorem 17 that the worst-case distortion when $\alpha \in [0, \frac{1}{2}]$ in that setting (only have agent's ordinal preferences over alternatives) is unbounded, and the same example shows $\alpha \in [0, \frac{1}{2}]$ in our setting is also unbounded. However, we are able to give a distortion of 3 for $\alpha \in [\frac{1}{2}, 1]$ in this paper, while for the setting in [2], the lower bound for distortion when $\alpha \in [\frac{1}{2}, \frac{2}{3}]$ is 5. The reason is that the ordinal preferences between alternatives are also available in our setting. We show in the full version that Algorithm 2 gives a distortion of at most 3 not only for the median objective, but also for the general α-percentile objective with $\frac{1}{2} \leq \alpha \leq 1$, because all the lemmas we used to prove the conclusion for the median objective can be generalized to α-percentile.

Theorem 9. *The distortion of Algorithm 2 for the α-PC objective social cost with $\frac{1}{2} \leq \alpha \leq 1$ is at most 3.*

Algorithm 2 and the Total Social Cost
Although Algorithm 2 is designed for the median objective, it also performs quite well for the sum objective. Interestingly, the distortion of this algorithm for the minimum total social cost is at most 5, which is the same as Copeland (the best known deterministic algorithm with no knowledge of candidate preferences). Thus this algorithm gives a distortion of 3 for median (and in fact for all α-percentile objectives) and distortion of 5 for sum simultaneously. In settings where we are not sure which objectives to optimize, or ones where we care both about the total social good, and about fairness, this social choice mechanism provides the best of both worlds. The following theorem is proved in the full version.

Theorem 10. *The distortion of Algorithm 2 for minimum total social cost is at most 5, and this bound is tight.*

3   Facility Assignment Problems

The mechanism we used for approximation of total social cost in Theorem 1 can be applied to far more general problems. In this section, we describe a set of facility assignment problems that fit in this framework. As before, let $\mathcal{A} = \{1, 2, \ldots, n\}$ be a set of agents, and $\mathcal{F} = \{F_1, F_2, \ldots, F_m\}$ be a set of facilities, with each agent i having a preference ranking σ_i over the facilities, and $\sigma = (\sigma_1, \ldots, \sigma_n)$.

As in the social choice model, we assume that there exists an arbitrary unknown metric $d : (\mathcal{A} \cup \mathcal{F})^2 \to \mathbb{R}_{\geq 0}$ on the set of agents and facilities. The distances $d(i, F_j)$ between agents and facilities are unknown, but the ordinal preferences σ and the distances l between facilities are given. Let $\mathcal{D}(\sigma, l)$ be the set of metrics consistent with σ and l, as defined previously in Sect. 2.

Unlike for social choice, our goal is now to choose which facilities to open, and which agents should be assigned to which facilities. Formally, we must choose an assignment $x : \mathcal{A} \to \mathcal{F}$, where $x(i)$ is the facility that i is assigned to. Every $i \in \mathcal{A}$ must be assigned to one (and only one) facility in $\mathcal{F}$; other than that, there could be arbitrary constraints on the assignment. Here are some examples of constraints which fall into our framework: each facility F_i has a capacity c_i, which is the maximum number of agents that can be assigned to F_i; at least (or at most) p facilities should have agents assigned to them; agents i and j must be (or must not be) assigned to the same facility, etc. The social choice model is a special case of this one with the constraint that exactly one facility must be opened, and all agents must be assigned to it. Note that the constraints are only on the assignment, and independent of the metric space d. An assignment x is valid if it satisfies all constraints. Let $\mathcal{X}$ be the set of all valid assignments.

The Cost Function of Assignments. The cost of an assignment x consists of two parts. The first part is the distance cost between agents and facilities. $\forall i \in \mathcal{A}$, let s_i denote the distance between i and the facility it is assigned to, i.e., $s_i = d(i, x(i))$. For a given metric d and assignment x, let $s(x, d)$ denote the vector of distances between each $i \in \mathcal{A}$ and $x(i)$, i.e., $s(x, d) = (s_1, s_2, \ldots, s_n)$. Let $c_d : \mathbb{R}_{\geq 0}^n \to \mathbb{R}_{\geq 0}$ be a cost function that takes a vector of distances as input. For example, this could simply sum up all the distances, take the maximum distance for an egalitarian objective, etc. To be as general as possible, instead of fixing a specific function c_d we consider the set of distance cost functions that are monotone nondecreasing and subadditive. Formally, c_d is monotonically nondecreasing means that for any vectors s and s' such that $s \leq s'$ componentwise, we have that $c_d(s) \leq c_d(s')$. Any reasonable cost function should satisfy this property if agents desire to be assigned to closer facilities. c_d being subadditive means that for any vectors s and s', we have that $c_d(s + s') \leq c_d(s) + c_d(s')$. While not all functions are subadditive, many important ones are, as they represent the concept of "economies of scale", a common property of realistic costs.

The second part of the assignment cost is the facility cost. Let $c_f(x)$ denote the facility cost for assignment x. c_f can be an *arbitrary* function over the assignments, for example, the opening cost of facilities, the penalty (or reward) for assigning certain agents to the same facility, etc. Our framework includes all such functions, and thus is quite general, as we discuss below. The main components needed for our framework to work is that the function c_f does not depend on the distances, only on x, and that the function c_d is subadditive.

The total cost $c(x, d)$ of an assignment x is the sum of the distance cost and the facility cost, i.e. $c(x, d) = c_d(s(x, d)) + c_f(x)$. We study algorithms to approximate the minimum cost assignment given only agents' ordinal preferences over facilities, and the distances between facilities, as described above.

Social Cost Distortion. As for social choice, we use the notion of distortion to measure the quality of an assignment in the worst case, similar to the notation in [10,27]. For any assignment x, we define the distortion of x as the ratio between the social cost of x and the optimal assignment:

$$dist(x, \sigma, l) = \sup_{d \in \mathcal{D}(\sigma, l)} \frac{c(x, d)}{\min_{x' \in \mathcal{X}} c(x', d)}$$

A social choice function f on $\mathcal{A}$ and $\mathcal{F}$ takes σ and l as input, and returns a valid assignment on $\mathcal{A}$ and $\mathcal{F}$. We say the distortion of f on σ and l is the same as the distortion of the assignment returned by f. In other words, the distortion of an assignment function f on a profile σ and facility distances l is the worst-case ratio between the social cost of $x = f(\sigma, l)$, and the social cost of the true optimal assignment, to obtain which we would need the true distances d.

Approximation Ratio of Omniscient Algorithms. Consider omniscient algorithms which know the true numerical distances between agents and facilities for the facility assignment problems, in other words, the metric d. In some sense, the goal of our work is to determine when algorithms with only limited information can compete with such omniscient algorithms. With the full distances information, we can of course obtain the optimal assignment using brute force, while for our algorithms with limited knowledge this is impossible even given unlimited computational resources. Nevertheless, we are also interested in what is possible to achieve if we restrict ourselves to polynomial time. To differentiate traditional approximation algorithms from algorithms with small distortion, suppose that an omniscient approximation algorithm $\tilde{f}$ returns assignment x. Then we denote the approximation ratio of a valid assignment x as:

$$ratio(x) = \frac{c(x, d)}{\min_{x' \in \mathcal{X}} c(x', d)}$$

Thus we say the approximation ratio of an omniscient algorithm $\tilde{f}$ is β if for any input of the problem, the assignment x returned by $\tilde{f}$ has $ratio(x) \leq \beta$.

3.1   Examples of Facility Assignment Problems

The total social cost problem we discussed in Sect. 2.1 is a special case of the facility assignment problem such that the constraint is only one facility (alternative) is chosen, and all agents are assigned to it. For any assignment x, the facility cost function $c_f(x) = 0$, and the distance cost function $c_d(s(x, d))$ is the sum of distances from the winning alternative to all agents in the metric d. c_d is monotone and additive (thus subadditive). Here are some other examples that fit into our framework:

Minimum Weight Metric Bipartite Matching. Given a set of agents $\mathcal{A}$ and a set of facilities $\mathcal{F}$ such that $|\mathcal{A}| = |\mathcal{F}| = n$. $G = (\mathcal{A}, \mathcal{F}, E)$ is an undirected complete bipartite graph. The facilities and agents lie in a metric space d. The weight of each edge $(i, F) \in E$ is the distance between i and F, $w(i, F) = d(i, F)$. The goal is to find a minimum weight perfect matching of the bipartite graph given only ordinal information. We can also consider Egalitarian bipartite matching, where the goal is to find a perfect matching such that maximum edge weight (instead of the total weight) in the matching is minimized [9]. Here $c_d(s(x, d))$ is the maximum edge weight in the assignment.

Metric Facility Location. In this problem, one is given a set of agents $\mathcal{A}$ and a set of facilities $\mathcal{F}$ such that $|\mathcal{A}| = n$, $|\mathcal{F}| = m$. The facilities and agents lie in a metric space d. Each facility $F_j \in \mathcal{F}$ has an opening cost f_j. Each agent is assigned to a facility; in different versions there may be capacities on the number of agents assigned to a facility, lower bounds on the number of agents assigned to a facility, or various other constraints [19]. The goal is to find a subset of facilities $\hat{\mathcal{F}} \subseteq \mathcal{F}$ to open, so that the sum of opening costs for facilities in $\hat{\mathcal{F}}$ and total distance of the assignment is minimized.

k-center and k-median. The goal in the classic k-center problem is to open a set of k facilities, with each agent assigned to the closest one. The optimal solution is the subset of $\hat{\mathcal{F}}$ which minimizes $\max_{i \in \mathcal{A}} d(i, x(i))$. In k-median, the goal is to minimize the sum of distances of agents to the facilities instead of the maximum distance. This problem can be thought of as a multi-winner election, in which we elect a committee of k candidates, and the quality of this committee for a voter depends on how well the closest candidate on the committee represents that voter.

3.2 Distortion of Facility Assignment Problems

Given agents $\mathcal{A} = \{1, 2, \ldots, n\}$ and facilities $\mathcal{F} = \{F_1, F_2, \ldots, F_m\}$, suppose facility F' is i's top choice in $\mathcal{F}$. We create a new agent $\tilde{i}$ at the location of F' in the metric space. Consequently, $\forall F \in \mathcal{F}$, $d(\tilde{i}, F) = d(F', F)$. Denote the set of the new agents as $\tilde{\mathcal{A}} = \{\tilde{1}, \tilde{2}, \ldots, \tilde{n}\}$.

The original assignment problem is on agents $\mathcal{A}$ and facilities $\mathcal{F}$, and only ordinal preferences of agents in $\mathcal{A}$ over facilities are given. The projected problem is on agents $\tilde{\mathcal{A}}$ and facilities $\mathcal{F}$, and we know the actual distances between agents in $\tilde{\mathcal{A}}$ and facilities $\mathcal{F}$, since we know the distances l between facilities. The constraints and costs c_d and c_f remain the same for both the original and the projected problem; the only difference is in the distances d. Our main result is that if we have a β-approximation assignment to the minimum assignment cost on the projected problem, then we can get an assignment that has a distortion of $2\beta + 1$ for the original problem in polynomial time.

Theorem 11. *Given a valid assignment $\tilde{x}$ for the projected problem on $\tilde{\mathcal{A}}$ and $\mathcal{F}$, with ratio($\tilde{x}$) $\leq \beta$, the assignment $x(i) = \tilde{x}(\tilde{i})$ has distortion of at most $(1 + 2\beta)$ for original assignment problem on $\mathcal{A}$ and $\mathcal{F}$.*

Proof. First, $\tilde{x}$ is a valid assignment for the projected problem on $\tilde{\mathcal{A}}$ and $\mathcal{F}$, so x must also be a valid assignment for the original problem on $\mathcal{A}$ and $\mathcal{F}$. This is because the constraints are only on the assignment, and are independent of the metric space d. For the same reason, the facility cost of x equals the facility cost of $\tilde{x}$, $c_f(x) = c_f(\tilde{x})$.

Now consider the distance cost of x. Let x^* denote the optimal assignment for the original problem. $\forall i \in \mathcal{A}$, let $s_i = d(i, x(i))$, $t_i = d(i, \tilde{i})$, $b_i = d(\tilde{i}, x(i))$. Similarly, let $s_i^* = d(i, x^*(i))$, $b_i^* = d(\tilde{i}, x^*(i))$.

For any agent i and facility $x(i)$, by triangle inequality,

$$s_i = d(i, x(i)) \leq d(i, \tilde{i}) + d(\tilde{i}, x(i)) = t_i + b_i$$

Because c_d is monotonically nondecreasing and subadditive,

$$c_d(s_1, s_2, \ldots, s_n) \leq c_d(t_1 + b_1, t_2 + b_2, \ldots, t_n + b_n)$$
$$\leq c_d(t_1, t_2, \ldots, t_n) + c_d(b_1, b_2, \ldots, b_n)$$

Therefore, the cost of our assignment x is bounded as follows:

$$c_f(x) + c_d(s(x, d)) = c_f(x) + c_d(s_1, s_2, \ldots, s_n)$$
$$= c_f(\tilde{x}) + c_d(s_1, s_2, \ldots, s_n)$$
$$\leq c_f(\tilde{x}) + c_d(t_1, t_2, \ldots, t_n) + c_d(b_1, b_2, \ldots, b_n)$$

Because $\tilde{i}$ is located at i's top choice facility, and $x^*(i)$ is a facility, we thus know that $t_i \leq s_i^*$, and by monotonicity $c_d(t_1, t_2, \ldots, t_n) \leq c_d(s_1^*, s_2^*, \ldots, s_n^*)$. Thus,

$$c_f(x) + c_d(s(x, d)) \leq c_f(\tilde{x}) + c_d(t_1, t_2, \ldots, t_n) + c_d(b_1, b_2, \ldots, b_n)$$
$$\leq c_f(\tilde{x}) + c_d(s_1^*, s_2^*, \ldots, s_n^*) + c_d(b_1, b_2, \ldots, b_n)$$

We know that $\tilde{x}$ is a β-approximation to the optimum assignment for the projected problem. Its total cost is exactly $c_f(\tilde{x}) + c_d(b_1, b_2, \ldots, b_n)$, since the distance from $\tilde{i}$ to $\tilde{x}(\tilde{i}) = x(i)$ is exactly b_i. Now consider another assignment for the projected problem, in which $\tilde{i}$ is assigned to $x^*(i)$. The cost of this assignment is $c_f(x^*) + c_d(b_1^*, b_2^*, \ldots, b_n^*)$, by definition of b_i^*. Since $\tilde{x}$ is a β-approximation, we therefore know that

$$c_f(\tilde{x}) + c_d(b_1, b_2, \ldots, b_n) \leq \beta c_f(x^*) + \beta c_d(b_1^*, b_2^*, \ldots, b_n^*),$$

and thus

$$c_f(x) + c_d(s(x, d)) \leq c_f(\tilde{x}) + c_d(s_1^*, s_2^*, \ldots, s_n^*) + c_d(b_1, b_2, \ldots, b_n)$$
$$\leq c_d(s_1^*, s_2^*, \ldots, s_n^*) + \beta c_f(x^*) + \beta c_d(b_1^*, b_2^*, \ldots, b_n^*)$$

For any agent i and facility $x^*(i)$ in x^*, by triangle inequality,

$$b_i^* = d(\tilde{i}, x^*(i)) \leq d(i, x^*(i)) + d(i, \tilde{i}) \leq 2d(i, x^*(i)) = 2s_i^*$$

$d(i, \tilde{i}) \leq d(i, x^*(i))$ above since $\tilde{i}$ is located at the closest facility to i. Because c_d is monotone and subadditive, we also have that

$$c_d(b_1^*, b_2^*, \ldots, b_n^*) \leq c_d(2s_1^*, 2s_2^*, \ldots, 2s_n^*) \leq 2c_d(s_1^*, s_2^*, \ldots, s_n^*)$$

Putting everything together,

$$c_f(x) + c_d(s(x, d)) \leq c_d(s_1^*, s_2^*, \ldots, s_n^*) + \beta c_f(x^*) + \beta c_d(b_1^*, b_2^*, \ldots, b_n^*)$$
$$\leq \beta c_f(x^*) + c_d(s_1^*, s_2^*, \ldots, s_n^*) + 2\beta c_d(s_1^*, s_2^*, \ldots, s_n^*)$$
$$= \beta c_f(x^*) + (1 + 2\beta)c_d(s_1^*, s_2^*, \ldots, s_n^*)$$
$$\leq (1 + 2\beta)(c_f(x^*) + c_d(s(x^*, d))) \qquad \square$$

Note that the above theorem immediately implies that if we are only concerned with what is possible to achieve given limited ordinal information in addition to distances between facilities, and are not worried about our algorithms running in polynomial time, then we can always form an assignment with distortion of at most 3 from knowing only σ and l. This is because we can solve the projected problem with brute force, and then we have $\beta = 1$. This bound of 3 is tight for many facility assignment problems: consider for example an instance of min-cost metric matching with two agents and two facilities, with both preferring F_1 to F_2. One of the agents has distance to F_1 of 0, and one is located halfway between F_1 and F_2, but since we only have ordinal information we do not know which agent is which. If we assign the wrong agent to F_1, then we end up with distortion of 3, and it is impossible to do better for any deterministic mechanism.

If on the other hand we want to form poly-time algorithms with small distortion, the above theorem gives a black-box reduction: if we have a β-approximation algorithm for the omniscient case, then we can form a $1 + 2\beta$-distortion algorithm for the ordinal case. Actually, we get a $1 + 2\beta$-distortion for the distance cost, and a β-distortion for the facility cost, which is shown in the second-to-last line of the proof for Theorem 11. This leads to the following (see the full version):

Corollary 12. *We can achieve the following distortion in polynomial time:*

1. *At most 3 for the minimum weight bipartite matching problem.*
2. *At most 3 for Egalitarian bipartite matching.*
3. *At most 3.976 for the facility location problem (1.488-approximation for the facility cost, and 3.976-approximation for the distance cost).*
4. *At most 5 for the k-center problem.*
5. *At most 6.35 for the k-median problem.*

Note that the median function, unlike sum and maximum, is not subadditive, and thus does not fit into our framework. In fact, while both min-cost and egalitarian matching problems have algorithms with small distortion in our setting, the same is not possible for forming a matching where the objective function is the cost of the *median* edge: see the full version for a lower bound.

Acknowledgements. We thank Onkar Bhardwaj for discussion of the lower bound example for the k-median problem. This work was partially supported by NSF award CCF-1527497.

References

1. Abramowitz, B., Anshelevich, E.: Utilitarians without utilities: maximizing social welfare for graph problems using only ordinal preferences. In: AAAI (2018)
2. Anshelevich, E., Bhardwaj, O., Postl, J.: Approximating optimal social choice under metric preferences. In: AAAI (2015)
3. Anshelevich, E., Postl, J.: Randomized social choice functions under metric preferences. In: IJCAI (2016)
4. Anshelevich, E., Sekar, S.: Blind, greedy, and random: algorithms for matching and clustering using only ordinal information. In: AAAI (2016)
5. Anshelevich, E., Sekar, S.: Truthful mechanisms for matching and clustering in an ordinal world. In: Cai, Y., Vetta, A. (eds.) WINE 2016. LNCS, vol. 10123, pp. 265–278. Springer, Heidelberg (2016). https://doi.org/10.1007/978-3-662-54110-4_19
6. Anshelevich, E., Zhu, W.: Tradeoffs between information and ordinal approximation for bipartite matching. In: Bilò, V., Flammini, M. (eds.) SAGT 2017. LNCS, vol. 10504, pp. 267–279. Springer, Cham (2017). https://doi.org/10.1007/978-3-319-66700-3_21
7. Benade, G., Nath, S., Procaccia, A.D., Shah, N.: Preference elicitation for participatory budgeting. In: AAAI (2017)
8. Bhalgat, A., Chakrabarty, D., Khanna, S.: Social welfare in one-sided matching markets without money. In: Goldberg, L.A., Jansen, K., Ravi, R., Rolim, J.D.P. (eds.) APPROX/RANDOM -2011. LNCS, vol. 6845, pp. 87–98. Springer, Heidelberg (2011). https://doi.org/10.1007/978-3-642-22935-0_8
9. Bogomolnaia, A., Moulin, H.: Random matching under dichotomous preferences. Econometrica **72**(1), 257–279 (2004)

10. Boutilier, C., Caragiannis, I., Haber, S., Lu, T., Procaccia, A.D., Sheffet, O.: Optimal social choice functions: a utilitarian view. Artif. Intell. **227**, 190–213 (2015)

11. Byrka, ., Pensyl, T., Rybicki, B., Srinivasan, A., Trinh, K.: An improved approximation for k-median, and positive correlation in budgeted optimization. In: SODA (2014)

12. Caragiannis, I., Filos-Ratsikas, A., Frederiksen, S.K.S., Hansen, K.A., Tan, Z.: Truthful facility assignment with resource augmentation: an exact analysis of serial dictatorship. In: Cai, Y., Vetta, A. (eds.) WINE 2016. LNCS, vol. 10123, pp. 236–250. Springer, Heidelberg (2016). https://doi.org/10.1007/978-3-662-54110-4_17

13. Caragiannis, I., Nath, S., Procaccia, A.D., Shah, N.: Subset selection via implicit utilitarian voting. J. Artif. Intell. Res. **58**, 123–152 (2017)

14. Cheng, Y., Dughmi, S., Kempe, D.: Of the people: voting is more effective with representative candidates. In: EC (2017)

15. Cheng, Y., Dughmi, S., Kempe, D.: On the distortion of voting with multiple representative candidates. In: AAAI (2018)

16. Christodoulou, G., Filos-Ratsikas, A., Frederiksen, S.K.S., Goldberg, P.W., Zhang, J., Zhang, J.: Social welfare in one-sided matching mechanisms. In: Osman, N., Sierra, C. (eds.) AAMAS 2016. LNCS (LNAI), vol. 10002, pp. 30–50. Springer, Cham (2016). https://doi.org/10.1007/978-3-319-46882-2_3

17. Enelow, J.M., Hinich, M.J.: The Spatial Theory of Voting: An Introduction. Cambridge University Press, New York (1984)

18. Fain, B., Goel, A., Munagala, K., Sakshuwong, S.: Sequential deliberation for social choice. In: Devanur, N.R., Lu, P. (eds.) WINE 2017. LNCS, vol. 10660, pp. 177–190. Springer, Cham (2017). https://doi.org/10.1007/978-3-319-71924-5_13

19. Farahani, R.Z., Hekmatfar, M.: Facility Location: Concepts, Models, Algorithms and Case Studies. Springer, Heidelberg (2009). https://doi.org/10.1007/978-3-7908-2151-2

20. Feldman, M., Fiat, A., Golomb, I.: On voting and facility location. In: EC (2016)

21. Filos-Ratsikas, A., Frederiksen, S.K.S., Zhang, J.: Social welfare in one-sided matchings: random priority and beyond. In: Lavi, R. (ed.) SAGT 2014. LNCS, vol. 8768, pp. 1–12. Springer, Heidelberg (2014). https://doi.org/10.1007/978-3-662-44803-8_1

22. Goel, A., Krishnaswamy, A.K., Munagala, K.: Metric distortion of social choice rules: lower bounds and fairness properties. In: EC (2017)

23. Gross, S., Anshelevich, E., Xia, L.: Vote until two of you agree: mechanisms with small distortion and sample complexity. In: AAAI (2017)

24. Hochbaum, D.S., Shmoys, D.B.: A best possible heuristic for the k-center problem. Math. Oper. Res. **10**(2), 180–184 (1985)

25. Hoefer, M., Kodric, B.: Combinatorial secretary problems with ordinal information. In: ICALP (2017)

26. Li, S.: A 1.488 approximation algorithm for the uncapacitated facility location problem. In: Aceto, L., Henzinger, M., Sgall, J. (eds.) ICALP 2011. LNCS, vol. 6756, pp. 77–88. Springer, Heidelberg (2011). https://doi.org/10.1007/978-3-642-22012-8_5

27. Procaccia, A.D., Rosenschein, J.S.: The distortion of cardinal preferences in voting. In: Klusch, M., Rovatsos, M., Payne, T.R. (eds.) CIA 2006. LNCS (LNAI), vol. 4149, pp. 317–331. Springer, Heidelberg (2006). https://doi.org/10.1007/11839354_23

28. Skowron, P.K., Elkind, E.: Social choice under metric preferences: scoring rules and STV. In: AAAI (2017)

Infinite-Duration Poorman-Bidding Games

Guy Avni[1]([✉]), Thomas A. Henzinger[1], and Rasmus Ibsen-Jensen[2]

[1] IST Austria, Klosterneuburg, Austria
guy.avni@ist.ac.at
[2] University of Liverpool, Liverpool, England

Abstract. In two-player games on graphs, the players move a token through a graph to produce an infinite path, which determines the winner or payoff of the game. Such games are central in formal verification since they model the interaction between a non-terminating system and its environment. We study *bidding games* in which the players bid for the right to move the token. Two bidding rules have been defined. In *Richman* bidding, in each round, the players simultaneously submit bids, and the higher bidder moves the token and pays the other player. *Poorman* bidding is similar except that the winner of the bidding pays the "bank" rather than the other player. While poorman reachability games have been studied before, we present, for the first time, results on *infinite-duration* poorman games. A central quantity in these games is the *ratio* between the two players' initial budgets. The questions we study concern a necessary and sufficient ratio with which a player can achieve a goal. For reachability objectives, such *threshold ratios* are known to exist for both bidding rules. We show that the properties of poorman reachability games extend to complex qualitative objectives such as parity, similarly to the Richman case. Our most interesting results concern quantitative poorman games, namely poorman mean-payoff games, where we construct optimal strategies depending on the initial ratio, by showing a connection with *random-turn based games*. The connection in itself is interesting, because it does not hold for reachability poorman games. We also solve the complexity problems that arise in poorman bidding games.

1 Introduction

Two-player infinite-duration games on graphs are a central class of games in formal verification [3] and have deep connections to foundations of logic [35]. They are used to model the interaction between a system and its environment, and the problem of synthesizing a correct system then reduces to finding a winning strategy in a graph game [33]. Theoretically, they have been widely studied.

This research was supported in part by the Austrian Science Fund (FWF) under grants S11402-N23 (RiSE/SHiNE), Z211-N23 (Wittgenstein Award), and M 2369-N33 (Meitner fellowship).

© Springer Nature Switzerland AG 2018
G. Christodoulou and T. Harks (Eds.): WINE 2018, LNCS 11316, pp. 21–36, 2018.
https://doi.org/10.1007/978-3-030-04612-5_2

For example, the problem of deciding the winner in a parity game is a rare problem that is in NP and coNP [22], not known to be in P, and for which a quasi-polynomial algorithm was only recently discovered [11].

A graph game proceeds by placing a token on a vertex in the graph, which the players move throughout the graph to produce an infinite path ("play") π. The game is zero-sum and π determines the winner or payoff. Two ways to classify graph games are according to the type of *objectives* of the players, and according to the *mode of moving* the token. For example, in *reachability games*, the objective of Player 1 is to reach a designated vertex t, and the objective of Player 2 is to avoid t. An infinite play π is winning for Player 1 iff it visits t. The simplest mode of moving is *turn based*: the vertices are partitioned between the two players and whenever the token reaches a vertex that is controlled by a player, he decides how to move the token.

We study a new mode of moving in infinite-duration games, which is called *bidding*, and in which the players *bid* for the right to move the token. The bidding mode of moving was introduced in [27,28] for reachability games, where two bidding rules were defined. The first bidding rule, called *Richman* rule (named after David Richman), is as follows: Each player has a budget, and before each move, the players submit bids simultaneously, where a bid is legal if it does not exceed the available budget. The player who bids higher wins the bidding, pays the bid to other player, and moves the token. The second bidding rule, which we focus on in this paper and which is called *poorman* bidding in [27], is similar except that the winner of the bidding pays the "bank" rather than the other player. Thus, the bid is deducted from his budget and the money is lost. Note that while the sum of budgets is constant in Richman games, in poorman games, the sum of budgets shrinks as the game proceeds.

Bidding for moving is a general concept that is relevant in any setting in which a scheduler needs to decide the order in which selfish agents perform actions. For example, in a multi-process system, a scheduler decides the order in which the processes execute. Allowing the processes to bid for moving is one method to resolve this conflict, and it ensures that processes never starve, a property that is called *fairness*. Systems that use internal currency to prevent free-riding are called "scrip systems" [24], and are popular in databases for example. Other examples in which bidding for moving can be used to determine agent ordering include multi-rounded negotiations [36], sequential auctions [29], and local search for Nash equilibria [19].

Poorman bidding is appropriate in modeling settings in which the agents pay the scheduler to gain priority. In order to accept payment, a scheduler needs to be a selfish entity. An example of such a scheduler appears in *Blockchain* technology like the one used in Bitcoin or Etherium. As another example, when the agents are buyers in a sequential auction, the scheduler models the auctioneer, and the agents' winning bids are its revenue [21]. An advantage of the poorman rule over the Richman rule is that their definition is easier to generalize to other important domains such as multi-player games, where the restriction of fixed sum of budgets in Richman bidding, is an obstacle.

A central quantity in bidding games is the *ratio* of the players' initial budgets. Formally, let $B_i \in \mathbb{R}_{\geq 0}$, for $i \in \{1, 2\}$, be Player i's initial budget. The *total initial budget* is $B = B_1 + B_2$ and Player i's *initial ratio* is B_i/B. The first question that arises in the context of bidding games is a necessary and sufficient initial ratio for a player to guarantee winning. For reachability games, it was shown in [27,28] that such *threshold ratios* exist in every Richman and poorman reachability game: for every vertex v there is a ratio $\text{Th}(v) \in [0, 1]$ such that (1) if Player 1's initial ratio exceeds $\text{Th}(v)$, he can guarantee winning, and (2) if his initial ratio is less than $\text{Th}(v)$, Player 2 can guarantee winning. This is a central property of the game, which is a form of *determinacy*, and shows that no ties can occur[1].

An interesting probabilistic connection was observed in [27,28] for reachability Richman games. For $r \in [0, 1]$, the *random turn-based game* that corresponds to a game $\mathcal{G}$ w.r.t. r, denoted $\text{RTB}^r(\mathcal{G})$, is a special case of stochastic game [17] in which, rather than bidding for moving, in each round, independently, Player 1 is chosen to move with probability r and Player 2 moves with the remaining probability of $1 - r$. The probabilistic connection is the following: the probability with which Player 1 can guarantee reaching his target in the uniform game $\text{RTB}^{0.5}(\mathcal{G})$ from a vertex v equals $1 - \text{Th}(v)$ in $\mathcal{G}$. Random-turn based games have been extensively studied (see for example the seminal paper [32]). For poorman reachability games, no such probabilistic connection is known. Moreover, such a connection is unlikely to exist since there are finite poorman games with irrational threshold ratios. The lack of a probabilistic connection makes poorman games technically more complicated.

More interesting, from the synthesis and logic perspective, are infinite winning conditions, but they have only been studied in the Richman setting previously [5]. We show, for the first time, existence of threshold ratios in qualitative poorman games with infinite winning conditions such as parity. The proof technique is similar to the one for Richman bidding: we show a linear reduction from poorman games with qualitative objectives to poorman reachability games.

Things get more interesting in poorman *mean-payoff games*, which are quantitative games; an infinite play π of the game is associated with a *payoff* $c \in \mathbb{R}_{\geq 0}$, which is Player 1's reward and Player 2's cost. Accordingly, we refer to the players in a mean-payoff game as Max and Min. The payoff of π is determined according to the weights it traverses and, as in the previous games, the bids are only used to determine whose turn it is to move. The central question in these games is: Given a value $c \in \mathbb{Q}$, what is the initial ratio that is necessary and sufficient for Max to guarantee a payoff of c? More formally, we say that c is the *value* with respect to a ratio $r \in [0, 1]$ if for every $\epsilon > 0$, we have (1) when Max's initial ratio is $r + \epsilon$, he can guarantee a payoff of at least c, and (2) intuitively, Max cannot hope for more: if Max's initial ratio is $r - \epsilon$, then Min can guarantee a payoff of at most c.

---

[1] When the initial budget of Player 1 is exactly $\text{Th}(v)$, the winner of the game depends on how we resolve draws in biddings, and our results hold for any tie-breaking mechanism.

Our most technically-involved contribution is a construction of optimal strategies in poorman mean-payoff games, which depend on the initial ratio $r \in [0, 1]$. The crux of the solution is reasoning about strongly-connected games: we first reason on the bottom strongly-connected components of a game graph and extend the solution by, intuitively, playing a reachability game in the rest of the graph. Before describing our solution, let us highlight an interesting difference between Richman and poorman rules. With Richman bidding, it is shown in [5] that a strongly-connected Richman mean-payoff game has a value that does not depend on the initial ratio and only on the structure of the game. It thus seems reasonable to guess that the initial ratio would not matter with poorman bidding as well. We show, however, that this is not the case; the higher Max's initial ratio is, the higher the payoff he can guarantee. We demonstrate this phenomenon with the following simple game. Technically, each vertex in a mean-payoff game is labeled by a weight. Consider an infinite play π. The *energy* of a prefix π^n of length n of π, denoted $E(\pi^n)$, is the sum of the weights it traverses. The payoff of π is $\liminf_{n \to \infty} E(\pi^n)/n$.

Example 1. Consider the mean-payoff bidding game that is depicted in Figure 1, where for convenience the weights are placed on the edges rather than the vertices. We take the viewpoint of Min in this example. We consider the case of $r = \frac{1}{2}$, and claim that the value with respect to $r = \frac{1}{2}$ is 0. Note that the players' choices upon winning a bid in the game are obvious, and the difficulty in devising a strategy is finding the right bids. Intuitively, Min copies Max's strategy. Suppose, for example, that Min starts with a budget of $1 + \epsilon$ and Max starts with 1, for some $\epsilon > 0$. A strategy for Min that ensures a payoff of 0 is based on a stack of numbers as follows: In round i, if the stack is empty Min bids $\epsilon \cdot 2^{-i}$, and otherwise the first number of the stack. If Min wins, he removes the first number on the stack (if non-empty). If Max wins, Min pushes Max's winning bid on the stack. It is not hard to show that Min never bids higher than the available budget. Also, we can show that every Max win is eventually matched, thus Min's queue empties infinitely often and the energy hits 0 infinitely often. Since we use $\liminf$ in the definition of the payoff, Min guarantees a non-positive payoff. Showing that Max can guarantee a non-negative payoff with an initial ratio of $\frac{1}{2} + \epsilon$ is harder, and a proof for the general case can be found in Sect. 4 (Theorem 3).

We show that the value c decreases with Max's initial ratio r. We set $r = \frac{1}{3}$. Suppose, for example, that Min's initial budget is $2 + \epsilon$ and Max's initial budget is 1. We claim that Min can guarantee a payoff of $-1/3$. His strategy is similar to the one above, only that whenever Max wins with b, Min pushes b to the stack twice. Now, every Max win is matched by two Min wins, and the claim follows.   □

In order to solve strongly-connected poorman mean-payoff games, we identify a probabilistic connection for these games. Consider such a game $\mathcal{G}$ and a ratio $r \in [0, 1]$. Recall that $\mathtt{RTB}^r(\mathcal{G})$ is a random turn-based game in which Max chooses the next move with probability r and Min with probability $1 - r$. The game

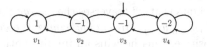

Fig. 1. A mean-payoff game.   **Fig. 2.** A second mean-payoff game.

$\text{RTB}^r(\mathcal{G})$ is a stochastic mean-payoff game, and its value, denoted $\text{MP}(\text{RTB}^r(\mathcal{G}))$, is the optimal expected payoff that the players can guarantee. The probabilistic connection we show is that the value in $\mathcal{G}$ with respect to r equals the value $\text{MP}(\text{RTB}^r(\mathcal{G}))$.

Reachability games tend to be simpler than mean-payoff games, thus we find the existence of a probabilistic connection in poorman mean-payoff games surprising given the inexistence of such a connection for poorman reachability games. A corollary of the result is that poorman mean-payoff games with initial ratio 0.5 are essentially equivalent to mean-payoff Richman games (see details in Remark 1). We are not aware of previous such connections between the two bidding rules.

Finally, we address, for the first time, complexity issues in poorman games; namely, we study the problem of finding threshold ratios in poorman games.

Due to lack of space, some proofs can be found in the full version [6].

Further Related Work. Beyond the works that are directly relevant to us, which we have compared to above, we list previous work on Richman games. To the best of our knowledge, since their introduction, poorman games have not been studied. Motivated by recreational games, e.g., bidding chess [9, 26], *discrete bidding games* are studied in [18], where the money is divided into chips, so a bid cannot be arbitrarily small unlike the bidding games we study. Non-zero-sum two-player Richman games were recently studied in [23].

2 Preliminaries

A graph game is played on a directed graph $G = \langle V, E \rangle$, where V is a finite set of vertices and $E \subseteq V \times V$ is a set of edges. The *neighbors* of a vertex $v \in V$, denoted $N(v)$, is the set of vertices $\{u \in V : \langle v, u \rangle \in E\}$, and we say that G has out-degree 2 if for every $v \in V$, we have $|N(v)| = 2$. A *path* in G is a finite or infinite sequence of vertices $v_1, v_2, \ldots$ such that for every $i \geq 1$, we have $\langle v_i, v_{i+1} \rangle \in E$.

Objectives. An objective O is a set of infinite paths. In reachability games, Player 1 has a target vertex v_R and an infinite path is winning for him if it visits v_R. In *parity* games each vertex has a parity index in $\{1, \ldots, d\}$, and an infinite path is winning for Player 1 iff the maximal parity index that is visited infinitely often is odd. We also consider games that are played on a weighted graph $\langle V, E, w \rangle$, where $w : V \to \mathbb{Q}$. Consider an infinite path $\pi = v_1, v_2, \ldots$. For $n \in \mathbb{N}$, we use π^n to denote the prefix of length n of π. We call the sum of weights that π^n traverses the *energy* of the game, denoted $E(\pi^n)$. Thus, $E(\pi^n) = \sum_{1 \leq j < n} w(v_j)$. In *energy games*, the goal of Player 1 is to keep the energy level positive, thus he wins an infinite path iff for every $n \in \mathbb{N}$, we have $E(\pi^n) > 0$. Unlike the previous objectives, a path in a *mean-payoff* game is associated with a payoff, which is Player 1's reward and Player 2's cost. Accordingly, in mean-payoff games, we refer to Player 1 as Min and Player 2 as Max. We define the payoff of π to be $\liminf_{n \to \infty} \frac{1}{n} E(\pi^n)$. We say that Max wins an infinite path of a mean-payoff game if the payoff is non-negative.

Strategies and Plays. A *strategy* prescribes to a player which *action* to take in a game, given a finite *history* of the game, where we define these two notions below. For example, in turn-based games, a strategy takes as input, the sequence of vertices that were visited so far, and it outputs the next vertex to move to. In bidding games, histories and strategies are more complicated as they maintain the information about the bids and winners of the bids. A strategy prescribes an action $\langle b, v \rangle$, where b is a bid that does not exceed the available budget and v is a vertex to move to upon winning. The winner of the bidding is the player who bids higher, where we assume there is some mechanism to resolve draws, and our results are not affected by what the mechanism is. More formally, for $i \in \{1, 2\}$, let B_i be the initial budgets of Player i, and, for a finite history π, let $W_i(\pi)$ be the sum of Player i winning bids throughout π. In Richman bidding, the winner of a bidding pays the loser, thus Player 1's budget following π is $B_1 - W_1 + W_2$. In poorman bidding, the winner pays the "bank", thus Player 1's budget following π is $B_1 - W_1$. Note that in poorman bidding, the loser's budget does not change following a bidding. An initial vertex together with two strategies for the players determine a unique infinite *play* π for the game. The vertices that π visits form an infinite path $path(\pi)$. Player 1 wins π according to an objective O iff $path(\pi) \in O$. We call a strategy f *winning* for Player 1 if for every strategy g of Player 2 the play they determine satisfies O. Winning strategies for Player 2 are defined dually. For more formal definitions see the full version.

Definition 1 (Initial ratio). *Suppose the initial budget of Player i is B_i, for $i \in \{1, 2\}$, then the total initial budget is $B = B_1 + B_2$ and Player i's initial ratio is B_i/B. We assume $B > 0$.*

The first question that arrises in the context of bidding games asks what is the necessary and sufficient initial ratio to guarantee an objective. We generalize the definition in [27,28]:

Definition 2 (Threshold ratios). *Consider a poorman or Richman game $\mathcal{G}$, a vertex v, and an initial ratio r and objective O for Player 1. The threshold ratio in v, denoted $Th(v)$, is a ratio in $[0, 1]$ such that if $r > Th(v)$, then Player 1 has a winning strategy that guarantees O is satisfied, and if $r < Th(v)$, then Player 2 has a winning strategy that violates O.*

Recall that we say that Max wins a mean-payoff game $\mathcal{G} = \langle V, E, w \rangle$ if the mean-payoff value is non-negative. Finding $Th(v)$ for a vertex v in $\mathcal{G}$ thus answers the question of what is the minimal ratio of the initial budget that guarantees winning. A more refined question asks what is the optimal payoff Max can guarantee with an initial ratio r. Formally, for a constant $c \in \mathbb{Q}$, let $\mathcal{G}^c$ be the mean-payoff game that is obtained from $\mathcal{G}$ by decreasing all weights by c.

Definition 3 (Mean-payoff values). *Consider a mean-payoff game* $\mathcal{G} = \langle V, E, w \rangle$ *and a ratio* $r \in [0,1]$. *The value of* $\mathcal{G}$ *with respect to* c, *denoted* $MP^r(\mathcal{G}, v)$, *is such that* $Th(v) = r$ *in* $\mathcal{G}^c$.

Random Turn-Based Games. In a 2.5-*player game* the vertices of the graph are partitioned between two players and a *nature* player. As in turn-based games, whenever the game reaches a vertex of Player i, for $i = 1, 2$, he choses how the game proceeds, and whenever the game reaches a vertex v that is controlled by nature, the next vertex is chosen according to a probability distribution that depends only on v. For more details see the full version.

Consider a game $\mathcal{G} = \langle V, E \rangle$. The *random-turn based game* with ratio $r \in [0,1]$ that is associated with $\mathcal{G}$ is a 2.5-player game that intuitively simulates the fact that Player 1 chooses the next move with probability r and Player 2 chooses with probability $1 - r$. Formally, we define $\text{RTB}^r(\mathcal{G}) = \langle V_1, V_2, V_N, E, \text{Pr}, w \rangle$, where each vertex in V is split into three vertices, each controlled by a different player, thus for $\alpha \in \{1, 2, N\}$, we have $V_\alpha = \{v_\alpha : v \in V\}$, nature vertices simulate the fact that Player 1 chooses the next move with probability r, thus $\text{Pr}[v_N, v_1] = r = 1 - \text{Pr}[v_N, v_2]$, and reaching a vertex that is controlled by one of the two players means that he chooses the next move, thus $E = \{\langle v_\alpha, u_N \rangle : \langle v, u \rangle \in E \text{ and } \alpha \in \{1, 2\}\}$. When $\mathcal{G}$ is weighted, then the weights of v_1, v_2, and v_N equal that of v.

The value of a 2.5-player game is a well-known to exist. We give an intuitive definition below and refer the reader for more details in the full version.

Definition 4 (Values). *Let* $r \in [0,1]$. *For a qualitative game* $\mathcal{G}$, *the value of* $RTB^r(\mathcal{G})$, *denoted* $val(RTB^r(\mathcal{G}))$, *is the probability that Player 1 wins when he plays optimally. For a mean-payoff game* $\mathcal{G}$, *the* mean-payoff value *of* $RTB^r(\mathcal{G})$, *denoted* $MP(RTB^r(\mathcal{G}))$, *is the maximal expected payoff Max obtains when he plays optimally.*

3   Poorman Parity Games

For qualitative objectives, poorman games have mostly similar properties to the corresponding Richman games. We start with reachability objectives, which were studied in [27,28]. The objective they study is slightly different than ours. We call their objective *double-reachability*: both players have targets and the game ends once one of the targets is reached. As we show below, for our purposes, the variants are equivalent since there are no draws in finite-state poorman and Richman double-reachability games.

Consider a double-reachability game $\mathcal{G} = \langle V, E, u_1, u_2 \rangle$, where, for $i = 1, 2$, the target of Player i is u_i. In both Richman and poorman bidding, trivially

Player 1 wins in u_1 with any initial budget and Player 2 wins in u_2 with any initial budget, thus $\mathrm{Th}(u_1) = 0$ and $\mathrm{Th}(u_2) = 1$. For $v \in V$, let $v^+, v^- \in N(v)$ be such that, for every $v' \in N(v)$, we have $\mathrm{Th}(v^-) \leq \mathrm{Th}(v') \leq \mathrm{Th}(v^+)$.

Theorem 1 [27,28]. *Threshold ratios exist in Richman and poorman reachability games. Moreover, consider a double-reachability game $\mathcal{G} = \langle V, E, u_1, u_2 \rangle$. In Richman bidding, for $v \in V \setminus \{u_1, u_2\}$, we have $\mathrm{Th}(v) = \frac{1}{2}\big(\mathrm{Th}(v^+) + \mathrm{Th}(v^-)\big)$, and it follows that $\mathrm{Th}(v) = val(\mathrm{RTB}^{0.5}(\mathcal{G}, v))$ and that $\mathrm{Th}(v)$ is a rational number. In poorman bidding, for $v \in V \setminus \{u_1, u_2\}$, we have $\mathrm{Th}(v) = \mathrm{Th}(v^+)/\big(1 - \mathrm{Th}(v^-) + \mathrm{Th}(v^+)\big)$.*

We continue to study poorman games with richer objectives.

Theorem 2. *Poorman parity games are linearly reducible to poorman reachability games. Specifically, threshold ratios exist in poorman parity games.*

Proof. The crux of the proof is to show that in a bottom strongly-connected component (BSCC, for short) of $\mathcal{G}$, one of the players wins with every initial budget. Thus, the threshold ratios for vertices in BSCCs are either 0 or 1. For the rest of the vertices, we construct a reachability game in which a player's goal is to reach a BSCC that is "winning" for him. The details can be found in the full version.   □

4   Poorman Mean-Payoff Games

This section consists of our most technically challenging contribution. We construct optimal strategies for the players in poorman mean-payoff games. The crux of the solution regards strongly-connected mean-payoff games, which we develop in the first three sub-sections.

Consider a strongly-connected game $\mathcal{G}$ and an initial ratio $r \in [0, 1]$. It is not hard to see that Theorem 2 implies that the value in $\mathcal{G}$ w.r.t. r does not depend on the initial vertex. We write $\mathrm{MP}^r(\mathcal{G})$ to denote the value of $\mathcal{G}$ w.r.t. r. We show the following probabilistic connection: the value $\mathrm{MP}^r(\mathcal{G})$ equals the value $\mathrm{MP}(\mathrm{RTB}^r(\mathcal{G}))$ of the random turn-based mean-payoff game $\mathrm{RTB}^r(\mathcal{G})$ in which Max chooses the next move with probability r and Min with probability $1 - r$.

4.1   Warm Up: Solving a Simple Game

In this section we solve a simple game through which we demonstrate the ideas of the general case. Recall that in an energy game, Min wins a finite play if the sum of weights it traverses, a.k.a. the energy, is 0 and Max wins an infinite play in which the energy stays positive throughout the play. Consider the game depicted in Fig. 1 and view the game as an energy game. It is shown in [27] that if the initial energy is $k \in \mathbb{N}$, then Max wins iff his initial ratio exceeds $\frac{k+2}{2k+2}$. We describe an alternative proof for the first implication.

We need several definitions. For $k \in \mathbb{N}$, let S_k be the square of area k^2. In Fig. 3, we depict S_5. We split S_k into unit-area boxes such that each of its sides contains k boxes. A diagonal in S_k splits it into a smaller black triangle and a larger white one. For $k \in \mathbb{N}$, we respectively denote by t_k and T_k the areas of the smaller black triangle and the larger white triangle of S_k. For example, we have $t_5 = 10$ and $T_5 = 15$, and in general $t_k = \frac{k(k-1)}{2}$ and $T_k = \frac{k(k+1)}{2}$.

	2	3	4	5	6
t_k	1	3	6	10	15
T_k	3	6	10	15	21

$\cdots$

Fig. 3. The square S_5 with area 25 and the sizes of some triangles.

Suppose the game starts with energy $\kappa \in \mathbb{N}$. We show that Max wins when his ratio exceeds $(\kappa + 2)/(2\kappa + 2)$, which equals $T_{\kappa+1}/(\kappa + 1)^2$. For ease of presentation, it is convenient to assume that the players' ratios add up to $1 + \epsilon_0$, Max's initial ratio is $\frac{T_{\kappa+1}}{(\kappa+1)^2} + \epsilon_0$, and Min's initial ratio is $\frac{t_{\kappa+1}}{(\kappa+1)^2}$. For $j \geq 0$, we think of ϵ_j as Max's slush fund in the j-th round of the game, though its role here is somewhat less significant than in Theorem 1. Consider a play π. We think of changes in energy throughout π and changes in budget ratio as representing two walks on two sequences. The *energy sequence* is $\mathbb{N}$ and the *budget sequence* is $\{t_k/S_k : k \in \mathbb{N}\}$, with the natural order in the two sets. We show a strategy for Max that maintains the invariant that whenever the energy is $k \in \mathbb{N}$, then Max's ratio is greater than $T_{k+1}/(k+1)^2$. That is, whenever Max wins a bidding, both sequences take a "step up" and when he loses, both sequences take a "step down".

We describe Max's strategy. Upon winning a bidding, Max chooses the $+1$ edge and we assume Min chooses the -1 edge. The challenge is to find the right bids. Suppose the energy level is k at the j-th round. Thus, Max and Min's ratio are respectively $T_{k+1}/(k+1)^2 + \epsilon_j$ and $t_{k+1}/(k+1)^2$. In other words, Min owns t_{k+1} boxes and Max owns a bit more than T_{k+1} boxes. Max's bid consists of two parts. Max bids $1/(k+1)^2 + \epsilon_j/2$, or in other words, a single box and half of his slush fund. We first show how the strategy maintains the invariant and then how it guarantees that an energy of 0 is never reached. Suppose first that Max wins the bidding. The total number of boxes decreases by one to $(k+1)^2 - 1$, his slush fund is cut by half, and Min's budget is unchanged. Thus, Max's ratio of the budget is more than $(T_{k+1} - 1)/((k+1)^2 - 1)$, which equals $T_{k+2}/(k+2)^2$. In other words, after normalization, Max owns more than T_{k+2} boxes and Min owns t_{k+2} boxes; the budget sequence takes a step up, matching the increase of 1 in the energy. The other case is when Min wins the bidding, the energy decreases by 1, and we show that the budget sequences takes a step down. Since Max bids more than one box, Min overbids, and in the worst case, he bids 1 box. Max's new ratio is more than $T_{k+1}/((k+1)^2 - 1) = T_k/k^2$. For example, let $k = 4$. Following a Max win, Max's ratio is $\frac{T_5-1}{t_5+T_5-1} = \frac{15-1}{25-1} = \frac{21}{36} = \frac{T_6}{t_6+T_6}$ and upon losing, Max's ratio is $\frac{15}{25-1} = \frac{10}{16} = \frac{T_4}{t_4+T_4}$.

We conclude by showing that the energy never reaches 0 by showing that the walk on the budget sequence never reaches the first element. Suppose the energy is $k = 1$ in the j-th round, thus according to the invariant, Max's ratio is $\frac{3}{4} + \epsilon_j$ and Min's ratio is $\frac{1}{4}$. Recall that Max bids $\frac{1}{(k+1)^2} + \epsilon_j/2$ at energy k, thus he bids $\frac{1}{4} + \epsilon_j/2$ at energy 1, and necessarily wins the bidding, implying that the energy increases.

4.2   Defining a Richer Budget Sequence

The game studied in the previous section is very simple. In this section we generalize the budget sequence that is used there so that we can handle arbitrary strongly-connected graphs. We proceed in two steps. Note that the budget sequence that is used in the previous section tends to 0.5 as the initial energy increases. It can thus be used for an initial ratio $r = 0.5$. The first generalization allows us to deal with $r \neq 0.5$. Recall the geometric intuition in the previous section. For some $k \in \mathbb{N}$, Min owns the smaller black triangle t_k and Max's owns the larger white triangle T_k. The total area of the square is $t_k + T_k$. Let $\mu, \nu \in \mathbb{Q}_{>0}$. We generalize the sequence by setting Min's budget to be μ black triangles and Max's budget to be ν white triangles. The total budget, or area, is thus $\mu \cdot t_k + \nu \cdot T_k$ and Max's initial ratio is $r = \frac{\nu \cdot T_k}{\nu \cdot T_k + \mu \cdot t_k}$. For example, set $k = 5, \mu = 2$, and $\nu = 3$. Then, Min has $2 \cdot t_5 = 2 \cdot 10$ boxes and Max has $3 \cdot T_5 = 3 \cdot 15$ boxes. It is nice to note the following property, which can of course be generalized: a Min win with a bid of 2 results in a step down, indeed $\frac{3T_5}{2t_5+3T_5-2} = \frac{3 \cdot 15}{2 \cdot 10+3 \cdot 15-2} = \frac{3 \cdot T_4}{2 \cdot t_4+3 \cdot T_4}$, and a Max win with a bid of 3 results in a step up, indeed $\frac{3 \cdot T_5-3}{2 \cdot t_5+3 \cdot T_5-3} = \frac{3 \cdot T_6}{2 \cdot t_6+3 \cdot T_6}$.

We make a second generalization. Rather than restricting to a discrete domain in which k gets values in $\mathbb{N}$, we replace k with a variable x whose domain is the real numbers. We define a function $\mathcal{R}^r : \mathbb{R} \to \mathbb{R}$ by $\mathcal{R}^r(x) = \frac{\nu \cdot T_x}{\mu \cdot t_x + \nu \cdot T_x}$. Note that $\lim_{x \to \infty} \mathcal{R}^r(x) = \frac{\nu}{\mu+\nu}$, and that the limit is reached from above.

We describe the intuition of how the following lemma is used. A play is going to induce a walk on a budget sequence $B \subseteq \mathbb{R}$. Max's strategy will ensure that whenever the walk reaches $x \in B$, Max's ratio is greater than $\mathcal{R}^r(x)$. In the first part of the lemma Min bids $\mu \cdot y$, wins the bidding, and the walk proceeds down y steps. In the second part, Max bids $\nu \cdot y$, wins the bidding, and the walk proceeds up y steps. The proof can be found in the full version.

Lemma 1. *Consider $\mu, \nu \in \mathbb{Q}_{>0}$ and $0 < y \leq 1$ such that $\mu > \nu \cdot y$ when $\nu > \mu$ and $\nu > \mu \cdot y$ when $\mu > \nu$. Then, for every $x \geq 1$ both of the following hold*

$$\frac{\nu \cdot T_x}{\mu \cdot t_x + \nu \cdot T_x - \mu \cdot y} \leq \frac{\nu \cdot T_{x-y}}{\mu \cdot t_{x-y} + \nu \cdot T_{x-y}} \quad and$$

$$\frac{\nu \cdot T_x - \nu \cdot y}{\mu \cdot t_x + \nu \cdot T_x - \nu \cdot y} \leq \frac{\nu \cdot T_{x+y}}{\mu \cdot t_{x+y} + \nu \cdot T_{x+y}}$$

4.3 The Potential and Strength of Vertices

In an arbitrary strongly-connected game the bids in the different vertices cannot be the same. In this section we develop a technique to determine the "importance" of a node v, which we call its *strength* and measures how high the bid should be in v compared with the other nodes.

Consider a strongly-connected game $\mathcal{G} = \langle V, E, w \rangle$ and $r \in [0, 1]$. Recall that $\text{RTB}^r(\mathcal{G})$ is a random-turn based game in which Max chooses the next move with probability r and Min with probability $1 - r$. A *positional strategy* is a strategy that always chooses the same action (edge) in a vertex. It is well known that there exist optimal positional strategies for both players in 2.5-player mean-payoff games.

Consider two optimal positional strategies f and g in $\text{RTB}^r(\mathcal{G})$, for Min and Max, respectively. For a vertex $v \in V$, let $v^-, v^+ \in V$ be such that $f(v_{Min}) = v^-$ and $g(v_{Max}) = v^+$. We assume w.l.o.g. that $\text{MP}(\text{RTB}^r(\mathcal{G})) = 0$. The *potential* of v, denoted $\text{Pot}^r(v)$, is a known concept in probabilistic models and its existence is guaranteed [34]. The *strength* of v is denoted $\text{St}^r(v)$.

$$\text{Pot}^r(v) = \frac{\nu \cdot \text{Pot}^r(v^+) + \mu \cdot \text{Pot}^r(v^-)}{\mu + \nu} + w(v) \text{ and}$$
$$\text{St}^r(v) = \nu\mu \cdot \frac{\text{Pot}^r(v^+) - \text{Pot}^r(v^-)}{\mu + \nu}$$

There are optimal strategies for which $\text{Pot}^r(v^-) \leq \text{Pot}^r(v') \leq \text{Pot}^r(v^+)$, for every $v' \in N(v)$, which can be found for example using the strategy iteration algorithm.

Consider a finite path $\pi = v_1, \ldots, v_n$ in $\mathcal{G}$. We intuitively think of π as a play, where for every $1 \leq i < n$, the bid of Max in v_i is $\text{St}(v_i)$ and he moves to v_i^+ upon winning. Thus, if $v_{i+1} = v_i^+$, we say that Max won in v_i, and if $v_{i+1} \neq v_i^+$, we say that Max lost in v_i. Let $W(\pi)$ and $L(\pi)$ respectively be the indices in which Max wins and loses in π. We call Max wins *investments* and Max loses *gains*, where intuitively he *invests* in increasing the energy and *gains* a higher ratio of the budget whenever the energy decreases. Let $G(\pi)$ and $I(\pi)$ be the sum of gains and investments in π, respectively, thus $G(\pi) = \sum_{i \in L(\pi)} \text{St}(v_i)$ and $I(\pi) = \sum_{i \in W(\pi)} \text{St}(v_i)$. Recall that the energy of π is $E(\pi) = \sum_{1 \leq i < n} w(v_i)$. The following lemma connects the strength, potential, and energy. Its proof can be found in the full version.

Lemma 2. *Consider a strongly-connected game $\mathcal{G}$, a ratio $r = \frac{\nu}{\mu + \nu} \in (0, 1)$ such that $\text{MP}(\text{RTB}^r(\mathcal{G})) = 0$, and a finite path π in $\mathcal{G}$ from v to u. Then, $\text{Pot}^r(v) - \text{Pot}^r(u) \leq E(\pi) + G(\pi)/\mu - I(\pi)/\nu$.*

4.4 Putting it All Together

In this section we combine the ingredients developed in the previous sections to solve arbitrary strongly-connected mean-payoff games.

Theorem 3. *Consider a strongly-connected poorman mean-payoff game $\mathcal{G}$ and a ratio $r \in [0, 1]$. The value of $\mathcal{G}$ with respect to r equals the value of the random-turn based mean-payoff game $RTB^r(\mathcal{G})$ in which Max chooses the next move with probability r, thus $MP^r(\mathcal{G}) = MP(RTB^r(\mathcal{G}))$.*

Proof. We assume w.l.o.g. that $MP(RTB^r(\mathcal{G})) = 0$ since otherwise we decrease this value from all weights. Also, the case where $r \in \{0, 1\}$ is easy since $RTB^r(\mathcal{G})$ is a graph and in $\mathcal{G}$, one of the players can win all biddings. Thus, we assume $r \in (0, 1)$. Recall that $MP(\pi) = \liminf_{n \to \infty} \frac{E(\pi^n)}{n}$. We show a Max strategy that, when the game starts from a vertex $v \in V$ and with an initial ratio of $r + \epsilon$, guarantees that the energy is bounded below by a constant, which implies $MP(\pi) \geq 0$. Showing such a strategy for Max suffices to prove $MP^r(\mathcal{G}) = 0$ since our definition for payoff favors Min.

Before we describe Max's strategy, we need several definitions. Let $S = \max_{v \in V} |St(v)|$ and $r = \frac{\nu}{\nu + \mu}$. We choose $0 < \beta \leq 1$ such that $\beta \cdot \nu \cdot S < 1$ and $\beta \cdot \mu \cdot \nu \cdot S < \frac{\mu}{\nu}$. Let $B = \{\beta \cdot i : i \in \mathbb{N}\}$. We choose $x_0 \in B$ such that Max's ratio is greater than $\mathcal{R}^r(x_0)$, which is possible since $\mathcal{R}^r$ tends to $1 - r$ from above. Suppose Max is playing according to the strategy we describe below and Min is playing according to some strategy. The play induces a walk on B, which we refer to as the *budget walk*. Max's strategy guarantees the following:

Invariant: Whenever the budget walk reaches an $x \in B$, then Max's ratio is greater than $\mathcal{R}^r(x)$.

The walk starts in x_0 and the invariant holds initially due to our choice of x_0. Suppose the token is placed on the vertex $v \in V$ and the walk reaches x. Max bids $St(v) \cdot \beta \cdot \mu \cdot \nu \cdot (D^r(x))^{-1}$, where $D^r(x)$ is the denominator of $\mathcal{R}^r(x)$, and he moves to v^+ upon winning. If Max loses, the walk proceeds down to $x - \nu \cdot St(v) \cdot \beta$, and by Lemma 1, the invariant is maintained. If Max wins, the walk proceeds up to $x + \mu \cdot St(v) \cdot \beta$, and by the other part of Lemma 1, and the invariant is maintained.

In the full version we show the following.

Claim. The budget walk is bounded: For every Min strategy, the budget walk never reaches $x = 1$.

Claim. The energy throughout a play is bounded from below. Formally, there exists a constant $c \in \mathbb{R}$ such that for every Min strategy and a finite play π, we have $E(\pi) \geq c$.

Consider a finite play π. We view π as a sequence of vertices in $\mathcal{G}$. Recall that the budget walk starts at x_0, and that $G(\pi)$ and $I(\pi)$ represent sums of strength of vertices. Suppose the budget walk reaches x following the play π, then $x = x_0 - G(\pi) \cdot \nu \cdot \beta + I(\pi) \cdot \mu \cdot \beta$. Recall that for every $v \in V$, we have $St(v) \geq -S$. Rephrasing Lemma 2, we have $\frac{-G(\pi) \cdot \nu + I(\pi) \cdot \mu}{\nu \cdot \mu} \leq 2S + E(\pi)$. Thus, $\frac{x - x_0}{\beta \mu \nu} \leq 2S + E(\pi)$. By the claim above $x \geq 1$. It follows that $\frac{1 - x_0}{\beta \mu \nu} - 2S \leq E(\pi)$, and we are done. $\square$

Remark 1. An interesting connection between poorman and Richman biddings arrises from Theorem 3. Consider a strongly-connected mean-payoff game $\mathcal{G}$. For an initial ratio $r \in [0,1]$, let $\text{MP}_\mathcal{P}^r(\mathcal{G})$ denote the value of $\mathcal{G}$ with respect to r with poorman bidding. It is shown in [5] that the value with Richman bidding does not depend on r, thus we denote it by $\text{MP}_\mathcal{R}(\mathcal{G})$. Moreover, $\text{MP}_\mathcal{R}(\mathcal{G})$ equals the value in the RTB in which the players are selected uniformly, thus $\text{MP}_\mathcal{R}(\mathcal{G}) = \text{MP}(\text{RTB}^{0.5}(\mathcal{G}))$. Our results show that poorman games with initial ratio 0.5 coincide with Richman games. Indeed, we have $\text{MP}_\mathcal{R}(\mathcal{G}) = \text{MP}_\mathcal{P}^{0.5}(\mathcal{G})$. To the best of our knowledge such a connection between the two bidding rules has not been identified before.

4.5   Extention to General Mean-Payoff Gamesf

We extend the solution in the previous sections to general graphs in a similar manner to the qualitative case; we first reason about the BSCCs of the graph and then construct an appropriate reachability game on the rest of the vertices. Formally, a *generalized reachability game* is $\mathcal{H} = \langle V, E, \langle u_i, r_i \rangle_{1 \leq i \leq m} \rangle$, where, for $1 \leq i \leq m$, we have $u_i \in V$. Player 1 wins a path in $\mathcal{H}$ iff it visits some u_i and when it visits u_i, Player 1's ratio is at least r_i. Consider a poorman mean-payoff game $\mathcal{G} = \langle V, E, w \rangle$. Recall that, for a vertex $v \in V$, the ratio $\text{Th}(v)$ is a necessary and sufficient initial ratio to guarantee a payoff of 0. Let S be a BSCC of $\mathcal{G}$ and let $\mathcal{G}_S$ be the restriction of $\mathcal{G}$ to S. If there is an $r \in [0,1]$ such that $\text{MP}^r(\mathcal{G}_S) = 0$, then by Theorem 3, for $v \in S$, we have $\text{Th}(v) = r$. Otherwise, either $\text{Th}(v) = 0$ or $\text{Th}(v) = 1$. We obtain the generalized reachability game that corresponds with $\mathcal{G}$ by replacing every BSCC S in $\mathcal{G}$ with a pair $\langle u_S, r \rangle$. It is not hard to generalize the proof of Theorem 1 to generalized reachability poorman games and obtain the following.

Theorem 4. *The threshold ratios in a poorman mean-payoff game $\mathcal{G}$ coincide with the threshold ratios in the generalized reachability game that corresponds to $\mathcal{G}$.*

5   Computational Complexity

We study the complexity of finding the threshold ratios in poorman games. We formalize this search problem as the following decision problem.

THRESH-BUD. Given a bidding game $\mathcal{G}$, a vertex v, and a ratio $r \in [0,1] \cap \mathbb{Q}$, decide whether $\text{Th}(v) \geq r$.

Theorem 5. *For poorman parity games, THRESH-BUD is in PSPACE. For poorman mean-payoff games, it is in PSPACE, for strongly-connected games, it is in NP and coNP, and for strongly-connected games with out-degree 2, THRESH-BUD is in P.*

Proof. To show membership in PSPACE for parity games, we guess the optimal moves for the two players. To verify the guess, we construct a program of the *existential theory of the reals* that uses the relation between the threshold ratios that is described in Theorem 1. Deciding whether such a program has a solution is known to be in PSPACE [12]. For mean-payoff games, membership in PSPACE is obtained similarly: We reduce the problem of solving general strongly-connected games, to solving 2.5-player mean-payoff games. The more challenging case is the solution for strongly-connected games with out-degree 2. There, we observe that the 2.5-player game is actually an MDP, which we can solve in polynomial time. See details in the full version.   □

6   Discussion

We studied for the first time infinite-duration poorman bidding games. We show the existence of threshold ratios for poorman games with qualitative objectives and give, to the best of our knowledge, the first complexity upper bounds on finding threshold ratios. For poorman mean-payoff games, we construct optimal strategies with respect to the initial ratio of the budgets and show a probabilistic connection for these games.

Historically, poorman bidding has been studied less than Richman bidding, but the reason was technical difficulty, not lack of motivation. On the contrary, we believe that poorman bidding is as motivated as Richman bidding, if not more so, particularly since they are easier to generalize. Poorman bidding has been less approachable since, e.g., poorman reachability games do not necessarily have rational threshold ratios. We expect that the structure we find here, namely the probabilistic connection for poorman bidding, will make these game more approachable and assist in introducing concepts like multiple-players, recharging stations, and partial information to bidding games, which are hard to add to Richman bidding.

This work belongs to a line of works that transfer concepts and ideas between the areas of formal verification and algorithmic game theory [31]. Examples of works in the intersection of the two fields include logics for specifying multi-agent systems [2,15,30], studies of equilibria in games related to synthesis and repair problems [1,13,14,20], non-zero-sum games in formal verification [10,16], and applying concepts from formal methods to *resource allocation games* such as rich specifications [8], efficient reasoning about very large games [4,25], and a dynamic selection of resources [7].

References

1. Almagor, S., Avni, G., Kupferman, O.: Repairing multi-player games. In: Proceedings of the 26th CONCUR, pp. 325–339 (2015)
2. Alur, R., Henzinger, T.A., Kupferman, O.: Alternating-time temporal logic. J. ACM **49**(5), 672–713 (2002)

3. Apt, K.R., Grädel, E.: Lectures in Game Theory for Computer Scientists. Cambridge University Press, Cambridge (2011)

4. Avni, G., Guha, S., Kupferman, O.: An abstraction-refinement methodology for reasoning about network games. In: Proceedings of the 26th IJCAI, pp. 70–76 (2017)

5. Avni, G., Henzinger, T.A., Chonev, V.: Infinite-duration bidding games. In: Proceedings of the 28th CONCUR, vol. 85 of LIPIcs, pp. 21:1–21:18 (2017)

6. Avni, G., Henzinger, T.A., Ibsen-Jensen, R.: Infinite-duration poorman-bidding games. CoRR, abs/1804.04372, (2018). arXiv:1804.04372

7. Avni, G., Henzinger, T.A., Kupferman, O.: Dynamic resource allocation games. In: Gairing, M., Savani, R. (eds.) SAGT 2016. LNCS, vol. 9928, pp. 153–166. Springer, Heidelberg (2016). https://doi.org/10.1007/978-3-662-53354-3_13

8. Avni, G., Kupferman, O., Tamir, T.: Network-formation games with regular objectives. Inf. Comput. **251**, 165–178 (2016)

9. Bhatt, J., Payne, S.: Bidding chess. Math. Intell. **31**, 37–39 (2009)

10. Brihaye, T., Bruyère, V., De Pril, J., Gimbert, H.: On subgame perfection in quantitative reachability games. Log. Methods Comput. Sci. **9**(1) (2012). https://doi.org/10.2168/LMCS-9(1:7)2013

11. Calude, C., Jain, S., Khoussainov, B., Li, W., Stephan, F.: Deciding parity games in quasipolynomial time. In: Proceedings of the 49th STOC (2017)

12. Canny, J.F.: Some algebraic and geometric computations in PSPACE. In: Proceedings of the 20th STOC, pp. 460–467 (1988)

13. Chatterjee, K.: Nash equilibrium for upward-closed objectives. In: Ésik, Z. (ed.) CSL 2006. LNCS, vol. 4207, pp. 271–286. Springer, Heidelberg (2006). https://doi.org/10.1007/11874683_18

14. Chatterjee, K., Henzinger, T.A., Jurdzinski, M.: Games with secure equilibria. Theor. Comput. Sci. **365**(1–2), 67–82 (2006)

15. Chatterjee, K., Henzinger, T.A., Piterman, N.: Strategy logic. Inf. Comput. **208**(6), 677–693 (2010)

16. Chatterjee, K., Majumdar, R., Jurdziński, M.: On nash equilibria in stochastic games. In: Marcinkowski, J., Tarlecki, A. (eds.) CSL 2004. LNCS, vol. 3210, pp. 26–40. Springer, Heidelberg (2004). https://doi.org/10.1007/978-3-540-30124-0_6

17. Condon, A.: On algorithms for simple stochastic games. In: Proceedings of the DIMACS, pp. 51–72 (1990)

18. Develin, M., Payne, S.: Discrete bidding games. Electron. J. Combin. **17**(1), R85 (2010)

19. Feldman, M., Snappir, Y., Tamir, T.: The efficiency of best-response dynamics. In: Bilò, V., Flammini, M. (eds.) SAGT 2017. LNCS, vol. 10504, pp. 186–198. Springer, Cham (2017). https://doi.org/10.1007/978-3-319-66700-3_15

20. Fisman, D., Kupferman, O., Lustig, Y.: Rational synthesis. In: Proceedings of the 16th TACAS, pp. 190–204 (2010)

21. Huang, Z., Devanur, N.R., Malec, D.: Sequential auctions of identical items with budget-constrained bidders. CoRR, abs/1209.1698 (2012)

22. Jurdzinski, M.: Deciding the winner in parity games is in up ∩ co-up. Inf. Process. Lett. **68**(3), 119–124 (1998)

23. Kalai, G., Meir, R., Tennenholtz, M.: Bidding games and efficient allocations. In: Proceedings of the 16th EC, pp. 113–130 (2015)

24. Kash, I.A., Friedman, E.J., Halpern, J.Y.: Optimizing scrip systems: crashes, altruists, hoarders, sybils and collusion. Distrib. Comput. **25**(5), 335–357 (2012)

25. Kupferman, O., Tamir, T.: Hierarchical network formation games. In: Legay, A., Margaria, T. (eds.) TACAS 2017. LNCS, vol. 10205, pp. 229–246. Springer, Heidelberg (2017). https://doi.org/10.1007/978-3-662-54577-5_13

26. Larsson, U., Wästlund, J.: Endgames in bidding chess. Games No Chance **5**, 70 (2018)

27. Lazarus, A.J., Loeb, D.E., Propp, J.G., Stromquist, W.R., Ullman, D.H.: Combinatorial games under auction play. Games Econ. Behav. **27**(2), 229–264 (1999)

28. Lazarus, A.J., Loeb, D.E., Propp, J.G., Ullman, D.: Richman games. Games No Chance **29**, 439–449 (1996)

29. Leme, R.P., Syrgkanis, V., Tardos, É.: Sequential auctions and externalities. In: Proceedings of the 23rd SODA, pp. 869–886 (2012)

30. Mogavero, F., Murano, A., Perelli, G., Vardi, M.Y.: Reasoning about strategies: on the model-checking problem. ACM Trans. Comput. Log. **15**(4), 34:1–34:47 (2014)

31. Nisan, N., Roughgarden, T., Tardos, E., Vazirani, V.: Algorithmic Game Theory. Cambridge University Press, Cambridge (2007)

32. Peres, Y., Schramm, O., Sheffield, S., Wilson, D.B.: Tug-of-war and the infinity laplacian. J. Am. Math. Soc. **22**, 167–210 (2009)

33. Pnueli, A., Rosner, R.: On the synthesis of a reactive module. In: Proceedings of the 16th POPL, pp. 179–190 (1989)

34. Puterman, M.L.: Markov Decision Processes: Discrete Stochastic Dynamic Programming. Wiley, New York (2005)

35. Rabin, M.O.: Decidability of second order theories and automata on infinite trees. Trans. AMS **141**, 1–35 (1969)

36. Winter, E.: Negotiations in multi-issue committees. J. Public Econ. **65**(3), 323–342 (1997)

Incentives and Coordination in Bottleneck Models

Moshe Babaioff[1] and Sigal Oren[2(✉)]

[1] Microsoft Research, Herzliya, Israel
moshe@microsoft.com
[2] Ben-Gurion University of the Negev, Be'er Sheva, Israel
sigal3@gmail.com

Abstract. We study a variant of Vickrey's classic bottleneck model. In our model there are n agents and each agent strategically chooses when to join a first-come-first-served observable queue. Agents dislike standing in line and they take actions in discrete time steps: we assume that each agent has a cost of 1 for every time step he waits before joining the queue and a cost of $w > 1$ for every time step he waits in the queue. At each time step a single agent can be processed. Before each time step, every agent observes the queue and strategically decides whether or not to join, with the goal of minimizing his expected cost.

In this paper we focus on symmetric strategies which are arguably more natural as they require less coordination. This brings up the following twist to the usual price of anarchy question: what is the main source for the inefficiency of symmetric equilibria? is it the players' strategic behavior or the lack of coordination?

We present results for two different parameter regimes that are qualitatively very different: (i) when w is fixed and n grows, we prove a tight bound of 2 and show that the entire loss is due to the players' selfish behavior (ii) when n is fixed and w grows, we prove a tight bound of $\Theta\left(\sqrt{\frac{w}{n}}\right)$ and show that it is mainly due to lack of coordination: the same order of magnitude of loss is suffered by any symmetric profile.

1 Introduction

William Vickrey is well known for his fundamental contributions to Mechanism Design, including the celebrated second price auction. However, his contributions were not limited to Mechanism Design. In a seminal paper from 1969, Vickrey [31] identifies bottlenecks as a significant reason for traffic congestion. Bottlenecks are short road segments with a fixed capacity. Once the capacity is reached a queue is formed. Vickrey presents a rush hour model – there are many employees that need to get to work around the same time, all need to cross the same bridge. It is assumed that they have some cost associated with each minute

We thank Refael Hassin, Moshe Haviv, Ella Segev and the participants of the "Queuing and Games" Seminar at TAU for useful comments on this manuscript. The full version of the paper (including all proofs) can be found at https://arxiv.org/abs/1808.00034.

G. Christodoulou and T. Harks (Eds.): WINE 2018, LNCS 11316, pp. 37–50, 2018.
https://doi.org/10.1007/978-3-030-04612-5_3

they arrive early to work and a potentially different cost associated with each minute they arrive late to work. Moreover, they have a different (and usually higher cost) for every minute they wait in traffic. Given these costs the employees need to decide when to leave for work in order to minimize their total cost. Similar timing decisions appear in many other situations: e.g., deciding when to go to the doctor or when to enter a traffic intersection.

Vickrey's paper has inspired a line of work analyzing variants of this model both in economics and transportation theory (for example, [1,3,5]). Vickrey, as common in the literature, assumes that the population is continuous. This assumption considerably simplifies the analysis and is motivated by the observation that in large populations the externalities that any single agent is imposing on the rest are negligible. The clear downside of this assumption is that it fails to model scenarios with relatively small population. Moreover, models of discrete and continuous populations can behave differently, as the analysis of the price of anarchy (PoA) in routing games with high-degree polynomials demonstrated. Specifically, while for continuous population the PoA is linear in the degree, it is exponential for discrete population ([4,28,29]).

Unlike Vickrey, we study a discrete population model. One of the choices we need to make is whether the agents observe the state of the traffic or not (this makes no difference in Vickrey's model where the population is modeled as a continuum). The few works that did study discrete variants of Vickrey's model (for example [23,26]), all made the assumption that no traffic information is provided (i.e., the queue is unobservable). However, technological advancements such as webcams that are installed over bridges, as well as mobile apps that provide information regarding traffic, call for focusing the analysis around the case that commuters *do* have some aggregate information on the state of traffic, e.g., they observe the length of the queue. Hence, in our model we assume that the agents do observe the traffic's state (i.e., observable queue).

Our Model. We study a stylized variant of Vickrey's model to allow us to focus on two issues that were not explored in the original model: a discrete population and an observable queue.[1] Formally, we have n agents that, starting at time 0, need to get a service which is offered by a first-come-first-served queue. Time progresses in discrete steps and in each step, each agent needs to decide whether to enter the queue at that time step or stay outside. When multiple agents decide to enter the queue simultaneously, they are ordered by a uniform random permutation. In our model, agents observe everything, and in particular, they observe the length of the queue and the set of agents that have not joined it yet.

For each agent, starting at time 0, the cost per time step for waiting before joining the queue is normalized to 1, and the cost per time step for waiting in the queue is $w > 1$ (agents dislike waiting in the queue). At each time step a single agent can be processed.

Our model is inspired by traffic related scenarios similar to Vickrey's rush hour scenario, such as the following one: the first game of the 2018 NBA finals

---

[1] Our minor simplifications of Vickrey's model include an assumption that the agents can only join the queue after some starting time.

at the Oracle Arena has just ended, and the audience wants to get home to San Francisco. To get home they should all cross the Bay Bridge that has a limited capacity (a bottleneck). While each person wants to get home as soon as possible, he dislikes standing in traffic. So, he should strategically decide when to leave the stadium and try to cross the bridge, aiming to minimize his discomfort. Hanging out around the stadium is an option that is costly, but not as much as standing in traffic. Fortunately for the audience, there are cameras installed over the bridge and apps that constantly broadcast the traffic state, and they can observe it and make their decision accordingly.

Solution Concept. As our game is symmetric and all agents are ex-ante the same, we focus on equilibria in symmetric randomized strategies that are anonymous. In such profiles all agents play the same randomized strategy that does not depend on the identities of the other agents. Observe that as agents are symmetric, equilibrium in asymmetric strategies requires different agents to play differently although they are ex-ante symmetric. This requires the agents to coordinate on which strategy each of them will play. Thus, symmetric strategies are arguably more natural than other, more general, strategies[2].

Moreover, our focus will be on *stationary strategies* that do not depend on the time step, but rather only on the state – the number of agents that are outside the queue as well as the number of agents that are in the queue. More formally, we consider symmetric Nash equilibria in anonymous stationary strategies, or *symmetric strategies* for short. Under such strategies, for any state there is some defined probability of entering the queue, and that probability is used by all agents. In the full version we prove that such equilibria always exist.

The discrete population and discrete time assumptions make the analysis of our model quite challenging. In particular computing equilibrium strategies for a game with n agents requires solving n polynomial equations of degrees increasing from 1 to n. Otsubo and Rapoport [26] are among the few that studied discrete population variants of Vickrey's model. They have only provided a complicated algorithm to numerically compute symmetric mixed Nash equilibrium rather than obtaining closed-form expressions.

Other approaches that were taken include analyzing the fluid limit of the system [18], and computing an equilibrium for the continuous time model by solving differential equations [19]. We take a different approach and instead of explicitly computing a symmetric equilibrium, present asymptotically tight upper and lower bounds on the social cost of any symmetric equilibrium.[3]

Our Results. We study the efficiency of symmetric equilibria in terms of the social cost, which is simply the sum of the players' costs. We present bounds that hold for any symmetric equilibria, any n and any w.[4] Thus, we establish

---

[2] Similar argument in favor of symmetric strategies is made in [26].

[3] Doing so alleviates the need to precisely compute symmetric equilibria and the need to determine if the game has a *unique* symmetric equilibrium or not.

[4] While we prove bounds that hold for any n and any w (see Theorems 11 and 12) our focus in the presentation is on the asymptotic social cost of symmetric equilibria, when either w or n gets large.

tight asymptotic price of stability and price of anarchy results for symmetric equilibria. Moreover, our lower bounds imply price of anarchy lower bounds for general Nash equilibria.[5]

We first analyze the ratio between the social cost of any symmetric equilibrium and the social cost of the optimal solution. We observe that whenever $1 < w \leq 2$ (the cost of waiting one time step in line is at most twice the cost of waiting outside), the unique symmetric equilibrium is for all agents to enter the queue immediately, and the total social cost is $w \cdot n \cdot (n-1)/2$. On the other hand, when agents enter sequentially, the social cost is only $n \cdot (n-1)/2$ (this is also the minimal social cost when we do not impose any restrictions on the strategies, in particular the strategies can be non-anonymous and non-stationary.)

Thus, in this case the ratio between the social costs of the unique symmetric equilibrium and the optimal solution is $w \leq 2$. Loosely speaking a similar bound of 2 also holds when we take a fixed $w << n$, but the proof is much more involved. A bit more formally, we prove the following result:

Theorem 1. *Fix $w > 2$. As n approaches infinity the ratio between the social costs of any symmetric equilibrium and the optimal solution is approaching 2.*

Usually in scenarios such as commuting to work or deciding when to head to the bridge after a game, the number of agents is relatively large while the normalized cost of waiting a unit of time in traffic, w, is relatively small. Theorem 1 tells us that in such cases the loss of efficiency in a symmetric Nash equilibrium is constant and relatively low, only 2.

We next consider the other extreme parameter regime, where the cost w is large relative to n. Such scenarios might arise in cases where either people do not care so much about getting the service early (for example, taking a routine medical check-up or running some non-urgent bureaucratic errand) or they have an arbitrary high cost for arriving simultaneously with others to receive the service. For example, one can think of a traffic intersection as providing a service for which simultaneous entry might cause an accident and has a very high cost. For the regime that n is small relative to w we obtain qualitatively different results than those for the regime that n is large relative to w. This provides an additional confirmation for the value of studying models of discrete population.

Theorem 2. *Fix n. As w approaches infinity the ratio between the social costs of any symmetric equilibrium and the (unrestricted) optimal solution is approaching $2 \cdot \sqrt{\frac{w}{n}}$, up to an additive term of $O(\frac{1}{\sqrt{n}})$.*

The theorem shows that when w is significantly greater than n, the multiplicative efficiency loss is very high (grows asymptotically as $\sqrt{w}$). Essentially, the issue with symmetric equilibria is that when the cost of standing in line is very high, the players are so horrified at the prospect of waiting in the queue that they enter the queue at a very low probability. Thus, the service is actually

---

[5] Similarly to the "Fully Mixed Nash Equilibrium Conjecture" [9] we suspect that in our game symmetric equilibria are in fact the worst equilibria.

idle most of the time. To reduce this kind of inefficiency, society came up with symmetry breaking mechanisms such as traffic lights or doctors appointments (see [6]), which provide a much needed coordination. Such a coordination mechanism induces a sequential order of entry, no player ever waits in the queue and the social cost is minimal. Moreover, for large w $(w > 2n)$, this is actually an (asymmetric) equilibrium.

Theorem 2 shows that symmetric equilibria are highly inefficient, but what is the main source of the inefficiency of symmetric equilibria: is it the players' strategic behavior or is it the lack of coordination imposed by symmetric strategies? As far as we know, no prior work has tried to separate between the loss of efficiency of symmetric equilibria due to strategic behavior and due to the symmetry requirement. To answer this question we bound the cost of the symmetric optimal solution and provide bounds on the ratio between the cost of any symmetric equilibrium and the symmetric optimal solution. We derive two different asymptotic bounds, depending on whether n or w are fixed.

Theorem 3. *Fix $w > 2$. As n approaches infinity the ratio between the social costs of any symmetric equilibrium and the* symmetric *optimal solution is approaching 2.*

In fact, as the ratio between the social costs of any symmetric equilibrium and the optimal solution is also approaching 2, this implies that as n increases the social cost of the symmetric optimal solution is approaching the social cost of the optimal solution, and both have essentially the same gap from any symmetric equilibrium. Thus, for this case, we conclude that the *main source of inefficiency of symmetric Nash equilibria is the strategic selfish behavior of the players.*

When considering the regime in which n is fixed but large, while w grows to infinity, a different picture emerges. Intuitively, since the cost of having two or more agents entering the queue simultaneously is so large, to avoid this cost, any profile of symmetric strategies must use entry probabilities that are low enough to ensure that the expected number of agents that join the queue at each step is much lower than 1. As a result, the agents will wait for a long time before anyone enters the queue, which implies a high social cost.

This creates a large gap between the social costs of the symmetric optimal solution and the (unrestricted) optimal solution. Interestingly, the gap is of the same magnitude as the gap between the worst symmetric equilibrium and the optimal solution. We show:

Theorem 4. *Fix n. As w approaches infinity the ratio between the social costs of any symmetric equilibrium and the* symmetric *optimal solution is approaching $\frac{3}{2\sqrt{2}} \approx 1.06$, up to an additive term of $O(\frac{1}{\sqrt{n}})$.*

Recall that Theorem 2 is showing that as w approaches infinity the ratio between the social costs of any symmetric equilibrium and the optimal solution is approaching $2 \cdot \sqrt{\frac{w}{n}}$. When we combine the two theorems we conclude that in the case that w goes to infinity, the *main source of inefficiency of the symmetric Nash equilibrium is the lack of coordination in symmetric randomized strategies.* Nevertheless, there is still some small constant loss that does not vanish and is due to incentives, yet it dwarfs compared to the loss of $2 \cdot \sqrt{\frac{w}{n}}$ (which tends to infinity as w grows to infinity) that is due to lack of coordination.

Related Work. Bottleneck models were studied in both the traffic science literature and the economics literature. Arnott et al. [1] provide economic analysis of Vickrey's bottleneck model and also consider how tolls can reduce the cost associated with strategic behavior in such models. Later papers extended the model to handle more general pricing schemes (for example [3,7]). In [2] Arnott et al. consider giving traffic information to commuters in a continuous population model in which commuters need to choose when to leave and which route to take. The new twist is that travel time is affected by unexpected events such as accidents or bad weather that the commuters can get information about. Arnott et al. reach the conclusion that providing the commuters perfect information about these unpredictable events can eliminate the inefficiency resulting from them. Other variants of the model that were studied more recently include: heterogeneous commuters [30] and the effects of congested bottlenecks on the roads leading to them [21].

The literature on strategic queuing is also related to the current paper. In his seminal paper, Naor was the first to introduce both economic and strategic considerations into the queuing literature [25]. Up till then queuing theory mainly focused on the efficiency of queues. The most well known model in classic queuing theory is the M/M/1 queue model, where there is one server and the jobs arrive according to a Poisson distribution and have an exponentially distributed service time. According to Naor's model, the service has a price and the "jobs" need to decide whether to join the queue or not. This gave rise to a new area called strategic queuing which studies the users' behavior in different queuing systems under various assumptions (see [12,13] for extensive surveys).

In general, the literature on strategic queuing has traditionally focused on models of unobservable queues as these are easier to analyze (see Chap. 2 of [12] for a survey of recent works on observable queues). Hassin and Roet-Green [15] bridge the gap between observable queues and non-observable queues by presenting and analyzing a natural model in which the agents have the option to pay to see the length of the queue. Most of the works in strategic queuing (both on observable and unobservable queues) consider games in which each player arrives at some time and needs to immediately decide whether to join the queue or not. In this setting, the paper of Kerner [20] applies a solution concept similar to ours in studying symmetric equilibrium joining probabilities for an M/G/1 observable queue.

The more elaborate model in which the player's strategy is to choose an arrival time with the goal of minimizing his waiting time was first suggested in [11] for unobservable queues. [11] studied a model in which agents can choose to join the queue before its opening time (early arrivals) while later [14] showed that the efficiency of the equilibrium can be sometimes increased by disallowing early arrivals. Discrete time and discrete population versions of this model were later studied in [22,27] that concentrated on symmetric mixed Nash equilibria for this unobservable queue model. A recent work [24] studies a setting in which while the queue is unobservable the service provider can observe the queue and give the agents some information regarding the queue.

A specific line of papers in strategic queuing which is similar both in intuition and in formalism to our model is on the so called "concert queuing game". This game was first defined in [18] and was later studied in follow up papers such as [17,19]. In this game concert attendants wish to get home as soon as possible once the concert ends but they dislike standing in traffic. The main distinctions between the concert queuing game and our model are in the assumptions on the observability of the queue and its processing time (in our model, processing time is fixed, while in the other model it is distributed according to some distribution). In each of [18] and [19] the authors use different analysis techniques to establish a bound of 2 on the price of anarchy for their model.

The strategic queuing literature includes a few papers dealing with price of anarchy. The first one was [16] which studied the price of anarchy of a multi server system where the players strategically choose which (unobservable) queue to join. In [10] Gilboa-Freedman et al. provide a price of anarchy bound for Naor's model [25] where the queue is observable but the strategy of each player is limited to the one-time decision whether to join the queue or not.

The paper of Fiat et al. [8] also considers strategic entry by selfish players – players that need to broadcast on a joint channel. The model in that paper is fundamentally different than ours, as simultaneous entry results in all players failing to enter the channel, rather than a formation of a queue as in our model.

Paper Outline. We start by formally presenting our model and some useful observations. Next, in Sect. 3 we study the two player case as a warm-up. We then present our upper and lower bounds on the cost of any symmetric equilibria in Sect. 4, Theorems 1 and 2 follow from these results. Finally, we present bounds for the social cost of symmetric optimal strategies in Sect. 5, Theorems 3 and 4 follow from these bounds.

2   Model and Preliminaries

There is a set N of n identical agents and time is discrete. At every time step t the following sequence takes place:

- Each agent decides whether or not to enter the queue (possibly using randomization).
- After the agents make their decisions, all agents that have decided to enter the queue are added to the end of the queue in a random order.
- If the queue is not empty then the first agent in the queue is processed. The rest of the agents in the queue incur a cost of $w > 1$.
- Every agent outside the queue incurs a cost of 1.

The goal of each agent is to minimize his expected cost. We use $G(n; w)$ to denote a game with $n \geq 2$ agents and waiting cost per unit of $w > 1$.

In general, a strategy of an agent needs to specify the probability of entry at each *history*, such a history specifies the time, the realized action of each agent at every prior time, and the randomization results whenever multiple agents that enter at the same time are ordered at the end of the queue. Our focus in this paper is on anonymous stationary strategies – strategies that do not depend on the time step or the identity of agents, such strategies will only depend on a summary statistics specified by the (anonymized) *state* of our game. A *state* of our game is defined as a pair (m, k), where there are $m \geq 1$ agents that are still outside the queue and $k \geq 0$ agents in the queue, and $m + k \leq n$. We formally define anonymous stationary strategies as follows:

Definition 1. (anonymous stationary strategies in $G(n; w)$). *A strategy of an agent is an anonymous stationary strategy if for any history it specifies a probability that an agent enters the queue that is only a function of the state (m, k). We denote that probability of entry at a state (m, k) by $q_{m,k} \in [0, 1]$ and the probability at state $(n, 0)$ by q_n.*

We use $S = (S_1, S_2, \ldots, S_n)$ to denote a profile of strategies. A profile S of anonymous stationary strategies is *symmetric* if all players use the same strategy ($S_i = S_j$ for every $i, j \in N$). We are interested in Nash equilibria of the game, in such equilibria each player minimizes his cost, given the strategies of the others. In the full version we discuss general strategies and show that if a profile is an equilibrium with respect to anonymous stationary strategies, it is also an equilibrium with respect to any arbitrary strategies that might depend on the time or on the identities of the agents. For this reason, for the rest of the paper we will only consider deviations to anonymous stationary strategies.

We denote the expected cost of every player in a symmetric equilibrium S in the game $G(n; w)$ by $c_{n,w}(S)$. When w and S are clear from the context we simplify the notation to c_n. We denote the social cost of strategy profile S by $C_{n,w}(S) = n \cdot c_{n,w}(S)$. As we will see, it is useful to extend this notation to sub-games as well. We denote the sub-game that starts with a state (m, k) by $G(m, k; w)$. For a symmetric equilibrium S we denote the cost of each of the m players that are outside the queue by $c_w(m, k; S)$. When S and w are clear from the context we will use the shorter notation $c(m, k)$. With this notation we have that $c_{n,w}(S) = c(n, 0)$.

2.1   Basic Observations

If at state (m, k) a player enters with a non-trivial probability $q_{m,k} \in (0, 1)$ then he must be indifferent between entering and waiting at that step. Otherwise, the player will choose to either enter the queue or wait with probability one. Our analysis of the social cost of symmetric Nash Equilibria heavily depends on this observation. Thus, it is useful to first work out the expressions for a cost of a player i that joins the queue with probability 1 at state (m, k) and the cost of player i that joins the queue with probability 0 at state (m, k). We define the two costs as $c^1(m, k; q)$ and $c^0(m, k; q)$ respectively, where we assume that in state

(m, k) all the players but i enter with probability q and at any other state all the players play according to some symmetric strategy profile S. We now give expressions for the two costs for every symmetric profile S:

Observation 5. *For every* $w > 1$, $q \in [0, 1]$, $m \geq 1$ *and* $k \geq 0$: $c^1(m, k; q) = \frac{m-1}{2} \cdot q \cdot w + k \cdot w$.

Observation 6. *For every* $q \in [0, 1]$, $m \geq 1$ *and* $k \geq 1$:

$$c^0(m, 0; q) = \frac{1}{1 - (1 - q)^{m-1}} + \frac{1}{1 - (1 - q)^{m-1}}$$
$$\cdot \sum_{i=1}^{m-1} \binom{m-1}{i} q^i \cdot (1 - q)^{m-1-i} \cdot c(m - i, i - 1; S)$$

$$c^0(m, k; q) = 1 + \sum_{i=0}^{m-1} \binom{m-1}{i} q^i \cdot (1 - q)^{m-1-i} \cdot c(m - i, k + i - 1; S)$$

Next, we claim that a symmetric equilibrium always exists. This is not a priori clear as our strategies cannot be easily defined as a mix of pure anonymous stationary strategies.

Theorem 7. *Fix any* $n \geq 2$ *and* $w > 1$. *There exists a symmetric equilibrium in anonymous stationary strategies in the game* $G(n; w)$.

We observe that for small values of w ($w \leq 2$) the unique symmetric equilibrium outcome is for all players to enter immediately.

Observation 8. *For any* $n \geq 2$. *If* $w \in [1, 2]$ *then in the game* $G(n; w)$ *the unique symmetric equilibrium with anonymous stationary strategies is for all agents to enter with probability* $q_n = 1$ *and the social cost is* $w \cdot \frac{n(n-1)}{2}$.

We focus our equilibrium analysis on the case that $w > 2$. We now show that when $w > 2$ it is no longer the case that a player in a symmetric equilibrium prefers to join the queue with probability 1. Furthermore, in case the queue is empty, deterministically not entering the queue is also not a best response to the strategies of others in a symmetric equilibrium.

Observation 9. *For any* $n \geq 2$, $w > 2$ *and for any* (m, k) *such that* $m + k \leq n$, $m \geq 2$ *and* $k \geq 0$, *in any symmetric equilibrium* $0 \leq q_{m,k} < 1$ *and* $0 < q_{m,0} < 1$.

The following observation characterizes the cost of symmetric Nash equilibria based on the observations above:

Observation 10. *Fix* $w > 2$ *and* $n \geq 2$. *Given a symmetric equilibrium in the game* $G(n; w)$, *the cost of some player* i *which is outside the queue in the state* (m, k) *satisfies the following:*

- *For* $m \leq n$ *it holds that* $c^1(m, 0; q_m) = c^0(m, 0; q_m)$.
- *For* $k \geq 1$, *if* $q_{m,k} \in (0, 1)$ *then* $c(m, k) = c^0(m, k; q_{m,k}) = c^1(m, k; q_{m,k})$. *Otherwise,* $q_{m,k} = 0$ *and* $c(m, k) = c^0(m, k; 0)$.

3   Warm-Up - The 2 Players Case

To get some intuition it is instructive to consider first the simple case of only 2 players. For this case we obtain an exact expression for the players' costs in the unique symmetric equilibrium. Note that a single player will always join the queue ($q_{1,0} = 1$) and for this reason $c(1,0) = 0$. Now, to compute a symmetric equilibrium we only need to compute the probability that the agents enter when both are still outside ($q_{2,0}$).

Claim. For $n = 2$ players, if $w > 2$, there exists a unique symmetric equilibrium S in anonymous stationary strategies in which each player plays the strategy: $q_{2,0} = \sqrt{2/w}$, $q_{1,0} = 1$. The social cost for both players is $C_{2,w}(S) = \sqrt{2w}$.

We compare the cost at Nash equilibrium against the cost of the optimal solution. For two players, the cost of the optimal solution is simply 1 as one of the players will enter first and pay a cost of 0 and the other will enter second and pay a cost of 1. This implies the following corollary:

Corollary 1. *In the game $G(2;w)$ the ratio between the social costs of the unique*[6] *symmetric Nash equilibrium and optimal solution is $\sqrt{2w}$.*

This relatively large gap that grows with w leads us to ask what is the source of this gap – is it due to strategic behavior, or to the lack of coordination imposed by symmetric strategies? To answer this question we compute the minimal cost when all agents are required to use the same strategy and use anonymous stationary strategies, which are not necessarily an equilibrium. We note that even just for 2 players computing the optimal symmetric solution is simple yet not completely trivial, as it requires computing the minimum of a function which is the ratio of two polynomials. As the number of players increases this becomes more complicated and hence instead of directly computing the optimal symmetric strategy we will compute bounds on its cost.

As in the Nash equilibrium, once an agent is the only one outside, he clearly enters immediately. Thus, we only need to compute the probability of each agent entering, assuming both agents are outside (denoted p_2).

Claim. For $n = 2$ players, if $w > 1$, the symmetric anonymous stationary strategy that minimizes the social cost is: $p_2 = \frac{\sqrt{2w-1}-1}{w-1}$ and $p_1 = 1$. The social cost for this profile is $OPT(2,w) = \frac{w+1}{\sqrt{2w-1}+1}$.

The following corollary is easily derived from the above claim:

Corollary 2. *In the game $G(2;w)$ the ratio between the social costs of the unique symmetric Nash equilibrium and optimal solution in symmetric strategies is approaching 2 as w approaches infinity.*

---

[6] Note that when $w > 2$ the two players game admits exactly three equilibria: the two optimal equilibria in which one player enters after the other, and the symmetric random equilibrium we discussed. Thus, our result is both a price of anarchy result for unrestricted equilibria and a price of stability result for symmetric equilibria.

We conclude that for large values of w, there is a huge loss for insisting on symmetric profiles: the optimal cost grows from 1 in an asymmetric optimum, to about $\sqrt{\frac{w}{2}}$ in the symmetric optimum. An additional, much smaller, loss of factor 2 comes from further requiring the symmetric profile to be an equilibrium. Our goal in this paper is to understand the source of inefficiency of symmetric Nash equilibria for any n. We present a separation between the cost ratio of symmetric Nash equilibria and the symmetric optimal solution when n is fixed and w is large and the case that w is fixed but n is large.

4 Bounds on the Cost of Symmetric Nash Equilibria

In this section we provide bounds on the cost of symmetric Nash equilibria in any profile of anonymous stationary strategies, these bounds hold for any n and any w. We present two types of bounds, each will be tight for a different regime of the parameters n and w, and we use these bounds to prove Theorems 1 and 2. We first present a bound that is useful when w is relatively small compared to n.

Theorem 11. *For every $w > 2$, $n \geq 2$ and symmetric equilibrium S:*

$$n - 1 \leq c_{n,w}(S) \leq n + w \cdot O(\ln(n))$$

Clearly the above bound is asymptotically tight whenever $w = o(\frac{n}{\ln(n)})$. This implies that when $w = o(\frac{n}{\ln(n)})$, for a sufficiently large value of n, the social cost of any symmetric equilibrium is about n^2. Denote by $SC(n,w)$ the social cost of the optimal solution. Recall that in the optimal solution the players enter sequentially and hence the cost of the optimal solution is $SC(n,w) = n(n-1)/2$. The ratio between the cost of any Nash equilibrium and the cost of an optimal solution is essentially 2, proving Theorem 1. Formally:

Corollary 3. *For every fixed $w > 2$ and every $\varepsilon > 0$ there exists $n_0^w(\varepsilon)$ such that for any $n > n_0^w(\varepsilon)$ for every symmetric equilibrium S it holds that*

$$2 \leq C_{n,w}(S)/SC(n,w) \leq 2 + \varepsilon$$

Next, we give a different bound which will be tight for the case of large enough n, and w that goes to infinity. We show that the social cost of any symmetric equilibrium is essentially approaching $n \cdot \sqrt{w \cdot n}$. Formally:

Theorem 12. *For every $w > 2$, $n \geq 2$ and any symmetric equilibrium S:*

$$(1 - \varepsilon_n(w))^{n-1} \cdot \sqrt{w \cdot n - w \cdot 2\ln n} \leq c_{n,w}(S) \leq \frac{e}{e-1} \cdot n + (1 + \varepsilon_n(w)) \cdot \sqrt{w \cdot n + 2w\sqrt{n-1}}$$

for some decreasing function $\varepsilon_n(w) \leq 1$ that for any fixed n, converges to 0.

In this case if $n = o(\sqrt{w})$ we get that the social cost of any symmetric equilibrium is about $n \cdot \sqrt{w \cdot n}$ and hence the ratio between the costs of any symmetric equilibrium and the optimal solution (which has cost of $n(n-1)/2$) is approaching $2 \cdot \sqrt{\frac{w}{n}}$, proving Theorem 2:

Corollary 4. *Fix any $\varepsilon > 0$. There exists $n_0(\varepsilon)$ such that for any $n > n_0(\varepsilon)$ there exist $w^n(\varepsilon)$ such that for any $w > w^n(\varepsilon)$ it holds that for any symmetric equilibrium S:*

$$(2 - \varepsilon)\sqrt{\frac{w}{n}} \leq \frac{C_{n,w}(S)}{SC(n,w)} \leq (2 + \varepsilon)\sqrt{\frac{w}{n}}$$

5   Bounds on the Cost of Symmetric Optimal Solutions

In this section we provide bounds on the *optimal symmetric cost*, this is the minimal cost when agents are restricted to play symmetric anonymous stationary strategies, but not necessarily equilibrium strategies. That is, for $w > 1$ and $n \geq 2$, the optimal cost $OPT(n,w) = \inf_S C_{n,w}(S)$, the infimum is taken over all profiles S of symmetric profiles of anonymous stationary strategies.

It is easy to see that since the cost of waiting in the queue is greater than the cost of waiting outside ($w > 1$) it is socially suboptimal to direct an agent to enter when the queue is not empty. Thus, when considering an optimal symmetric strategy in the game $G(n; w)$ we can restrict ourselves to strategies that define an entrance probability p_m for every number of players $1 \leq m \leq n$ such that the queue is empty and there are m agents outside the queue.

Providing a closed form expression for the optimal symmetric cost for the game $G(n; w)$ is very challenging as it requires minimizing a function which is the ratio of two polynomials each of degree $n - 1$. Hence, we compute lower and upper bounds on the optimal symmetric cost instead. As in the case of the symmetric equilibrium we provide different bounds for the case that w is fixed and n goes to infinity and for the case that n is fixed and w goes to infinity:

Theorem 13. *$OPT(n, w)$ is bounded as follows:*

- *Fix any $w > 2$ and any $\varepsilon > 0$. There exists $n_0^w(\varepsilon)$ such that for any $n > n_0^w(\varepsilon)$ it holds that*

$$\frac{n(n-1)}{2} \leq OPT(n, w) \leq (1 + \varepsilon)\frac{n(n-1)}{2}$$

- *Fix any $n \geq 2$ and any $\varepsilon > 0$. There exists $w_1^n(\varepsilon)$ such that for any $w > w_1^n(\varepsilon)$ it holds that*

$$(1 - \varepsilon)\sqrt{2w} \cdot \frac{2}{3}(n-1)\sqrt{n-1} \leq OPT(n, w) \leq (1 + \varepsilon)\sqrt{2w} \cdot \left(\frac{2}{3}n\sqrt{n} + \sqrt{n}\right)$$

From the theorem we derive two corollaries about the asymptotic cost of $OPT(n, w)$. Our first corollary shows that for fixed $w > 2$, the symmetric optimal cost grows asymptotically the same as the cost of the optimal schedule, when n grows large. This implies that in this case the source of the inefficiency of the symmetric equilibrium is *only* due to strategic behavior, and not lack of coordination.

Corollary 5. *For every fixed $w > 2$ and every $\varepsilon > 0$ there exists $n_0^w(\varepsilon)$ such that for any $n > n_0^w(\varepsilon)$ it holds that $1 \leq OPT(n,w)/SC(n,w) \leq 1 + \varepsilon$.*

Combining the previous corollary together with Corollary 3 establishes the proof of Theorem 3. Thus, we have that for every fixed $w > 2$ and every $\varepsilon > 0$ there exists $n_0^w(\varepsilon)$ such that for any $n > n_0^w(\varepsilon)$ it holds that $2 - \varepsilon \leq C_{n,w}(S)/OPT(n,w) \leq 2 + \varepsilon$.

For the case that n is large enough and w is greater than n we get that the cost of a symmetric optimal solution is about $\frac{2\sqrt{2}}{3} \cdot n \cdot \sqrt{w \cdot n}$. Recall that for the same case by Theorem 12 we have that the social cost of any symmetric Nash equilibrium is about $n \cdot \sqrt{w \cdot n}$. Thus, we have that in this case the ratio between the cost of any symmetric equilibrium and the symmetric optimal cost converges to $\frac{3}{2\sqrt{2}} \approx 1.06$, which proves Theorem 4. This means that in the case that w is relatively larger than n the lack of coordination is playing a major role in deteriorating the efficiency of symmetric equilibria. This explains why in such cases, it is common that measures are taken to increase coordination and help boost social welfare. Formally, In the full version we show that:

Corollary 6. *Fix any $\delta > 0$. There exists $n_0(\delta)$ such that for any $n > n_0(\delta)$ there exist $w^n(\delta)$ such that for any $w > w^n(\delta)$ it holds that for any symmetric equilibrium S*

$$(1 - \delta)\frac{3}{2\sqrt{2}} \leq \frac{C_{n,w}(S)}{OPT(n,w)} \leq (1 + \delta)\frac{3}{2\sqrt{2}}$$

References

1. Arnott, R., De Palma, A., Lindsey, R.: Economics of a bottleneck. J. Urban Econ. **27**(1), 111–130 (1990)
2. Arnott, R., De Palma, A., Lindsey, R.: Does providing information to drivers reduce traffic congestion? Transp. Res. Part A: Gen. **25**(5), 309–318 (1991)
3. Arnott, R., De Palma, A., Lindsey, R.: A structural model of peak-period congestion: a traffic bottleneck with elastic demand. Am. Econ. Rev. **83**(1), 161–179 (1993)
4. Awerbuch, B., Azar, Y., Epstein, A.: The price of routing unsplittable flow. SIAM J. Comput. **42**(1), 160–177 (2013)
5. Ben-Akiva, M., De Palma, A., Isam, K.: Dynamic network models and driver information systems. Transp. Res. Part A: Gen. **25**(5), 251–266 (1991)
6. Cayirli, T., Veral, E.: Outpatient scheduling in health care: a review of literature. Prod. Oper. Manag. **12**(4), 519–549 (2003)
7. Daganzo, C.F., Garcia, R.C.: A pareto improving strategy for the time-dependent morning commute problem. Transp. Sci. **34**(3), 303–311 (2000)
8. Fiat, A., Mansour, Y., Nadav, U.: Efficient contention resolution protocols for selfish agents. In: Proceedings of the Eighteenth Annual ACM-SIAM Symposium On Discrete Algorithms (SODA), pp. 179–188 (2007)
9. Gairing, M., Lücking, T., Mavronicolas, M., Monien, B., Spirakis, P.: Extreme nash equilibria. In: Blundo, C., Laneve, C. (eds.) ICTCS 2003. LNCS, vol. 2841, pp. 1–20. Springer, Heidelberg (2003). https://doi.org/10.1007/978-3-540-45208-9_1

10. Gilboa-Freedman, G., Hassin, R., Kerner, Y.: The price of anarchy in the markovian single server queue. IEEE Trans. Autom. Control **59**(2), 455–459 (2014)
11. Glazer, A., Hassin, R.: ?/M/1: on the equilibrium distribution of customer arrivals. Eur. J. Oper. Res. **13**(2), 146–150 (1983)
12. Hassin, R.: Rational Queueing. Chapman and Hall/CRC, London (2016)
13. Hassin, R., Haviv, M.: To Queue or Not to Queue: Equilibrium Behavior in Queueing Systems, 59. Springer, Heidelberg (2003). https://doi.org/10.1007/978-1-4615-0359-0
14. Hassin, R., Kleiner, Y.: Equilibrium and optimal arrival patterns to a server with opening and closing times. IIE Trans. **43**(3), 164–175 (2010)
15. Hassin, R., Roet-Green, R.: The impact of inspection cost on equilibrium, revenue, and social welfare in a single-server queue. Oper. Res. **65**(3), 804–820 (2017)
16. Haviv, M., Roughgarden, T.: The price of anarchy in an exponential multi-server. Oper. Res. Lett. **35**(4), 421–426 (2007)
17. Honnappa, H., Jain, R.: Strategic arrivals into queueing networks: the network concert queueing game. Oper.Res. **63**(1), 247–259 (2015)
18. Jain, R., Juneja, S., Shimkin, N.: The concert queueing game: to wait or to be late. Discrete Event Dyn. Syst. **21**(1), 103–138 (2011)
19. Juneja, S., Shimkin, N.: The concert queueing game: strategic arrivals with waiting and tardiness costs. Queueing Syst. **74**(4), 369–402 (2013)
20. Kerner, Y.: Equilibrium joining probabilities for an M/G/1 queue. Games Econ. Behav. **71**(2), 521–526 (2011)
21. Lago, A., Daganzo, C.F.: Spillovers, merging traffic and the morning commute. Transp. Res. Part B: Methodol. **41**(6), 670–683 (2007)
22. Lariviere, M.A., Van Mieghem, J.A.: Strategically seeking service: how competition can generate poisson arrivals. Manuf. Serv. Oper. Manag. **6**(1), 23–40 (2004)
23. Levinson, D.: Micro-foundations of congestion and pricing: a game theory perspective. Transp. Res. Part A: Policy Pract. **39**(7), 691–704 (2005)
24. Lingenbrink, D., Iyer, K.: Optimal signaling mechanisms in unobservable queues with strategic customers. In: Proceedings of the 2017 ACM Conference on Economics and Computation (EC), pp. 347–347. ACM (2017)
25. Naor, P.: The regulation of queue size by levying tolls. Econometrica **37**(1), 15–24 (1969)
26. Otsubo, H., Rapoport, A.: Vickrey's model of traffic congestion discretized. Transp. Res. Part B: Methodol. **42**(10), 873–889 (2008)
27. Rapoport, A., Stein, W.E., Parco, J.E., Seale, D.A.: Equilibrium play in single-server queues with endogenously determined arrival times. J. Econ. Behav. Organ. **55**(1), 67–91 (2004)
28. Roughgarden, T.: Selfish routing with atomic players. In: Proceedings of the Sixteenth Annual ACM-SIAM Symposium on Discrete Algorithms (SODA), pp. 1184–1185 (2005)
29. Roughgarden, T., Tardos, É.: How bad is selfish routing? J. ACM (JACM) **49**(2), 236–259 (2002)
30. van den Berg, V., Verhoef, E.T.: Congestion tolling in the bottleneck model with heterogeneous values of time. Transp. Res. Part B: Methodol. **45**(1), 60–78 (2011)
31. Vickrey, W.S.: Congestion theory and transport investment. Am. Econ. Rev. **59**, 251–260 (1969)

Strategy-Proof Incentives for Predictions

Amir Ban[✉]

Faculty of Industrial Engineering and Management, Technion,
Israel Institute of Technology, Haifa, Israel
amirban@netvision.net.il

Abstract. Our aim is to design mechanisms that motivate all agents to reveal their predictions truthfully and promptly. For myopic agents, proper scoring rules induce truthfulness. However, when agents have multiple opportunities for revealing information, and take into account long-term effects of their actions, deception and reticence may appear. Such situations have been described in the literature. No simple rules exist to distinguish between the truthful and the untruthful situations, and a determination has been done in isolated cases only. This is of relevance to prediction markets, where the market value is a common prediction, and more generally in informal public prediction forums, such as stock-market estimates by analysts. We describe three different mechanisms that are strategy-proof with non-myopic considerations, and show that one of them, a discounted market scoring rule, meets all our requirements from a mechanism in almost all prediction settings. To illustrate, we extensively analyze a prediction setting with continuous outcomes, and show how our suggested mechanism restores prompt truthfulness where incumbent mechanisms fail.

1 Introduction

Mechanisms that motivate all agents to reveal their information truthfully and promptly are desirable in many situations.

Consider, for example, the estimation of company earnings by stock-market analysts, a longstanding Wall Street institution. Publicly-traded companies announce their earnings for the latest quarter or year, on dates set well in advance. Each company is typically covered by several stock-market analysts, the larger ones by dozens. These analysts issue reports containing predictions of a company's future earnings. The timing of such predictions ranges from several years to days before earnings announcement, and every analyst typically updates his prediction several times in the interval. A consensus calculated from all predictions in force may be viewed on several popular finance websites. Not least, the analysts themselves are aware of, and are no doubt influenced by the actions and opinions of their peers.

In essence, this earnings estimation functions as a public prediction forum with an evolving consensus, that terminates when a company announces its true earnings for the forecast period, which we call the *outcome*. It acts as a sort of

© Springer Nature Switzerland AG 2018
G. Christodoulou and T. Harks (Eds.): WINE 2018, LNCS 11316, pp. 51–65, 2018.
https://doi.org/10.1007/978-3-030-04612-5_4

advisory forum for the public of investors, and this public's interest is best served if analysts share their information and judgement truthfully and promptly.

Prediction markets are public prediction forums organized as markets. A de-facto standard for organizing prediction markets is due to Hanson (2003), using a *market scoring rule*. In such a market, probability estimates are rewarded by a *proper scoring rule* an amount $S(\boldsymbol{p}, r)$, where $\boldsymbol{p}$ is a probability distribution of the outcome, and r is the outcome. A trader in Hanson's markets not only makes her probability estimate public, she changes the market price to it. She then stands to be rewarded by the *market maker* for her prediction (when the outcome becomes known), but she also commits to compensate the *previous* trader for his prediction. Her total compensation is therefore the difference $S(\boldsymbol{p}, r) - S(\boldsymbol{p}', r)$ where $\boldsymbol{p}'$ is the replaced market prediction. When the logarithmic scoring rule $(S(\boldsymbol{p}, r) = \log p_r)^1$ is used in a prediction market, the mechanism is called LMSR (Logarithmic Market Scoring Rule). Hanson also demonstrated how a market maker can facilitate such an LMSR market and provide liquidity by selling and buying shares of each outcome.

Proper scoring rules, are, by definition, incentive compatible for *myopic* agents. That is, an agent maximizes her expected score by announcing her true belief, provided longer-term effects of the prediction, if any, are ignored. The incremental, market variation of the scoring rule does not affect this incentive compatibility, because the previous agent's score does not depend on the current prediction.

When an agent is *not* myopic, and *does* take into account all consequences of her action, truthfulness will, in many cases, *not* be her optimal strategy, and incentive compatibility is lost. Such scenarios have been described in the literature, and our paper adds many further examples. As we will show, the damage to incentive compatibility caused by long-term strategic considerations is extensive, and it is *a priori* unclear whether any remedy is available.

Our aim is the design of mechanisms for rewarding predictions that are strategy-proof. We demonstrate the problem with proper scoring rules, as often leading to reticence or deception. We formulate criteria for determining which prediction settings are truthful, and apply these criteria for a complete classification of the important class of prediction settings with normal and lognormal signals. We suggest three strategy-proof mechanisms, and identify one of them, *discounting*, as having all desirable properties. We prove the applicability and effectiveness of the discounting mechanism.

1.1   The Problem with Scoring Rules

A scoring rule $S : \Delta(\mathbb{R}) \times \mathbb{R} \mapsto \mathbb{R}$ scores a prediction $\boldsymbol{p}$, representing a probability distribution of the outcome, a value $S(\boldsymbol{p}, r)$ when the outcome is r. An agent whose belief of the outcome distribution is $\boldsymbol{q}$ has score expectation $S(\boldsymbol{p}, \boldsymbol{q}) := \mathbb{E}_{r \sim \boldsymbol{q}} S(\boldsymbol{p}, r)$ for prediction $\boldsymbol{p}$. A *proper* scoring rule is one for which $S(\boldsymbol{q}, \boldsymbol{q}) \geq S(\boldsymbol{p}, \boldsymbol{q})$ for every $\boldsymbol{p}, \boldsymbol{q} \in \Delta(\mathbb{R})$, so that predicting one's true belief has maximal

---

[1] We use the notation p_x for the density of distribution $\boldsymbol{p}$ at x.

score expectation. A *strictly proper* scoring rule is one where the inequality is tight only for $\boldsymbol{p} = \boldsymbol{q}$. The *logarithmic* scoring rule $S(\boldsymbol{p}, r) = \log p_r$, and the *quadratic* (a.k.a. *Brier*) scoring rule $S(\boldsymbol{p}, r) = 2p_r - \boldsymbol{p} \cdot \boldsymbol{p} - 1$ are examples of strictly proper scoring rules. More background on scoring rules may be found, e.g., in Gneiting and Raftery (2007).

The following generic example illustrates the problem when non-myopic considerations apply.

Example 1. The public wants to predict a variable, whose outcome is x. Every signal of x is, i.i.d., $x + \epsilon$ with probability $1/2$, and $x - \epsilon$ with probability $1/2$. ϵ is unknown. There is an expert, who gets private signals. The public gets public signals.

On Sunday, expert gets a signal. On Monday, the public gets a signal. On Tuesday, expert gets another signal. On Wednesday, outcome x is revealed.

Question: Should expert reveal his information truthfully on Sunday?

Answer: No. Whoever sees two different truthful signals is able to calculate the outcome $x = (x + \epsilon)/2 + (x - \epsilon)/2$ exactly. For any distribution of ϵ, and for the logarithmic and almost[2] every other scoring rule, the expert should not tell the truth on Sunday. This prevents the 50% probability that the market will know x on Monday, preserving a 75% probability that the expert can announce x on Tuesday.

The canonical case, to which this example belongs, is "Alice-Bob-Alice", where Alice speaks before and after Bob's single speaking opportunity, both are awarded by a proper scoring rule for each prediction, and both maximize their total score. Chen et al. (2010) as well as Chen and Waggoner (2016) studied situations where several agents, each having private information, are given more than one opportunity to make a public prediction. The situations are reducible to the Alice-Bob-Alice game. The proper scoring rule assures that each will tell the truth on their last prediction, and the open question is whether Alice, when going first, will tell the truth, lie, or keep her silence. Chen et al. (2010) make the key observation that truthfulness is optimal if, in a different setup, namely, a single-prediction Alice-Bob game where Alice chooses whether to go first or second, she will always prefer going first. Building on that insight, Chen and Waggoner (2016) show that when the players' information is what they define as "perfect informational substitutes", they will predict truthfully and as early as allowed, when they are "perfect informational complements", they will predict truthfully and as *late* as allowed, while when players are neither substitutes nor complements, untruthfulness can and will occur. While this characterization is helpful, few concrete cases have been settled. The most significant of those was to show that when signals are independent conditional on the outcome, and the logarithmic scoring rule (LMSR) is used, the signals are informational substitutes, meaning that in such a case, Alice will reveal all her information truthfully in her first round.

---

[2] The example is true for the logarithmic scoring rule because its scores are unbounded. Every scoring rule that values exact predictions over inexact ones sufficiently will do.

A strategy-proof mechanism of the Alice-Bob-Alice setting easily generalizes to a strategy-proof mechanism for any number of experts and prediction order, because whenever an expert (call her Alice) makes more than one prediction, one can roll together all experts making predictions between Alice's successive predictions into one expert (call him Bob), who shares their information[3]. This is formally proved in Proposition 5.

1.2   Goals of the Mechanism

We seek a mechanism with several desirable traits.

1. TRUTHFULNESS: The mechanism should motivate all experts to make truthful predictions, that is, to reveal their true subjective distributions of the outcome. At minimum, this means that truth-telling should be a best-response to truth-telling by all other experts, according to the player's beliefs at the time of prediction, i.e., it is a perfect Bayesian equilibrium.[4]
An untruthful mechanism may still be *locally truthful* by which we mean that infinitesimal variations from the truth are suboptimal, but telling a sufficiently big lie may be advantageous.
2. FULL DISCLOSURE: All information possessed by the experts should be disclosed. This means that every expert makes a (truthful) prediction some time after getting his last signal. Otherwise, the information on the outcome possessed in that last signal would never reach the public.
3. PROMPTNESS: Experts should reveal their signals by a truthful public prediction as soon as the prediction schedule allows, and make an updated prediction whenever receiving a new signal, again, at the earliest opportunity. We shall require a *strong* preference for promptness. Indifference to timing shall not count as prompt.
4. BOUNDED LOSS: Hanson (2003) notes that in his market scoring rule mechanism, the market maker effectively subsidizes traders to motivate their truthfulness, and he shows that market maker's expected loss due to that is *bounded*. We seek mechanisms that achieve this property.

1.3   Our Results

We propose three different incentive mechanisms, all of which are based on proper scoring rules, all of which achieve truthfulness, and the third and last also achieves promptness. They are

1. GROUP PREDICTION: All agents receive the final prediction's (non-incremental) score. Since all agents have a stake in the final prediction, all will

---

[3] Chen et al. (2010) use the same construction to generalize from Alice-Bob-Alice to a finite-players game.

[4] Note that the ideal of truth-telling as dominant strategy is not attainable here, because if a player is aware of another player's distortion, the correct Bayesian response is to compensate for the distortion.

reveal their information truthfully. On the negative side, they are not moti- vated to be prompt about it. Another problem is freeloaders, since agents with no information can participate and gain without contributing anything.

2. ENFORCE SINGLE PREDICTION: Score each of agent's prediction with an incre- mental scoring rule, and award each agent the minimum score. Agents are therefore motivated to predict once only, since having made a prediction, a further prediction can only lower their reward expectation. With a proper scoring rule, this assures incentive compatibility with truthfulness. Agents are not motivated to be prompt, but instead need to find the optimal timing to make their single prediction. A major drawback is that when agents receive a time-varying signal, they will not reveal all their information.

3. DISCOUNTING: Discount each of agent's incremental prediction scores by a monotonically increasing factor of the time. The idea is that if signals are not informational substitutes, they will become ones if a sufficiently steep negative time gradient is applied. When successful, this mechanism achieves the ideal result of motivating all agents to reveal their information truthfully and promptly, including when they receive time-varying signals.

   We show that, under some light conditions, discounting will always work unless signals are perfectly correlated, i.e., have a deterministic relation given the outcome (as in Example 1).

Table 1 summarizes how the incumbent mechanism, and our three proposed mechanisms, measure up against each of the traits we described as desirable in the previous section.

We thoroughly investigate the Alice-Bob-Alice game (and, by extension, multi-player, multi-signal games) with both the logarithmic and the quadratic (Brier) scoring rule, when player signals have a multivariate normal distribution or a multivariate lognormal distribution.

These distributions are among the best-known continuous distributions and naturally arise in many situations. They are characterized, *inter alia*, by the correlation coefficient ($\rho \in [-1, 1]$) between Alice's and Bob's signals. When the logarithmic scoring rule is used, we find that when these signals are too well-correlated (whether positively or negatively), prompt truthfulness is *not* optimal. On the other hand, if the correlation is low, the game will be truthful and prompt. This includes the case $\rho = 0$, where, as is well-known, the sig- nals are conditionally independent, confirming the Chen et al. (2010) result for conditionally independent signals.

However, when the quadratic scoring rule, one of the oldest and most com- monly used scoring rules, is used with these multivariate distributions, it is *never* truthful for repeated predictions.

In all settings with either the logarithmic or the quadratic scoring rules, we show that our discounting mechanism restores prompt truthfulness, with the single exception of perfect correlation ($|\rho| = 1$) of the players' signals.

We make the observation that information aggregation works differently in the presence of a common-knowledge prior than where the prior is unknown. (The fact that agents have a common prior does not necessarily mean that they know what it is). Whenever there exists a public prediction forum, as for earnings estimates described above, or a market, the initial running consensus or the initial market price serves as a common-knowledge prior. On the other hand, if players do not know what their common prior is, even if they have one, their behavior is as if they have no common prior.

We show that the discounting mechanism can be effectively implemented with an automated market maker, as in Hanson's markets, thus showing that it may be practically applied in prediction markets.

1.4   Related Literature

Scoring rules have a long history, going back to De Finetti (1937), Brier (1950) and Good (1952). Proper scoring rules are often used for incentive-compatible belief elicitation of risk-neutral agents. Market scoring rules for prediction markets were introduced by Hanson (2003).

The role of Chen et al. (2010) (which is based on earlier papers Chen et al. (2007) and Dimitrov and Sami (2008)) and Chen and Waggoner (2016) in investigating the strategy-proofness of prediction markets was already described. Gao et al. (2013) and Kong and Schoenebeck (2018) resolve some more scenarios. Conitzer (2009) embarks on a program similar to ours, citing mechanism design as a guiding principle. Accordingly, he strives to achieve the Revelation Principle, where all experts announce their private information to some organizing entity that makes the appropriate Bayesian aggregation. As we discuss in Sect. 2.2 below, we do not share that vision: Experts often do not know what part of their belief stems from truly private information, and even when they do, they cannot afford to go on record with a prediction which is not their best judgement. His "Group-Rewarding Information Mechanism" is similar to our Group Prediction mechanism, and its lack of fairness is pointed out. Conitzer does not propose a mechanism that achieves prompt truthfulness.

Chen et al. (2010) also suggest discounting, that "reduces the opportunity for bluffing", in their words, but does not prevent it (Sect. 9.1), so their discounting mechanism does not achieve our basic requirement of truthfulness. The reason is that their formulation is different from ours, applying same discount to *before* and *after* scores. On the other hand, we discount every prediction score according to the time its prediction was made. The difference is crucial, because theirs does not result in a true market scoring rule, as defined by Hanson (2003). In consequence, our Sect. 2.4, on which our results rest, as well as our Sect. 5.2, do not apply to their formulation.

We shall occasionally rely on well-known facts of the normal and the multivariate normal distributions. The reader will find the basis for these in, e.g., Tong (2012).

To keep this paper short, we omitted some proofs, and other details. The reader will find a full version, including all proofs, in Ban (2018). The rest of this paper is organized as follows: In Sect. 2 we formulate the problem. In Sect. 3 we investigate which predictions settings are already truthful and prompt. In

Sect. 4 we offer strategy-proof mechanisms, and show how they can solve the gaps we have found. In Sect. 5 we summarize and offer concluding remarks.

2 Problem Formulation

2.1 Basics

Two players, Alice and Bob, make public predictions of a real parameter λ, whose prior distribution is π. The *outcome* x will be revealed after all predictions have been made. A prediction consists of revealing one's belief of the distribution of λ. Assume all agents take others' predictions as truthful. All agents (Alice, Bob, and the public) are Bayesian, and each prediction causes them to update their beliefs, i.e. the posterior distribution, of λ.[5]

The posterior beliefs are distributions of the parameter λ which are inferred from priors and likelihood functions using Bayesian techniques. In our discussion we find it more convenient and succinct to represent beliefs, without loss of generality, by a real number, rather than by a probability distribution. We use the fact that, in Bayesian analysis, when the likelihood functions belong to some family of distributions (e.g. exponential), all posterior beliefs belong to another family of distributions (Gamma distribution for the exponential family) $Q(\boldsymbol{Y}) \in \Delta(\lambda)$, called the *conjugate prior* of the first family. $\boldsymbol{Y}$ is a set of real *parameters* of the inferred distribution Q. We will assume models where, one, and only one of these parameters is dependent on previous predictions, while the rest $\boldsymbol{Y} \setminus \{x\}$ is known from the model, the timing and the identity of the believer, but *does not depend on any previous prediction*. An example illustrates this:

Example 2. Assume Alice's belief of λ to be normally distributed $N(\mu_A, 1/\tau_A)$ where μ_A is the mean and τ_A the accuracy (i.e. inverse of the variance), and Bob's is $N(\mu_B, 1/\tau_B)$ and independent of Alice's. τ_A and τ_B are set by the model and are commonly known. Assume an uninformative prior. Using well-known aggregation rules for independent normal observations, if Alice announces μ_A, Bob's belief changes to the normally distributed $N(\mu_{AB}, 1/\tau_{AB})$, where

$$\mu_{AB} = \frac{\tau_A \mu_A + \tau_B \mu_B}{\tau_A + \tau_B}$$

$$\tau_{AB} = \tau_A + \tau_B$$

Notice that τ_{AB} can be calculated without knowing any of the means μ_A, μ_B, while μ_{AB} can be evaluated once μ_A and μ_B is known.

In this context we are therefore able to describe a prediction by a single real number (the mean) rather than by a probability distribution. We shall say

---

[5] This formulation is different from the mechanism of prediction markets, but equivalent to it. In prediction markets, an agent replaces the current market prediction by his own. In our formulation, the agent merely announces a prediction, which, assuming the agent is truthful, becomes the market prediction by Bayesian inference. This is because all rational agents reach the same beliefs from the same data.

that Alice's prior belief A_1 is μ_A and Bob's prior belief B_1 is μ_B. After Alice makes her prediction, Bob's belief changes to μ_{AB}. After Alice and Bob both make a prediction, the public's belief is μ_{AB}. In context, these statements are unambiguously equivalent to specifying the probability distributions in full.

The prior π may be uninformative, assigning equal probabilities to all possibilities[6], or, if not, as also representable by a parameter. For example, if Alice and Bob participate in a prediction market, the prior parameter is the market value before Alice's first prediction.

Time is discrete, $t = 0, 1, 2, \ldots, T$, with T the time the outcome is known. $A_t, B_t, C_t \in \mathbb{R}$ are, respectively, Alice's, Bob's and the public's (or market's) beliefs at time t. At $t = 1$, A_1, B_1, C_1 are their respective prior beliefs. Any prediction takes place at $t > 1$. $t = 0$ is "pre-prior" time, when players beliefs are equal to their *private* signals, so that A_0, B_0 are respectively, Alice and Bob's private signals. At $t = 1$, each player is additionally aware of the public prior C_1, so that A_1 is an inference from A_0 and C_1 while B_1 is an inference from B_0 and C_1. If the public prior is uninformative, then we have $A_0 = A_1$ and $B_0 = B_1$. In other words, the players' priors equal their private signals. For completeness, we define $C_0 = C_1$.

To avoid degenerate exceptions, we assume the players' signals are *informative*. This means that $A_1 \neq C_1$ and $B_1 \neq C_1$.

The signals have a common-knowledge joint distribution $f(a, b; \lambda)$ conditional on the parameter

$$f(a, b; \lambda) := \Pr(A_1 = a, B_1 = b | \lambda)$$

The order of predictions is Alice, Bob, then Alice again, and then the outcome x is revealed.

A twice-differentiable, w.r.t. λ, proper scoring rule $S(\boldsymbol{p}, \lambda)$ incrementally rewards each prediction made, i.e., if a player's prediction changed the public's belief from $\boldsymbol{p}'$ to $\boldsymbol{p}$, the player's reward for this prediction, calculated when x is known, is $S(\boldsymbol{p}, x) - S(\boldsymbol{p}', x)$. Each player seeks to maximize their total reward.

As the scoring rule is proper, Bob will tell the truth on his only prediction, and Alice will tell the truth on her second and last prediction. The remaining question is whether Alice will tell the truth on her first prediction. More accurately, the question is of equilibrium: If Bob is truthful, and Bob and the public take Alice's predictions as truthful, is truth-telling Alice's best response?

2.2   Knowledge Model

As will be shown, the players behavior is affected by their *common-knowledge prior*, by which we mean a belief distribution which is explicitly known to both players and from which each inferred his or her current belief. That the common prior is commonly known is significant, because it is quite possible, and even

---

[6] Technically, an uninformative prior may be envisioned as the limit of a uniform or normal distribution as the variance goes to infinity.

likely, that the players share a prior but do not know what it is. For example, in predicting a poll, Alice and Bob may be basing themselves on knowledge of how their acquaintances voted, but they may not know which acquaintances they have in common. Or, they both may be basing themselves on a paper they read, but neither is aware that the other has read it. If the players do not know what their common prior is, they cannot infer anything from having one, and their behavior is as if they have an uninformative prior.

In prediction markets, and more generally in public prediction forums where all communication is done in public, the common-knowledge prior is known. In the context of this paper, it is the initial market value C_1 (or, equivalently, distribution π), when the Alice-Bob-Alice game starts.

2.3   Inferences from Predictions

Assume that inference functions are invertible, so that if a player's prediction is known, her signal can be computed. If Alice announces $A_1 = a$, the posterior outcome distribution can be calculated from her marginal distribution, $f_A(a; \lambda) := \Pr(A_1 = a | \lambda) = \int_{-\infty}^{\infty} f(a, b'; \lambda) db'$. Mark it $g(a)$.

$$g(a)_\lambda = \Pr(\lambda | A_1 = a) = \frac{\pi_\lambda f_A(a; \lambda)}{\int_{-\infty}^{\infty} \pi_{\lambda'} f_A(a; \lambda') d\lambda'} \tag{1}$$

Similarly, if Alice announces $A_1 = a$, and Bob privately observes $B_1 = b$, Bob's posterior outcome is inferred from $f(a, b; \lambda)$. It will become the public prediction when Bob announces it. Mark it $h(a, b)$.

$$h(a, b)_\lambda = \Pr(\lambda | A_1 = a, B_1 = b) = \frac{\pi_\lambda f(a, b; \lambda)}{\int_{-\infty}^{\infty} \pi_{\lambda'} f(a, b; \lambda') d\lambda'} \tag{2}$$

2.4   Maximizing the Reward

How does Alice maximize her total reward for both her predictions? And is this maximum achieved by telling the truth on both predictions? We will show that Alice maximizes her reward by *minimizing* Bob's reward, and therefore is truthful if truth minimizes Bob's reward.

Proposition 1. *Alice maximizes her expected total reward by making a first prediction that minimizes Bob's expected reward, where expectations are taken according to Alice's beliefs on her first prediction.*

Proof. The proof of this proposition is in the full version Ban (2018).

Corollary 1. *If a truthful first prediction minimizes Alice's expectation of Bob's reward, i.e., if*

$$\underset{b \sim B_1 | (A_1 = a)}{\mathrm{E}} \; \underset{\lambda \sim h(a, b)}{\mathrm{E}} \; \Pi_B(\lambda; a, \hat{a}, b) = \left\{ \underset{b \sim B_1 | (A_1 = a)}{\mathrm{E}} s(h(\hat{a}, b), h(a, b)) \right\} - s(g(\hat{a}), g(a))$$

is minimized at $\hat{a} = a$, then truth is Alice's best policy.

3   Which Prediction Settings Are Already Truthful?

We described in Example 1 an elementary setting that is not truthful, and a handful of other settings have been settled either way in the literature. But in the landscape of prediction settings that are of interest, the coverage has been very sparse. Beyond Chen and Waggoner (2016)'s criterion of "Informational Substitutes", which does not amount to an explicit algorithm[7], we have no procedure to settle any given case, and the problem remains opaque.

With the results we derived in Sect. 2.4, we now have such procedures. In the full version Ban (2018) we apply them to classify settings belonging to two of the most commonly-met continuous distributions: the multivariate normal distribution and the multivariate lognormal distribution, and the two most commonly used scoring rules: the Logarithmic and the Brier/Quadratic.

See details and results in Sect. 3 of the full version Ban (2018).

4   Strategy-Proof Mechanisms

4.1   Group Prediction

Our first strategy-proof mechanism scores the last prediction made by a proper scoring rule, and awards the score to *each* of the participating experts.

Proposition 2. *Let x be the outcome, and let the last prediction made before the outcome is revealed be distribution $\boldsymbol{p}$. Then the mechanism that awards $S(\boldsymbol{p}, x)$ to each participating player, where S is a proper scoring rule, is truthful. Furthermore, the mechanism elicits full disclosure.*

Proof. The proof of this proposition is in the full version Ban (2018).

The mechanism does not motivate promptness, which we defined as a strong preference: Players may predict as late as possible without harming their welfare.

Another drawback is unfairness: The mechanism awards all experts the same, regardless of their contribution. Indeed, a so-called expert who has no information of his own may reap the same reward as other experts by simply repeating the current public prediction. This means that, unless the number of experts is bounded, the mechanism is not loss-bounded. Attempts to fix this would be counterproductive, compromising truthfulness. For example, rewarding nothing to "predictions" that merely repeat the public prediction, will motivate an uninformed expert to pretend knowledge by "tweaking" the current public prediction.

---

[7] A submodularity property is required of the signal lattice, in a context described in their article.

Table 1. Mechanism score card

Mechanism	Truthful	Full disclosure	Prompt	Bounded loss
Market scoring rule	✗	✓	✗	✓
Group prediction	✓	✓	✗	✗
Enforce single prediction	✓	✗	✗	✓
Discounting	✓	✓	✓	✓

4.2 Enforce Single Prediction

Since multiple predictions are the source of the potential for manipulation, a mechanism that prevents that would restore general truthfulness.

Proposition 3. *Let S be a proper scoring rule. The mechanism that scores each prediction with the increment $S(\boldsymbol{p}, x) - S(\boldsymbol{p}', x)$, where $\boldsymbol{p}$ is the predicted distribution and $\boldsymbol{p}'$ the previous public prediction, and rewards each expert with the minimum score out of all her predictions, is truthful.*

Proof. Once an expert has made a prediction, a further prediction may only lower her reward expectation and so is not optimal. Since incremental, proper scoring rules are truthful with a single prediction, the mechanism is truthful. □

While this mechanism is truthful, and loss-bounded, it does not motivate promptness. Every expert needs to figure out the best time to make his single prediction, given other players' strategies, resulting in a Bayes-Nash equilibrium. As Azar et al. (2016) show, this can be complex. Furthermore, full disclosure is not assured, since experts may choose to make a prediction before getting their final signal.

4.3 Discounting

The last, and, we will argue, the most successful mechanism we suggest for incentive compatibility is *discounting*.

Discounting essentially uses a proper market scoring rule. As explained in the Introduction, for a proper scoring rule $S(\boldsymbol{p}, r)$, the market scoring rule scores a prediction $\boldsymbol{p}$ $S(\boldsymbol{p}, r) - S(\boldsymbol{p}', r)$ where $\boldsymbol{p}'$ is the outcome distribution that was replaced by $\boldsymbol{p}$, and r is the outcome. Now any scoring rule can be scaled by an arbitrary constant k and remain proper. Furthermore, $kS(\boldsymbol{p}, r) - k'S(\boldsymbol{p}', r)$ where k, k' may be different, is also proper. Generally, we can employ a time-varying scale factor $k(t) \in \mathbb{R}_{>0}$, and use the proper scoring rule $k(t)S(\boldsymbol{p}, r)$, where t is the time (which may be the elapsed time from some base, or an integer counter of events) of announcement of $\boldsymbol{p}$. Discounting means choosing $k(t)$ that is weakly decreasing in t, whence we get a discounted scoring rule $D(\boldsymbol{p}, r, t) := k(t)S(\boldsymbol{p}, r)$, where t is the time of prediction $\boldsymbol{p}$. When $t' < t$ is the time replaced prediction $\boldsymbol{p}'$ was made, the discounted market scoring rule for prediction $\boldsymbol{p}$ is

$$D(\boldsymbol{p}, r, t) - D(\boldsymbol{p}', r, t') = k(t)S(\boldsymbol{p}, r) - k(t')S(\boldsymbol{p}', r)$$

The idea of discounting is that, if without discounting Alice's optimal strategy is to hide or distort information on her first prediction, in order to reap a bigger benefit on her second prediction, her calculus will change if a sufficiently steep discount, motivating earlier predictions, is imposed on her reward.

We must be careful to use non-positive scoring rules with the discounting mechanism. Otherwise a possibility exists that the discounted scoring rule will have negative expectation for a prediction, creating a situation in which a player will prefer *not* predicting to making a truthful prediction. On the other hand, for non-positive scoring rules, $S(\boldsymbol{p}, r) \leq 0$, so that, whenever $0 \leq k(t) \leq k(t')$

$$k(t)S(\boldsymbol{p}, r) - k(t')S(\boldsymbol{p}', r) \geq k'(t)[S(\boldsymbol{p}, r) - S(\boldsymbol{p}', r)]$$

So, if our original scoring rule had positive expectation, so does our discounted one.

Note that the logarithmic scoring rule $\log p_r$ and quadratic scoring rule $2p_r - \boldsymbol{p} \cdot \boldsymbol{p} - 1$, which we have analyzed, are non-positive. Any scoring rule can be made non-positive by affine transform.

Remark 1. Non-positive scoring rules suffer from an artefact that is the mirror image of positive ones: Experts have positive score expectation ($[k(t) - k(t')]S(p, r)$) for merely repeating the current prediction, and so will do so even if they have no information. In fact, the mechanism can offer them this "reward" automatically, sparing them the need to make an empty prediction. While this is ugly, its effect is minor, and the expected loss is still bounded by $-k(t)S(p, r)$ at any time t. As in other cases, attempts to mend this would be counterproductive. E.g., if we deduct the "unearned" $[k(t) - k(t')]S(p, r)$ from every score, we will compromise truthfulness, as this reverts to the discounting mechanism proposed in Chen et al. (2010).

Proposition 4. *Using the notation of Sect. 2.3, if S is a non-positive scoring rule, and there exists K such that*

$$K \geq \frac{\underset{b \sim B_1 | (A_1 = a)}{\mathbb{E}} \left[S(h(a, b), h(a, b) - S(h(\hat{a}, b), h(a, b)) \right]}{S(g(a), g(a)) - S(g(\hat{a}), g(a))} \tag{3}$$

for every $A_1 = a$ and every possible $\hat{a}$, then the game is promptly truthful with a discount factor $k(t)$ that satisfies $k(t_1)/k(t_2) \geq K$, where t_1 is the time of Alice's first prediction and t_2 is the time of the second.

Proof. If (3) is satisfied, scoring rule $k(t)S(p, r)$ satisfies Corollary 1.   □

From Proposition 4, we state sufficient conditions for discounting to succeed.

Corollary 2. *If*

1. *S is a non-positive strictly proper scoring rule, and*
2. $\mathbb{E}_{b\sim B_1|(A_1=a)}\left[S(h(a,b),h(a,b)) - S(h(\hat{a},b),h(a,b))\right]$ *is bounded for every $a,\hat{a}$*
   - *In particular, if S is the quadratic scoring rule, or any other bounded scoring rule, and*
3. *The right-hand side of (3) is bounded when $\hat{a}\to\pm\infty$, and*
4. *$lim_{\hat{a}\to a}\dfrac{\frac{\partial^2}{\partial\hat{a}^2}S(h(\hat{a},b),h(a,b))}{\frac{\partial^2}{\partial\hat{a}^2}S(g(\hat{a}),g(a))}$ exists for every a,b.*

Then there exists a discount factor effective at restoring truthfulness.

Proof. As S is strictly proper, and the nominator of (3) is bounded, (3) can be unbounded only at infinity or at $\hat{a} = a$. At $\hat{a} = a$, (3) evaluates to $\frac{0}{0}$, so we invoke L'Hôpital's rule to find the limit. By the definition of a proper score rule, first derivatives again evaluate to $\frac{0}{0}$, so invoke L'Hôpital again for second derivatives.
□

In Sect. 4.3 of the full version Ban (2018), we show that discounting can restore truthfulness in the untruthful settings found in Sect. 3.

Finally, we can formulate a proposition for the prompt truthfulness of a general prediction forum.

Proposition 5. *The public and a set of experts predict a parameter λ. Let there be a fixed schedule in $[0, T]$ specifying when experts and public receive signals, and experts may make a prediction. Agents may receive multiple signals, and experts may have multiple prediction opportunities.*

Then the forum is generally truthful and prompt with a discount function $k(t) \in \mathbb{R}_{>0}$ that makes all Alice-Bob-Alice subgames in the schedule truthful and prompt. These subgames are all occurrences where any expert, identified as "Alice", has two consecutive prediction opportunities, and all experts who makes predictions in between are rolled into a single player identified as "Bob".

Proof. The proof of this proposition is in the full version Ban (2018).

5 Discussion

5.1 Conclusions

We showed that using proper scoring rules for rewarding predictions often leads to reticence or deception. We formulated criteria for determining which prediction settings are truthful, and made a complete classification of the class of prediction settings with normal and lognormal signals, under the logarithmic and quadratic scoring rules. We suggested three new strategy-proof mechanisms, and identified one of them, discounting, as having all desirable properties, and proved the applicability and effectiveness of the discounting mechanism.

5.2   A Market Maker for the Discounting Mechanism

Hanson (2003) has shown that every market scoring rule can be implemented by an automated market maker, who provides liquidity, and is willing to make any fair trade in shares of the form "pay \$1 if outcome is r", for every r. This applies to our discounting mechanism for the logarithmic scoring rule, which, as noted, takes the form of a market scoring rule for $S(\boldsymbol{p}, r) = k(t) \log p_r$.

Applying Hanson's explanations to our case, if at t there is an inventory $\boldsymbol{s}$ of shares s_r for every r, the instantaneous share price is

$$m(s_r) = e^{\frac{s_r}{k(t)}} \Big/ \int_{-\infty}^{\infty} e^{\frac{s_x}{k(t)}} \, dx$$

Assuming an infinitesimal trade path, this induces a cost function $C(\boldsymbol{s}, t)$. The cost of a trade changing the inventory from $\boldsymbol{s}$ to $\boldsymbol{s}'$ at t', is

$$C(\boldsymbol{s}', t') - C(\boldsymbol{s}, t) := k(t') \log \int_{-\infty}^{\infty} e^{\frac{s'_x}{k(t')}} \, dx - k(t) \log \int_{-\infty}^{\infty} e^{\frac{s_x}{k(t)}} \, dx$$

Since $k(t)$ is weakly decreasing in t, differentiating the above expression with respect to t' shows that the more any trade with the automated market maker is delayed, the higher its cost.

References

Azar, Y., Ban, A., Mansour, Y.: When should an expert make a prediction? In: Proceedings of the 2016 ACM Conference on Economics and Computation, pp. 125–142. ACM (2016)

Ban, A.: Strategy-proof incentives for predictions. arXiv preprint arXiv:1805.04867 [cs.GT] (2018)

Brier, G.W.: Verification of forecasts expressed in terms of probability. Weather Rev. **78**(1950), 1–3 (1950)

Chen, Y., et al.: Gaming prediction markets: equilibrium strategies with a market maker. Algorithmica **58**(4), 930–969 (2010)

Chen, Y., Reeves, D.M., Pennock, D.M., Hanson, R.D., Fortnow, L., Gonen, R.: Bluffing and strategic reticence in prediction markets. In: Deng, X., Graham, F.C. (eds.) WINE 2007. LNCS, vol. 4858, pp. 70–81. Springer, Heidelberg (2007). https://doi.org/10.1007/978-3-540-77105-0_10

Chen, Y., Waggoner, B.: Informational substitutes. In: 2016 IEEE 57th Annual Symposium on Foundations of Computer Science (FOCS), pp. 239–247. IEEE (2016)

Conitzer, V.: Prediction markets, mechanism design, and cooperative game theory. In: Proceedings of the Twenty-Fifth Conference on Uncertainty in Artificial Intelligence, pp. 101–108. AUAI Press (2009)

De Finetti, B.: La prevision: Ses lois logiques, ses sources subjectives. Ann. Inst. Henri Poincaré **7**(1937), 1–68 (1937)

Dimitrov, S., Sami, R.: Non-myopic strategies in prediction markets. In: Proceedings of the 9th ACM conference on Electronic commerce, pp. 200–209. ACM (2008)

Gao, X.A., Zhang, J., Chen, Y.: What you jointly know determines how you act: strategic interactions in prediction markets. In: Proceedings of the Fourteenth ACM Conference on Electronic Commerce, pp. 489–506. ACM (2013)

Gneiting, T., Raftery, A.E.: Strictly proper scoring rules, prediction, and estimation. J. Amer. Statist. Assoc. **102**(477), 359–378 (2007). https://doi.org/10.1198/016214506000001437. http://amstat.tandfonline.com/

Good, I.J.: Rational decisions. J. Roy. Stat. Soc. Ser. B (Methodol.) **14**(1), 107–114 (1952)

Hanson, R.: Combinatorial information market design. Inf. Syst. Frontiers **5**(1), 107–119 (2003)

Kong, Y., Schoenebeck, G.: Optimizing Bayesian information revelation strategy in prediction markets: the Alice Bob Alice case. In: LIPIcs-Leibniz International Proceedings in Informatics, vol. 94. Schloss Dagstuhl-Leibniz-Zentrum fuer Informatik (2018)

Tong, Y.L.: The Multivariate Normal Distribution. Springer, Heidelberg (2012)

Revealed Preference Dimension via Matrix Sign Rank

Shant Boodaghians[(⊠)]

Department of Computer Science, University of Illinois at Urbana-Champaign,
Urbana, IL, USA
boodagh2@illinois.edu

Abstract. Given a data-set of consumer behaviour, the Revealed Preference Graph succinctly encodes inferred relative preferences between observed outcomes as a directed graph. Not all graphs can be constructed as revealed preference graphs when the market dimension is fixed. This paper solves the open problem of determining exactly which graphs are attainable as revealed preference graphs in d-dimensional markets. This is achieved via an exact characterization which closely ties the feasibility of the graph to the Matrix Sign Rank of its signed adjacency matrix. The paper also shows that when the preference relations form a partially ordered set with order-dimension k, the graph is attainable as a revealed preference graph in a k-dimensional market.

Keywords: Revealed preference · Matrix sign rank · Partial order

1 Introduction

In standard economic analysis and mechanism design, it is often assumed that the agents' valuation functions are known *a priori*, or more commonly, a probability distribution over possible agent types is assumed to be known. However, in practice, we may only observe the prices which are set, and the subsequent behaviour of the agents. Assuming the agents act rationally, and that their utility functions are restricted to some well-defined class, information about the relative values attributed to various outcomes may be inferred by simply observing the agents' behaviour at various prices. This idea was first pioneered by Samuelson in 1938 [23], and a large body of work has followed. See [24] for a thorough survey on the subject. The concept came to be known as *revealed preference*: when an agent chooses different outcomes given different prices, she is (under some assumptions) revealing that one is preferable to the other.

Though this may seem natural at first, the implementation of these ideas has required much mathematical development for the description, characterization, and computation of the revealed preferences. In the model originally studied by Samuelson, the agent is assumed to have an underlying valuation function and a fixed budget. She seeks to choose the collection of goods which has the largest value while satisfying her budget constraint at the current prices. This

G. Christodoulou and T. Harks (Eds.): WINE 2018, LNCS 11316, pp. 66–79, 2018.
https://doi.org/10.1007/978-3-030-04612-5_5

is the most common formulation, though others exist (see *e.g.* [11]). This paper, however, deals only with the standard model.

The market consists of d distinct, separable goods, and a collection of goods (or *bundle*) is denoted as a vector $\boldsymbol{x} \in \mathbb{R}^d_{\geq 0}$, where the i-th coordinate of $\boldsymbol{x}$ represents the quantity of the i-th good. The prices are linear, and are described by a vector $\boldsymbol{p} \in \mathbb{R}^d_{\geq 0}$ such that the price of a bundle $\boldsymbol{x}$ is given by the inner product $\langle \boldsymbol{p}, \boldsymbol{x} \rangle$. In this model, a bundle $\boldsymbol{x}$ is *revealed preferred* to a bundle $\boldsymbol{y}$ if at current prices $\boldsymbol{p}$, the agent chose $\boldsymbol{x}$, but $\boldsymbol{y}$ was more affordable, *i.e.* $\langle \boldsymbol{p}, \boldsymbol{x} \rangle \geq \langle \boldsymbol{p}, \boldsymbol{y} \rangle$. Since the agent is assumed to be maximizing value within a budget constraint, $\boldsymbol{y}$ must be less valuable than $\boldsymbol{x}$.

Samuelson originally asked whether revealed preferences may be used to verify whether an agent's behaviour is consistent with the model assumptions: if an agent's behaviour is contradictory, then one must conclude that the model assumption is incorrect. He first conjectured a simple test, which was confirmed to be correct in special cases by Rose [21], but disproved in the general case. Houthakker [19] proposed a stronger test of "cyclical consistency" and proved its correctness non-constructively in our setting. The most famous result is given by Afriat [1], where he shows a slightly more general result, and gives explicit constructions of the valuation function as a certificate for consistency. The notion of cyclical consistency has since been called the "Strong/Generalized Axiom of Revealed Preference" (SARP/GARP), and is used to this day in many empirical settings [16,25,26], including as bidding rules in combinatorial auction mechanisms, to deter non-truthful bidding practices [3,4,12,17]. We will not delve into the details of the axioms of revealed preference, but a thorough survey may be found at [24]. It is however helpful to be familiar with the concept of a *revealed preference graph*, which is defined below. A more familiar treatment to economic audiences is given in [10].

Revealed Preference Graphs: Recall that in Samuelson's model, an agent may choose goods from the *consumption space* $\mathbb{R}^d_{\geq 0}$, and seeks to choose the bundle $\boldsymbol{x}^*$ which maximizes her valuation $v(\boldsymbol{x})$ subject to the budget constraint $\langle \boldsymbol{p}, \boldsymbol{x} \rangle \leq 1$, up to re-scaling. Suppose now that we make n observations of this agent at *different* price points $\boldsymbol{p}_1, \boldsymbol{p}_2, \ldots, \boldsymbol{p}_n \in \mathbb{R}^d_{\geq 0}$, and that at prices $\boldsymbol{p}_i$, her optimal bundle was $\boldsymbol{x}_i$. If her behaviour is rational, then whenever $\langle \boldsymbol{p}_i, \boldsymbol{x}_i \rangle \geq \langle \boldsymbol{p}_i, \boldsymbol{x}_j \rangle$, she must value $\boldsymbol{x}_i$ greater than $\boldsymbol{x}_j$ since the latter was affordable when the former was chosen. Thus, she has revealed that $\boldsymbol{x}_i$ is preferable to $\boldsymbol{x}_j$.

These preference relations may be modelled as a directed graph: let G be a graph on vertex set $\{1, \ldots, n\}$, and add an edge directed from i to j if $\boldsymbol{x}_i$ is revealed preferred to $\boldsymbol{x}_j$. Thus, $(i, j) \in G$ if and only if $\langle \boldsymbol{p}_i, \boldsymbol{x}_i \rangle \geq \langle \boldsymbol{p}_i, \boldsymbol{x}_j \rangle$. This graph is implicit in the proofs of Afriat, and the notion of cyclical consistency is equivalent to requiring that G not contain any directed cycles. In general, a preference graph inferred from observations need not be acyclic.

Most uses of revealed preference as bidding rules in combinatorial auctions (cited above) rely on testing properties of this revealed preference graph. In past work [7,8], we have asked whether such tests, *e.g.* the *minimum feedback vertex set*, are efficiently computable. We concluded that for this one test, its computational complexity is in fact dependent on the market dimension d: when the dimension of the p_i and x_j vectors is fixed, but the number of observations is unbounded, then the class of observable graphs is restricted, and the computational complexity of some problems may depend on the parameter d. For example, it was shown by Deb and Pai [13] that when $d = n$, every directed graph on n vertices is observable over $\mathbb{R}^n$, but that for all fixed d, there exist exponentially large graphs which can not be observed in d dimensions.

This past work has led us to asking whether one could characterize the class of preference graphs observable in the market $\mathbb{R}^d$ for some fixed d. In fact, this question had been posed as an open problem by Echenique [10].

Question: For a fixed dimension d, which directed graphs may be observed as revealed-preference graphs on bundles in $\mathbb{R}^d$?

This paper answers this question by giving an exact characterization of all graphs observable over $\mathbb{R}^d$, given in terms of the *matrix sign rank* of signed adjacency matrices. The notion of matrix sign rank deals with the existence of low-rank matrices whose entries satisfy certain sign constraints; see [6] for an introduction. More formally, the matrix sign rank of a *sign matrix* $S \in \{0, +, -\}^{n \times m}$ is defined as the least rank real-valued matrix $M \in \mathbb{R}^{n \times m}$ whose entries have the same signs as the entries in S. Matrix sign rank has been influential in many fields, including complexity theory and learning theory. Seminal lower bounds in communication complexity [15] and circuit complexity [20] rely on using the matrix sign rank of a problem to measure its hardness, whereas linear classification algorithms benefit from low-dimensional embeddings of classification problems, as given by sign-rank [2, 9, 22].

Summary of the Results. To answer the above question, we introduce the notion of *RP dimension*, defined as follows: Given a directed graph G, what is the least d such that G may be observed as a preference graph over $\mathbb{R}^d$?

In this paper, we give an exact characterization of the set of graphs with RP dimension d, for all $d \geq 1$. We show that, for a given graph, it can be realized as preference observations over $\mathbb{R}^d$ if and only if its signed adjacency matrix has sign-rank at most $d + 1$. Thus, the RP dimension of the graph G is exactly the sign-rank of its signed adjacency matrix, minus one. In fact, this paper shows that determining the RP dimension of directed graphs is equivalent to determining the sign rank of a large class of sign matrices, a problem which is known to be NP-hard [6].

This paper also considers the special case of directed graphs which represent *partially ordered sets*, or posets. We show that the RP dimension is at most the *order-dimension* of the poset, and that this bound is tight when the order-dimension is at most 3. However, there exist posets of arbitrarily large order-dimension, which can be realized in $\mathbb{R}^3$.

2 Model, Preliminaries, and Summary of Results

This section formally lays out the concepts introduced above. As discussed in the introduction, it is assumed that an agent is observed repeatedly in a market. Faced with price-vector $p \in \mathbb{R}_{\geq 0}^d$, she chooses the item which maximizes her valuation subject to a budget constraint. Thus, she chooses x^* as the bundle which maximizes $v(x)$ subject to $\langle p, x \rangle \leq 1$. Assume that on the i-th observation, the agent was faced with prices p_i, and chose the bundle x_i. Then we have that x_i is *revealed preferred* to x_j if $\langle p_i, x_i \rangle \geq \langle p_i, x_j \rangle$, since x_j must have been affordable when x_i was chosen. Given a collection of observations $(p_1, x_1), (p_2, x_2), \ldots, (p_n, x_n)$, we may construct a directed *preference graph G* on vertex set $\{1, \ldots, n\}$ with an edge from i to j if x_i is revealed preferred to x_j.

This paper, however, does not deal with constructing preference graphs from data sets, but rather of constructing data sets from preference graphs. Thus, we introduce the notion of the *realization* of a preference graph:

Definition 1 (RP realization). Let $G = (V, E)$ be a directed graph with vertices labelled $\{1, \ldots, n\}$, and let $\mathcal{X} := \{(p_1, x_1), \ldots, (p_n, x_n)\}$ be pairs of vertices in $\mathbb{R}^d$ such that $p_i \geq 0$ for all $i \leq n$. Then $\mathcal{X}$ is said to *RP-realize* G if for all $1 \leq i, j \leq n$, the directed edge (i, j) is present in G if and only if $\langle p_i, x_i \rangle > \langle p_i, x_j \rangle$.

Note that we require strict inequality to induce preference. This is purely for mathematical convenience, and is not standard in the definitions of revealed preference. Since we are only considering the existence of the realization, rather than the realization itself, this is assumed without loss of generality. We also define a notion of *weak* RP realization, which we will show is equivalent.

Definition 2 (Weak RP realization). As above, let $G = (V, E)$ be a directed graph with vertices labelled $\{1, \ldots, n\}$, and let $\mathcal{X} := \{(p_1, x_1), \ldots, (p_n, x_n)\}$ be pairs of vertices in $\mathbb{R}^d$ such that $\langle p_i, \mathbf{1} \rangle > 0$ for all $i \leq n$, where $\mathbf{1} = (1, 1, 1, \ldots)$. Then $\mathcal{X}$ is said to *weakly RP-realize* G if for all $1 \leq i, j \leq n$, the directed edge (i, j) is present in G if and only if $\langle p_i, x_i \rangle > \langle p_i, x_j \rangle$.

The difference between RP realization and weak RP realization is the restriction on the possible p vectors. We will show below that these two notions of RP realization are equivalent in the following sense: given a graph G and an integer d, there exists an RP realization of G in $\mathbb{R}^d$ if and only if there exists a weak RP realization in $\mathbb{R}^d$.

It is natural to ask whether an RP realization is possible, and whether this depends on the value of d. In fact, it was shown by Dep and Pai [13] that when $d = n$, a realization is always possible. (Simply set x_i to be the i-th standard basis vector, and set $(p_i)_j = 0$ if $(i, j) \in G$, 1 if $i = j$, and 2 if $(i, j) \notin G$.) We wish to determine the minimum value of d for which this is possible. Thus, we introduce the notion of RP dimension.

Definition 3 (RP dimension). Let $G = (V, E)$ be a directed graph on n vertices. Then the *RP dimension* of G, (denoted $\mathsf{RPDim}(G)$) is the minimum d

such that there exists an RP realization of G in $\mathbb{R}^d$. We denote as REVPREFDIM the computational problem of computing $\mathsf{RPDim}(G)$ for a given di-graph G.

As mentioned above, we will show that the RP dimension does not change if the RP realization is allowed to be weak. We will be characterizing the RP dimension of candidate preference graphs by the sign-rank of an associated sign matrix. We define below the notion of realization for sign matrices, and define sign rank as the minimum rank of a realization:

Definition 4 (Matrix sign realization). Let $M \in \{+, -, 0\}^{n \times m}$ be an $n \times m$ matrix whose entries are given by the symbols $+$, $-$, and 0. Let $A \in \mathbb{R}^{n \times m}$ be an $n \times m$ real-valued matrix. Then A is a *matrix sign realization* of M if $A_{ij} > 0$ whenever $M_{ij} = +$, $A_{ij} < 0$ whenever $M_{ij} = -$, and $A_{ij} = 0$ whenever $M_{ij} = 0$.

Definition 5 (Matrix sign-rank). Let $M \in \{+, -, 0\}^{n \times m}$, then the *sign-rank* of M, (denoted $\mathsf{SgnRnk}(M)$) is the minimum r such that there exists a matrix sign realization A of M with rank r. We denote as MATSGNRNK the computational problem of finding $\mathsf{SgnRnk}(M)$ for a given matrix M.

Finally we will show that when the preference graph is induced by a *partially ordered set* or *poset*, the RP dimension of the graph is related to properties of the poset. Partially ordered sets have been the subject of much study, and many textbooks on the matter make for a good introduction (see *e.g.* [18]). Below is a formal definition, included for completeness.

Definition 6 (Partially-ordered set (Poset)). Let S be some (finite) ground set, and let $\succ$ be a transitive, acyclic, and irreflexive, binary relation on S. That is, for all $a, b, c \in S$, $a \nsucc a$, and if $a \succ b$ and $b \succ c$, then $a \succ c$. [1] Then the pair $(S, \succ)$ is termed a *partially-ordered set, or poset*. The poset may also be seen as a directed graph $G = (S, \succ)$, which must be acyclic and transitively closed. A *total order* is a poset whose underlying undirected graph is complete. Alternatively, a total order is the poset induced by a ranking of the elements of S.

Every partial order is the *intersection* of some total orders. The *order dimension* of a poset captures the least number of total orders needed to realize it.

Definition 7 (Order dimension). Let $(S, \succ)$ be a poset, and $\mathcal{O} = \{\succ_1, \ldots, \succ_k\}$ be k distinct total orders on S. Then $\mathcal{O}$ *realizes* the poset $(S, \succ)$ if for all $a, b \in S$, $a \succ b$ if and only if $a \succ_i b$ for all $1 \leq i \leq k$. The *order dimension* of $(S, \succ)$ is the minimum k such that there exists a collection of k total orders which realize $\succ$.

2.1 Results

We formally state here the results of this paper and outlines of their proofs. The proofs in full technical detail will be presented in the next section. The main

---

[1] These two conditions imply that the relation must be acyclic.

goal is to show that for each directed graph G, there exists an associated sign matrix M such that $\mathsf{RPDim}(G) = \mathsf{SgnRnk}(M) - 1$.

For simplicity of notation, though, we extend all directed graphs with a fully-dominated and fully-dominating node as follows: given a directed graph G, let G^+ be the graph obtained by adding two nodes s and t to G, and for all $v \in G$, adding an edge (s,v) and an edge (v,t) to G, plus the edge (s,t).

Finally, for a directed graph G, we define its signed adjacency matrix $M(G)$ as follows: $M(G)_{ij}$ is 0 if $i = j$, $+1$ if $(i,j) \in G$, and -1 otherwise. This allows us to formally state the first, and most important result:

Theorem 1. *For all G, $\mathsf{RPDim}(G) = \mathsf{RPDim}(G^+) = \mathsf{SgnRnk}(M(G^+)) - 1$.*

To show this, we first constructively show that any RP realization in d dimensions implies a sign-rank realization with rank $d+1$. Next, we argue that a sign-rank realization of $M(G^+)$ with rank r implies a weak RP realization in $r-1$ dimensions. Using the following lemma, we conclude the theorem:

Lemma 1. *A graph G admits an RP realization with vectors in $\mathbb{R}^d$ if and only if it admits a weak RP realization with vectors in $\mathbb{R}^d$.*

As a special case, we consider the case of partially ordered sets.

We begin by showing first that the order-dimension of the poset is an upper-bound on the RP dimension.

Theorem 2. *Let $(S, \succ)$ be a poset with order-dimension k, a and let $G = (S, \succ)$ be the associated directed graph. Then $\mathsf{RPDim}(G) \leq k$.*

This is shown by first introducing the "dominance ordering" interpretation of order dimension, and then embedding the elements of the partial order in such a way as to achieve such an ordering. A dominance ordering is simply a partial order on points in $\mathbb{R}^k$, where $\boldsymbol{x} \succ \boldsymbol{y}$ if and only if $x_i \geq y_i$ for all $1 \leq i \leq k$. This naturally has order dimension at most k, since it suffices to sort the points by their positing in each of the k dimensions.

Unfortunately, the converse does not hold. We show first that for k sufficiently small, order dimension and RP dimension are equal, but bad examples exist for large order dimension.

Theorem 3. *Let $(S, \succ)$ be a poset with order-dimension k, a and let $G = (S, \succ)$ be the associated directed graph. Then $\mathsf{RPDim}(G) \geq \min\{k, 3\}$. Furthermore, for all $k \geq 3$, there exist posets $G = (S, \succ)$ of order-dimension k, but $\mathsf{RPDim}(G) = 3$.*

3 RP Dimension and Sign Rank – Proof of Theorem 1

In this section, we prove Theorem 1, as was outlined in Sect. 2.1. We begin by showing the equivalence of RP realization and weak RP realization, given by Definitions 1 and 2, respectively. Thus, we wish to prove Lemma 1. The proof of this lemma effectively reduces to transforming a cone into the cone spanned by the standard basis vectors, though we have included a careful analysis, which allows us to greatly reduce the technical burden of proving Theorem 1.

Lemma 1. *A graph G admits an RP realization with vectors in $\mathbb{R}^d$ if and only if it admits a weak RP realization with vectors in $\mathbb{R}^d$.*

Proof. An RP realization is by definition a weak RP realization, so one direction of the implication follows trivially. It suffices to show that the existence of a weak RP realization implies the existence of an RP realization, which we will do constructively. Let $\mathcal{X} := \{(\boldsymbol{p}_1, \boldsymbol{x}_1), \ldots, (\boldsymbol{p}_n, \boldsymbol{x}_n)\}$ be the d-dimensional vectors which *weakly* realize G. In other words, $(i, j) \in G$ if and only if $\langle \boldsymbol{p}_i, \boldsymbol{x}_i \rangle > \langle \boldsymbol{p}_i, \boldsymbol{x}_j \rangle$. Let $\boldsymbol{b}_1, \boldsymbol{b}_2, \ldots, \boldsymbol{b}_{d-1}$ be a basis for the space orthogonal to the all-ones vector $\boldsymbol{1}$. Thus, $\langle \boldsymbol{1}, \boldsymbol{b}_i \rangle = 0$ for all i, and $\mathcal{B} = \{\boldsymbol{1}, \boldsymbol{b}_1, \boldsymbol{b}_2, \ldots, \boldsymbol{b}_{d-1}\}$ is a basis for $\mathbb{R}^d$.

We wish to express the $\boldsymbol{p}$ vectors as positive combinations of the $\boldsymbol{b}$ vectors, and thus restrict them to lie in a cone. This will allow us to map the rays of the cone to the standard basis vectors, and get the desired "strong" RP realization. Since $\mathcal{B}$ is a basis, we can express $\boldsymbol{p}_i = \sum_{j=0}^{d-1} \alpha_j^i \boldsymbol{b}_j$ for all i and j, where $\boldsymbol{b}_0 = \boldsymbol{1}$. We have chosen the $\boldsymbol{b}_i$'s as orthogonal to $\boldsymbol{1}$, and by assumption $\langle \boldsymbol{1}, \boldsymbol{p}_i \rangle > 0$ for all i. Hence, $\alpha_0^i > 0$ for all i, and we define $\epsilon = \min\{\alpha_0^1, \ldots, \alpha_0^n\} > 0$. However, for all $j \neq 0$, we may have α_j^i negative. To this end, define $\lambda_j = \min\{-1, \alpha_j^1, \alpha_j^2, \ldots, \alpha_j^n\}$ for all $1 \leq j \leq d - 1$. Hence, $\alpha_j^i - \lambda_j \geq 0$ for all i and j. We now define a slightly modified basis $\widehat{\mathcal{B}} = \{\boldsymbol{b}_1, \ldots, \boldsymbol{b}_{d-1}, \widehat{\boldsymbol{b}}_d\}$, where $\widehat{\boldsymbol{b}}_d = \epsilon \boldsymbol{1} + \sum_{j=1}^{d-1} \lambda_j \boldsymbol{b}_j$. In this new basis, we can express

$$\boldsymbol{p}_i = \frac{\alpha_0^i}{\epsilon} \cdot \widehat{\boldsymbol{b}}_d + \sum_{j=1}^{d-1} \boldsymbol{b}_j \cdot \left(\alpha_j^i - \frac{\alpha_0^i \lambda_j}{\epsilon} \right) \tag{1}$$

Recall that $\lambda_j < 0$ and $\alpha_0^i \geq \epsilon$, so $\frac{\alpha_0^i \lambda_j}{\epsilon} \leq \lambda_j$, and thus $\alpha_j^i - \frac{\alpha_0^i \lambda_j}{\epsilon} \geq \alpha_j^i - \lambda_j > 0$, by construction. Thus, not only is $\widehat{\mathcal{B}}$ a basis for $\mathbb{R}^d$, but the $\boldsymbol{p}_i$ vectors are *non-negative* combinations of the basis vectors.

It remains, then to construct a linear map which goes between the standard basis and the basis $\widehat{\mathcal{B}}$. Let B be the matrix whose columns are the vectors of $\widehat{\mathcal{B}}$, and note that for the j-th standard basis vector $\boldsymbol{e}_j$, we have $B\boldsymbol{e}_j = \boldsymbol{b}_j$, for all $1 \leq j \leq d$, setting $\boldsymbol{b}_d := \widehat{\boldsymbol{b}}_d$. Therefore, $B^{-1}\boldsymbol{b}_j = \boldsymbol{e}_j$ for all $1 \leq j \leq d$. Since we have shown that the $\boldsymbol{p}$ vectors are non-negative combinations of the $\widehat{\mathcal{B}}$ vectors, we may conclude that for all i, $B^{-1}\boldsymbol{p}_i$ has all non-negative entries. Furthermore,

$$\langle B^{-1}\boldsymbol{p}_i, B^{\mathsf{T}}\boldsymbol{x}_j \rangle = (B^{-1}\boldsymbol{p}_i)^{\mathsf{T}} B^{\mathsf{T}} \boldsymbol{x}_j = \boldsymbol{p}_i^{\mathsf{T}} (B^{-1})^{\mathsf{T}} B^{\mathsf{T}} \boldsymbol{x}_j = \langle \boldsymbol{p}_i, \boldsymbol{x}_j \rangle \tag{2}$$

Therefore, setting $\widehat{\boldsymbol{p}}_i = B^{-1}\boldsymbol{p}_i$, and $\widehat{\boldsymbol{x}}_j = B^{\mathsf{T}}\boldsymbol{x}_j$, we have that $\langle \widehat{\boldsymbol{p}}_i, \widehat{\boldsymbol{x}}_j \rangle > 0$ if and only if $\langle \boldsymbol{p}_i, \boldsymbol{x}_j \rangle > 0$, so $\widehat{\mathcal{X}} := \{(\widehat{\boldsymbol{p}}_1, \widehat{\boldsymbol{x}}_1), \ldots, (\widehat{\boldsymbol{p}}_n, \widehat{\boldsymbol{x}}_n)\}$ are d-dimensional vectors which *strongly* realize G, as desired.   □

To complete the proof of the theorem, it remains to construct low-rank sign matrices for preference graphs which have low-dimensional RP realizations, and construct low-dimensional *weak* RP realizations when the *augmented* directed graph has a sign-incidence matrix with low sign rank. We begin by recalling the definition of the augmented preference graph: For any directed graph G, let

G^+ be constructed by appending two nodes s and t to G, adding the directed edge (s,t), and for all $v \in G$, adding the directed edges (s,v) and (v,t). We begin by observing that the addition of these two extra nodes does not affect the RP dimension of the graph:

Claim. $\mathsf{RPDim}(G) = \mathsf{RPDim}(G^+)$.

Proof. Let $d := \mathsf{RPDim}(G)$ and $d' := \mathsf{RPDim}(G^+)$. Clearly, $\mathsf{RPDim}(G) \leq d'$, since it suffices to remove the vectors representing the s and t nodes from any realization of G^+ in d' dimensions. It remains to show $\mathsf{RPDim}(G^+) \leq d$. We say a vector $\boldsymbol{x} = (x_1, \ldots, x_d)$ *dominates* $\boldsymbol{y} = (y_1, \ldots, y_d)$ if $x_i \geq y_i$ for all $1 \leq i \leq d$. Now, let $\mathcal{X} := \{(\boldsymbol{p}_1, \boldsymbol{x}_1), \ldots, (\boldsymbol{p}_n, \boldsymbol{x}_n)\}$ be the d-dimensional vectors which realize G. Assume that the realization is a standard realization, as in Definition 1, as opposed to a weak one. Then we must have that if $\boldsymbol{x}_i$ dominates $\boldsymbol{x}_j$, there is an (i,j) edge in G.

Now, the collection $\boldsymbol{x}_1, \ldots, \boldsymbol{x}_n$ is finite, and so there must exist vectors $\boldsymbol{x}_s$ and $\boldsymbol{x}_t$ such that $\boldsymbol{x}_s$ dominates $\boldsymbol{x}_i$ and $\boldsymbol{x}_i$ dominates $\boldsymbol{x}_t$ for all $1 \leq i \leq n$. It suffices to set $\boldsymbol{p}_s = \boldsymbol{p}_t = \mathbf{1}$, and this gives a d-dimensional realization of G^+, as desired. □

Now that we have shown that G and G^+ have the same sign-rank, we may introduce the signed adjacency matrix. For any graph G on the vertex set $\{1, \ldots, s\}$, let $M(G)$ be defined as

$$M(G)_{ij} = \begin{cases} 0 & \text{if } i = j \\ + & \text{if } (i,j) \in G \\ - & \text{if } (i,j) \notin G \end{cases} \tag{3}$$

In what follows, we show that if $\mathsf{RPDim}(G^+) = d$, then $\mathsf{SgnRnk}(M(G^+)) \leq d+1$, and if $\mathsf{SgnRnk}(M(G^+)) = r$, then there is a weak RP realization for G^+ in $r-1$ dimensions. Both of those directions will be shown constructively, following a similar construction. We begin by showing the first direction:

Lemma 2. $\mathsf{SgnRnk}(M(G^+)) \leq \mathsf{RPDim}(G^+) + 1$.

Proof. Let $\mathcal{X} := \{(\boldsymbol{p}_1, \boldsymbol{x}_1), \ldots, (\boldsymbol{p}_n, \boldsymbol{x}_n)\}$ be the d-dimensional vectors which realize G^+, that is for all $i, j \in V(G)$, $(i,j) \in E(G)$ if and only if $\langle \boldsymbol{p}_i, \boldsymbol{x}_i \rangle > \langle \boldsymbol{p}_i, \boldsymbol{x}_j \rangle$. We construct the following matrix: let $A(\mathcal{X})$ be the $n \times n$-dimensional matrix whose entries are given by $A(\mathcal{X})_{ij} = \langle \boldsymbol{p}_i, \boldsymbol{x}_i - \boldsymbol{x}_j \rangle$ for all $i, j \leq n$. Observe that we have chosen the entries of $M(G)$ to be exactly the signs of the entries of $A(\mathcal{X})$. Thus, it suffices to show that $A(\mathcal{X})$ has rank at most $d+1$, which will imply that $M(G)$ has sign-rank at most $d+1$. Indeed,

$$A(\mathcal{X}) = \underbrace{\begin{bmatrix} \leftarrow \boldsymbol{p}_1 \rightarrow & \langle \boldsymbol{p}_1, \boldsymbol{x}_1 \rangle \\ \leftarrow \boldsymbol{p}_2 \rightarrow & \langle \boldsymbol{p}_2, \boldsymbol{x}_2 \rangle \\ \vdots & \vdots \\ \leftarrow \boldsymbol{p}_n \rightarrow & \langle \boldsymbol{p}_n, \boldsymbol{x}_n \rangle \end{bmatrix}}_{n \times (d+1)} \cdot \underbrace{\begin{bmatrix} \uparrow & \uparrow & & \uparrow \\ -\boldsymbol{x}_1 & -\boldsymbol{x}_2 & \cdots & -\boldsymbol{x}_n \\ \downarrow & \downarrow & & \downarrow \\ 1 & 1 & \cdots & 1 \end{bmatrix}}_{(d+1) \times n} \tag{4}$$

Since the inner-dimension of the product is $d + 1$, this implies that $A(\mathcal{X})$ has rank at most $d + 1$, as desired.   □

We will use this same construction to show the converse. The extension G^+ is required to ensure that this is possible, and that the vectors do indeed form a weak RP realization.

Lemma 3. $\mathsf{RPDim}(G^+) \leq \mathsf{SgnRnk}(M(G^+)) - 1$.

Proof. Let A be some rank-r realization of $M(G^+)$. Assume without loss of generality that the first and second rows and columns of $M(G^+)$ are associated to the dominating and dominated vertices, respectively. Thus, $M(G^+)$ has the form

$$\begin{bmatrix} 0 & + & + & + & \cdots \\ - & 0 & - & - & - & \cdots \\ - & + & 0 & * & * & \cdots \\ - & + & * & 0 & * & \cdots \\ \vdots & \vdots & \vdots & \vdots & \ddots & \ddots \end{bmatrix}$$

Thus, letting A_i be the i-th row of A, we have that $A_1 - A_2$ is an all-positive vector. Since A has rank r, we may set $\boldsymbol{a}_1 = A_1 - A_2$, and extend it to a basis $\mathcal{A} = \{\boldsymbol{a}_1, \boldsymbol{a}_2, \ldots, \boldsymbol{a}_r\}$ for the rows of A. Let R be the $r \times n$ matrix whose rows are the vectors of $\mathcal{A}$, and let L be the matrix of coefficients such that $A = LR$. Note that scaling the columns of A is the same as scaling the columns of R, and this scaling process does not affect the rank of the matrix. Furthermore, if the scaling factors are positive, then the sign pattern is unaffected. Thus, we may assume without loss of generality that $\boldsymbol{a}_1 = \boldsymbol{1}$, by rescaling column j by $1/(\boldsymbol{a}_1[j]) > 0$, for all j. (We are using square brackets to denote the entries of the vector.) Thus, we may interpret the entries of L and R as in Eq. (4). Now, if the i-th row of L is given by the vector $\boldsymbol{\ell}_i$, we set $\boldsymbol{p}_i := \boldsymbol{\ell}_i[1..r-1]$, and assume $\langle \boldsymbol{p}_i, \boldsymbol{x}_i \rangle = \boldsymbol{\ell}_i[r]$. Furthermore, if the j-th column of R is given by $(\boldsymbol{r}_j, 1)$, then we set $\boldsymbol{x}_j = -\boldsymbol{r}_j$. Since the diagonal entries of $M(G^+)$ are zero, we must have that $1 \cdot \langle \boldsymbol{p}_i, \boldsymbol{x}_i \rangle + \langle \boldsymbol{p}_i, -\boldsymbol{x}_i \rangle = 0$, which is consistent.

Thus, we have vectors $\boldsymbol{p}_i$ and $\boldsymbol{x}_j$ in $\mathbb{R}^{r-1}$ such that $\langle \boldsymbol{p}_i, \boldsymbol{x}_i \rangle > \langle \boldsymbol{p}_i, \boldsymbol{x}_j \rangle$ if and only if $(i, j) \in G^+$. It suffices to transform these vectors to ensure $\langle \boldsymbol{1}, \boldsymbol{p}_i \rangle > 0$ for all i. Recall that we have assumed that G^+ contains the edges $(1, i)$ and $(i, 2)$ for all $3 \leq i \leq n$. Thus, we must have $\langle \boldsymbol{p}_i, \boldsymbol{x}_i \rangle \leq \langle \boldsymbol{p}_i, \boldsymbol{x}_1 \rangle$ and $\langle \boldsymbol{p}_i, \boldsymbol{x}_i \rangle > \langle \boldsymbol{p}_i, \boldsymbol{x}_2 \rangle$. Therefore, $\langle \boldsymbol{p}_i, \boldsymbol{x}_1 - \boldsymbol{x}_2 \rangle > 0$ for all $i \geq 3$. Similarly to the proof of Lemma 1, we will use this to find an appropriate linear transformation for the $\boldsymbol{p}$ and $\boldsymbol{x}$ vectors. Let Q be any invertible matrix such that $Q(\boldsymbol{x}_1 - \boldsymbol{x}_2) = \boldsymbol{1}$. Then, setting $\widehat{\boldsymbol{p}}_i := (Q^{-1})^\mathsf{T} \boldsymbol{p}_i$, and $\widehat{\boldsymbol{x}}_j := Q\boldsymbol{x}_j$, we have $\langle \boldsymbol{p}_i, \boldsymbol{x}_j \rangle = \langle \widehat{\boldsymbol{p}}_i, \widehat{\boldsymbol{x}}_j \rangle$, and $0 < \langle \widehat{\boldsymbol{p}}_i, \widehat{\boldsymbol{x}}_1 - \widehat{\boldsymbol{x}}_2 \rangle = \langle \widehat{\boldsymbol{p}}_i, \boldsymbol{1} \rangle$. Thus, we have constructed a weak RP realization for G^+ in $r - 1$ dimensions.   □

We claim that this completes the proof of Theorem 1: Lemma 1 ensures that a weak realization is possible if and only if a "strong" one is, the above Claim ensures that $\mathsf{RPDim}(G^+) = \mathsf{RPDim}(G)$, these two facts along with Lemma 2 imply that $\mathsf{RPDim}(G) \leq \mathsf{SgnRnk}(M(G^+)) - 1$, and finally, Lemma 3 implies that $\mathsf{RPDim}(G) \geq \mathsf{SgnRnk}(M(G^+)) - 1$, from which we conclude Theorem 1.

4   RP Dimension and Order Dimension – Proofs of Theorems 2 and 3

In this section, we prove Theorems 2 and 3, as was outlined in Sect. 2.1. We begin by defining the notion of a dominance order, and noting the natural interpretation of dimension as order dimension. Recall from Sect. 3 the notion of vector dominance: where we say x dominates y if it is at least as great in each coordinate. This is denoted $x \geq y$. A standard form of geometrically-defined partial orders is a vector-dominance partial order: Given a set of vectors $x_1, \ldots, x_n$, we set $i \succ j$ if and only if $x_i \geq x_j$. It is easy to check that this relation is transitive and acyclic. We remark that the vector-dominance poset induced by points in $\mathbb{R}^d$ has order dimension at most d: For all $1 \leq j \leq d$, set the j-th total order to be the ordering of the n points with respect to their j-th coordinate. Then $i \succ j$ if and only if all d total orders agree on the relative ordering of x_i and x_j. The converse also holds: if a partial order has order-dimension k, then it can be expressed as a vector-dominance poset in $\mathbb{R}^k$. We will formalize this fact and extend it to show Theorem 2.

Proof (of Theorem 2). Let $(S, \succ)$ be a poset with order-dimension k, a and let $G = (S, \succ)$ be the associated directed graph. We wish to show $\mathsf{RPDim}(G) \leq k$.

Let $\succ_1, \ldots, \succ_k$ be the k total orders which realize $\succ$. Note that each total order $\succ_i$ induces a ranking σ_i on the elements of S, such that $a \succ_i b$ if and only if $\sigma_i(a) > \sigma_i(b)$. Assume without loss of generality that σ_i maps the elements of S to $\{1, \ldots, |S|\}$. Then the usual dominance embedding of $(S, \succ)$ is given by mapping each $a \in S$ to $\phi(a) := \sigma(a) := (\sigma_1(a), \sigma_2(a), \ldots, \sigma_k(a))$. Now, $a \succ b$ if and only if $\phi(a)$ dominates $\phi(b)$.

For the purposes of RP-dimension, we need a rescaled embedding. If $k = 1$, then $(S, \succ)$ is a total order, and setting $x_i = \sigma(i)$, $p_i = 1$ will suffice. Otherwise, define $\psi(a) := (k^{\sigma_1(a)}, k^{\sigma_2(a)}, \ldots, k^{\sigma_k(a)})$. Since $k \geq 2$, this maintains the dominance ordering of ϕ. Now, if we set $x_i = \psi(i)$, and

$$p_i = \left(\frac{1}{k^{\sigma_1(i)}}, \frac{1}{k^{\sigma_2(i)}}, \ldots, \frac{1}{k^{\sigma_k(i)}} \right)$$

then we get $\langle p_i, x_i \rangle = k$. Furthermore, letting $y_j^i = (\ldots, 0, k^{\sigma_j(i)+1}, 0, \ldots)$, we have $\langle p_i, y_j^i \rangle = k$ for all $1 \leq j \leq k$. Therefore, the hyperplane normal to p_i, passing through x_i, will also pass through y_j^i for all j. Since the σ values are presumed to be positive integers, this means that $x_i = \psi(i)$ can only be revealed-preferred to $x_j = \psi(j)$ if x_i dominates x_j.

Thus, we have constructed a set of vectors in $\mathbb{R}^k$ which realizes $G = (S, \succ)$, which allows us to conclude that $\mathsf{RPDim}(G) \leq k$. □

It remains to prove Theorem 3, that is that for $k = 1, 2$, or 3, posets of order dimension k have RP dimension k, but that for all $k \geq 3$, there exists a poset of order dimension k but order dimension 3. We begin by showing this first part:

Lemma 4. *A poset has RP dimension 1 or 2 if and only if it has order dimension 1 or 2, respectively.*

Proof. Note that an RP realization in $\mathbb{R}^1$ is simply a total order on the players, since we have $(i, j) \in G$ if and only if $p_i x_i > p_i x_j$. Since we need $p_i > 0$, we have $(i, j) \in G$ if and only if $x_i > x_j$, and therefore, the values of $x_1, \ldots, x_n$ induce a total order on the elements. Thus, any graph has RP dimension 1 if and only if it is a poset of order dimension 1, namely, a total order.

Thus, we conclude that a poset of order dimension 2 must have RP dimension equal to 2. It remains to show that a poset with RP dimension 2 must have order dimension 2. It is known that a poset has order dimension 2 if and only if the complement of its *comparability graph* is also a comparability graph [5]. In our terms, a poset $(S, \succ)$ has order dimension 2 if and only if (a) at least one pair of elements is not comparable, *i.e.* there is some $x, y \in S$ such that neither $x \succ y$ nor $y \succ x$, and (b) there exists a partial order $\succ'$ on S whose comparable pairs are exactly those which are non-comparable in $(S, \succ)$. Thus, for any two $x, y \in S$, we must have exactly one of $x \prec y$, $x \succ y$, $x \prec' y$, and $x \succ' y$ hold. Therefore, to show that a partial order with RP dimension 2 must have order dimension 2, we must construct a partial order on its non-comparable pairs.

Let $\mathcal{X} := \{(\boldsymbol{p}_1, \boldsymbol{x}_1), \ldots, (\boldsymbol{p}_n, \boldsymbol{x}_n)\}$ be the 2-dimensional vectors which realize $(S, \succ)$, that is for all $i, j \in S$, $i \succ j$ if and only if $\langle \boldsymbol{p}_i, \boldsymbol{x}_i \rangle > \langle \boldsymbol{p}_i, \boldsymbol{x}_j \rangle$. Furthermore, denote $\boldsymbol{x}_i = (x_i, y_i)$ and $\boldsymbol{p}_i = (p_i, q_i)$ for all $1 \leq i \leq n$. Recall also that we have assumed $p_j^i \geq 0$ for all $1 \leq i \leq n$ and $j = 1, 2$. For every pair i, j such that neither $i \succ j$ nor $j \succ i$, we say $i \succ' j$ if $x_i > x_j$, and $j \succ' i$ otherwise. This is clearly acyclic, it remains to show that the relation is transitive.

Let $\boldsymbol{x}_1, \boldsymbol{p}_1, \boldsymbol{x}_2, \boldsymbol{p}_2, \boldsymbol{x}_3, \boldsymbol{p}_3$ be such that $\langle \boldsymbol{p}_i, \boldsymbol{x}_j \rangle \geq \langle \boldsymbol{p}_i, \boldsymbol{x}_i \rangle$ for all $j \neq i$, and $x_1 > x_2 > x_3$. This implies that $1 \succ' 2 \succ' 3$. We wish to show that $1 \succ' 3$. It is clear that $x_1 > x_3$, so it remains to show that both $\langle \boldsymbol{p}_1, \boldsymbol{x}_3 \rangle \geq \langle \boldsymbol{p}_1, \boldsymbol{x}_1 \rangle$, and $\langle \boldsymbol{p}_3, \boldsymbol{x}_1 \rangle \geq \langle \boldsymbol{p}_3, \boldsymbol{x}_3 \rangle$. We may assume without loss of generality that $p_i > 0$ for $i = 1, 2, 3$, since we may slightly rotate the space, and so we may assume without loss of generality that $p_i = 1$ for $i = 1, 2, 3$, since scaling $\boldsymbol{p}$ does not affect the induced preferences. Thus, $\langle \boldsymbol{p}_i, \boldsymbol{x}_j \rangle = x_j + q_i y_j$ for all i, j. Now, $x_1 > x_2$, but $x_2 + q_1 y_2 > x_1 + q_1 y_1$, so we must have $y_1 < y_2$. Similarly, we get $y_2 < y_3$. Furthermore, we have

$$
\begin{array}{r}
x_1 + q_2 y_1 > x_2 + q_2 y_2 \\
- \quad x_1 + q_1 y_1 < x_2 + q_1 y_2 \\
\hline
\Rightarrow (q_2 - q_1) y_1 > (q_2 - q_1) y_2
\end{array}
\tag{5}
$$

but $y_1 < y_2$, so we conclude $q_2 < q_1$. Similarly, we have $q_3 < q_2$. Now, since $q_1 > q_2$ but $y_2 < y_3$, we have that

$$
x_3 + q_2 y_3 \geq x_2 + q_2 y_2 \quad \Longrightarrow \quad x_3 + q_1 y_3 \geq x_2 + q_1 y_2
\tag{6}
$$

But we know that $x_2 + q_1 y_2 \geq x_1 + q_1 y_1$, so we have that $\langle \boldsymbol{p}_1, \boldsymbol{x}_3 \rangle \geq \langle \boldsymbol{p}_1, \boldsymbol{x}_1 \rangle$. The converse inequality is shown similarly, and thus we may conclude that the relation $\succ'$ is a partial order.

Since any two elements are comparable in $\succ$ if and only if they are not comparable in $\succ'$, we conclude that the order dimension of $\succ$ is at most 2. Since its RP dimension is not 1, it must have order dimension exactly 2, as desired. □

With this lemma in hand, we conclude that if a poset has order dimension 3, it must have RP dimension 3, as its RP dimension is at most 3, but it cannot have dimension 1 or 2. Thus, we have proved the first half of Theorem 3. It remains to show that for all $k \geq 3$, there exist posets with order dimension k, but RP dimension 3.

The family of *standard* posets S_2, S_3, ... is a sequence of posets such that S_k has order dimension k, and ground set of size $2k$. (See *e.g.* [14]) They are defined as follows: The ground set for S_k is labelled 1, 2, ..., k, 1', 2', ..., k', and we have $i' \succ j$ for all $i \neq j$. No pair i', j' or i, j is comparable.

We will show that $G_k = (S_k, \succ_k)$ has RP dimension 3 for all $k \geq 3$. Let $H := \{\boldsymbol{x} \in \mathbb{R}^3 : \langle \boldsymbol{1}, \boldsymbol{x} \rangle = 0\}$, the plane normal to the all-ones vector in $\mathbb{R}^3$, and let $\mathbb{S}_H$ be the unit circle in H centered at the origin. Thus, $\mathbb{S}_H = \{\boldsymbol{x} \in H : \|\boldsymbol{x}\|_2 = 1\}$. Finally, let $\boldsymbol{a}_1$, ..., $\boldsymbol{a}_n$ be n equally spaced points along the circumference of $\mathbb{S}_H$. We will use these to construct our realization of $(S_k, \succ_k)$ in $\mathbb{R}^3$. For all $i \leq k$, set $\boldsymbol{x}_i = (2+\epsilon)\boldsymbol{a}_i$, and $\boldsymbol{x}_{i'} = 1-\boldsymbol{a}_i$, where $\epsilon > 0$ will be chosen later. Set $\boldsymbol{p}_i = 1-\boldsymbol{a}_i$, and $\boldsymbol{p}_{i'} = 1 + \boldsymbol{a}_i$. Since the $\boldsymbol{a}_i$'s are unit vectors, we have that $\langle \boldsymbol{a}_i, \boldsymbol{a}_j \rangle < 1$ if $i \neq j$, and $= 1$ if $i = j$.

We have $\langle \boldsymbol{p}_i, \boldsymbol{x}_j \rangle = (2+\epsilon) \langle \boldsymbol{1}, \boldsymbol{a}_j \rangle - (2+\epsilon) \langle \boldsymbol{a}_i, \boldsymbol{a}_j \rangle$. The left hand term is 0, and the right hand term is minimized when $i = j$. Thus, $\boldsymbol{x}_i$ is not revealed preferred to $\boldsymbol{x}_j$ for all $j \neq i$. Furthermore, $\langle \boldsymbol{p}_i, \boldsymbol{x}_{j'} \rangle = \langle 1 - \boldsymbol{a}_i, 1 - \boldsymbol{a}_j \rangle = \langle \boldsymbol{1}, \boldsymbol{1} \rangle + \langle \boldsymbol{a}_i, \boldsymbol{a}_j \rangle$, since $\langle \boldsymbol{1}, \boldsymbol{a}_j \rangle = 0$ for all j. Thus, $\boldsymbol{x}_i$ is not revealed preferred to $\boldsymbol{x}_{j'}$ for all j.

Now, $\langle \boldsymbol{p}_{i'}, \boldsymbol{x}_{i'} \rangle = \langle 1 + \boldsymbol{a}_i, 1 - \boldsymbol{a}_i \rangle = \langle \boldsymbol{1}, \boldsymbol{1} \rangle - \langle \boldsymbol{a}_i, \boldsymbol{a}_i \rangle = 2$, whereas $\langle \boldsymbol{p}_{i'}, \boldsymbol{x}_j \rangle = (2+\epsilon) \langle 1 + \boldsymbol{a}_i, \boldsymbol{a}_j \rangle = (2+\epsilon) \langle \boldsymbol{a}_i, \boldsymbol{a}_j \rangle$. Thus, if we choose $\epsilon > 0$ sufficiently small, we have that $\boldsymbol{x}_{i'}$ is revealed preferred to $\boldsymbol{x}_j$ for all $j \neq i$, but not to $\boldsymbol{x}_i$. Furthermore, $\langle \boldsymbol{p}_{i'}, \boldsymbol{x}_{j'} \rangle = 3 - \langle \boldsymbol{a}_i, \boldsymbol{a}_j \rangle$, which is minimized when $i = j$, so $\boldsymbol{x}_{i'}$ is not revealed preferred to $\boldsymbol{x}_{j'}$ for all $j \neq i$.

Therefore, we have shown that our choice of $\boldsymbol{p}_i$'s and $\boldsymbol{x}_j$'s is a valid RP realization of $(S_k, \succ_k)$ in $\mathbb{R}^3$ for all $k \geq 3$. Thus, we have demonstrated the existence of partial orders with order dimension k but RP dimension 3, for all $k \geq 3$, hence concluding the proof of Theorem 3.

5   Further Work

This paper does not address the computational complexity of computing the RP dimension of a given graph, and this is left as an open problem for future work. Below is a summary of the computational complexity of matrix sign rank, and what this implies for RP dimension.

Complexity of Matrix Sign Rank. Recall that the problem of computing RP dimension is denoted REVPREFDIM, and the problem of computing matrix sign rank, MATSGNRNK. It is known [6] that MATSGNRNK in full generality is complete for the *existential theory of the reals*: the problem of determining whether a system of polynomial equalities and inequalities has a feasible solution over the reals. This complexity class, often denoted "∃ℝ", is known to lie between NP and PSPACE. In fact, it is ∃ℝ-complete to determine whether a matrix has sign rank at most 3. However, this hardness result only holds when the sign matrix is allowed to have arbitrarily many zero entries in each row and column. When sign matrices are constrained to have no 0 entries, MATSGNRNK is known only to be NP-hard. (Again, [6]). It is not known whether MATSGNRNK lies in NP in this restricted setting, though we think this is unlikely.

This paper shows that the REVPREFDIM problem is equivalent to computing the sign rank of signed adjacency matrices, which are a (large) subclass of sign matrices with exactly one zero in each row and column. Note that replacing a row containing a single zero entry with two copies, replacing the zero with a + in one copy, and a − in the other, does not affect the sign rank. Therefore, REVPREFDIM is a special case of MATSGNRNK in the restricted setting, and thus it cannot be a harder problem.

We leave as an open problem determining whether REVPREFDIM is itself NP-hard, and whether it is equivalent to MATSGNRNK on +, − matrices.

Acknowledgements. I would like to thank Ruta Mehta, Adrian Vetta, and Siddharth Barman, for their insightful discussion in the initial stages of work.

References

1. Afriat, S.N.: The construction of utility functions from expenditure data. Int. Econ. Rev. **8**(1), 67–77 (1967)
2. Alon, N., Moran, S., Yehudayoff, A.: Sign rank versus VC dimension. In: Conference on Learning Theory, pp. 47–80 (2016)
3. Ausubel, L.M., Baranov, O.V.: Market design and the evolution of the combinatorial clock auction. Am. Econ. Rev. **104**(5), 446–51 (2014)
4. Ausubel, L.M., Cramton, P., Milgrom, P.: The clock-proxy auction: a practical combinatorial auction design (2006)
5. Baker, K.A., Fishburn, P.C., Roberts, F.S.: Partial orders of dimension 2. Networks **2**(1), 11–28 (1972)
6. Bhangale, A., Kopparty, S.: The complexity of computing the minimum rank of a sign pattern matrix. arXiv preprint arXiv:1503.04486 (2015)
7. Boodaghians, S., Vetta, A.: The combinatorial world (of auctions) according to GARP. In: Hoefer, M. (ed.) SAGT 2015. LNCS, vol. 9347, pp. 125–136. Springer, Heidelberg (2015). https://doi.org/10.1007/978-3-662-48433-3_10
8. Boodaghians, S., Vetta, A.: Testing consumer rationality using perfect graphs and oriented discs. In: Markakis, E., Schäfer, G. (eds.) WINE 2015. LNCS, vol. 9470, pp. 187–200. Springer, Heidelberg (2015). https://doi.org/10.1007/978-3-662-48995-6_14

9. Boser, B.E., Guyon, I.M., Vapnik, V.N.: A training algorithm for optimal margin classifiers. In: Proceedings of the Fifth Annual Workshop on Computational Learning Theory, pp. 144–152. ACM (1992)
10. Chambers, C.P., Echenique, F.: Rational demand: 3.1. Revealed preference graphs, Chap. 3. In: Revealed Preference Theory. Cambridge University Press, Cambridge (2016)
11. Chambers, C.P., Echenique, F., Shmaya, E.: General revealed preference theory. Theor. Econ. **12**(2), 493–511 (2017)
12. Cramton, P.: Spectrum auction design. Rev. Ind. Organ. **42**(2), 161–190 (2013)
13. Deb, R., Pai, M.M.: The geometry of revealed preference. J. Math. Econ. **50**, 203–207 (2014)
14. Fishburn, P.C., Trotter, W.T.: Geometric containment orders: a survey. Order **15**(2), 167–182 (1998)
15. Forster, J.: A linear lower bound on the unbounded error probabilistic communication complexity. J. Comput. Syst. Sci. **65**(4), 612–625 (2002)
16. Gross, J.: Testing data for consistency with revealed preference. Rev. Econ. Stat. **77**, 701–710 (1995)
17. Harsha, P., Barnhart, C., Parkes, D.C., Zhang, H.: Strong activity rules for iterative combinatorial auctions. Comput. Oper. Res. **37**(7), 1271–1284 (2010)
18. Harzheim, E.: Ordered Sets. Springer, Boston (2006). https://doi.org/10.1007/b104891
19. Houthakker, H.S.: Revealed preference and the utility function. Economica **17**(66), 159–174 (1950)
20. Razborov, A.A., Sherstov, A.A.: The sign-rank of AC^0. SIAM J. Comput. **39**(5), 1833–1855 (2010)
21. Rose, H.: Consistency of preference: the two-commodity case. Rev. Econ. Stud. **25**(2), 124–125 (1958)
22. Rosenblatt, F.: The perceptron, a Perceiving and Recognizing Automaton Project Para. Cornell Aeronautical Laboratory, Buffalo (1957)
23. Samuelson, P.A.: A note on the pure theory of consumer's behavior. Economica **5**, 62–71 (1938)
24. Varian, H.R.: Revealed preference. In: Samuelsonian Economics and the Twenty-First Century, pp. 99–115 (2006)
25. Varian, H.R.: Position auctions. Int. J. Ind. Organ. **25**(6), 1163–1178 (2007)
26. Varian, H.R.: Revealed preference and its applications. Econ. J. **122**(560), 332–338 (2012)

Timing Matters: Online Dynamics in Broadcast Games

Shuchi Chawla[1], Joseph (Seffi) Naor[2], Debmalya Panigrahi[3(✉)], Mohit Singh[4], and Seeun William Umboh[5]

[1] Computer Sciences Department, University of Wisconsin - Madison, Madison, USA
shuchi@cs.wisc.edu
[2] Department of Computer Science, Technion, Haifa, Israel
naor@cs.technion.ac.il
[3] Department of Computer Science, Duke University, Durham, USA
debmalya@cs.duke.edu
[4] H. Milton Stewart School of Industrial and Systems Engineering, Georgia Institute of Technology, Atlanta, USA
mohitsinghr@gmail.com
[5] School of Information Technologies, The University of Sydney, Sydney, Australia
william.umboh@sydney.edu.au

Abstract. This paper studies the equilibrium states that can be reached in a network design game via *natural* game dynamics. First, we show that an arbitrarily interleaved sequence of arrivals and departures of players can lead to a polynomially inefficient solution at equilibrium. This implies that the central controller must have some control over the timing of agent arrivals and departures in order to ensure efficiency of the system at equilibrium. Indeed, we give a complementary result showing that if the central controller is allowed to restore equilibrium after every set of arrivals/departures via *improving moves*, the eventual equilibrium states reached have exponentially better efficiency.

1 Introduction

In multi-agent systems where different agents have competing objectives, it is well-known that selfish behavior leads to suboptimal system performance at *equilibrium*. The *Price of Anarchy* (PoA) and the *Price of Stability* (PoS), which respectively correspond to the worst and best equilibrium states, are widely used in the literature to quantify this suboptimality relative to an optimal solution designed by a central authority. If these two measures are close to each other, they

Part of this work was done when all the authors were visiting Microsoft Research - Redmond. Partial support for this work was provided by the following grants: S. Chawla from NSF grants CCF-1101429 and CCF-1320854; S. Naor from ISF grant 1585/15 and BSF grant 2014414; D. Panigrahi from NSF grants CCF-1527084 and CCF-1535972, an NSF CAREER Award CCF-1750140, and faculty research awards from Google and Yahoo; M. Singh from NSF grant CCF-1717947; S. Umboh from ERC consolidator grant 617951 and NSF grant CCF-1320854.

© Springer Nature Switzerland AG 2018
G. Christodoulou and T. Harks (Eds.): WINE 2018, LNCS 11316, pp. 80–95, 2018.
https://doi.org/10.1007/978-3-030-04612-5_6

provide a satisfactory understanding of the quality of stable states the system is expected to reach. However when these measures differ significantly, the system can exhibit multiple equilibria with highly varying performance. But, which of these equilibria can be achieved in actual game dynamics? More generally, what is the minimal guidance by a central authority that can guarantee near-optimal system performance in equilibrium?

In this paper, we study these questions in the context of a game that exhibits a particularly rich set of equilibria, namely the *broadcast game*. A broadcast game is defined on a rooted undirected graph with costs on edges. Every vertex in the graph has an agent whose goal is to select a routing path to the root that minimizes her own cost. The cost of every edge is shared equally among all agents using it, and the cost of an agent is the sum of her *cost shares* along the edges in her path to the root. The system is in Nash equilibrium (or NE) if no agent can lower her own cost by unilaterally changing her routing path. The cost of an equilibrium is the total cost of all edges used by at least one agent. The quality of equilibria is measured with respect to the social optimum, which for broadcast games is the minimum spanning tree (MST) of the graph.

Broadcast games are a kind of potential games and the existence of NE in any instance can be proved through a potential function argument, originally given by Rosenthal [34] (see also Monderer and Shapley [30]). Anshelevich et al. [4] observed that different NE in broadcast games can exhibit vastly different performance: the PoA can be as large as $\Omega(n)$ whereas the PoS (a concept they introduced to show this gap) is bounded by $O(\log n)$; here n denotes the number of vertices in the graph.[1] A long line of work (e.g., [10,19,27,28]) subsequently improved the PoS bound to $O(1)$.

Given this divergence of bounds, Chekuri et al. [13] posed the question of analyzing the quality of equilibria that are actually *reachable* via natural dynamics— a sequence of single agent moves where the moving agent always chooses a new path that strictly decreases her cost relative to her current path. We call such moves "improving moves" or "best response moves", depending on whether they merely lower the agent's cost or are optimal for the agent given the current state of the system. It follows from the potential function argument of Rosenthal [34] that *any* such sequence of moves will eventually converge to NE.

Chekuri et al. [13] considered the following *restricted* two-stage process: in the first stage, starting with an empty graph, agents arrive sequentially in arbitrary order and choose their respective best response paths upon arrival; in the second stage, agents make improving moves[2] in arbitrary order until they reach equilibrium. They showed that the equilibria reachable through this process have a cost of $O(\sqrt{n}\log^2 n)$ times the MST, a significant improvement over the PoA bound. This bound was subsequently improved to $O(\log^3 n)$ for the same two-stage process by Charikar et al. [11].

---

[1] The full version of this paper [12] provides examples illustrating these bounds.

[2] Observe that when an agent arrives or makes an improving move, paths of other agents may become suboptimal for them.

The Dynamic Price of Stability. These previous works motivate extending the notion of the price of stability to online dynamics. In the static (or "one shot") version of our problem, in which players are initially in an empty configuration, the central planner can force the players into any configuration, in particular the one realizing the price of stability. In the dynamic case, however, the central planner cannot do so since some players have already chosen a route. Thus, the central planner has to offer existing players a better strategy, so as to incentivize changes. Informally, the *dynamic price of stability* is the cost of a solution in equilibrium resulting from online dynamics, while allowing for algorithmic intervention by the central planner. The notion of dynamic price of stability can be applied to any game in which one needs to characterize which equilibria can be reached via online dynamics, while minimizing the power of intervention of the central planner. It would be very interesting to find further applications of this new notion.

One way to restate Charikar et al.'s result is that the dynamic PoS is poly-logarithmic when all arrivals happen before any improving moves. But, what if some agents make improving moves before all of the other agents have arrived, i.e., the sequence of improving moves is interleaved with arrivals? Unfortunately, the analyses presented in [13] and [11] strongly build on the fact that all agents arrive upfront and remain in the system thereafter, and agents must wait for everyone to arrive before making any changes to their strategies. Charikar et al. posed the question of analyzing dynamics in which arrivals and improving moves are interleaved as a "tantalizing and difficult" open problem. In the decade following their work, in spite of tremendous progress in PoS bounds for broadcast games, no progress has been made on understanding more general game dynamics.[3]

More General Dynamics. Since the work of Chekuri et al., our work is the first to study more general dynamics of the broadcast game. We consider two kinds of extensions to the two-stage process. First, we consider systems with churn where agents arrive as well as depart over time. Second, we allow multiple interleaved stages of arrivals, departures, and improving moves. Our first result shows that if we make a minimal change to the two-stage dynamics studied above, namely adding departures to the first stage, then it is possible to reach an equilibrium that is polynomial (in n) worse than the social optimum, placing it in the same regime as the PoA bound. To the best of our knowledge, this is the first polynomial lower bound for *any* game dynamics for broadcast games.

Theorem 1. *For any large enough integer n, there exists an instance of the broadcast game with n vertices and a sequence of arrivals and departures that terminates in an NE of cost $\Omega(n^{1/3})$ times that of the minimum spanning tree on all the vertices.*

---

[3] Charikar et al. also studied a variant where arrivals happen in uniformly random order and are interleaved with adversarially ordered best response moves. For this setting, they were able to prove an upper bound of $O(\sqrt{n}\operatorname{polylog} n)$ on the quality of the equilibria reached, but did not present any lower bounds.

It is important to observe that since we allow departures, not all vertices have agents at the end of the game. This creates two candidates for OPT: the optimal Steiner tree on the remaining agents, or the MST on all vertices.[4] The former leads to trivial and uninteresting lower bounds (see the full version); so, we use the MST as OPT in this paper. This choice of a weaker optimum makes for a stronger lower bound.

The Power of Intervention. Given the above lower bound, a natural question is whether some limited intervention from a central planner can lead to a better outcome for the game. At one extreme, if the central planner is allowed to suggest a strategy to every player simultaneously, then any NE, in particular the best one corresponding to the PoS of $O(1)$, can be achieved. This level of control is unrealistic. A more reasonable level of control is for the central planner to suggest improving moves to players one by one; importantly, any such move should lower the corresponding agent's current cost share, otherwise the player has no incentive to follow the planner's suggestion.

What about the timing of such interventions? As our lower bound demonstrates, if the timing of interventions is completely adversarial, in particular if no interventions are allowed during the initial arrival/departure phase, the system can end up in a poor NE. To get around this lower bound, we consider dynamics where the central planner is allowed to make a series of improving moves after every adversarial arrival/departure. Observe that the sequence of arrivals and departures can still be ordered adversarially, and indeed can depend on the previous algorithmic interventions. We call such dynamics *equilibrium-preserving* (EQ-P) dynamics because the central planner restores the system to a good equilibrium after every adversarial arrival/departure.

Specifically, the EQ-P dynamics starts from an empty configuration and continues in epochs. At the beginning of each epoch the system is at equilibrium. The epoch begins with an arrival or departure, followed by a series of improving moves to restore equilibrium. Once equilibrium is restored, the epoch ends. Our analysis, in fact, allows for multiple simultaneous arrivals[5] at the beginning of an epoch, and multiple departures at any point of time during the epoch. Formally, we define three different kinds of moves within an epoch:

1. (**Arrivals.**) A set of new players arrive and each picks a best response path with respect to the configuration reached at the end of the previous epoch. (The choice of the set of arrivals is adversarial.)
2. (**Departures.**) A set of players departs the system. (Choice of departing players is adversarial.)

---

[4] Another bound is the optimal Steiner tree on all vertices for which an agent arrived at some point in the dynamics. Since we can assume metric costs, we can restrict our attention to these vertices and then MST cost is within a factor of two of the cost of optimal Steiner tree.

[5] Note that although arrivals within a single epoch are simultaneous in that every arriving player picks a best response path with respect to the equilibrium state at the beginning of the epoch, arrivals in different epochs are sequential. In this sense our model captures sequential arrivals with interleaved improving moves.

3. (**Equilibrium Restoration.**) The central authority offers players strategies that can improve their (shared) connection costs. This step continues till equilibrium is restored to the system.

Our second result shows that this limited level of central intervention is sufficient to guarantee a NE with exponentially better performance:

Theorem 2. *Every instance of the broadcast game using* EQ-P *dynamics converges to an* NE *of cost* $O(\log n)$ *times that of the minimum spanning tree on all the vertices.*

Observe that, as for our lower bound result, we compare the performance of EQ-P dynamics to the MST on all vertices[6] and not the optimal Steiner tree on the vertices that remain in the system. The two benchmarks are identical when there are no departures but the MST benchmark can potentially be much weaker when there are many departures. However, as mentioned earlier, the Steiner tree benchmark is not interesting because it leads to trivial polynomial lower bounds (see the full version [12]). Furthermore that the guarantee provided by the above theorem holds at the end of *every* epoch as compared against the MST over vertices that have arrived up to the end of that epoch, not including future arrivals. A natural open question is whether a polylogarithmic dynamic PoS can be achieved through less algorithmic intervention relative to EQ-P dynamics, for example, by allowing players to make best response moves instead of improving moves.

Technical Challenges. The broadcast game exhibits a rich set of equilibria and a far richer set of intermediate states of the system. For example, whereas the set of agent strategies (paths) in any equilibrium of the game always forms a tree,[7] intermediate states, even those reached by a series of best response moves, can contain a complex structure of interconnected cycles. A major impediment to analyzing dynamics is that it is extremely challenging to maintain *any* structural invariant on intermediate states. Our work overcomes this challenge by algorithmically maintaining such a structural invariant. Whenever the structural invariant is broken by arrivals or departures, we restore it algorithmically. Importantly, we show that this can always be achieved through a sequence of *improving* moves.

Our structural invariant is a charging of the cost of a state (i.e. collection of paths) against a family of partitions of the underlying graph. Each partition corresponds to a solution to the dual of the standard MST linear program. As such, our charging scheme can be interpreted as a *dual fitting* approach. One challenge in carrying out this approach is that as agents arrive and leave, our analysis must allow for the dual to become grossly infeasible at intermediate states, which in turn results in very expensive intermediate (non-equilibrium)

---

[6] Because of this comparison against the MST, we prefer the term "broadcast game" for this setting, rather than the "multicast game".

[7] The existence of a cycle would imply that one of the agents can improve her cost share by switching to a different path and the current state is not an equilibrium.

states. At the crux of our argument is a careful construction of improving moves that ensures that the system cycles between a small set of states of which the stable ones correspond to feasible duals.

Related Work. We have already mentioned the long line of work on improving PoS bounds for broadcast games [4,10,19,27], and the game dynamics studied by Chekuri et al. [13] and Charikar et al. [11]. A different approach was taken by Balcan et al. [6], who considered the problem of influencing the dynamics of broadcast games so as to achieve socially efficient equilibria. In their model, players use expert learning, choosing between a best response expert and a central authority expert suggesting (near-)optimal global behavior. Broadcast games belong to a broader class called network design games (see, e.g., [2,4,9,10,14,15,18,20,26,28]), which in turn, are a special case of the widely studied congestion and potential games (see, e.g., [1,7,17,24,29,30,33–35]).

The analysis of game dynamics in this paper crucially relies on the construction of a hierarchial family of multiple dual solutions. This method of analysis has been highly influential in designing online algorithms for network design problems. Implicit use of this method dates back to the work of Imase and Waxman [25] on online Steiner trees and a subsequent line of work of Awerbuch et al. [5], Berman and Coulston [8], Naor et al. [31]. More recently, this method has been explicitly employed in solving a range of node and edge-weighted Steiner network design problems in the online setting [3,16,21–23,32]. In terms of the exact techniques, perhaps the closest to our work is that of Umboh [36], who uses hierarchical tree embeddings to analyze greedy-like online algorithms for network design problems. In contrast to these applications in competitive analysis where decisions are irrevocable, game dynamics allows temporary overcharging of dual solutions, which we crucially use in this paper.

Organization of the Paper. We present a proof of our lower bound (Theorem 1) in Sect. 2, and a proof of our upper bound (Theorem 2) in Sect. 3. Due to lack of space, we defer most of the proofs to the full version [12].

2   Lower Bound

In this section, we will show that if arrivals and departures are allowed at non equilibrium states, then no dynamics can lead to a good equilibrium (Theorem 1).

We construct a family of lower bound instances parameterized by an integer $m \geq 1$. The mth instance uses the metric induced by weighted graph G_m (see Fig. 1). The vertex set of this graph consists of a root r and $m + 1$ layers $V^0, \ldots, V^m$. For $1 \leq i \leq m$, layer V^i consists of m clusters $C_1^i, \ldots, C_m^i$, each of which is a clique over m vertices. We use $v_{j,k}^i$ to denote the k-th vertex of C_j^i; recall that each of i, j, and k take on integral values in $[m]$. Layer V^0 also consists of m^2 vertices, which are labeled $v_{j,k}^0$ for $j, k \in [m]$, but there are no edges between these vertices. The vertices of V^m are called *end* vertices,

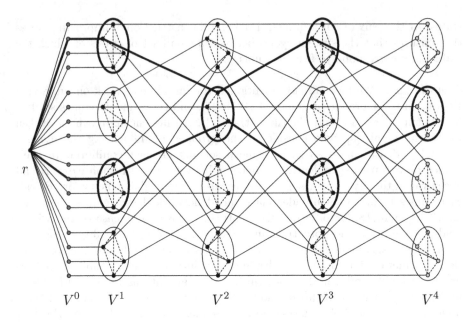

$$V^0 \quad V^1 \qquad\qquad V^2 \qquad\qquad V^3 \qquad\qquad V^4$$

Fig. 1. Example for $m = 4$. Auxiliary vertices are in red, end vertices are in blue. Ovals represent clusters. Intra-cluster edges are shown as dashed edges. The two bold paths starting from the same cluster successively diverge into different clusters and converge into the same cluster on their way to the root.

and those of V^0 are called *auxiliary* vertices. Observe that the graph G_m has $n = m^2(m + 1) + 1$ vertices in all.

Next, we describe the edges. Each pair of vertices within the same cluster C_j^i is connected by an edge of length $1/m$ for all layers except V^0. The remaining edges in the graph connect vertices in neighboring layers and are all of length 1. Each auxiliary vertex $v_{j,k}^0$ in V_0 is connected to the root and to its corresponding vertex $v_{j,k}^1$ in layer 1. For $1 \leq i \leq m - 1$, we have an edge $(v_{j,k}^i, v_{k,j}^{i+1})$ for each $j, k \in [m]$. In other words, the vertices of the j-th cluster in layer i are connected to the k-th vertices of the clusters in layer $i + 1$; in particular, the k-th vertex of the j-th cluster in layer i is connected to the j-th vertex of the k-th cluster in layer $i + 1$. For example, see the edges leaving the first (top) cluster of V_1 in Fig. 1. Observe that there are exactly $m^2(m + 1)$ inter-layer edges, and exactly $m^3(m - 1)/2$ intra-cluster edges.

Observe that each end vertex $v_{j,k}^m$ has a unique path $P_{j,k}$ to the root that consists of only inter-layer edges (see Fig. 1). We call these paths *canonical paths*. Note that each inter-layer edge belongs to exactly one canonical path. In other words, the set of inter-layer edges is a disjoint union of all the canonical paths $P_{j,k}$.

The Cost of the Final Equilibrium. Before describing the sequence of arrivals and departures of terminals, let us analyze the final equilibrium state and its cost relative to the optimal cost. Let OPT denote the cost of the minimum spanning tree over all vertices in G_m. Observe that this is an upper bound on the cost of any optimal solution at the end. The final state following our sequence of arrivals and departures, denoted $\mathscr{F}$, consists of m players situated at every end vertex $v_{j,k}^m$ in layer m; each player uses the canonical path $P_{j,k}$ to route to the root. The following lemma shows that this is an equilibrium state with a polynomially larger cost relative to OPT.

Lemma 1. *State $\mathscr{F}$ is an equilibrium and the cost of $\mathscr{F}$ is $\Omega(m)$ OPT.*

Sequence of Arrivals and Departures. The sequence is constructed in m *phases*, each phase consisting of m^2 *rounds*, one per end vertex $v_{j,k}^m$, and indexed by (j,k). Informally, the objective of each phase is to add one more terminal at each of the end vertices $v_{j,k}^m$. Within round (j,k) in a phase, we use a set of "temporary" terminals whose sole aim is to force the terminal at $v_{j,k}^m$ that arrives at the end of the round to choose the canonical path as its best response. The temporary terminals are introduced at intermediate vertices along the canonical path during the round, and removed at the end of the round.

Formally, let $\prec$ be an arbitrary total order on the pairs (j,k). The sequence σ is constructed to maintain the following invariant: at the end of round (j,k) of phase ℓ, there will be ℓ players on $v_{j',k'}^m$ for $(j',k') \prec (j,k)$, and $\ell-1$ players on the remaining end vertices. Furthermore, each player on $v_{j,k}^m$ uses the path $P_{j,k}$.

We now specify the subsequence for each round. Consider round (j,k) of phase ℓ. For simplicity of notation, we use v^i to denote the vertex of V^i on $P_{j,k}$. We also use P^i to denote the segment of $P_{j,k}$ starting at v^i and ending at the root. The round consists of $m+1$ *iterations*. In iteration $0 \leq i \leq m-1$, m^2 players arrive at v^i. In iteration $i = m$, one player arrives at v^m. Finally, the players on $v^0, \ldots, v^{m-1}$ depart. We can now show using induction over the terminal arrivals, that for every terminal the best-response path on arrival is the segment of the canonical path connecting it to the root.

Lemma 2. *Consider a terminal arriving at vertex v^i in iteration i of round (j,k) in phase ℓ. The best-response path of the terminal to the root is the segment of its canonical path P^i.*

Lemma 2 shows that the sequence of arrivals and departures above terminates in the final state $\mathscr{F}$, which costs $\Omega(m)$ OPT by Lemma 1. Since m is polynomial in the number of vertices, Theorem 1 follows.

3   EQ-P **Dynamics**

In this section and the next, we describe and analyze EQ-P dynamics for the broadcast game. We first set up our notation and terminology, and prove some basic structural properties that are used in the rest of the paper. Let $G = (V, E)$ be a complete graph, $|V| = n$, with metric costs $c : V \times V \to \mathbb{R}_+$ defined on the edges. We assume without loss of generality that every vertex has a unique agent (a.k.a. terminal) residing at it. Agents arrive and depart over time. Since edge costs satisfy the triangle inequality, before an agent arrives, no other agents route their paths via the vertex corresponding to this agent. Indeed, we assume that our intervention algorithm has no knowledge of vertices corresponding to agents that are yet to arrive. However, if an agent already in the system departs, other agents may continue to route their paths via its vertex, and the vertex remains in the graph. At any point of time during the process, our algorithm only considers the subgraph induced over vertices whose agents have arrived prior to that time. We call a vertex active if the agent at that vertex is still in the system.

The graph G is revealed via an online process that is divided into *epochs* (indexed by time t). At the start of epoch t, the set of vertices in G that have already appeared is denoted by V_t. We denote the set of active terminals among them by $A_t \subseteq V_t$. Each terminal $v \in A_t$ has a current routing path p_v connecting it to the common root r. The cost share of v along this path is the sum of v's cost share over the edges in the path, where the cost of an edge is equally shared between all terminals currently using it. In the EQ-P scenario, we further enforce the invariant that the set of paths p_v are in NE, i.e., no terminal has an incentive to unilaterally deviate to a different routing path.

The *routing* at any time t is defined to be the set of routing paths $(p_v)_{v \in A_t}$. A *best response path* of a terminal v with respect to a routing, denoted p_v^*, is a path from v to r with the minimum shared cost if v were to move to this path. If there are multiple such paths, we break ties in favor of paths having fewer edges with no terminal other than v using them. Note that this may not break all ties, in which case, any of these paths can be designated as the best response path. A terminal $v \in A$ is said to have an *improving move* with respect to a routing if by moving from its current path p_v to a new path q_v strictly decreases v's cost share. Given a routing, its potential [34] is defined to be $\Phi = \sum_{e \in E} \sum_{i=1}^{N_e} c_e/i$, where N_e is the number of agents using e. A standard argument shows that any improving move decreases the potential by a value which is uniformly bounded away from zero, resulting in a finite convergence of our dynamics. The following well-known lemma states that in equilibrium, the routing paths form a tree.

Lemma 3. *In equilibrium, the routing paths of a broadcast game form a tree.*

Each epoch t is divided into several phases. The first phase consists of an arrival or departure event. In the former case, a new set of terminals $U_t \subseteq V \setminus V_t$ arrive, and the cost of all edges incident on terminals in U_t is revealed. Each new terminal $u \in U_t$ chooses a *best response* routing path p_u. In the latter case, a set of terminals leave, thereby removing the corresponding vertices from the set of terminals A_t. (Note that the corresponding vertices remain in V_t.) Lemma 6 establishes that the structure of the set of routing paths after arrivals or departures remains a tree.

Both arrival and departure events lead to changes in the cost shares of edges. In the EQ-P scenario, this might lead to a violation of the equilibrium state that was being previously maintained. In this case, the system performs a sequence of *improving moves*, in each of which a terminal changes its routing path in order to reduce its cost share.

Improving moves may temporarily create cycles in the collection of routing paths $\{p_v\}_{v \in A_t}$. We order and group improving moves into contiguous blocks or phases such that every phase ends with the routing paths forming a tree. Furthermore, the trees at the beginning and end of the phase differ in a single pair of edges. The collection of moves in each such phase is called a *tree-follow* move.

Definition 1 (Tree-follow move). *A tree-follow move from u to v in T is a sequence of improving moves that start with routing tree T and end with routing tree $T' = T \setminus (u, parent(u)) \cup (u, v)$, where $parent(u)$ is the parent vertex of u in T. Observe that each terminal in the subtree rooted at u in T reroutes its path to the root according to T'.*

Because of departure events, the routing tree may contain non-terminal vertices as Steiner vertices. It is convenient to extend the notion of an improving move to vertices that are not terminals. Let $w \notin A$ be a non-terminal vertex. We say that w has an improving move if the following properties hold: (1) There exists a terminal v whose routing path p_v includes w; let p_w denote the segment of p_v between w and r; (2) There exists a path q_w between w and r such that if v were to retain its current routing path from v to w but move from p_w to q_w, then the cost share of v would strictly decrease.

A priori, it is not clear whether improving moves can always be grouped into tree-follow moves. In Lemma 7, we show that in every routing tree T which is not in equilibrium, there exists a sequence of improving moves that collectively form the tree-follow move from u to v for some vertices u and v. When there are multiple such moves, we use a careful charging scheme to identify the order in which tree-follow moves should be implemented. (See Algorithm SELECT TREE MOVE defined at the end of this section.)

Since every vertex in a tree has a unique path to the root, it suffices to specify the tree itself in lieu of all of the routing paths. Henceforth, we will use T_t to denote the tree induced by $\{p_v\}_{v \in A_t}$ without explicitly specifying the paths themselves.

EQ-P **Dynamics**

1. *Initialization.* $t = 1$, $V_0 = \{r\}$, $T_0 = \{r\}$, $A_0 = \emptyset$.
2. *For $t = 1, 2, \ldots$*
   - (**Arrivals.**) Let U_t be the set of terminals arriving. Let $A_t \leftarrow A_{t-1} \cup U_t$. For each $v \in U_t$, let $p_v = p_v^*$ where p_v^* is the best response path of v with respect to T_{t-1}. Let $T_t = T_{t-1} \cup_{v \in U_t} p_v^*$.
   - (**Departures.**) Let D_t be the set of terminals departing. Let $A_t = A_t \setminus D_t$. Let $T_t = \cup_{v \in A_t} p_v$.
   - (**Tree Follow Moves.**) While T_t is not in equilibrium:
     Use algorithm SELECT TREE MOVE to determine a tree-follow move to implement in T_t; let this be a move from u to v, and let parent(u)

denote the parent of u in T_t. Implement the sequence of improving moves for this tree-follow move to obtain the new routing tree $T_t \leftarrow T_t \setminus (u, \text{parent}(u)) \cup (u, v)$.

Charging Scheme. In proving the upper bound for EQ-P dynamics, we use a dual charging scheme to bound the cost of the routing tree. We define the dual and the corresponding lower bound on the optimal cost next. We call a partition $P = (S_1, \cdots, S_m)$ of the vertex set V a *level-j* dual for an integer j if it satisfies the following:

- P is a partition: $\cup_{S \in P} S = V$, and for any $S_a, S_b \in P$, $S_a \cap S_b = \emptyset$.
- The components have bounded diameter: for any $S \in P$, and any vertices $x, y \in S$, $c(x, y) < 2^j$.
- The components are far from each other: there exists a "center" s_i in each component S_i, such that for all $S_a, S_b \in P$, $c(s_a, s_b) \geq 2^{j-1}$.

We use the term *cuts* to denote the components S of the partition. The lemma below follows immediately from the observation that any spanning tree over V must connect the centers of all cuts in a level-j dual P.

Lemma 4. *For any level-j dual P, the cost of the minimum spanning tree* OPT *is at least* $2^{j-1}(|P| - 1)$.

In order to bound the cost of an equilibrium resulting from EQ-P, we relate the cost of the edges used in the solution to a family of duals. Let $\Pi = \{P_j\}_{j \in \mathbb{Z}}$ denote a family of partitions, where P_j is a level-j dual.

Our charging scheme for routing solutions that form a tree proceeds as follows. Every vertex in the routing tree is responsible for the cost of its parent edge. Consider an edge $e = (v, \text{parent}(v))$ with length in $[2^{j+2}, 2^{j+3})$ for some $j \in \mathbb{Z}$. We charge the cost of this edge to the cut in the level-j dual that contains v: $S \in P_j$ such that $v \in S$. Our goal is to show that every cut gets charged a small number of times, and use the following well-known property (see the full version for a proof).

Lemma 5. *Suppose that our charging scheme charges each cut in the family Π at most once. Then the cost of the solution is at most* $O(\log n)$OPT.

For much of our analysis, we will assume that the dual family Π is provided to us. In the full version, we discuss how to construct this family algorithmically as terminals arrive online.

Classification of a Tree Routing. We classify the tree routings reachable via EQ-P dynamics into one of four states depending on the charging structure defined by the solution. We remark that not all tree routings are reachable via EQ-P dynamics, indeed even the set of equilibria obtained is smaller than the set of all equilibria. Let T be a routing tree for some set of active terminals A. We say a vertex u is a leaf (non-leaf) if it is a leaf (non-leaf) in T. Note that all leaves must be terminals, but a non-leaf vertex may or may not be a terminal.

1. BALANCED-EQUILIBRIUM: In this state, no terminal (and therefore, no non-terminal vertex in T) has an improving move. Furthermore, every cut is charged at most once. (Note that not every NE is a BALANCED-EQUILIBRIUM state.)
2. BALANCED: In this state, some terminals (and potentially non-terminals) may have improving moves, but every cut is charged at most once.
3. LEAF-UNBALANCED: In this state, every cut is charged by at most one non-leaf vertex (and any number of leaf terminals). (Recall that leaf vertices in the routing tree are necessarily terminals.)
4. NON-LEAF-UNBALANCED: In this state, all but one of the cuts are charged by at most one non-leaf vertex (and any number of leaf terminals). The exceptional cut, that we denote by S^*, is charged by at most two non-leaf vertices, say u and v (and any number of leaf terminals). One of these, u or v, must be the last vertex to have made a (tree-follow) move.

Note: BALANCED-EQUILIBRIUM $\subseteq$ BALANCED $\subseteq$ LEAF-UNBALANCED $\subseteq$ NON-LEAF-UNBALANCED, where $A \subseteq B$ implies that a routing tree in state A is also in state B.

Selecting a Tree-Follow Move. To define the tree-follow move performed in a non-equilibrium tree state T, we establish a system of priorities among the improving tree moves based on the current state of the routing tree. A tree follow move of u to v is said to be a *leaf move* if v is a leaf in T, and a *non-leaf move* otherwise.

Algorithm SELECT TREE MOVE

1. BALANCED-EQUILIBRIUM: No terminal has an improving move. The system can deviate from an equilibrium state only via arrivals or departure events.
2. BALANCED: In this state, for any vertex u that has an improving tree move, move u to the closest vertex to which it has an improving move.
3. LEAF-UNBALANCED:
   (a) If there exists a leaf terminal u with a non-leaf move, then make any such move for u.
   (b) Else, if there exists a non-leaf vertex u with a non-leaf move then move u to the closest such non-leaf v.
   (c) Else, if there exists a non-leaf vertex u and a leaf terminal v such that u and v are charging the same dual cut, then move u to v. If there are multiple such leaf terminals v, then make any such move.
   (d) Else, make any improving move. (This will necessarily be a leaf-to-leaf move by exclusion of the previous three cases.)

4. NON-LEAF-UNBALANCED: Let u and v be the non-leaf vertices that are charging the special cut S^*. If u has an improving move to v then move u, else move v, in either case to the closest vertex to which they have an improving move.

The validity of the algorithm depends on two claims. The first shows that whenever a cut is being charged by a leaf and a non-leaf, at least one of these two vertices has an improving move to the other. In this case, we can find a valid tree-move for Step (3c) of SELECT TREE MOVE. The second shows that in a NON-LEAF-UNBALANCED state, whenever a cut is being charged by two non-leaves, at least one of these two vertices has an improving move to the other; we can then find a valid tree-move for Step (4) of SELECT TREE MOVE.

3.1   Analysis of EQ-P Dynamics

We now given an outline of the proof of Theorem 2. Our argument hinges on a closure property: the epoch starts with the routing tree being in the BALANCED-EQUILIBRIUM state; Lemma 7 argues that whenever the current routing tree is not in equilibrium, at least one improving move exists, and we can use algorithm SELECT TREE MOVE to make a move; Lemma 8 then shows that for the moves made by algorithm SELECT TREE MOVE, the routing tree remains in one of the four states defined above, in particular, it is always in a NON-LEAF-UNBALANCED state. The epoch ends when the routing tree re-enters a BALANCED-EQUILIBRIUM state. At this point, by definition, each dual cut is charged at most once, and therefore, by Lemma 5 the cost of the routing tree is bounded, and Theorem 2 follows. We must also argue termination of the sequence of moves, but this follows directly from a standard potential argument based on the fact that all our moves are improving moves. The following lemmas capture the essence of our argument.

Observation 1. *In EQ-P dynamics the routing paths at the end of every phase form a tree.*

Lemma 6. *After the arrival or departure of a set of terminals in an BALANCED-EQUILIBRIUM state, the routing tree T remains in a LEAF-UNBALANCED state.*

Lemma 7. *If the routing tree is not in equilibrium, then at least one improving tree-follow move exists.*

Lemma 8. *Let T be the routing tree for which we make an improving tree-move in Step (3) of algorithm EQ-P.*

(i) *If T is in a BALANCED state but not in a BALANCED-EQUILIBRIUM state, then after the move selected in Step (2) of SELECT TREE MOVE, the resulting tree is in a NON-LEAF-UNBALANCED state.*

(ii) *If* T *is in a* LEAF-UNBALANCED *state, then after the move selected in Step (3) of* SELECT TREE MOVE, *the resulting tree is in a* NON-LEAF-UNBALANCED *state.*

(iii) *If* T *is in a* NON-LEAF-UNBALANCED *state, then after the move selected in Step (4) of* SELECT TREE MOVE, *the resulting tree is in a* NON-LEAF-UNBALANCED *state.*

References

1. Aland, S., Dumrauf, D., Gairing, M., Monien, B., Schoppmann, F.: Exact price of anarchy for polynomial congestion games. SIAM J. Comput. **40**(5), 1211–1233 (2011)
2. Albers, S.: On the value of coordination in network design. SIAM J. Comput. **38**(6), 2273–2302 (2009)
3. Alon, N., Awerbuch, B., Azar, Y., Buchbinder, N., Naor, J.: A general approach to online network optimization problems. ACM Trans. Algorithms **2**(4), 640–660 (2006)
4. Anshelevich, E., Dasgupta, A., Kleinberg, J.M., Tardos, É., Wexler, T., Roughgarden, T.: The price of stability for network design with fair cost allocation. SIAM J. Comput. **38**(4), 1602–1623 (2008)
5. Awerbuch, B., Azar, Y., Bartal, Y.: On-line generalized steiner problem. Theor. Comput. Sci. **324**(2–3), 313–324 (2004)
6. Balcan, M.-F., Blum, A., Mansour, Y.: Circumventing the price of anarchy: leading dynamics to good behavior. SIAM J. Comput. **42**(1), 230–264 (2013)
7. Beier, R., Czumaj, A., Krysta, P., Vöcking, B.: Computing equilibria for a service provider game with (im)perfect information. ACM Trans. Algorithms **2**(4), 679–706 (2006)
8. Berman, P., Coulston, C.: On-line algorithms for steiner tree problems. In: Proceedings of the Twenty-Ninth Annual ACM Symposium on Theory of Computing, pp. 344–353. ACM (1997)
9. Bilò, V., Bove, R.: Bounds on the price of stability of undirected network design games with three players. J. Interconnect. Netw. **12**(1–2), 1–17 (2011)
10. Bilò, V., Flammini, M., Moscardelli, L.: The price of stability for undirected broadcast network design with fair cost allocation is constant. In: FOCS, pp. 638–647 (2013)
11. Charikar, M., Karloff, H.J., Mathieu, C., Naor, J., Saks, M.E.: Online multicast with egalitarian cost sharing. In: SPAA, pp. 70–76 (2008)
12. Chawla, S., Naor, J., Panigrahi, D., Singh, M., Umboh, S.W.: Timing matters: online dynamics in broadcast games. CoRR, abs/1611.07745 (2016). http://arxiv.org/abs/1611.07745
13. Chekuri, C., Chuzhoy, J., Lewin-Eytan, L., Naor, J.(Seffi), Orda, A.: Non-cooperative multicast and facility location games. IEEE J. Sel. Areas Commun. **25**(6), 1193–1206 (2007)
14. Chen, H.-L., Roughgarden, T.: Network design with weighted players. Theory Comput. Syst. **45**(2), 302–324 (2009)
15. Christodoulou, G., Chung, C., Ligett, K., Pyrga, E., van Stee, R.: On the price of stability for undirected network design. In: Bampis, E., Jansen, K. (eds.) WAOA 2009. LNCS, vol. 5893, pp. 86–97. Springer, Heidelberg (2010). https://doi.org/10.1007/978-3-642-12450-1_8

16. Ene, A., Chakrabarty, D., Krishnaswamy, R., Panigrahi, D.: Online buy-at-bulk network design. In: IEEE 56th Annual Symposium on Foundations of Computer Science, FOCS 2015, Berkeley, CA, USA, 17–20 October, 2015, pp. 545–562 (2015)
17. Fabrikant, A., Papadimitriou, C.H., Talwar, K.: The complexity of pure Nash equilibria. In: Proceedings of the 36th Annual ACM Symposium on Theory of Computing, Chicago, IL, USA, 13–16 June 2004, pp. 604–612 (2004)
18. Fanelli, A., Leniowski, D., Monaco, G., Sankowski, P.: The ring design game with fair cost allocation. Theor. Comput. Sci. **562**, 90–100 (2015)
19. Fiat, A., Kaplan, H., Levy, M., Olonetsky, S., Shabo, R.: On the price of stability for designing undirected networks with fair cost allocations. In: Bugliesi, M., Preneel, B., Sassone, V., Wegener, I. (eds.) ICALP 2006. LNCS, vol. 4051, pp. 608–618. Springer, Heidelberg (2006). https://doi.org/10.1007/11786986_53
20. Freeman, R., Haney, S., Panigrahi, D.: On the Price of stability of undirected multicast games. In: Cai, Y., Vetta, A. (eds.) WINE 2016. LNCS, vol. 10123, pp. 354–368. Springer, Heidelberg (2016). https://doi.org/10.1007/978-3-662-54110-4_25
21. Gupta, A., Ravi, R., Talwar, K., Umboh, S.W.: LAST but not least: online spanners for buy-at-bulk. In: Proceedings of the Twenty-Eighth Annual ACM-SIAM Symposium on Discrete Algorithms, SODA 2017, 16–19 January, Hotel Porta Fira, Barcelona, Spain, pp. 589–599 (2017)
22. Hajiaghayi, M.T., Liaghat, V., Panigrahi, D.: Online node-weighted steiner forest and extensions via disk paintings. In: FOCS, pp. 558–567 (2013)
23. Hajiaghayi, M.T., Liaghat, V., Panigrahi, D.: Near-optimal online algorithms for prize-collecting steiner problems. In: Esparza, J., Fraigniaud, P., Husfeldt, T., Koutsoupias, E. (eds.) ICALP 2014. LNCS, vol. 8572, pp. 576–587. Springer, Heidelberg (2014). https://doi.org/10.1007/978-3-662-43948-7_48
24. Harks, T., Klimm, M.: On the existence of pure nash equilibria in weighted congestion games. Math. Oper. Res. **37**(3), 419–436 (2012)
25. Imase, M., Waxman, B.M.: Dynamic steiner tree problem. SIAM J. Discrete Math. **4**(3), 369–384 (1991)
26. Kawase, Y., Makino, K.: Nash equilibria with minimum potential in undirected broadcast games. Theor. Comput. Sci. **482**, 33–47 (2013)
27. Lee, E., Ligett, K.: Improved bounds on the price of stability in network cost sharing games. In: EC, pp. 607–620 (2013)
28. Li, J.: An $O(\log(n)/\log(\log(n)))$ upper bound on the price of stability for undirected shapley network design games. Inf. Process. Lett. **109**(15), 876–878 (2009)
29. Milchtaich, I.: Congestion games with player-specific payoff function. Games Econ. Behav. **13**, 111–124 (1996)
30. Monderer, D., Shapley, L.S.: Potential games. Games Econ. Behav. **14**, 124–143 (1996)
31. Naor, J., Panigrahi, D., Singh, M.: Online node-weighted steiner tree and related problems. In: FOCS, pp. 210–219 (2011)
32. Qian, J., Umboh, S.W., Williamson, D.P.: Online constrained forest and prize-collecting network design. Algorithmica **80**(11), 3335–3364 (2018)
33. Law-Yone, N., Holzman, R.: Strong equilibrium in congestion games. Games Econ. Behav. **21**, 85–101 (1997)

34. Rosenthal, R.W.: A class of games possessing pure-strategy Nash equilibria. Int. J. Game Theory **2**(1), 65–67 (1973)
35. Ui, T.: A shapley value representation of potential games. Games Econ. Behav. **31**(1), 121–135 (2000)
36. Umboh, S.: Online network design algorithms via hierarchical decompositions. In: Proceedings of the Twenty-Sixth Annual ACM-SIAM Symposium on Discrete Algorithms, pp. 1373–1387. SIAM (2015)

A Simple Mechanism
for a Budget-Constrained Buyer

Yu Cheng[1], Nick Gravin[2], Kamesh Munagala[1], and Kangning Wang[1(✉)]

[1] Duke University, Durham, USA
{yucheng,kamesh,knwang}@cs.duke.edu
[2] Shanghai University of Finance and Economics, Shanghai, China
nikolai@mail.shufe.edu.cn

Abstract. We study a classic Bayesian mechanism design setting of monopoly problem for an additive buyer in the presence of budgets. In this setting a monopolist seller with m heterogeneous items faces a single buyer and seeks to maximize her revenue. The buyer has a budget and additive valuations drawn independently for each item from (non-identical) distributions. We show that when the buyer's budget is publicly known, the better of selling each item separately and selling the grand bundle extracts a constant fraction of the optimal revenue. When the budget is private, we consider a standard Bayesian setting where buyer's budget b is drawn from a known distribution B. We show that if b is independent of the valuations and distribution B satisfies monotone hazard rate condition, then selling items separately or in a grand bundle is still approximately optimal. We give a complementary example showing that no constant approximation simple mechanism is possible if budget b can be interdependent with valuations.

1   Introduction

Revenue maximization is one of the fundamental problems in auction theory. The well-celebrated result of Myerson [43] characterized the revenue-maximizing mechanism when there is only one item for sale. Specifically, in the single buyer case, the optimal solution is to post a take-it-or-leave-it price. Since Myerson's work, the optimal mechanism design problem has been studied extensively in computer science literature and much progress has been made [2,12–15,26]. The problem of finding the optimal auction turned out to be so much more complex than the single-item case. Unlike the Myerson's single-item auction, the optimum can use randomized allocations and price bundles of items already for two items and a single buyer. It is also known that the gap between the revenue of the optimal randomized and optimal deterministic mechanism can be arbitrarily large [11,38], the optimal mechanism may require a menu with infinitely many options [27,42], and the revenue of the optimal auction may decrease when the buyer's valuation distributions move upwards (in the stochastic dominance sense).

© Springer Nature Switzerland AG 2018
G. Christodoulou and T. Harks (Eds.): WINE 2018, LNCS 11316, pp. 96–110, 2018.
https://doi.org/10.1007/978-3-030-04612-5_7

In light of these negative results for optimal auction design, many recent papers focused on the design of *simple* mechanisms that are *approximately* optimal. One such notable line of work initiated by Hart and Nisan [39] concerns a basic and natural setting of monopoly problem for the buyer with item values drawn independently from given distributions $D_1, \ldots, D_m$ and whose valuation for the sets of items is additive[1] (linear). A remarkable result by Babaioff et al. [4] showed that the better mechanism of either selling items separately, or selling the grand bundle extracts at least $(1/6)$-fraction of the optimal revenue. It was also observed [4,38,45] that the independence assumption on the items is essentially necessary and without it no simple (any deterministic) mechanism cannot be approximately optimal.

Auction design with budget constraints is an even harder problem. Because buyer's utility is no longer quasi-linear, many standard concepts do not carry over[2]. For example, even for one buyer and one item, the optimal mechanism may require randomization when the budget is public [21], and may need an exponential-size menu when the budget is private [30]. Despite many efforts [1,7–10,18,21,23,24,28–31,34,35,40,46], the theory of optimal auction design with budgets is still far behind the theory without budgets.

In this paper, we investigate the effectiveness of simple mechanisms in the presence of budgets. Our work is motivated by the following questions:

How powerful are simple mechanisms in the presence of budgets? In particular, is there a simple mechanism that is approximately optimal for a budget-constrained buyer with independent valuations?

To this end we consider one of the most basic and natural settings of extensively studied monopoly problem for an additive buyer. In this setting, a monopolistic seller sells m items to a single buyer. The buyer has additive valuations drawn independently for each item from an arbitrary (non-identical) distribution. We study two different budget settings: the *public budget* case where the buyer has a fixed budget known to the seller, and the *private budget* case where the buyer's budget is drawn from a distribution. The seller wishes to maximize her revenue by designing an auction subject to individual rationality, incentive compatibility, and budget constraints. We consider the Bayesian setting where the buyer knows his budget and his values for each item, but the seller only knows the prior distributions.

1.1 Our Results and Techniques

Our first result is that simple mechanisms remain approximately optimal when the buyer has a public budget.

---

[1] A buyer has additive valuations if his value for a set of items is equal to the sum of his values for the items in the set.

[2] E.g., the classic VCG mechanism may not be implementable and social efficiency may not be achievable in the budgeted-setting [46].

Theorem 1. *For an additive buyer with a known public budget and independent valuations, the better of selling each item separately and selling the grand bundle extracts a constant fraction of the optimal revenue.*

Theorem 1 is among the few positive results in budget-constrained settings that hold for arbitrary distributions. Before our work, it is not clear that any mechanism extracting a constant fraction of the optimal revenue can be computed in polynomial time.

In Sects. 3 and 4, we present two different approaches to prove Theorem 1. Both approaches truncate the valuation distribution V according to the budget b (in different ways) and then relate the revenues of the optimal/simple mechanisms on the truncated distribution to the revenues on the original valuations. The first approach uses the main result of [4] in a black-box way, and the second approach adapts the duality-based framework developed in [16].

It is worth pointing out that many of our structural lemmas hold for correlated valuations as well. Using these lemmas, we can generalize Theorem 1 with minimum effort to allow the buyer to have weakly correlated valuations. We call a distribution $\widehat{V}$ *weakly correlated* if it is the result of conditioning an independent distribution V on the sum of $v \sim V$ being at most c: $\widehat{V} = V_{|(\sum v_i \leq c)}$ (See Definition 4 for the formal definition).

Corollary 2. *Let $\widehat{V}$ be a weakly correlated distribution. For an additive buyer with a public budget and valuations drawn from $\widehat{V}$, the better of selling separately and selling the grand bundle extracts a constant fraction of the optimal revenue.*

In Sect. 5, we examine the private budget setting. The budget b is no longer fixed but is drawn from a distribution B. The seller only knows the prior distribution B but not the value of b. We first show that if the valuations can be correlated with the budget, the problem is at least as hard as budget-free mechanism design with correlated valuations, where simple mechanisms are known to be ineffective. In light of this negative result, we focus on the setting where the budget distribution B is independent of the valuations V. In this setting, we show that simple mechanisms are approximately optimal when the budget distribution satisfies the monotone hazard rate (MHR) condition.

Theorem 3. *When the budget distribution B is MHR, the better mechanism of pricing items separately and selling a grand bundle achieves a constant fraction of the optimal revenue.*

We will show that it is sufficient to pretend the buyer has a public budget $b^* = \mathbb{E}_{b \sim B}[b]$. The proof of Theorem 3 uses the MHR condition, as well as the fact that for a public budget b, the (budget-constrained) optimal revenue is nondecreasing in b, but optimal revenue divided by b is nonincreasing in b.

1.2 Related Work

The most closely related to ours are the following two lines of work.

Simple Mechanisms. In a line of work initiated by Hart and Nisan [4,39,41], [4] first showed that for an additive buyer with independent valuations, either selling separately or selling the grand bundle extracts a constant fraction of the optimal revenue. This was later extended to multiple buyers [49], as well as buyers with more general valuations (e.g., sub-additive [45], valuations with a common-value component [6], and valuations with complements [33]). Others have studied the trade-off between the complexity and approximation ratio of an auction, along with the design of small-menu mechanisms in various settings [3, 25,32,38,48].

Auctions for Budget-Constrained Buyers. There has been a lot of work studying the impact of budget constraints on mechanism design. Most of the earlier work required additional assumptions on the valuations distributions, like regularity or monotone hazard rate [9,23,40,44]. We mention a few results that work for arbitrary distributions. For public budgets, [21] designed approximately optimal mechanisms for several single-parameter settings and multi-parameter settings with unit-demand buyers. For private budgets, [30] characterized the structure of the optimal mechanism for one item and one buyer. [28] gave a constant-factor approximation for additive bidders whose private budgets can be correlated with their values. However, they require the buyers' valuation distribution to be given explicitly, which is of exponential size in our setting. There are also approximation and hardness results in the prior-free setting [1, 10,29], as well as designing Pareto optimal auctions [31,34].

Other Related Work. Our work concerns revenue maximization for additive buyer. Another natural and basic scenario extensively studied in the literature concerns buyers with unit-demand preferences [19,20,22]. Our work studies monopoly problem for additive budgeted buyer in the standard Bayesian approach. In this framework, the prior distribution is known to the seller and typically is assumed to be independent. Parallel to this framework, the (budgeted) additive monopoly problem has been studied in a new robust optimization framework [17,36]. Another group of papers on budget feasible mechanism design [7,24,46,47] studies different reverse auction settings and are concerned with value maximization.

2 Preliminaries

2.1 Optimal Mechanism Design

We study the design of optimal auctions with one buyer, one seller, and m heterogeneous items labeled by $[m] = \{1, \ldots, m\}$. There is exactly one copy of each item, and the items are indivisible. The buyer has additive valuation $(v(S) = \sum_{j \in S} v(\{i\})$ for any set $S \subseteq [m])$ and a publicly known budget b^3.

---

[3] In this paper, we mostly focus on the public budget case. So we define notations and discuss backgrounds assuming the buyer has a public budget.

We use $v \in \mathbb{R}^m$ to denote the buyer's valuations, where v_j is the buyer's value for item j. We consider the Bayesian setting of the problem, in which the buyer's values are drawn from a discrete[4] distribution V. Let $T = \text{supp}(V)$ be the set of all possible valuation profiles in V. We use $f(t)$ for any $t \in T$ to denote the probability mass function of V: $f(t) = \text{Pr}_{v \sim V}[v = t]$. Let $T_j = \text{supp}(V_j)$. We say the valuation distribution V is independent across items if it can be expressed as $V = \times_j V_j$.

We assume the buyer is risk-neutral and has quasi-linear utility when the payment does not exceed his budget. Let $\pi : T \to [0,1]^m$ and $p : T \to \mathbb{R}$ denote the allocation and payment rules of a mechanism respectively. That is, when the buyer reports type t, the probability that he will receive item j is $\pi_j(t)$, and his expected payment is $p(t)$ (over the randomness of the mechanism). Thus, if the buyer has type t, his (expected) value for reporting type t' is exactly $\pi(t')^\top t,$[5] and his (expected) utility for reporting type t' is

$$u(t, t') = \begin{cases} \pi(t')^\top t - p(t') & \text{if } p(t') \le b, \text{ and} \\ -\infty & \text{otherwise.} \end{cases}$$

By the revelation principle, it is sufficient to consider mechanisms that are incentive compatible (i.e., "truthful"). A mechanism $M = (\pi, p)$ is (interim) incentive-compatible (IC) if the buyer is incentivized to tell the truth (over the randomness of mechanism), and (interim) individually rational (IR) if the buyer's expected utility is non-negative whenever he reports truthfully. We use $\varnothing$ for the option of not participating in the auction ($\pi(\varnothing) = 0, p(\varnothing) = 0$), and let $T^+ = T \cup \{\varnothing\}$. Then, the IC and IR constraints can be unified as follows:

$$u(t, t) \ge u(t, t') \quad \forall t \in T, t' \in T^+.$$

To summarize, when the seller faces a single buyer with budget b and valuation drawn from V, the optimal mechanism $M^* = (\pi^*, p^*)$ is the optimal solution to the following (exponential-size) linear program (LP):

$$
\begin{aligned}
\text{maximize } & \sum_{t \in T} f(t) p(t) \\
\text{subject to } & \pi(t')^\top t - p(t') \le \pi(t)^\top t - p(t), \forall t \in T, t' \in T^+. \\
& 0 \le \pi_j(t) \le 1, & \forall t \in T, j \in [m]. \\
& p(t) \le b, & \forall t \in T. \\
& \pi(\varnothing) = 0, \; p(\varnothing) = 0.
\end{aligned}
\tag{1}
$$

A mechanism is called ex-post IC, ex-post IR, or ex-post budget-preserving respectively, if the corresponding constraints hold for all possible outcomes, without averaging over the randomness in the mechanism. We will show the better of pricing each item separately and pricing the grand bundle, which is a deterministic ex-post mechanism, can extract a constant fraction of the revenue of any interim mechanism.

---

[4] Like previous work on simple and approximately optimal mechanisms, our results extend to continuous types as well (see, e.g., [16] for a more detailed discussion).

[5] We use $x^\top y = \sum_{i=1}^m x_i y_i$ to denote the inner product of two vectors x and y.

2.2 Simple Mechanisms

For a buyer with valuation distribution V, we frequently use the following notations in our analysis:

- REV(V): the revenue of the optimal truthful mechanism.
- SREV(V): the maximum revenue obtainable by pricing each item separately.
- BREV(V): the maximum revenue obtainable by pricing the grand bundle.
- REV^b(V): the revenue of the optimal truthful mechanism, when the buyer has a budget b.
- SREV^b(V): the maximum revenue that can be extracted by pricing each item separately, when the buyer has a public budget b.
- BREV^b(V): the maximum revenue that can be extracted by pricing the grand bundle, when the buyer has a public budget b.

We know that SREV(V) is obtained by running Myerson's optimal auction separately for each item, and BREV(V) is obtained by running Myerson's auction viewing the grand bundle as one item. Similarly, BREV^b(V) is a single-parameter problem as well, with the minor change that the posted price is at most b.

The case of SREV^b(V) is more complicated. For example, when a budgeted buyer of type $t \in \mathbb{R}^m$ participates in an auction with posted price p_j for each item j, he will maximize his utility by solving a KNAPSACK problem. There exists a poly-time computable mechanism that extracts a constant fraction of SREV^b(V) (e.g., [8]). We focus on the structural result that the better of SREV^b(V) and BREV^b(V) is a constant approximation of REV^b(V). A better approximation for SREV^b(V) is an interesting open problem that is beyond the scope of this paper.

2.3 Weakly Correlated Distributions

We call a distribution like $\widehat{V}$ *weakly correlated* if the only condition causing the correlation is a cap on its sum.

Definition 4. *For an m-dimensional independent distribution V and a threshold $c > 0$, we remove the probability mass on any $t \in \mathrm{supp}(V)$ with $\|t\|_1 > c$ and renormalize. Let $\widehat{V} := V_{|(\|v\|_1 \leq c)}$ denote the resulting distribution. Formally,*

$$\Pr_{\widehat{v} \sim \widehat{V}}[\widehat{v} = t] = \Pr_{v \sim V}[v = t \mid \|v\|_1 \leq c], \ \forall t \in \mathrm{supp}(V).$$

Weakly correlated distributions arise naturally in our analysis. We will show that if the buyer's valuations are weakly correlated, then the better of selling separately and selling the grand bundle is approximately optimal, and this holds with or without a (public) budget constraint.

2.4   First-Order Stochastic Dominance

Stochastic dominance is a partial order between random variables. A random variable X with $\text{supp}(X) \subseteq \mathbb{R}$ *(weakly) first-order stochastically dominates* another random variable Y with $\text{supp}(Y) \subseteq \mathbb{R}$ if and only if

$$\Pr[X \geq a] \geq \Pr[Y \geq a] \text{ for all } a \in \mathbb{R}.$$

This notion of stochastic dominance can be extended to multi-dimensional distributions. In this paper, we use the notion of coordinate-wise dominance.

Definition 5. *Given two m-dimensional distributions D_1 and D_2, we say D_1 coordinate-wise stochastically dominates D_2 ($D_1 \succeq D_2$ or $D_2 \preceq D_1$) if there exists a randomized mapping $f : \text{supp}(D_1) \to \text{supp}(D_2)$ such that $f(x) \sim D_2$ when $x \sim D_1$, and $f(x) \leq x$ coordinate-wise for all $x \in \text{supp}(D_1)$ with probability 1.*

This notion helps us express the monotonicity of optimal revenues in some cases. For example, we can show that $\text{SREV}(V_1) \geq \text{SREV}(V_2)$ when $V_1 \succeq V_2$. The mapping f allows us to couple the draws $v_1 \sim V_1$ and $v_2 \sim V_2$, so that for a set of fixed prices, if the buyer buys an item under v_2, he will also buy it under v_1.

3   Public Budget

In this section, we focus on the public budget case and prove our main result (Theorem 1). The buyer has a fixed budget b and valuations drawn from an independent distribution V.

Theorem 1. $\text{REV}^b(V) \leq 8\text{SREV}^b(V) + 24\text{BREV}^b(V)$.

It follows that the better of $\text{SREV}^b(V)$ and $\text{BREV}^b(V)$ is at least $\frac{\text{REV}^b(V)}{32}$.[6]

Overview of Our Approach. Instead of taking the Lagrangian dual of LP (1) to derive an upper bound on the optimal objective value $\text{REV}^b(V)$, we adopt a more combinatorial approach. Intuitively, we come up with a charging argument that splits $\text{REV}^b(V)$ and charges each part to either $\text{SREV}^b(V)$ or $\text{BREV}^b(V)$.

First, we partition the buyer types $t \in \text{supp}(V)$ into two sets: *high-value* types where $\|t\|_\infty \geq b$ and *low-value* types where $\|t\|_\infty < b$. Note that we can already charge the revenue of high-value types to $\text{BREV}^b(V)$: If we sell the grand bundle at price b, all high-value types will exhaust their budgets.

We now examine the low-value types. Let V' denote the valuation distribution conditioned on the buyer having a low-value type. Observe that V' is independent because it is defined using ℓ_∞-norm, and we can remove the budget to upper

---

[6] We do not optimize the constants in our proofs. In Sect. 4, we will give an alternative proof of Theorem 1 that shows $\text{REV}^b(V) \leq 5\text{SREV}^b(V) + 6\text{BREV}^b(V)$, thus improving this constant from 32 to 11.

bound its revenue. For a budget-free additive buyer with independent valuations, we can apply the main result of [4], which states that either selling separately or grand bundling works for V': $\text{REV}(V') = O(\text{SREV}(V') + \text{BREV}(V'))$.

Next, we will relate $\text{SREV}(V'), \text{BREV}(V')$ to $\text{SREV}^b(V'), \text{BREV}^b(V')$. We can assume the sum of $v' \sim V'$ is usually much smaller than b. Similar to standard tail bounds, if the sum $\|v'\|_1$ is often small and the random variables are independent and bounded (each v'_j is at most b), then $\|v'\|_1$ must have an exponentially decaying tail. Therefore, we can add back the budget, because the sum $\|v'\|_1$, which upper bounds the buyer's payment, is rarely very large.

Finally, we will show that $\text{SREV}^b(V') = O(\text{SREV}^b(V))$ and $\text{BREV}^b(V') \leq \text{BREV}^b(V)$. The BREV statement is easy to verify, but the SREV statement is more tricky. The monotonicity of $\text{SREV}(V)$ in the budget-free case (see Sect. 2.4) no longer holds when there is a budget. Fortunately, we can pay a factor of two and circumvent this non-monotonicity due to budget constraints.

We will now make our intuitions formal and present three key lemmas. Throughout the paper, we will always use $V' = V_{\|v\|_\infty \leq b}$ as defined below.

Definition 6. *Fix an m-dimensional distribution $V = \times V_j$. Let V' be the independent distribution where every coordinate of V is capped at b. That is, $V' = \times_j V'_j$, and V'_j is given by $\Pr_{V'_j}[x] = \Pr_{v_j \sim V_j}[\min(v_j, b) = x]$.*

Lemma 7. $\text{REV}^b(V) \leq \text{REV}(V') + \text{BREV}^b(V)$.

Lemma 8. *Assume $\text{BREV}^b(V') < \frac{b}{10}$. Then, $\text{BREV}(V') \leq 3\text{BREV}^b(V')$ and $\text{SREV}(V') \leq \text{SREV}^b(V') + 4\text{BREV}^b(V')$.*

Lemma 9. $\text{BREV}^b(V') \leq \text{BREV}^b(V)$ and $\text{SREV}^b(V') \leq 2\text{SREV}^b(V)$.

We defer the proofs of these lemmas to the full version of this paper, and first use them to prove Theorem 1.

Proof (of Theorem 1). If $\text{BREV}^b(V') \geq \frac{b}{10}$, then the theorem holds because the optimal revenue $\text{REV}^b(V)$ is at most the budget b. By Lemma 9, $\text{BREV}^b(V) \geq \text{BREV}^b(V') \geq \frac{b}{10} \geq \frac{\text{REV}^b(V)}{10}$.

We now assume $\text{BREV}^b(V') < \frac{b}{10}$. The theorem follows straightforwardly from Lemmas 7, 8, 9, and a black-box use of the main result of [4].

$$\text{REV}^b(V) \leq \text{REV}(V') + \text{BREV}^b(V) \qquad \text{(Lemma 7)}$$
$$\leq 4\text{SREV}(V') + 2\text{BREV}(V') + 2\text{BREV}^b(V) \qquad \text{([4])}$$
$$\leq 4\text{SREV}^b(V') + 22\text{BREV}^b(V') + 2\text{BREV}^b(V). \qquad \text{(Lemma 8)}$$
$$\leq 8\text{SREV}^b(V) + 24\text{BREV}^b(V). \qquad \text{(Lemma 9)}$$

$\square$

4 Public Budget and Weakly Correlated Valuations

In this section, we present an alternative approach to prove our main result (Theorem 1). Recall that the buyer has a public budget b and valuations drawn from an independent distribution V.

Theorem 1. $\text{REV}^b(V) \leq 5\text{SREV}^b(V) + 6\text{BREV}^b(V)$.

Overview of Our Approach. We will truncate the input distribution V in a different way: instead of truncating $v \sim V$ in ℓ_∞-norm (as in Sect. 3), we will truncate v in ℓ_1-norm. This truncation produces a correlated distribution $\widehat{V}$. The upshot of truncating in ℓ_1-norm is that we always have $\|\widehat{v}\|_1 \leq b$, so $\widehat{V}$ can ignore the budget. In addition, as in Sect. 3, we can relate the optimal revenue to the revenue of $\widehat{V}$, and we can relate the revenue of simple mechanisms on $\widehat{V}$ back to revenue of simple mechanisms on V.

We still need to argue that simple mechanisms work well for $\widehat{V}$. This is the main challenge in this approach. Because $\widehat{V}$ is correlated, we cannot apply the result of [4] in a black-box way. Instead, we need to modify the analysis of previous work [4,16,41] and build on the key ideas like "core-tail" decomposition. More specifically, we generalize the duality-based framework developed in [16] to handle the specific type of correlation $\widehat{V}$ has.

Weakly Correlated Valuations. It is worth mentioning that our structural lemmas do not require the input distribution to be independent. This is why our techniques can be applied to more general settings. For example, in this section, we will generalize Theorem 1 with minimum effort to handle weakly correlated valuations (see Definition 4 for the formal definition).

Corollary 2. Let $\widehat{V}$ be a weakly correlated distribution (Definition 4). We have $\text{REV}^b(\widehat{V}) \leq 5\text{SREV}^b(\widehat{V}) + 6\text{BREV}^b(\widehat{V})$.

Our main contribution in this section is Lemma 10. Lemma 10 shows that for any weakly correlated distribution $\widehat{V}$ (see Definition 4), the better of $\text{SREV}(\widehat{V})$ and $\text{BREV}(\widehat{V})$ is a constant approximation to the optimal revenue $\text{REV}(\widehat{V})$.

Lemma 10. Fix $c > 0$. Let $\widehat{V} = V_{|(\|v\|_1 \leq c)}$ for an independent distribution V. We have $\text{REV}(\widehat{V}) \leq 5\text{SREV}(\widehat{V}) + 4\text{BREV}(\widehat{V})$.

We defer the proof of Lemma 10 to the full version. We first use these lemmas to prove Theorem 1.

Proof (of Theorem 1). If $\min_{v \sim V} \|v\|_1 \geq b/2$, then the seller can price the grand bundle at $b/2$ and the buyer always buys it. In this case, the revenue is $b/2$ and $\text{REV}^b(V) \leq b \leq 2\text{BREV}^b(V)$. Thus, we focus on the more interesting case where $\text{Pr}_{v \sim V}[\|v\|_1 \leq b/2] > 0.$[7]

---

[7] Throughout the paper, when we consider the conditional distribution $\widehat{V} := V_{|(\|v\|_1 \leq c)}$, we will always have $c > \min_{v \in \text{supp}(V)} \|v\|_1$, so that the event we condition on happens with non-zero probability.

Let $\widehat{V} := V_{|(\|v\|_1 \leq c)}$ for $c = b/2$. The proof outline goes as:

$$\begin{aligned}
\mathrm{REV}^b(V) &\leq (b/c) \cdot \mathrm{BREV}^b(V) + \mathrm{REV}(\widehat{V}) \\
&\leq 2\mathrm{BREV}^b(V) + 5\mathrm{SREV}(\widehat{V}) + 4\mathrm{BREV}(\widehat{V}) & \text{(Lemma 10)} \\
&= 2\mathrm{BREV}^b(V) + 5\mathrm{SREV}^c(\widehat{V}) + 4\mathrm{BREV}^c(\widehat{V}) & (\|\widehat{v}\|_1 \leq c) \\
&\leq 2\mathrm{BREV}^b(V) + 5\mathrm{SREV}^b(V) + 4\mathrm{BREV}^b(V) \\
&= 5\mathrm{SREV}^b(V) + 6\mathrm{BREV}^b(V).
\end{aligned}$$

The details are proved in the full version.   □

5 Private Budget

In this section, we consider the case where the budget b is no longer fixed but instead drawn from a distribution B. One natural model is that the buyer's budget b is first drawn from B, and then depending on the value of b, the buyer's valuations are drawn independently for each item.

We show that in this case, the problem is at least as hard as finding (approximately) optimal mechanisms for correlated valuations in the budget-free setting. Consider an instance in which all possible budgets are larger than $\max_{v \sim V} \|v\|_1$ so they are irrelevant. However, the budget can still be used as a signal (or a correlation device) to produce correlated valuations. It is known that for correlated distributions, the better of selling separately and bundling together [37], or even the best partition-based mechanism [4], does not offer a constant approximation.

This negative result motivates us to study the private budget setting when the budget distribution B is independent of the valuation distributions V.

5.1 Monotone-Hazard-Rate Budgets

We focus on the case where the budget is independent of valuations, and it is drawn from a continuous[8] monotone-hazard-rate (MHR) distribution. Let $g(\cdot)$ and $G(\cdot)$ be the probability density function and cumulative distribution function of B. The MHR condition says $\frac{g(b)}{1-G(b)}$ is non-decreasing in b.

Lemma 11. *Let b^* be the expectation of an MHR distribution B. Let M^* be the optimal mechanism for a buyer with a public budget b^*. Then in expectation, M^* extracts at least $\frac{1}{2e}$-fraction of the expected optimal revenue when the buyer has a private budget drawn from B.*

Proof. Let $R(b, V)$ denote the expected revenue of M^* when the buyer has a public budget b and valuations drawn from V. Let $R(B, V) = \mathbb{E}_{b \sim B}[R(b, V)]$

---

[8] If the distribution is a discrete MHR distribution, similar results still hold. For discrete distributions we have $\mathrm{Pr}_{b \sim B}[b \geq \lfloor b^* \rfloor] \geq e^{-1}$ instead of $\mathrm{Pr}_{b \sim B}[b \geq b^*] \geq e^{-1}$.

denote the expected revenue of M^* when the buyer's budget is drawn from B.

$$R(B, V) = \int_b g(b)R(b, V)\mathrm{d}b \geq \int_{b \geq b^*} g(b)R(b, V)\mathrm{d}b$$

$$= \int_{b \geq b^*} g(b)R(b^*, V)\mathrm{d}b \geq e^{-1} \cdot R(b^*, V).$$

The second last step uses $R(b, V) = R(b^*, V)$ when $b \geq b^*$, because M^* provides a menu of allocation/payment pairs for the buyer to choose from; A buyer with budget $b \geq b^*$ can afford any option on the menu so he will choose the same option as if he had budget b^*. The last inequality comes from the fact that for any MHR distribution B, $\Pr_{b \sim B}[b \geq b^*] \geq e^{-1}$ (see, e.g., [5]).

Let $\mathrm{REV}^B(V)$ denote the optimal revenue we can extract when the buyer has private budgets drawn from B.

$$\mathrm{REV}^B(V) \leq \int_{b < b^*} g(b)\mathrm{REV}^b(V)\mathrm{d}b + \int_{b \geq b^*} g(b)\mathrm{REV}^b(V)\mathrm{d}b$$

$$\leq \int_{b < b^*} g(b)\mathrm{REV}^{b^*}(V)\mathrm{d}b + \int_{b \geq b^*} g(b) \cdot \frac{b}{b^*} \cdot \mathrm{REV}^{b^*}(V)\mathrm{d}b$$

$$\leq \mathrm{REV}^{b^*}(V) + \frac{\int_b g(b)b\,\mathrm{d}b}{b^*} \cdot \mathrm{REV}^{b^*}(V) = 2\mathrm{REV}^{b^*}(V).$$

The first line is because the seller can only do better if she knows the buyer's budget b. The third line is because $b^* = \mathbb{E}[b]$. The second line uses the fact that $\mathrm{REV}^b(V) \leq \mathrm{REV}^{b^*}(V)$ when $b < b^*$ and $\mathrm{REV}^b(V) \leq \frac{b}{b^*}\mathrm{REV}^{b^*}(V)$ when $b > b^*$.

We have $\mathrm{REV}^b(V) \leq \mathrm{REV}^{b^*}(V)$ when $b < b^*$ because a buyer with budget b^* can afford all options from the menu that achieves $\mathrm{REV}^b(V)$. When $b > b^*$, consider the menu that achieves $\mathrm{REV}^b(V)$ and cap all prices at b^*. A buyer with budget $b > b^*$ either chooses the same option as if he had budget b^*, or chooses a different option whose price must be b^*, and therefore $\mathrm{REV}^b(V) \leq \frac{b}{b^*}\mathrm{REV}^{b^*}(V)$.

By definition $R(b^*, V) = \mathrm{REV}^{b^*}(V)$. Therefore, $R(B, V) \geq \frac{1}{2e}\mathrm{REV}^B(V)$. □

Theorem 3. *When the budget distribution B is MHR, the better of pricing items separately and bundling them together achieves a constant fraction of the optimal revenue.*

Proof. By pretending the budget is b^*,

$$\mathrm{SREV}^B(V) \geq \int_{b \geq b^*} g(b)\mathrm{SREV}^{b^*}(V)\mathrm{d}b \geq \frac{1}{e}\mathrm{SREV}^{b^*}(V).$$

Similarly, $\mathrm{BREV}^B(V) \geq \frac{1}{e}\mathrm{BREV}^{b^*}(V)$. Therefore, by Theorem 1 and Lemma 11, $\mathrm{SREV}^B(V) + \mathrm{BREV}^B(V) = \Omega(\mathrm{SREV}^{b^*}(V) + \mathrm{BREV}^{b^*}(V)) = \Omega(\mathrm{REV}^{b^*}(V)) = \Omega(\mathrm{REV}^B(V))$. □

6   Conclusion and Future Directions

In this paper, we investigated the effectiveness of simple mechanisms in the presence of budgets, and showed that for an additive buyer with independent valuations and a public budget, either selling separately or selling the grand bundle gives a constant approximation to optimal revenue.

The area of designing simple and approximately optimal auctions with budget constraints is still largely unexplored. Our work leaves many natural follow-up questions. We only considered selling to a single buyer. An immediate open question is whether our results can be extended to multiple bidders. A generalization to multiple bidders is known in the budget-free case [16,49].

Question 1. *Is there a simple mechanism that is approximately optimal for multiple additive buyers, when each buyer has the same public budget b?*

For private budgets where the budget is independent of the valuations, we showed that if the budget distribution satisfies monotone hazard rate, then we can extract a constant fraction of the revenue. The general case with arbitrary budget distributions appears to be nontrivial and is an interesting avenue for future work.

Question 2. *Is there a simple mechanism that is approximately optimal for an additive buyer with private budgets, when the budget distribution is independent of the valuations?*

Acknowledgements. Yu Cheng is supported by NSF grants CCF-1527084, CCF-1535972, CCF-1637397, CCF-1704656, IIS-1447554, and NSF CAREER Award CCF-1750140. Kamesh Munagala is supported by NSF grants CCF-1408784, CCF-1637397, and IIS-1447554; and by an Adobe Data Science Research Award. Kangning Wang is supported by NSF grants CCF-1408784 and CCF-1637397.

References

1. Abrams, Z.: Revenue maximization when bidders have budgets. In: Proceedings of 17th ACM-SIAM Symposium on Discrete Algorithms, pp. 1074–1082 (2006)
2. Alaei, S., Fu, H., Haghpanah, N., Hartline, J.D., Malekian, A.: Bayesian optimal auctions via multi- to single-agent reduction. In: Proceedings of 13th ACM Conference on Electronic Commerce, p. 17 (2012)
3. Babaioff, M., Gonczarowski, Y.A., Nisan, N.: The menu-size complexity of revenue approximation. In: Proceedings of 49th ACM Symposium on Theory of Computing, pp. 869–877 (2017)
4. Babaioff, M., Immorlica, N., Lucier, B., Weinberg, S.M.: A simple and approximately optimal mechanism for an additive buyer. In: Proceedings of 55th IEEE Symposium on Foundations of Computer Science, pp. 21–30 (2014)
5. Barlow, R.E., Marshall, A.W.: Tables of bounds for distributions with monotone hazard rate. J. Am. Stat. Assoc. **60**(311), 872–890 (1965)
6. Bateni, M.H., Dehghani, S., Hajiaghayi, M.T., Seddighin, S.: Revenue maximization for selling multiple correlated items. In: Bansal, N., Finocchi, I. (eds.) ESA 2015. LNCS, vol. 9294, pp. 95–105. Springer, Heidelberg (2015). https://doi.org/10.1007/978-3-662-48350-3_9

7. Bei, X., Chen, N., Gravin, N., Lu, P.: Budget feasible mechanism design: from prior-free to Bayesian. In: Proceedings of 44th ACM Symposium on Theory of Computing, pp. 449–458 (2012)
8. Bhattacharya, S., Conitzer, V., Munagala, K., Xia, L.: Incentive compatible budget elicitation in multi-unit auctions. In: Proceedings of 21st ACM-SIAM Symposium on Discrete Algorithms, pp. 554–572 (2010)
9. Bhattacharya, S., Goel, G., Gollapudi, S., Munagala, K.: Budget constrained auctions with heterogeneous items. In: Proceedings of 42nd ACM Symposium on Theory of Computing, pp. 379–388 (2010)
10. Borgs, C., Chayes, J.T., Immorlica, N., Mahdian, M., Saberi, A.: Multi-unit auctions with budget-constrained bidders. In: Proceedings of 6th ACM Conference on Electronic Commerce, pp. 44–51 (2005)
11. Briest, P., Chawla, S., Kleinberg, R., Weinberg, S.M.: Pricing randomized allocations. In: Proceedings of 21st ACM-SIAM Symposium on Discrete Algorithms, pp. 585–597 (2010)
12. Cai, Y., Daskalakis, C., Weinberg, S.M.: An algorithmic characterization of multidimensional mechanisms. In: Proceedings of 44th ACM Symposium on Theory of Computing, pp. 459–478 (2012)
13. Cai, Y., Daskalakis, C., Weinberg, S.M.: Optimal multi-dimensional mechanism design: reducing revenue to welfare maximization. In: Proceedings of 53rd IEEE Symposium on Foundations of Computer Science, pp. 130–139 (2012)
14. Cai, Y., Daskalakis, C., Weinberg, S.M.: Reducing revenue to welfare maximization: approximation algorithms and other generalizations. In: Proceedings of 24th ACM-SIAM Symposium on Discrete Algorithms, pp. 578–595 (2013)
15. Cai, Y., Daskalakis, C., Weinberg, S.M.: Understanding incentives: mechanism design becomes algorithm design. In: Proceedings of 54th IEEE Symposium on Foundations of Computer Science, pp. 618–627 (2013)
16. Cai, Y., Devanur, N.R., Weinberg, S.M.: A duality based unified approach to Bayesian mechanism design. In: Proceedings of 48th ACM Symposium on Theory of Computing, pp. 926–939 (2016)
17. Carroll, G.: Robustness and separation in multidimensional screening. Econometrica **85**(2), 453–488 (2017)
18. Chakrabarty, D., Goel, G.: On the approximability of budgeted allocations and improved lower bounds for submodular welfare maximization and GAP. SIAM J. Comput. **39**(6), 2189–2211 (2010)
19. Chawla, S., Hartline, J.D., Kleinberg, R.D.: Algorithmic pricing via virtual valuations. In: Proceedings of 8th ACM Conference on Electronic Commerce, pp. 243–251 (2007)
20. Chawla, S., Hartline, J.D., Malec, D.L., Sivan, B.: Multi-parameter mechanism design and sequential posted pricing. In: Proceedings of 42nd ACM Symposium on Theory of Computing, pp. 311–320 (2010)
21. Chawla, S., Malec, D.L., Malekian, A.: Bayesian mechanism design for budget-constrained agents. In: Proceedings of 12th ACM Conference on Electronic Commerce, pp. 253–262 (2011)
22. Chawla, S., Malec, D.L., Sivan, B.: The power of randomness in Bayesian optimal mechanism design. Games Econ. Behav. **91**, 297–317 (2015)
23. Che, Y., Gale, I.L.: The optimal mechanism for selling to a budget-constrained buyer. J. Econ. Theory **92**(2), 198–233 (2000)
24. Chen, N., Gravin, N., Lu, P.: On the approximability of budget feasible mechanisms. In: Proceedings of 22nd ACM-SIAM Symposium on Discrete Algorithms, pp. 685–699 (2011)

25. Cheng, Y., Cheung, H.Y., Dughmi, S., Emamjomeh-Zadeh, E., Han, L., Teng, S.: Mixture selection, mechanism design, and signaling. In: Proceedings of 56th IEEE Symposium on Foundations of Computer Science, pp. 1426–1445 (2015)
26. Daskalakis, C.: Multi-item auctions defying intuition? SIGecom Exch. **14**(1), 41–75 (2015)
27. Daskalakis, C., Deckelbaum, A., Tzamos, C.: Mechanism design via optimal transport. In: Proceedings of 14th ACM Conference on Electronic Commerce, pp. 269–286 (2013)
28. Daskalakis, C., Devanur, N.R., Weinberg, S.M.: Revenue maximization and ex-post budget constraints. In: Proceedings of 16th ACM Conference on Economics and Computation, pp. 433–447 (2015)
29. Devanur, N.R., Ha, B.Q., Hartline, J.D.: Prior-free auctions for budgeted agents. In: Proceedings of 14th ACM Conference on Electronic Commerce, pp. 287–304 (2013)
30. Devanur, N.R., Weinberg, S.M.: The optimal mechanism for selling to a budget constrained buyer: the general case. In: Proceedings of 18th ACM Conference on Economics and Computation, pp. 39–40 (2017)
31. Dobzinski, S., Lavi, R., Nisan, N.: Multi-unit auctions with budget limits. Games Econ. Behav. **74**(2), 486–503 (2012)
32. Dughmi, S., Han, L., Nisan, N.: Sampling and representation complexity of revenue maximization. In: Liu, T.-Y., Qi, Q., Ye, Y. (eds.) WINE 2014. LNCS, vol. 8877, pp. 277–291. Springer, Cham (2014). https://doi.org/10.1007/978-3-319-13129-0_22
33. Eden, A., Feldman, M., Friedler, O., Talgam-Cohen, I., Weinberg, S.M.: A simple and approximately optimal mechanism for a buyer with complements. In: Proceedings of 18th ACM Conference on Economics and Computation, p. 323 (2017)
34. Goel, G., Mirrokni, V.S., Leme, R.P.: Polyhedral clinching auctions and the adwords polytope. J. ACM **62**(3), 18:1–18:27 (2015)
35. Goldberg, A.V., Hartline, J.D., Wright, A.: Competitive auctions and digital goods. In: Proceedings of 12th ACM-SIAM Symposium on Discrete Algorithms, pp. 735–744 (2001)
36. Gravin, N., Lu, P.: Separation in correlation-robust monopolist problem with budget. In: Proceedings of 29th ACM-SIAM Symposium on Discrete Algorithms, pp. 2069–2080 (2018)
37. Hart, S., Nisan, N.: Approximate revenue maximization with multiple items. In: Proceedings of 13th ACM Conference on Electronic Commerce, p. 656 (2012)
38. Hart, S., Nisan, N.: The menu-size complexity of auctions. In: Proceedings of 14th ACM Conference on Electronic Commerce, pp. 565–566 (2013)
39. Hart, S., Nisan, N.: Approximate revenue maximization with multiple items. J. Econ. Theory **172**, 313–347 (2017)
40. Laffont, J.-J., Robert, J.: Optimal auction with financially constrained buyers. Econ. Lett. **52**(2), 181–186 (1996)
41. Li, X., Yao, A.C.: On revenue maximization for selling multiple independently distributed items. Proc. Natl. Acad. Sci. **110**(28), 11232–11237 (2013)
42. Manelli, A.M., Vincent, D.R.: Multidimensional mechanism design: revenue maximization and the multiple-good monopoly. J. Econ. Theory **137**(1), 153–185 (2007)
43. Myerson, R.B.: Optimal auction design. Math. Oper. Res. **6**(1), 58–73 (1981)
44. Pai, M.M., Vohra, R.: Optimal auctions with financially constrained buyers. J. Econ. Theory **150**, 383–425 (2014)
45. Rubinstein, A., Weinberg, S.M.: Simple mechanisms for a subadditive buyer and applications to revenue monotonicity. In: Proceedings of 16th ACM Conference on Economics and Computation, pp. 377–394 (2015)

46. Singer, Y.: Budget feasible mechanisms. In: Proceedings of 51st IEEE Symposium on Foundations of Computer Science, pp. 765–774 (2010)
47. Singla, A., Krause, A.: Truthful incentives in crowdsourcing tasks using regret minimization mechanisms. In: WWW, pp. 1167–1178 (2013)
48. Tang, P., Wang, Z.: Optimal mechanisms with simple menus. J. Econ. Theory **69**, 54–70 (2017)
49. Yao, A.C.: An n-to-1 bidder reduction for multi-item auctions and its applications. In: Proceedings of 26th ACM-SIAM Symposium on Discrete Algorithms, pp. 92–109 (2015)

The Communication Complexity
of Graphical Games on Grid Graphs

Jen-Hou Chou and Chi-Jen Lu[(✉)]

Institute of Information Science, Academia Sinica, Taipei, Taiwan
{jhchou,cjlu}@iis.sinica.edu.tw

Abstract. We consider the problem of deciding the existence of pure Nash equilibrium and the problem of finding mixed Nash equilibrium in graphical games defined on the two dimensional $d \times m$ grid graph. Unlike previous works focusing on the computational complexity of centralized algorithms, we study the communication complexity of distributed protocols for these problems, in the setting that each player initially knows only his private input of constant length describing his utility function and each player can only communicate directly with his neighbors. For the pure Nash equilibrium problem, we show that in any protocol, the players in some game must communicate a total of at least $\Omega(dm^2)$ bits when $d \geq \log m$ and at least $\Omega(d2^d m)$ bits when $d < \log m$. For the mixed Nash equilibrium problem, we show that in any protocol, the players in some game must communicate at least $\Omega(d^2m^2)$ bits in total, and moreover, every player must communicate at least $\Omega(dm)$ bits. We also provide protocols with matching or almost matching upper bounds.

Keywords: Nash equilibrium · Communication complexity

1 Introduction

Game theory has become an important topic in computer science, as it provides an appropriate framework for studying the behavior of a network of non-cooperative players (or agents) who have their own interests. To model the interaction among players, Kearns, Littman, and Singh [12] proposed the notion of graphical games, in which each player is represented by a node of a graph and his utility depends only on his action and the actions of his neighbors on the graph. Much effort has been devoted to the problem of deciding the existence of pure Nash equilibrium (PNE) and the problem of computing mixed Nash equilibrium (NE), but no efficient algorithm is known so far for either problem in such games on general graphs. In fact, there are evidences showing that no such algorithm may exist. More precisely, for general graphical games, the problem of deciding the existence of PNE was shown in [9,14] to be NP-complete, while the problem of computing NE was shown in [4] to be PPAD-complete even when each player has at most two actions and three neighbors. On the other hand, there are polynomial-time algorithms for games on special graphs. For games on

© Springer Nature Switzerland AG 2018
G. Christodoulou and T. Harks (Eds.): WINE 2018, LNCS 11316, pp. 111–125, 2018.
https://doi.org/10.1007/978-3-030-04612-5_8

graphs with maximum degree 2 in which each player has two actions, such an algorithm was given in [7] for computing NE. For games on graphs with logarithmic tree-width and constant degree in which each player has a constant number of actions, such algorithms were given in [6], for deciding the existence of PNE and for computing approximate NE. For games on a k-dimensional torus (or grid) in which all players have the same action set and the same utility function (the input thus consists of only that utility function and the number of nodes in each dimension), it was shown in [5] that the problem of deciding the existence of PNE falls in P when $k = 1$ but becomes NEXP-complete when $k \geq 2$.

Note that most of the previous works focused on the centralized setting, which corresponds to the scenario in which there is a centralized authority who knows every player's utility function, does all the computation, and then tells every player what to do. In such a setting, the dominant question is the computational complexity of the centralized algorithm. Although results in this direction have been fruitful, they may not tell the whole story, since many systems in use today operate in a distributed fashion with participants having their own inputs and working autonomously, instead of being controlled by a central authority. Therefore, we would like to argue that another direction worth exploring is to consider the distributed setting. In fact, different types of questions naturally arise from such a setting, which may provide different perspectives and help us gain better understanding of some fundamental problems in game theory. As a start, we consider the issue of communication, which is one of the main issues in a distributed system. This issue was studied by Hart and Mansour [10], who considered the case of graphical games on a complete graph. They showed that for such games with n players each having two actions, $\Omega(2^n)$ bits of communication are required to decide the existence of PNE or to compute NE. Note that in such games, each player's utility depends on the actions of all players, which means that the size of each player's input, his utility function, is at least 2^n. Recently, Babichenko and Rubinstein [2] showed that such an lower bound even holds for finding approximate NE, with some small enough constant approximation error.

We are more interested in graphical games on sparse graphs, in which each player has direct interaction with only a small number of players. That is, each player's utility depends only on the actions of his small number of neighbors, and moreover, each player can only communicate directly with his small number of neighbors. We believe that this captures more accurately many social or physical networks in our daily life, and many interesting questions may be waiting to be asked and answered. As a first step, following [5], we start by considering a class of highly regular graphs: the two dimensional grids. A $d \times m$ grid is a graph with dm nodes, aligned in d rows and m columns, with each node connected to at most four other nodes, at its left, right, top, and bottom. In games on such a graph, each player's utility only depends on the actions of at most four other players. We consider the setting in which players initially know only their own utility functions, given as their private inputs, and they must then communicate with others in order to collectively solve the PNE problem or the NE problem. We consider the communication model in which each player can only communicate directly with his neighbors on the graph; thus, if a player wants to send one bit of message to a player with k edges away from him, he needs the players in between to relay the message, and this costs k bits, instead of one bit, of communication complexity in total. This seems to be the model often used in the areas of networking and distributed computing, but is different from what is usually considered in the area of communication complexity [8], including [2,10], which allows any player to communicate directly with others. We adopt this model because we believe it to be a more appropriate model for graphical games, as it corresponds to the case that the neighbors who affect your utility are those you have communication with. To avoid obscuring the picture, we focus on the case that each player's input is short, with $O(1)$ bits, which means that the total input length is $O(dm)$, and we ignore the issue of computation complexity. We obtain the following results, assuming without loss of generality that $d \leq m$ (by rotating the grid if necessary).

First, we consider the problem of deciding the existence of PNE in such games. In the case with $d \geq \log m$, we show that in any protocol, the players must communicate at least $\Omega(dm^2)$ bits in total, and thus most players must each communicate at least $\Omega(m)$ bits. Note that these are the largest lower bounds a decision problem can possibly have, because matching upper bounds can be achieved by a simple protocol, which lets all players send their inputs to one particular player who then computes and broadcasts the answer. In the case with $d < \log m$, we show that any protocol must communicate at least $\Omega(d2^d m)$ bits in total, and we also provide a protocol which communicates at most $O(2^{2d} m)$ bits in total. When $d = O(1)$, our upper bound matches our lower bound, and each player in our protocol only communicates $O(1)$ bits.

Next, we consider the problem of finding NE in such games. We show that in any protocol, the players must communicate at least $\Omega(d^2 m^2)$ bits in total, and furthermore, every player must communicate at least $\Omega(dm)$ bits. These are also the largest lower bounds a search problem can possibly have, because matching upper bounds can be achieved by another simple protocol, which lets each player send his input to all other players and then lets each player compute his answer by himself. Our lower bounds also imply that every player's distribution in an NE is in fact influenced by the utility functions of almost all the players. Furthermore, our lower bounds can be extended to the more general problem of finding approximate NE, and we show that the same lower bounds hold even when the approximation error ε is at most $2^{-\Omega(dm)}$, while the lower bounds become $\Omega(dm \log(1/\varepsilon))$ in total and $\Omega(\log(1/\varepsilon))$ for each player when $\varepsilon \geq 2^{-O(dm)}$.

Let us make some remarks about our results. First, observe that our bounds for the problem of finding NE are larger than those for the problem of deciding the existence of PNE. For example, in the case with $d = O(1)$, each player can communicate $O(1)$ bits for the PNE problem but must communicate $\Omega(m)$ bits for the NE problem. One reason may be that while the PNE problem is a decision problem, the NE is a search problem whose solution may need many bits to describe. Next, our results can be seen as giving an example of graphical games in which each player's utility depends only on the actions of a small number of players, but long range influence from many other players still exists. Therefore, the degree of graphs alone is not the parameter which determines the communication complexity of graphical games. Moreover, although we discuss only games on grids, our lower bounds in fact apply to games on a much broader class of graphs, those which can embed such grids (formally, those which have such grids as minors). In fact, our lower bound for the NE problem even works for graphs which can embed a subtree of the $2 \times m$ grid (as shown in Fig. 2(1)). Finally, our communication lower bounds are unconditional, which show that the PNE problem and the NE problem are indeed hard in terms of communication complexity, since the lower bounds are the largest possible for decision problems and search problems, respectively. This is in contrast to the current status on their computation complexity: the negative results currently known for them are only of the form of NP-hardness and PPAD-hardness, respectively.

2   Preliminaries

Let $\mathbb{N}$ denote the set of positive integers and $\mathbb{R}$ the set of real numbers. For $n \in \mathbb{N}$, let $[n]$ denote the set $\{1, \ldots, n\}$. All logarithms in the paper will have base two. The entropy of a distribution $\mathcal{X}$ is defined as $\sum_x \Pr[\mathcal{X} = x] \log(1/\Pr[\mathcal{X} = x])$.

Games and Nash Equilibria. In a game, there are some number n of players indexed by $1, \ldots, n$. Each player i has a finite set of actions A_i, as well as a utility function $u_i : \prod_{j \in [n]} A_j \to \mathbb{R}$. One can allow each player i to play according to some distribution σ_i over his actions. We say that a sequence of independent distributions $(\sigma_1, \ldots, \sigma_n)$ is an ε-Nash equilibrium, if for any player i, changing his distribution σ_i to a different σ_i' unilaterally cannot improve his expected utility by more than ε. More precisely, we have the following.

Definition 1. *A sequence of independent distributions $\sigma = (\sigma_1, \ldots, \sigma_n)$ is an ε-Nash equilibrium, or ε-NE for short, if for any player i and for any sequence of independent distributions σ^* which differs from σ only in its i'th distribution, $\mathrm{E}_{a \in \sigma^*}[u_i(a)] \le \mathrm{E}_{a \in \sigma}[u_i(a)] + \varepsilon$. An ε-NE with $\varepsilon = 0$ is called a Nash Equilibrium, or NE for short. An NE in which each player plays some action with probability 1 is called a pure Nash equilibrium, or PNE for short.*

Graphical Games and Communication Model. A graphical game on a graph is a game in which every player is represented by a node of the graph and the utility of a player only depends on the joint action of himself and his neighbors on the graph. The graphs which we will focus on in this paper are two-dimensional grid graphs. For $d, m \in \mathbb{N}$, a $d \times m$ grid graph has dm nodes, aligned in d rows and m columns, with each node connecting to at most four neighbors, on its left, right, top, and bottom.

Definition 2. *Let $G_{d\times m}$ denote the set of graphical games on a $d \times m$ grid graph, in which each player is represented by a node on the graph, has a binary action set $\{0, 1\}$, and has a utility function mapping from the joint action taken by himself and his neighbors to a value in the set $U = \{0, 1/2, 1\}$.*[1]

Note that each game in $G_{d\times m}$ is determined by the dm utility functions of the players, which can be specified by $O(dm)$ bits since each utility function has only a constant number of possibilities. We study two problems: a decision problem which decides if a PNE exists, and a search problem which outputs an ε-NE. We consider the setting in which each player initially knows only his utility function as his private input, and there is an underlying communication network, which is the grid graph, such that each player can only communicate directly with his neighbors on the communication network. Thus, on the grid graph, each player can only communicate directly with at most four players. If a player wants to send one bit of message to a player with k edges away from him, he needs the players in between to relay the message, and this costs k bits, instead of one bit, of communication complexity in total because there are k players each sending one bit. Note that this is different from what is usually considered in the area of communication complexity, which allows any player to communicate directly with others (or equivalently, the communication network is a complete graph). For a given input, the communication complexity of a player is the number of bits he sends and receives, and the communication complexity of a protocol is the total number of bits sent and received by its players. The maximum communication complexity over all possible inputs is taken as the communication complexity of a protocol for that problem. The goal is to have protocols with small communication complexity for the PNE problem and the ε-NE problem.

Definition 3. *For the problem of deciding if a PNE exists, we say that a protocol succeeds on an input if every player knows the final answer (true or false) of whether or not a PNE exists, while for the problem of finding an ε-NE, we say that a protocol succeeds on an input if the players agree on some ε-NE and each player knows his distribution in that ε-NE.*

We allow a protocol to be randomized, which can use a random string to help its computation, and this makes our lower bound results stronger. In fact, our lower bounds even work for the stronger model known as the public coins model, in which the random string is generated publicly and broadcasted to all players without counting its communication complexity. We say that a protocol is a δ-error randomized protocol for a problem if for any input, the protocol succeeds with probability at least $1 - \delta$.

---

[1] Our algorithms actually work for any set U of constant size. We make this restriction on U to make our lower bound results stronger—the problems remain hard even when specialized to such a set U. In fact, our lower bound in Sect. 3 even holds when U is restricted to $\{0, 1\}$.

Set-Disjointness Problem. We will prove a communication lower bound for our PNE problem by reducing to it a hard communication problem. The hard problem we choose is the set-disjointness problem, denoted as Φ_n, which takes the input $(x, y) = ((x_1, \ldots, x_n), (y_1, \ldots, y_n)) \in \{0,1\}^n \times \{0,1\}^n$ and outputs $\bigvee_{i \in [n]}(x_i \wedge y_i)$, where $\wedge$ is the AND operation and $\vee$ is the OR operation. Note that $\Phi_n(x, y) = \bigvee_{i \in [n]} \Phi_1(x_i, y_i)$. This problem is known to have a high communication complexity for any two-party protocol, with one party having x and the other having y. We will use the following stronger result.

Lemma 1. [1] *For $n \in \mathbb{N}$, there is a distribution $\mathcal{D}_n = ((\mathcal{X}_1, \ldots, \mathcal{X}_n), (\mathcal{Y}_1, \ldots, \mathcal{Y}_n))$ over $\Phi_n^{-1}(0) \subseteq \{0,1\}^n \times \{0,1\}^n$ such that the following two conditions hold:*

- *For each $i \in [n]$, the pair of distributions $(\mathcal{X}_i, \mathcal{Y}_i)$ over $\{0,1\} \times \{0,1\}$ is independent from all other pairs of distributions $(\mathcal{X}_j, \mathcal{Y}_j)$ for $j \neq i$.*
- *Given any δ-error randomized two-party protocol for Φ_n, with a constant $\delta < 1/4$, the expected length of its messages, over the inputs sampled from $\mathcal{D}_n$ and the randomness it uses, must be at least $\Omega(n)$.*

Note that what [1] actually proved is that the distribution of messages sent by a protocol must have entropy at least $\Omega(n)$, but this implies that the expected length of the messages must also be at least $\Omega(n)$. The reason is that one can convert the messages into a prefix-free encoding of their distribution with their lengths increased only by a constant factor, and it is well-known that the expected length of any prefix-free encoding of a distribution is at least the entropy of the distribution (see, e.g., Theorem 5.3.1. of [3]).

3 Pure Nash Equilibria

In this section, we consider the communication complexity of deciding the existence of pure Nash equilibria in $G_{d \times m}$. Note that we can assume without loss of generality that $d \leq m$, by rotating the grid if necessary. Our main result is the following, which we will prove in Subsect. 3.1.

Theorem 1. *Suppose $d \geq c$, for a large enough constant c, and $0 \leq \delta < 1/4$. Then for the problem of deciding the existence of PNE in $G_{d \times m}$, any δ-error randomized protocol must have total communication complexity at least $\Omega(dm^2)$ when $d \geq \log m$ and at least $\Omega(d2^d m)$ when $d < \log m$.*

We also have the following upper bounds, which we will prove in Subsect. 3.2. Note that our lower bounds match our upper bounds when $d \geq \log m$ or when $d \leq O(1)$.

Theorem 2. *For the problem of deciding the existence of PNE in $G_{d \times m}$, there is a deterministic protocol which has total communication complexity at most $O(dm^2)$ in general and at most $O(2^{2d}m)$ when $d \leq O(\log m)$.*

3.1 Proof of Theorem 1

First, let us consider the case with $d \geq \log m$. To have a cleaner presentation, we will prove an $\Omega(dm^2)$ lower bound for graphical games on an $O(d) \times O(m)$ grid, instead of on a $d \times m$ grid, which clearly implies our result.

The basic idea behind our proof is the following. To show a communication lower bound for our problem, we take a hard problem which is known to have a high communication lower bound, and reduce it to our problem. The hard problem we choose is the set-disjointness problem Φ_{dm}, defined in Sect. 2, with inputs from $\{0,1\}^{dm} \times \{0,1\}^{dm}$, which is known to have a communication complexity of $\Omega(dm)$ for any two-party protocol. More precisely, given any input $(x, y) \in \{0,1\}^{dm} \times \{0,1\}^{dm}$ of Φ_{dm}, we will map it to a graphical game on an $O(d) \times O(m)$ grid, such that $\Phi_{dm}(x, y) = 1$ if and only if the graphical game has a PNE. For this, we will first construct a combinatorial circuit for Φ_{dm} which can be placed on an $O(d) \times O(m)$ grid in such a way that gates and wires of the circuit are placed on nodes and edges, respectively, of the graph. Then we will design the utility functions for players so that the actions taken by them will match the values of the corresponding gates on them. Furthermore, we will add two extra players to force the condition that the game has a PNE if and only if the circuit has output 1. We will have dm input gates for the input x as well as dm input gates for the input y, and they will be placed on two sides of the graph separable by a cut in the middle. Then by seeing players on two sides of the graph as two parties, any protocol for our problem gives a two-party protocol for Φ_{dm}. Thus, the two-party communication lower bound for Φ_{dm} implies a communication lower bound for our problem.

However, there seems to be an obstacle in this approach. If we take a deterministic circuit for Φ_{dm}, there must be at least $\Omega(dm)$ wires crossing any cut that separates x and y. This is because otherwise one can obtain a two-party protocol for Φ_{dm} by simulating the circuit which only needs $o(dm)$ bits of communication, contradicting the known lower bound of $\Omega(dm)$. Therefore, it is impossible to embed any deterministic circuit for Φ_{dm} on an $O(d) \times O(m)$ grid. We overcome this obstacle by using a non-deterministic circuit, instead of a deterministic one, for Φ_{dm}, where a non-deterministic circuit, unlike a deterministic one, has some non-deterministic gates each of which can be assigned any binary value, and the circuit accepts an input if it outputs one for some assignment of the non-deterministic gates. As we will see, a non-deterministic circuit can indeed have much fewer crossing wires and can thus be placed on the grid.

Now we proceed to give the formal proof. First, we show how to construct a non-deterministic circuit C to compute Φ_{dm}. Let us arrange the input $x \in \{0,1\}^{dm}$ of Φ_{dm} as a $d \times m$ array and write $x = (x_1, \ldots, x_m)$ with $x_b = (x_{b,1}, \ldots, x_{b,d}) \in \{0,1\}^d$ denoting the b'th column of the array; the other input $y \in \{0,1\}^{dm}$ is treated similarly. Then we have

$$\Phi(x, y) = 1 \iff \bigvee_{b \in [m]} \bigvee_{j \in [d]} (x_{b,j} \wedge y_{b,j}) = 1 \iff \exists b \in [m] : \Phi_d(x_b, y_b) = 1.$$

What our non-deterministic circuit C does is to guess $b \in [m]$, select x_b and y_b, and then compute $\Phi_d(x_b, y_b)$. Note that $\Phi_d(x_b, y_b) = \bigvee_{j \in [d]} (x_{b,j} \wedge y_{b,j})$, which can be easily realized by a small deterministic circuit with $O(d)$ gates and wires. The difficult part is to select x_b and y_b. For this, we use the guessed b to produce a mask $a \in \{0,1\}^m$ with $a_b = 1$ and $a_i = 0$ for $i \neq b$. Then for every $i \in [m]$, we produce the string $(a_i \wedge x_{i,1}, \ldots, a_i \wedge x_{i,d})$, which gives the string x_b if $i = b$ and the all-0 string otherwise, and then by taking bit-wise OR of these m string, we obtain the string x_b. Similarly, we do this for selecting y_b. In summary, our non-deterministic circuit C consists of the following.

1. There are $\ell = \log m$ non-deterministic gates $(b_1, \ldots, b_\ell)$ for guessing the binary representation of the index b.
2. There are two copies of a sub-circuit C_a, each of which takes $(b_1, \ldots, b_\ell)$ as its input and outputs the m-bit string $a = (a_1, \ldots, a_m)$ such that $a_b = 1$ and $a_i = 0$ for $i \neq b$.
3. There is a sub-circuit C_x which contains x as input gates, takes the string a from C_a, and outputs the d-bit string x_b. There is a similar sub-circuit C_y for y_b.
4. There is a sub-circuit C_Φ which takes x_b from C_x as well as y_b from C_y, and outputs $\Phi_d(x_b, y_b)$.

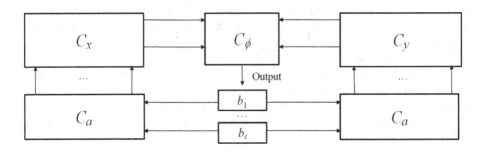

Fig. 1. The layout of the circuit C.

We compose these sub-circuits in the way as shown in Fig. 1 to obtain the nondeterministic circuit C for Φ_{dm}. To implement the circuit, we adopt the binary $\wedge$ gate for the AND of two inputs, the binary $\vee$ gate for the OR of two inputs, and the binary $\equiv$ gate for comparing if two inputs are equal. In addition, to embed the circuit on the grid, we adopt the binary $\oplus$ gate for the XOR of two inputs, which enables us to simulate the crossing of two wires, and we also adopt the unary $\rightarrow$ gate for passing its input to its output, which enables us to simulate a long wire. Lemma 2 below guarantees the feasibility of such an implementation. The proof is somewhat straightforward following the discussion above, which we omit due to the page limit.

Lemma 2. *The circuit C can be implemented using gates from $\{\wedge, \vee, \equiv, \oplus, \rightarrow\}$, as well as input gates and non-deterministic gates. Furthermore, it can be placed on an $O(d) \times O(m)$ grid such that every gate is on a different node and every wire is on a different edge of the grid.*

Recall that each node of the grid corresponds to a player of the graphical game, so each gate is now associated with a player of the graphical game. Next, we show how to set the utility functions for players so that they simulate their associated gates of the circuit. In particular, for a player who is associated with some gate, given the actions taken by his neighbors which correspond to the input values for that gate, we will force the player to play the action equal to the output value of that gate. Note that we can force a player to play a particular action by giving him utility 1 for playing that action and 0 otherwise. More precisely, we do the following.

1. For a player who simulates a nondeterministic gate b_i, we allow him to play any action by always giving him utility 1.
2. For a player who simulates an input gate $x_{i,j}$ (or $y_{i,j}$), we force him to play action $x_{i,j}$ (or $y_{i,j}$), by giving him utility 1 if he does so and 0 otherwise.
3. For a player who simulates the unary $\rightarrow$ gate, we force him to play the same action as his neighbor who corresponds to the input. That is, we give him utility 1 if he plays the same action as that neighbor and 0 otherwise.
4. For a player who simulates a binary gate $\odot \in \{\vee, \wedge, \equiv, \oplus\}$, we force him to play action $\alpha_1 \odot \alpha_2$ whenever his two neighbors, who correspond to the two inputs, play actions α_1 and α_2. That is, we give him utility 1 if he plays action $\alpha_1 \odot \alpha_2$ and 0 otherwise.

So far, by setting utility functions in this way, one can show that for any assignment to the non-deterministic gates $(b_1, \ldots, b_\ell)$, the graphical game always has a PNE which corresponds to the evaluation of the circuit. However, what we actually want is to have a graphical game which has a PNE if and only if there is an assignment to the non-deterministic gates which makes the circuit output 1. To achieve this, we would like the player p who represents the output gate to play action 1, by adding two extra players p_1, p_2 and set their utility functions in the following way. We force p_2 to play the same action as p_1, but we force p_1 to play the same action as p_2 if and only if p plays 1. Thus, p_2 and p_1 can both be happy if and only if p plays 1, which accomplishes what we want. Note that to incorporate these two new players, we only need to add a new row of nodes to the grid, so we still have an $O(d) \times O(m)$ grid. This completes the reduction.

Now, we are ready to show the communication lower bound for our problem. Let us focus on those graphical games constructed above. In these games, for each $i \in [m]$ and $j \in [d]$, the player with the input gate $x_{i,j}$ has two possible utility functions as his input, and so is the player with the input gate $y_{i,j}$, while each of the remaining players has a fixed utility function. Thus, there are 2^{2dm} such games, one for each possible (x, y). Let us abuse the notation and also use $\mathcal{D}_{dm}$ to denote the distribution of graphical games with (x, y) sampled from the distribution $\mathcal{D}_{dm}$ in Lemma 1 and with the utility functions of other players fixed in the way discussed before. Now consider any randomized protocol Π_G for deciding the existence of a PNE. Note that for any $i \in [m]$, the players with the input gates $x_{i,1}, \ldots, x_{i,d}$ are all located on the same column of the grid, which is different from those with other input gates. Thus for any $i \in [m]$, there is a cut on the graph which separates those with input gates for $x_1, \ldots, x_i$ from those with other input gates, and these m cuts share no edge. Then from Lemma 1, we have the following, as one can derive from Π_G a two party-protocol for computing $\Phi_{di}((x_1, \ldots, x_i), (y_1, \ldots, y_i))$. We omit the proof due to the page limit.

Lemma 3. *For any $i \in [m]$, over the input sampled from the distribution $\mathcal{D}_{dm}$ and the random string used by Π_G, the expected length of the messages sent across the i'th cut by Π_G must be at least $\Omega(di)$.*

Finally, by summing the bound in Lemma 3 over i, we can obtain a lower bound on the total communication complexity. More precisely, using the fact that the m cuts share no edge as well as the linearity of expectation, we know that the messages sent across the m cuts altogether must have an expected total length of at least $\sum_{i \in [m]} \Omega(di) = \Omega(dm^2)$. This completes the proof for the case with $d \geq \log m$.

For the case with $d < \log m$, let us divide the $d \times m$ grid into $m/2^d$ copies of a $d \times 2^d$ grid (assuming for simplicity of presentation that 2^d divides m). For each copy of the $d \times 2^d$ grid, we sample an input for players on it from an independent copy of the distribution $\mathcal{D}_{d2^d}$ discussed before. We claim that if we sample an input from $G_{d \times m}$ according to such a distribution, then the messages communicated within each copy of the $d \times 2^d$ grid by any protocol must have an expected total length of $\Omega(d2^{2d})$. This is because otherwise it would give a protocol for $G_{d \times 2^d}$ violating the lower bound we have shown for the case with $d \geq \log m$. The argument is similar to that for proving Lemma 3, so we omit it here. As a result, by summing over the $m/2^d$ copies, the expected total length of all messages must be at least $\Omega(d2^{2d}) \cdot (m/2^d) = \Omega(d2^d m)$. This completes the proof of Theorem 1.

3.2   Proof of Theorem 2

An $O(dm^2)$ upper bound can be achieved by the straightforward deterministic protocol: every player sends his utility function to the first player, who decides if a PNE exists and then tells each player the answer.

For the case with $d \leq O(\log m)$, we can do the following instead, which is inspired by the protocols in [6,12]. Let us start with some preparation. For $i \in [m]$ and $j \in [d]$, let $c_{i,j}$ denote the player on column i and row j of the $d \times m$ grid. For each column $i \in [m]$, let us see the d players in that column as a meta-player c_i, who has 2^d meta-actions from $\{0,1\}^d$ corresponding to the joint-actions of those d players. For convenience, let us add a column 0 in front and a column $m+1$ at the end, and consider two corresponding imaginary meta-players c_0 and c_{m+1}, who are always happy (with utility 1) for taking any meta-actions. For any $i \in [m]$, we say that a meta-action α_i is *good* for c_i with respect to a meta-action α_{i+1} of c_{i+1} if they have the potential of being part of a PNE in the sense that there is a meta-action α_{i-1} of c_{i-1} such that the following two conditions hold:

- The meta-action α_{i-1} is good for c_{i-1} (we assume that any meta-action is good for c_0) with respect to the meta-action α_i of c_i.
- The meta-action α_i is *stable* for c_i with respect to c_{i-1}'s meta-action α_{i-1} and c_{i+1}'s meta-action α_{i+1}, in the sense that for any $j \in [d]$, the player $c_{i,j}$ cannot improve his utility by unilaterally changing his action in α_i.

Then our protocol does the following. For any $i \in [m]$, we let the first player in column i represent the meta-player c_i by collecting all the utility functions in that column; this only costs $O(d^2 m)$ bits of communication in total. Then our protocol enters the following $m-1$ iterations. For i going from 1 up to $m-1$, let c_i first receive the message from c_{i-1} and then send c_{i+1} the following message: for every meta-action α_{i+1} of c_{i+1}, which meta-actions are good for c_i with respect to α_{i+1}. Finally, after receiving the message from c_{m-1}, the meta-player c_m declares that a PNE exists if and only if there is a meta-action α_m which is good for c_m with respect to some meta-action of c_{m+1}.

For the correctness of our protocol, a simple induction shows that c_m has a meta-action α_m which is good with some meta-action α_{m+1} of c_{m+1} if and only if there exist meta-actions $\alpha_0, \alpha_1, \ldots, \alpha_m, \alpha_{m+1}$ such that for any $i \in [m]$, α_i is stable with respect to α_{i-1} and α_{i+1}. Since the imaginary meta-players c_0 and c_{m+1} are always happy, the condition that these meta-actions are stable is equivalent to the condition that no player $c_{i,j}$ can improve his utility by unilaterally changing his distribution in α_i, which means the existence of a PNE.

To bound the communication complexity of our protocol, note that the message sent in each iteration can be described by a $2^d \times 2^d$ Boolean matrix, which costs 2^{2d} bits of communication. Therefore, the total communication complexity of our protocol is $O(d^2 m + 2^{2d} m) = O(2^{2d} m)$, which completes the proof.

4   Mixed Nash Equilibria

In this section, we prove a communication lower bound for finding (approximate) mixed Nash equilibria in $G_{d \times m}$. Our result is the following.

Theorem 3. *For the problem of finding ε-NE in $G_{d \times m}$, any δ-error randomized protocol, with $0 \le \delta \le 1/4$, must have the following communication lower bounds.*

- *The communication complexity of any player is at least $\Omega(dm)$ for $\varepsilon \le 2^{-\Omega(dm)}$ and at least $\Omega(\log(1/\varepsilon))$ for $\varepsilon \ge 2^{-O(dm)}$.*
- *The total communication complexity of all players is at least $\Omega(d^2 m^2)$ for $\varepsilon \le 2^{-\Omega(dm)}$ and at least $\Omega(dm \log(1/\varepsilon))$ for $\varepsilon \ge 2^{-O(dm)}$.*

Before proving the theorem, let us make some remarks. First, for the task of finding an ε-NE with $\varepsilon \le 2^{-\Omega(dm)}$ (in particular, finding an NE, with $\varepsilon = 0$), our lower bounds are tight, as matching upper bounds can be achieved by the straightforward deterministic protocol: every player sends his utility function to all other players and then every player computes his distribution in the first (according to some order) NE. Another straightforward algorithm, similar to that in Theorem 2, is to have one particular player collecting all utility functions, computing an NE, and telling each player his distribution in that NE. However, this does not give a smaller communication complexity because each player's distribution in general may need much more than a constant number of bits to specify. This may explain why the communication complexity for finding an NE given in Theorem 3 is larger than that for deciding the existence of a PNE given in Theorem 1. Finally, note that the first item of the theorem does not imply the second item. This is because different players may have their high communication complexity on different inputs, so it may be possible that every input has only few players with high communication complexity. Nevertheless, the second item of the theorem guarantees that there indeed exists some input on which many players all have high communication complexity.

4.1   Proof of Theorem 3

The basic idea behind our proof is to show the existence of $2^{\Omega(dm)}$ different inputs which all have different NE. This means that to distinguish among these many different NE, many bits must be communicated. We will first show this for games in $G_{2 \times 2n}$, which are graphical games on the $2 \times 2n$ grid. Then we will show that by embedding the $2 \times 2n$ grid into the $d \times m$ grid, with $n = \Theta(dm)$, we can obtain lower bounds for $G_{d \times m}$ from lower bounds for $G_{2 \times 2n}$.

Let us start by considering graphical games in $G_{2 \times 2n}$. Among the $4n$ players, the $2n$ players $g_1, \ldots, g_n, v_1, \ldots, v_n$ will be the main ones for us, the n players $w_1, \ldots, w_n$ will be the auxiliary ones, and the remaining n players will be ignored. Figure 2(1) shows their positions on the grid. In our games, each player z has only two actions, denoted by 0 and 1, and we let $p[z]$ denote the probability that z plays the action 1. For each player v_j, we would like $p[v_j]$ in any NE to have exactly two possibilities, either 0 or 1, which can be achieved by forcing him to play either action 0 or 1, by giving him utility 1 for playing that action and 0 otherwise. Furthermore, we would like to set the utility functions for the players $g_1, \ldots, g_m$ so that in any NE, each $p[g_i]$ is uniquely determined by the values of $p[v_1], \ldots, p[v_i]$. In particular, we will make

$$p[g_i] = \frac{p[v_i]}{2} + \frac{p[g_{i-1}]}{2} = \frac{p[v_i]}{2} + \frac{p[v_{i-1}]}{2^2} + \cdots + \frac{p[v_1]}{2^i}.$$

For this, we will rely on the following building block, which is a simplified version from [4]. We will consider the more general case of an ε-NE, and we will use the notation $u = p \pm \varepsilon$ for the condition $p - \varepsilon \leq u \leq p + \varepsilon$.

Proposition 1. [4] *Let g', v, w, g be the players on the graph shown in Fig. 2(2). Then there are fixed utility functions U_g and U_w for players g and w, respectively, with utility values in $\{0, 1/2, 1\}$, such that for any utility functions of players g' and v, $p[g] = \frac{1}{2}p[v] + \frac{1}{2}p[g'] \pm \varepsilon$ in any ε-NE. In particular, $p[g] = \frac{1}{2}p[v] + \frac{1}{2}p[g']$ in any NE.*

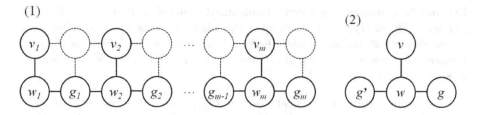

Fig. 2. (1) The graph of $G_{2\times 2n}$. (2) The graph of Proposition 1.

In our games, we give our players their utility functions in the following way. For convenience, let us have an imaginary player g_0 who is connected to w_1 and always plays the action 0. For $i \in [n]$, we give the players g_i and w_i the fixed utility functions U_g and U_w of Proposition 1, respectively. For $i \in [n]$, we give the player v_i two possible utility functions, specified by an input variable $x_i \in \{0, 1\}$, to force him to play the action x_i (by giving him utility 1 for playing x_i and 0 otherwise), and this implies that $\mathsf{p}[v_i] = x_i$ in any NE. Other players do not matter and we fixed their utility functions arbitrarily. In summary, each player v_i, for $i \in [n]$, has two possible utility functions as his input, indicated by the input variable $x_i \in \{0, 1\}$, and all other players have their utility functions fixed. That is, our games have exactly 2^n different inputs, and let S denote this set of inputs. Then we have the following.

Lemma 4. *In any ε-NE, any player g_i, for $i \in [n]$, has*

$$\mathsf{p}[g_i] = \frac{x_i}{2} + \frac{x_{i-1}}{2^2} + \cdots + \frac{x_1}{2^i} \pm 3\varepsilon.$$

Proof. Consider any ε-NE. Note that for any player v_i, we have $\mathsf{p}[v_i] = x_i \pm \varepsilon$. Then for any player g_i with $i \in [n]$, we know from Proposition 1 that

$$\mathsf{p}[g_i] = \frac{1}{2}\mathsf{p}[v_i] + \frac{1}{2}\mathsf{p}[g_{i-1}] \pm \varepsilon = \frac{1}{2}x_i + \frac{1}{2}\mathsf{p}[g_{i-1}] \pm \frac{3}{2}\varepsilon,$$

where we let $\mathsf{p}[g_0] = 0$, and a simple induction shows that $\mathsf{p}[g_i]$ equals

$$\frac{x_i}{2} + \frac{1}{2}\left(\frac{x_{i-1}}{2} + \frac{x_{i-1}}{2^2} + \cdots + \frac{x_1}{2^{i-1}} \pm 3\varepsilon\right) \pm \frac{3}{2}\varepsilon = \frac{x_i}{2} + \frac{x_{i-1}}{2^2} + \cdots + \frac{x_1}{2^i} \pm 3\varepsilon.$$

$\square$

In the case of an NE, with $\varepsilon = 0$, the lemma shows that different values of $(x_1, \ldots, x_i)$ give rise to different values of $\mathsf{p}[g_i]$. However, in the case of an ε-NE, this is no longer true, but we will show that there are still many different values of $\mathsf{p}[g_i]$ from many different values of $(x_1, \ldots, x_i)$. Then to distinguish the many possibilities, the player g_i must communicate many bits. Formally, we have the following. We omit the proof here due to the page limit.

Lemma 5. *Consider any δ-error randomized protocol, with $0 \leq \delta \leq 1/4$, for finding ε-NE in $G_{2\times 2n}$, and consider any player g_i with $i \in [n]$. Then over the random input sampled uniformly from S and the random string used by the protocol, the expected number of bits received by g_i must be at least $(i^* - 2)/2$, where $i^* = \min\{i, \lfloor \log(1/7\varepsilon)\rfloor\}$.*

Now let us return to graphical games in $G_{d\times m}$. We will focus on a subset of them which correspond to the graphical games in $G_{2\times 2n}$ discussed above. Note that one can easily embed the $2 \times 2n$ grid into the $d \times m$ grid, with $n = \Theta(dm)$, in several ways. We give the utility functions of the players in $G_{2\times 2n}$ to the corresponding players in $G_{d\times m}$. The remaining players in $G_{d\times m}$ will have no effect, and we fix their utility functions arbitrarily. Thus, we can see our games in $G_{d\times m}$ simply as games in $G_{2\times 2n}$.

For the first item of the theorem, note that we can choose an embedding such that the player in $G_{d\times m}$ corresponds to some player g_i in $G_{2\times 2n}$ with $i \geq n/2$. Then the first item follows immediately from Lemma 5.

For the second item of the theorem, we know from the linearity of expectation that over the random input sampled from S and the random string used by the protocol, the expected number of bits received by the players $g_1, \ldots, g_n$ equals the sum, over $i \in [n]$, of the expected number of bits received by player g_i. By Lemma 5, for the case with $\varepsilon \leq 2^{-\Omega(n)}$, this is at least $\sum_{i=O(1)}^{\Omega(n)} \Omega(i) \geq \Omega(n^2) \geq \Omega(d^2 m^2)$, while for the case with $\varepsilon \geq 2^{-O(n)}$, this is at least $\sum_{i=n/2}^{n} \Omega(\log(1/7\varepsilon)) \geq \Omega(n\log(1/\varepsilon)) \geq \Omega(dm\log(1/\varepsilon))$.

References

1. Bar-Yossef, Z., Jayram, T.S., Kumar, R., Sivakumar, D.: An information statistics approach to data stream and communication complexity. J. Comput. Syst. Sci. **68**(4), 702–732 (2004)
2. Babichenko, Y., Rubinstein, A.: Communication complexity of approximate Nash equilibria. In: Proceedings of 49th Annual ACM SIGACT Symposium on Theory of Computing, pp. 878–889 (2017)
3. Cover, T., Thomas, J.: Elements of Information Theory. Wiley, Hoboken (1991)
4. Daskalakis, C., Goldberg, P.W., Papadimitriou, C.H.: The complexity of computing a Nash equilibrium. SIAM J. Comput. **39**(1), 195–259 (2009)
5. Daskalakis, K., Papadimitriou, C.H.: The complexity of games on highly regular graphs. In: Brodal, G.S., Leonardi, S. (eds.) ESA 2005. LNCS, vol. 3669, pp. 71–82. Springer, Heidelberg (2005). https://doi.org/10.1007/11561071_9
6. Daskalakis, C., Papadimitriou, C.H.: Computing pure Nash equilibria in graphical games via Markov random fields. In: Proceedings of ACM 7th Conference on Electronic Commerce, pp. 91–99 (2006)
7. Elkind, E., Goldberg, L.A., Goldberg, P.W.: Nash equilibria in graphical games on trees revisited. In: Proceedings of 7th ACM Conference on Electronic Commerce, pp. 100–109 (2006)
8. Kushilevitz, E., Nisan, N.: Communication Complexity. Cambridge University Press, Cambridge (1996)

9. Gottlob, G., Greco, G., Scarcello, F.: Pure Nash equilibria: hard and easy games. J. Artif. Intell. Res. **24**, 357–406 (2005)
10. Hart, S., Mansour, Y.: How long to equilibrium? the communication complexity of uncoupled equilibrium procedures. Games Econ. Behav. **69**(1), 107–126 (2010)
11. Kakade, S., Kearns, M., Langford, J., Ortiz, L.: Correlated equilibria in graphical games. In: Proceeding of 4th ACM Conference on Electronic Commerce, pp. 42–47 (2003)
12. Kearns, M., Littman, M., Singh, S.: Graphical models for game theory. In: Proceedings of 17th Conference in Uncertainty in Artificial Intelligence, pp. 253–260 (2001)
13. Papadimitriou, C.H., Roughgarden, T.: Computing correlated equilibria in multi-player games. J. ACM **55**(3), 14 (2008)
14. Schoenebeck, G., Vadhan, S.: The computational complexity of Nash equilibria in concisely represented games. In: Proceedings of 7th ACM Conference on Electronic Commerce, pp. 270–279 (2006)

Approximating the Existential Theory of the Reals

Argyrios Deligkas[1,2], John Fearnley[1], Themistoklis Melissourgos[1(✉)], and Paul G. Spirakis[1,3]

[1] Department of Computer Science, University of Liverpool, Liverpool, UK
{Argyrios.Deligkas,John.Fearnley,T.Melissourgos,
P.Spirakis}@liverpool.ac.uk
[2] Leverhulme Research Centre for Functional Materials Design, Liverpool, UK
[3] Computer Engineering and Informatics Department, University of Patras,
Patras, Greece

Abstract. The existential theory of the reals (ETR) consists of existentially quantified boolean formulas over equalities and inequalities of real-valued polynomials. We propose the approximate existential theory of the reals (ϵ-ETR), in which the constraints only need to be satisfied approximately. We first show that unconstrained ϵ-ETR = ETR, and then study the ϵ-ETR problem when the solution is constrained to lie in a given convex set. Our main theorem is a sampling theorem, similar to those that have been proved for approximate equilibria in normal form games. It states that if an ETR problem has an exact solution, then it has a k-uniform approximate solution, where k depends on various properties of the formula. A consequence of our theorem is that we obtain a quasi-polynomial time approximation scheme (QPTAS) for a fragment of constrained ϵ-ETR. We use our theorem to create several new PTAS and QPTAS algorithms for problems from a variety of fields.

1 Introduction

Sampling Techniques. The Lipton-Markakis-Mehta algorithm (LMM) is a well known method for computing approximate Nash equilibria in normal form games [28]. The key idea behind their technique is to prove that there exist approximate Nash equilibria where both players use *simple* strategies.

Suppose that we have a convex set $C = \text{conv}(c_1, c_2, \dots, c_l)$ defined by vectors c_1 through c_l. A vector $x \in C$ is *k-uniform* if it can be written as a sum of the form $(\beta_1/k) \cdot c_1 + (\beta_2/k) \cdot c_2 + \cdots + (\beta_l/k) \cdot c_l$, where each β_i is a non-negative integer and $\sum_{i=1}^{l} \beta_i = k$.

A. Deligkas—The work of this author was supported by Leverhulme Trust and the Leverhulme Research Centre for Functional Materials Design.
J. Fearnley—The work of this author was supported by the EPSRC grant EP/P020909/1.
P. G. Spirakis—The work of this author was supported by the EPSRC Grant EP/P02002X/1, the Leverhulme Research Centre, and the ERC Project ALGAME.

© Springer Nature Switzerland AG 2018
G. Christodoulou and T. Harks (Eds.): WINE 2018, LNCS 11316, pp. 126–139, 2018.
https://doi.org/10.1007/978-3-030-04612-5_9

Since there are at most $l^{O(k)}$ k-uniform vectors, one can enumerate all k-uniform vectors in $l^{O(k)}$ time. For approximate equilibria in $n \times n$ bimatrix games, Lipton, Markakis, and Mehta showed that for every $\epsilon > 0$ there exists an ϵ-Nash equilibrium where both players use k-uniform strategies where $k \in O(\log n / \epsilon^2)$, and so they obtained a quasi-polynomial approximation scheme (QPTAS) for finding an ϵ-Nash equilibrium.

Their proof of this fact uses a sampling argument. Every bimatrix game has an exact Nash equilibrium (NE), and each player's strategy in this NE is a probability distribution. If we sample from each of these distributions k times, and then construct new k-uniform strategies using these samples, then when $k \in O(\log n / \epsilon^2)$ there is a positive probability the new strategies form an ϵ-NE. So by the probabilistic method, there must exist a k-uniform ϵ-NE.

The sampling technique has been widely applied. It was initially used by Althöfer [1] in zero-sum games, before being applied to non-zero sum games by Lipton, Markakis, and Mehta [28]. Subsequently, it was used to produce algorithms for finding approximate equilibria in normal form games with many players [3], sparse bimatrix games [4], tree polymatrix [5], and Lipschitz games [21]. It has also been used to find constrained approximate equilibria in polymatrix games with bounded treewidth [19].

At their core, each of these results uses the sampling technique in the same way as the LMM algorithm: first take an exact solution to the problem, then sample from this solution k times, and finally prove that with positive probability the sampled vector is an approximate solution to the problem. The details of the proofs, and the value of k, are often tailored to the specific application, but the underlying technique is the same.

The Existential Theory of the Reals. In this paper we ask the following question: *is there a broader class of problems to which the sampling technique can be applied?* We answer this by providing a sampling theorem for the existential theory of the reals. The existential theory of the reals consists of existentially quantified formulae using the connectives $\{\wedge, \vee, \neg\}$ over polynomials compared with the operators $\{=, \leq, <, \geq, >\}$. For example, each of the following is a formula in the existential theory of the reals.

$$\exists x \exists y \exists z \cdot (x = y) \wedge (x > z) \qquad \exists x \cdot (x^2 = 2)$$
$$\exists x \exists y \cdot \neg(x^{10} = y^{100}) \vee (y \geq 4) \qquad \exists x \exists y \exists z \cdot (x^2 + y^2 = z^2)$$

Given a formula in the existential theory of the reals, we must decide whether the formula is *true*, that is, whether there do indeed exist values for the variables that satisfy the formula.

The complexity class ETR is defined to be all problems that can be reduced in polynomial time to the existential theory of the reals. It is known that ETR $\subseteq$ PSPACE [12], and NP $\subseteq$ ETR since the problem can easily encode Boolean satisfiability. However, the class is not known to be equal to either PSPACE or NP, and it seems to be a distinct class of problems between the two. Many problems are now known to be ETR-complete, including various problems involving constrained equilibria in normal form games with at least three players [6–9, 23].

Our Contribution. In this paper we propose the *approximate* existential theory of the reals (ϵ-ETR), where we seek a solution that approximately satisfies the constraints of the formula. We show a subsampling theorem for a large fragment of ϵ-ETR, which can be used to obtain PTASs and QPTASs for the problems that lie within it. We believe that this will be useful for future research: instead of laboriously reproving subsampling results for specific games, it now suffices to simply write a formula in ϵ-ETR and then apply our theorem to immediately get the desired result. To exemplify this, we prove several new QPTAS and PTAS results using our theorem.

Our first result is actually that ϵ-ETR = ETR, meaning that the problem of finding an approximate solution to an ETR formula is as hard as finding an exact solution. However, this result crucially relies on the fact that ETR formulas can have solutions that are doubly-exponentially large. This motivates the study of *constrained* ϵ-ETR, where the solutions are required to lie within a given convex set.

Our main theorem (Theorem 2) gives a subsampling result for constrained ϵ-ETR. It states that if the formula has an exact solution, then it also has a k-uniform approximate solution, where the value of k depends on various parameters of the formula, such as the number of constraints and the number of variables. The theorem allows for the formula to be written using *tensor* constraints, which are a type of constraint that is useful in formulating game-theoretic problems.

The consequence of the main theorem is that, when various parameters of the formula are constant (see Corollary 1), we are able to obtain a QPTAS for approximating the existential theory of the reals. Specifically, this algorithm either finds an approximate solution of the constraints, or verifies that no exact solution exists. In many game theoretic applications an exact solution always exists, and so this algorithm will always find an approximate solution.

It should be noted that we are not just applying the well-known subsampling techniques in order to derive our main theorem. Our main theorem incorporates a new method for dealing with polynomials of degree d, which prior subsampling techniques were not able to deal with.

Our theorem can be applied to a wide variety of problems. In the game theoretic setting, we prove new results for constrained approximate equilibria in normal form games, and approximating the value vector of a Shapley game. We also show optimization results. Specifically, we give approximation algorithms for optimizing polynomial functions over a convex set, subject to polynomial constraints. We also give algorithms for approximating eigenvalues and eigenvectors of tensors. Finally, we apply the theorem to some problems from computational geometry.

2   The Existential Theory of the Reals

Let $x_1, x_2, \ldots, x_q \in \mathbb{R}$ be distinct variables, which we will treat as a vector $x \in \mathbb{R}^q$. A *term* of a multivariate polynomial is a function $T(x) := a \cdot x_1^{d_1} \cdot x_2^{d_2} \cdot \cdots \cdot x_q^{d_q}$, where $d_1, d_2, \ldots, d_q$ are non negative integers and $a \in \mathbb{R}$. A multivariate polynomial is a function $p(x) := T_1(x) + T_2(x) + \cdots + T_t(x) + c$, where each T_i is a term as defined above, and $c \in \mathbb{R}$ is a constant.

We now define *Boolean formulae* over multivariate polynomials. The atoms of the formula are polynomials compared with $\{<, \leq, =, \geq, >\}$, and the formula itself can use the connectives $\{\wedge, \vee, \neg\}$.

Definition 1. *The existential theory of the reals consists of every true sentence of the form $\exists x_1 \exists x_2 \ldots \exists x_q \cdot F(x)$, where F is a Boolean formula over multivariate polynomials of x_1 through x_q.*

Given a Boolean formula F, the ETR problem is to decide whether F is a true sentence in the existential theory of the reals. We will say that F has m constraints if it uses m operators from the set $\{<, \leq, =, \geq, >\}$ in its definition.

The Approximate ETR. In the *approximate* existential theory of the reals, we replace the operators $\{<, \leq, \geq, >\}$ with their approximate counterparts. We define the operators $<_\epsilon$ and $>_\epsilon$ with the interpretation that $x <_\epsilon y$ holds if and only if $x < y + \epsilon$ and $x >_\epsilon y$ if and only if $x > y - \epsilon$. The operators $\leq_\epsilon$ and $\geq_\epsilon$ are defined analogously.

We do not allow equality tests in the approximate ETR. Instead, we require that every constraint of the form $x = y$ should be translated to $(x \leq y) \wedge (y \leq x)$ before being weakened to $(x \leq_\epsilon y) \wedge (y \leq_\epsilon x)$.

We also do not allow negation in Boolean formulas. Instead, we require that all negations are first pushed to atoms, using De Morgan's laws, and then further pushed into the atoms by changing the inequalities. So the formula $\neg((x \leq y) \wedge (a \geq b))$ would first be translated to $(x \geq y) \vee (a \leq b)$ before then being weakened to $(x \geq_\epsilon y) \vee (a \leq_\epsilon y)$.

Definition 2. *The approximate existential theory of the reals consists of every true sentence of the form $\exists x_1 \exists x_2 \ldots \exists x_q \cdot F(x)$, where F is a negation-free Boolean formula using the operators $\{<_\epsilon, \leq_\epsilon, \geq_\epsilon, >_\epsilon\}$ over multivariate polynomials of x_1 through x_q.*

Given a Boolean formula F, the ϵ-ETR problem asks us to decide whether F is a true sentence in the approximate existential theory of the reals, where the operators $\{<_\epsilon, \leq_\epsilon, \geq_\epsilon, >_\epsilon\}$ are used.

Unconstrained ϵ-ETR. Our first result is that if no constraints are placed on the value of the variables, that is if each x_i can be arbitrarily large, then ϵ-ETR = ETR for *all* values of $\epsilon \in \mathbb{R}$. We show this via a two way reduction between ϵ-ETR and ETR. The reduction from ϵ-ETR to ETR is trivial, since we can just rewrite each constraint $x <_\epsilon y$ as $x < y + \epsilon$, and likewise for the other operators.

For the other direction, we show that the ETR-complete problem Feas, which asks us to decide whether a system of multivariate polynomials $(p_i)_{i=1,\ldots,k}$ has a shared root, can be formulated in ϵ-ETR. Here we rely on a result of Schaefer and Stefankovic [29], which showed that Feas has a solution if and only if there is a point x such that $|p_i(x)| < 2^{-2^{L+5}}$ for all i, where L is the number of bits used to represent the polynomials. To formulate the problem in ϵ-ETR, we blow-up the instance by multiplying each polynomial by a doubly-exponentially large number t that is bigger than $\epsilon \cdot 2^{2^{L+5}}$. The number t can be constructed by a polynomially-sized formula that uses repeated squaring. So if we write down the constraint $t \cdot p_i(x) \leq_\epsilon 0$ in ϵ-ETR, then this implies that $t \cdot p_i(x) \leq \epsilon$ and therefore $p_i(x) < 2^{-2^{L+5}}$. Thus, via the lemma of Schaefer and Stefankovic, we can formulate Feas in the ϵ-ETR. The full details of this reduction are given in the full version of this paper [18].

Theorem 1. ϵ-ETR = ETR for all $\epsilon \in \mathbb{R}$.

Constrained ϵ-ETR. In our negative result for unconstrained ϵ-ETR, we abused the fact that variables could be arbitrarily large to construct the doubly-exponentially large number t. So, it makes sense to ask whether ϵ-ETR gets easier if we *constrain* the problem so that variables cannot be arbitrarily large.

In this paper, we consider ϵ-ETR problems that are constrained by a convex set in $\mathbb{R}^q$. For vectors $c_1, c_2, \ldots, c_l \in \mathbb{R}^q$ we use $\mathrm{conv}(c_1, c_2, \ldots, c_l)$ to denote the set containing every vector that lies in the convex hull of c_1 through c_l. In the constrained ϵ-ETR, we require that the solution of the ϵ-ETR problem should also lie in the convex hull of c_1 through c_l.

Definition 3. *Given a Boolean formula F and vectors $c_1, c_2, \ldots, c_l \in \mathbb{R}^q$, the constrained ϵ-ETR problem asks us to decide whether*

$$\exists x_1 \exists x_2 \ldots \exists x_q \cdot \big(x \in \mathrm{conv}(c_1, c_2, \ldots, c_l) \wedge F(x)\big).$$

Note that, unlike the constraints used in F, the convex hull constraints are not weakened. So the resulting solution $x_1, x_2, \ldots, x_q$, must actually lie in the convex set.

3   Approximating Constrained ϵ-ETR

Polynomial Classes. To state our main theorem, we will use a certain class of polynomials where the coefficients are given as a tensor. This will be particularly useful when we apply our theorem to certain problems, such as normal form games. To be clear though, this is not a further restriction on the constrained ϵ-ETR problem, since all polynomials can be written down in this form.

The variables of the polynomials we will study will be p-dimensional vectors denoted as $x_1, x_2, \ldots, x_n$, where $x_j(i)$ will denote the i-th element of vector x_j. The coefficients of the polynomials will be a tensor denoted by A. Given a $\times_{j=1}^n p$ tensor A, we denote by $a(i_1, \ldots, i_n)$ its element with coordinates $(i_1, \ldots, i_n)$ on the tensor's dimensions $1, \ldots, n$, respectively, and by α we denote the maximum absolute value of these elements. We define the following two classes of polynomials.

- **Simple tensor multivariate.** We will use $\mathrm{STM}(A, x_1^{d_1}, \ldots, x_n^{d_n})$ denote an STM polynomial with n variables where each variable x_j, $j \in [n]$ is applied d_j times on tensor A that defines the coefficients. Tensor A has $\sum_{j=1}^{n} d_j$ dimensions with p indices each. We will say that an STM polynomial is of maximum degree d, if $d = \max_j d_j$. Here is an example of a degree 2 simple tensor polynomial with two variables:

$$\mathrm{STM}(A, x^2, y) = \sum_{i=1}^{p} \sum_{j=1}^{p} \sum_{k=1}^{p} x(i) \cdot x(j) \cdot y(k) \cdot a(i, j, k) + 10.$$

This polynomial itself is written as follows.

$$\mathrm{STM}(A, x_1^{d_1}, \ldots, x_n^{d_n}) =$$

$$\sum_{i_{1,1} \in [p]} \cdots \sum_{i_{n,d_n} \in [p]} (x_1(i_{1,1})) \cdot \ldots \cdot (x_1(i_{1,d_1})) \cdot \ldots \cdot (x_n(i_{n,1})) \cdot \ldots \cdot (x_n(i_{n,d_n})) \cdot$$

$$\cdot a(i_{1,1}, \ldots, i_{1,d_1} \cdots, i_{n,1}, \ldots, i_{n,d_n}) + a_0.$$

- **Tensor multivariate.** A tensor multivariate (TMV) polynomial is the sum over a number of simple tensor multivariate polynomials. We will use $\mathrm{TMV}(x_1, \ldots, x_n)$ to denote a tensor multivariate polynomial with n vector variables, which is formally defined as

$$\mathrm{TMV}(x_1, \ldots, x_n) = \sum_{i \in [t]} \mathrm{STM}(A_i, x_1^{d_{i1}}, \ldots, x_n^{d_{in}}),$$

where the exponents $d_{i1}, \ldots, d_{in}$ depend on i, and t is the number of simple multivariate polynomials. We will say that $\mathrm{TMV}(x_1, \ldots, x_n)$ has length t if it is the sum of t STM polynomials, and that it is of degree d if $d = \max_{i \in [t], j \in [n]} d_{ij}$.

ϵ-ETR with Tensor Constraints. We focus on ϵ-ETR instances F where all constraints are of the form $\mathrm{TMV}(x_1, \ldots, x_n) \bowtie 0$, where $\bowtie$ is an operator from the set $\{<_\epsilon, \leq_\epsilon, >_\epsilon, \geq_\epsilon\}$. Recall that each TMV constraint considers vector variables. We consider the number of variables used in F (denoted as n) to be the number of vector variables used in the TMV constraints. So the value of n used in our main theorem may be constant if only a constant number of vectors are used, even if the underlying ϵ-ETR instance actually has a non-constant number of variables. For example, if x and y and w are p-dimensional probability distributions and A_1 and A_2 are $p \times p$ tensors, the TMV constraint $x^T A_1 y + w^T A_2 x > 0$ has three variables, degree 1, length two; though the underlying problem has $3 \cdot p$ variables.

Note that every ϵ-ETR constraint can be written as a TMV constraint, because all multivariate polynomials can be written down as a TMV polynomial.

Every term of a TMV can be written as a STM polynomial where the tensor entry is non zero for exactly the combination of variables used in the term, and 0 otherwise. Then a TMV polynomial can be constructed by summing over the STM polynomial for each individual term.

The Main Theorem. Given an ϵ-ETR formula F, we define $\text{exact}(F)$ to be a Boolean formula in which every approximate constraint is replaced with its exact variant, meaning that every instance of $x \leq_\epsilon y$ is replaced with $x \leq y$, and likewise for the other operators.

Our main theorem is as follows.

Theorem 2. *Let F be an ϵ-ETR instance with n vector variables and m multivariate-polynomial constraints each one of maximum length t and maximum degree d, constrained by a convex set defined by $c_1, c_2, \ldots, c_l \in \mathbb{R}^{np}$. Let α be the maximum absolute value of the coefficients of constraints of F, and let $\gamma = \max_i \|c_i\|_\infty$. If $\text{exact}(F)$ has a solution in $\text{conv}(c_1, c_2, \ldots, c_l)$, then F has a k-uniform solution in $\text{conv}(c_1, c_2, \ldots, c_l)$ where*

$$k = \frac{48 \cdot \alpha^6 \cdot \gamma^{2d+2} \cdot d^5 \cdot t^4 \cdot n^6 \cdot \ln(2 \cdot \alpha \cdot \gamma \cdot d \cdot t \cdot n \cdot m)}{\epsilon^4}.$$

Consequences of the Main Theorem. Our main theorem gives a QPTAS for approximating a fragment of ϵ-ETR. The total number of k-uniform vectors in a convex set $C = \text{conv}(c_1, c_2, \ldots, c_l)$ is $l^{O(k)}$. So, if the parameters α, γ, d, t, and n are all constant, then our main theorem tells us that the total number of k-uniform vectors is $l^{O(\log m)}$, where m is the number of constraints. So if we enumerate each k-uniform vector x, we can check whether F holds, and if it does, we can output x. If no k-uniform vector exists that satisfies F, then we can determine that $\text{exact}(F)$ has no solution. This gives us the following result.

Corollary 1. *Let F be an ϵ-ETR instance constrained by the convex set defined by $c_1, c_2, \ldots, c_l$. If α, γ, d, t, and n are constant, and l is polynomial, then we have an algorithm that runs in time $l^{O(\log m)}$ that either finds a solution to F, or determines that $\text{exact}(F)$ has no solution.*

If m is constant and l is polynomial then this gives a PTAS, while if m and l are polynomial, then this gives a QPTAS.

In Sect. 5 we will show that the problem of approximating the best social welfare achievable by an approximate Nash equilibrium in a two-player normal form game can be written down as a constrained ϵ-ETR formula where α, γ, d, and m are constant. It has been shown that, assuming the exponential time hypothesis, this problem cannot be solved faster than quasi-polynomial time [11, 20], so this also implies that constrained ϵ-ETR where α, γ, d, and m are constant cannot be solved faster than quasi-polynomial time unless the exponential time hypothesis is false.

Many ϵ-ETR problems are naturally constrained by sets that are defined by the convex hull of exponentially many vectors. The cube $[0, 1]^n$ is a natural example of one such set. Brute force enumeration does not give an efficient algorithm for these problems, since we need to enumerate $l^{O(k)}$ vectors, and l is already exponential. However, our main theorem is able to provide non-deterministic polynomial time algorithms for these problems.

This is because each k-uniform vector is, by definition, the convex combination of at most k of the vectors in the convex set, and this holds even if l is exponential. So, provided that k is polynomial, we can guess the subset of vectors that are used, and then verify that the formula holds. This is particularly useful for problems where exact(F) always has a solution, which is often the case in game theory applications, since it places the approximation problem in NP, whereas computing the exact solution may be ETR-complete.

Corollary 2. *Let F be an ϵ-ETR instance constrained by the convex set defined by $c_1, c_2, \ldots, c_l$. If α, γ, d, t, n, are polynomial, then there is a non-deterministic polynomial time algorithm that either finds a solution to F, or determines that exact(F) has no solution. Moreover, if exact(F) is guaranteed to have a solution, then the problem of finding an approximate solution for F is in NP.*

A Theorem for Non-tensor Formulas. One downside of Theorem 2 is that it requires that the formula is written down using tensor constraints. We have argued that every ETR formula can be written down in this way, but the translation introduces a new vector-variable for each variable in the ETR formula. When we apply Theorem 2 to obtain PTASs or QPTASs we require that the number of vector variables is at most polylogarithmic, and so this limits the application of the theorem to ETR formulas that have at most polylogarithmically many variables.

The following theorem is a sampling result for ϵ-ETR with non-tensor constraints, which is proved in the full version of this paper [18].

Theorem 3. *Let F be an ϵ-ETR instance constrained over the convex set defined by $c_1, c_2, \ldots, c_l \in \mathbb{R}^q$. Let m be the number of constraints used in F, Let $\gamma = \max_i \|c_i\|_\infty$, let α be the largest constant coefficient used in F, let t be the number of terms used in F, and let d be the maximum degree of the polynomials in F. If exact(F) has a solution in $\mathrm{conv}(c_1, c_2, \ldots, c_l)$, then F has a k-uniform solution in $\mathrm{conv}(c_1, c_2, \ldots, c_l)$ where*

$$k = \alpha^2 \cdot \gamma^{2d-2} \cdot (2^d - 1)^2 \cdot t^2 \cdot \log l / \epsilon^2.$$

The key feature here is that the number of variables does not appear in the formula for k, which allows the theorem to be applied to some formulas for which Theorem 2 cannot. However, since the theorem does not allow tensor constraints, its applicability is more limited because the number of terms t will be much larger in non-tensor formulas. For example, as we will see in Sect. 5, we can formulate bimatrix games using tensor constraints over constantly many vector variables, and this gives a result using Theorem 2. No such result can be obtained via Theorem 3, because when we formulate problem without tensor constraints, the number of terms t used in the inequalities becomes polynomial.

4   The Proof of the Main Theorem

In this section we prove Theorem 2. Before we proceed with the technical results let us provide a roadmap. We begin by considering two special cases, which when combined will be the backbone of the proof of the main theorem.

Firstly, we will show how to deal with problems where every constraint of the Boolean formula is a *multilinear polynomial*, which we will define formally later. We deal with this kind of problems using Hoeffding's inequality and the union bound, which is similar to how such constraints have been handled in prior work.

Then, we study problems where the Boolean formula consists of a *single* degree d polynomial constraint. We reduce this kind of problems to a constrained $\epsilon/2$-ETR problem with multilinear constraints, so we can use our previous result to handle the reduced problem. Degree d polynomials have not been considered in previous work, and so this reduction is a novel extension of sampling based techniques to a broader class of ϵ-ETR formulas.

Finally, we deal with the main theorem: we reduce the original ETR problem with multivariate constraints to a set of $\epsilon'-$ETR problems with a single standard degree d constraint, and then we use the last result to derive a bound on k. The proof of the theorem is given in the full version of the paper [18].

Problems with Multilinear Constraints. We begin by considering constrained ϵ-ETR problems where the Boolean formula F consists of tensor-multilinear polynomial constraints. We will use $\mathrm{TML}(A, x_1, \ldots, x_n)$ to denote a tensor-multilinear polynomial with n variables and coefficients defined by tensor A of size $\times_{j=1}^{n} p$. Formally,

$$\mathrm{TML}(A, x_1, \ldots, x_n) = \sum_{i_1 \in [p]} \cdots \sum_{i_n \in [p]} x_1(i_1) \cdot \ldots \cdot x_n(i_n) \cdot a(i_1, \ldots, i_n) + c.$$

We will use α to denote the maximum entry of tensor A in the absolute value sense and γ to denote the infinite norm of the convex set that constrains the variables.

The following lemma is proved in the full version of this paper [18]. The proof uses Hoeffding's inequality and the union bound, and is similar to previous applications of the sampling technique.

Lemma 1. *Let F be a Boolean formula with n variables and m tensor-multilinear polynomial constraints and let $\mathcal{Y}$ be a convex set in the variables space. If the constrained ETR problem defined by exact(F) and $\mathcal{Y}$ has a solution, then the constrained ϵ-ETR problem defined by F and $\mathcal{Y}$ has a k uniform solution where*

$$k = \frac{2 \cdot \alpha^2 \cdot \gamma^2 \cdot n^2 \cdot \ln(2 \cdot n \cdot m)}{\epsilon^2}.$$

Problems with a Standard Degree d Constraints. We now consider constrained ϵ-ETR problems with *exactly one* tensor polynomial constraint of standard degree d. We will use $\mathrm{TSD}(A, x, d)$ to denote a standard degree d tensor-polynomial with coefficients defined by the $\times_{j=1}^{d} l$ tensor A. Here, d identical vectors x are applied on A. Formally,

$$\text{TSD}(A, x, d) = \sum_{i_1 \in [p]} \cdots \sum_{i_d \in [p]} x(i_1) \cdot \ldots \cdot x(i_d) \cdot a(i_1, \ldots, i_d) + c.$$

The following lemma is proved in the full version of this paper [18]. To prove the lemma we consider the variable x to be defined as the average of $r = O(\frac{\alpha^2 \cdot \gamma^d \cdot d^2}{\epsilon})$ variables. This allows us to "break" the standard degree d tensor polynomial to a sum of multilinear tensor polynomials and to a sum of not-too-many multivariate polynomials. Then, the choice of r allows us to upper bound the error occurred by the multivariate polynomials by $\frac{\epsilon}{2}$. Then, we observe that in order to prove the lemma we can write the sum of multilinear tensor polynomials as an $\frac{\epsilon}{2}$-ETR problem with r variables and roughly r^d multilinear constraints. This allows us to use Lemma 1 to complete the proof.

Lemma 2. *Let F be a Boolean formula with variable x and one tensor-polynomial constraint of standard degree d, and let $\mathcal{Y}$ be a convex set. If the constrained ETR problem defined by* $\text{exact}(F)$ *and $\mathcal{Y}$ has a solution, then the constrained ϵ-ETR problem defined by F and $\mathcal{Y}$ has a k-uniform solution where*

$$k = \frac{24 \cdot \alpha^6 \cdot \gamma^{2d+2} \cdot d^5 \cdot \ln(2 \cdot \alpha \cdot \gamma \cdot d)}{\epsilon^4}.$$

Problems with Simple Multivariate Constraints. We now assume that we are given a constraint-ϵ-ETR problem defined by a Boolean formula F of tensor simple multilinear polynomial constraints and a convex set $\mathcal{Y}$. As before $\gamma = \|\mathcal{Y}\|_\infty$ and let α be the maximum absolute value of the coefficients of the constraints. We will say that the constraints are of maximum degree d if d is the maximum degree among all variables. The following lemma is shown in the full version of this paper [18]. The idea is to rewrite the problem as an equivalent problem with standard degree d constraints and then apply Lemmas 2 and 1 to derive the bound for k.

Lemma 3. *Let F be a Boolean formula with n variables and m simple tensor-multivariate polynomial constraints of maximum degree d and let $\mathcal{Y}$ be a convex set in the variables space. If the constrained ETR problem defined by* $\text{exact}(F)$ *and $\mathcal{Y}$ has a solution, then the constrained ϵ-ETR problem defined by F and $\mathcal{Y}$ has a k uniform solution where*

$$k = \frac{48 \cdot \alpha^6 \cdot \gamma^{2d+2} \cdot d^5 \cdot n^6 \cdot \ln(2 \cdot \alpha \cdot \gamma \cdot d \cdot n \cdot m)}{\epsilon^4}.$$

5 Applications

We now show how our theorems can be applied to derive new approximation algorithms for a variety of problems.

Constrained Approximate Nash Equilibria. A *constrained* Nash equilibrium is a Nash equilibrium that satisfies some extra constraints, like specific bounds on the payoffs of the players. Constrained Nash equilibria attracted the attention of many authors, who proved NP-completeness for two-player games [6,14,24] and ETR-completeness for three-player games [6–9,23] for constrained *exact* Nash equilibria.

Constrained approximate equilibria have been studied, but so far only lower bounds have been derived [2,11,19,20,26]. It has been observed that sampling methods can give QPTASs for finding constrained approximate Nash equilibria for certain constraints in two player games [20].

By applying Theorem 2, we get the following result for games with a constant number of players: *Any property of an approximate equilibrium that can be formulated in ϵ-ETR where α, γ, d, t and n are constant has a QPTAS.* This generalises past results to a much broader class of constraints, and provides results for games with more than two players, which had not previously been studied in this setting. The details of this result are given in the full version of the paper [18].

Shapley Games. Shapley's stochastic games [30] describe a two-player infinite-duration zero-sum game. The game consists of N *states*. Each state specifies a two-player $M \times M$ matrix game where the players compete over: (1) a reward (which may be negative) that is paid by player two to player one, and (2) a probability distribution over the next state of the game. So each round consists of the players playing a bimatrix game at some state s, which generates a reward, and the next state s' of the game. The reward in round i is discounted by λ^{i-1}, where $0 < \lambda < 1$ is a *discount factor*. The overall payoff to player 1 is the discounted sum of the infinite sequence of rewards generated during the course of the game.

Shapley showed that these games are determined, meaning that there exists a value vector v, where v_s is the value of the game starting at state s. A polynomial-time algorithm has been devised for computing the value vector of a Shapley game when the number of states N is constant [25]. However, since the values may be irrational, this algorithm needs to deal with algebraic numbers, and the *degree* of the polynomial is $O(N)^{N^2}$, so if N is even mildly super-constant, then the algorithm is not polynomial.

Shapley showed that the value vector is the unique solution of a system of polynomial optimality equations, which can be formulated in ETR. Any approximate solution of these equations gives an approximation of the value vector, and applying Theorem 2 gives us a QPTAS. This algorithm works when $N \in O(\sqrt[6]{\log M})$, which is a value of N that prior work cannot handle. The downside of our algorithm is that, since we require the solution to be bounded by a convex set, the algorithm only works when the value vector is reasonably small. Specifically, the algorithm takes a constant bound $B \in \mathbb{R}$, and either finds the approximate value of the game, or verifies that the value is strictly greater than B. The algorithm's details are given in the full version of the paper [18].

Optimization Problems. Our framework can provide approximation schemes for optimization problems with one vector variable $x \in \mathbb{R}^p$ with polynomial constraints over bounded convex sets. Formally,

$$\max \quad h(x)$$
$$s.t. \quad h_1(x) > 0, \ldots, h_m(x) > 0$$
$$x \in \text{conv}(c_1, \ldots, c_l)$$

where $h(x), h_1(x), \ldots, h_m(x)$ are polynomials with respect to vector x; for example $h(x) = x^T A x$, where A is an $p \times p$ matrix, subject to $h_1(x) = x^T x > \frac{1}{10}$ and $x \in \Delta^p$. We will call the polynomials h_i *solution-constraints*. Optimization problems of this kind received a lot of attention over the years [15–17, 22].

For optimization problems, we sample from the solution that achieves the maximum when we apply Theorem 2, in order to prove that there is a k-uniform solution that is close to the maximum. Our algorithm enumerates all k-uniform profiles, and outputs the one that maximizes the objective function. Using this technique, Theorem 2 gives the following results.

1. There is a PTAS if $h(x)$ is a STM polynomial of maximum degree independent of p, the number of solution-constraints is independent of p, and $l = \text{poly}(p)$.
2. There is QPTAS if $h(x)$ is an STM polynomial of maximum degree up to $\text{poly} \log p$, the number of solution-constraints is $\text{poly}(p)$, and $l = \text{poly}(p)$.

To the best of our knowledge, the second result is new. The first result was already known, however it was proven using completely different techniques: in [10] it was proven for the special case of degree two, in [22] it was extended to any fixed degree, and alternative proofs of the fixed degree case were also given in [16, 17]. We highlight that in all of the aforementioned results solution constraints were not allowed. Note that unless NP=ZPP there is no FPTAS for quadratic programming even when the variables are constrained in the simplex [15]. Hence, our results can be seen as a partial answer to the important question posed in [15]: *"What is a complete classification of functions that allow a PTAS?"*

Tensor Problems. Our framework provides quasi-polynomial time algorithms for deciding the existence of approximate eigenvalues and approximate eigenvectors of tensors in $\mathbb{R}^{p \times p \times p}$, where the elements are bounded by a constant, where the solutions are required to be in a convex set. In [27] it is proven that there is no PTAS for these problems when the domain is unrestricted. To the best of our knowledge this is the first positive result for the problem even in this, restricted, setting. The algorithm's details are given in the full version of this paper [18].

Computational Geometry. Finally, we note that our theorem can be applied to problems in computational geometry, although the results are not as general as one may hope. Many problems in this field are known to be ETR-complete, including, for example, the Steinitz problem for 4-polytopes, inscribed polytopes and Delaunay triangulations, polyhedral complexes, segment intersection graphs, disk intersection graphs, dot product graphs, linkages, unit distance graphs, point visibility graphs, rectilinear crossing number, and simultaneous graph embeddings. We refer the reader to the survey of Cardinal [13] for further details.

All of these problems can be formulated in ϵ-ETR, and indeed our theorem does give results for these problems. However, our requirement that the bounding convex set be given explicitly limits their applicability. Most computational geometry problems are naturally constrained by a cube, so while Corollary 2 does give NP algorithms, we do not get QPTASs unless we further restrict the convex set. In the full version of the paper [18] we formulate QPTASs for the segment intersection graph and the unit disk intersection graph problems when the solutions are restricted to lie in a simplex. While it is not clear that either problem has natural applications that are restricted in this way, we do think that future work may be able to derive sampling theorems that are more tailored towards the computational geometry setting.

Acknowledgements. P. Spirakis wishes to dedicate this paper to the memory of his late father in law Mathematician and Professor Dimitrios Chrysofakis, who was among the first in Greece to work on tensor analysis.

References

1. Althöfer, I.: On sparse approximations to randomized strategies and convex combinations. Linear Algebra Appl. **199**, 339–355 (1994). Special Issue Honoring Ingram Olkin
2. Austrin, P., Braverman, M., Chlamtac, E.: Inapproximability of NP-complete variants of nash equilibrium. Theo. Comput. **9**, 117–142 (2013)
3. Babichenko, Y., Barman, S., Peretz, R.: Empirical distribution of equilibrium play and its testing application. Math. Oper. Res. **42**(1), 15–29 (2016)
4. Barman, S.: Approximating Nash equilibria and dense bipartite subgraphs via an approximate version of Caratheodory's theorem. In: Proceedings of STOC, pp. 361–369 (2015)
5. Barman, S., Ligett, K., Piliouras, G.: Approximating Nash equilibria in tree polymatrix games. In: Hoefer, M. (ed.) SAGT 2015. LNCS, vol. 9347, pp. 285–296. Springer, Heidelberg (2015). https://doi.org/10.1007/978-3-662-48433-3_22
6. Bilò, V., Mavronicolas, M.: The complexity of decision problems about Nash equilibria in win-lose games. In: Serna, M. (ed.) SAGT 2012. LNCS, pp. 37–48. Springer, Heidelberg (2012). https://doi.org/10.1007/978-3-642-33996-7_4
7. Bilò, V., Mavronicolas, M.: Complexity of rational and irrational Nash equilibria. Theo. Comput. Syst. **54**(3), 491–527 (2014)
8. Bilò, V., Mavronicolas, M.: A catalog of EXISTS-R-complete decision problems about Nash equilibria in multi-player games. In: 33rd Symposium on Theoretical Aspects of Computer Science, STACS 2016, Orléans, France, 17–20 February 2016, pp. 17:1–17:13 (2016). https://doi.org/10.4230/LIPIcs.STACS.2016.17
9. Bilò, V., Mavronicolas, M.: Existential-R-complete decision problems about symmetric Nash equilibria in symmetric multi-player games. In: 34th Symposium on Theoretical Aspects of Computer Science, STACS 2017, Hannover, Germany, 8–11 March 2017, pp. 13:1–13:14 (2017). https://doi.org/10.4230/LIPIcs.STACS.2017.13

10. Bomze, I.M., De Klerk, E.: Solving standard quadratic optimization problems via linear, semidefinite and copositive programming. J. Global Opt. **24**(2), 163–185 (2002)
11. Braverman, M., Kun-Ko, Y., Weinstein, O.: Approximating the best Nash equilibrium in $n^{o(\log n)}$-time breaks the exponential time hypothesis. In: Proceedings of SODA, pp. 970–982 (2015)
12. Canny, J.: Some algebraic and geometric computations in PSPACE. In Proceedings of STOC, pp. 460–467. ACM, New York (1988)
13. Cardinal, J.: Computational geometry column 62. ACM SIGACT News **46**(4), 69–78 (2015)
14. Conitzer, V., Sandholm, T.: New complexity results about Nash equilibria. Games Econ. Behav. **63**(2), 621–641 (2008)
15. De Klerk, E.: The complexity of optimizing over a simplex, hypercube or sphere: a short survey. CEJOR **16**(2), 111–125 (2008)
16. De Klerk, E., Laurent, M., Parrilo, P.A.: A PTAS for the minimization of polynomials of fixed degree over the simplex. Theoret. Comput. Sci. **361**(2–3), 210–225 (2006)
17. de Klerk, E., Laurent, M., Sun, Z.: An alternative proof of a PTAS for fixed-degree polynomial optimization over the simplex. Math. Program. **151**(2), 433–457 (2015)
18. Deligkas, A., Fearnley, J., Melissourgos, T., Spirakis, P.: Approximating the existential theory of the reals. ArXiv e-prints, October (2018)
19. Deligkas, A., Fearnley, J., Savani, R.: Computing constrained approximate equilibria in polymatrix games. In: Bilò, V., Flammini, M. (eds.) SAGT 2017. LNCS, vol. 10504, pp. 93–105. Springer, Cham (2017). https://doi.org/10.1007/978-3-319-66700-3_8
20. Deligkas, A., Fearnley, J., Savani, R.: Inapproximability results for constrained approximate Nash equilibria. Inf. Comput. **262**(Part), 40–56 (2018). https://doi.org/10.1016/j.ic.2018.06.001
21. Deligkas, A., Fearnley, J., Spirakis, P.: Lipschitz continuity and approximate equilibria. In: Gairing, M., Savani, R. (eds.) SAGT 2016. LNCS, vol. 9928, pp. 15–26. Springer, Heidelberg (2016). https://doi.org/10.1007/978-3-662-53354-3_2
22. Faybusovich, L.: Global optimization of homogeneous polynomials on the simplex and on the sphere. In: Floudas, C.A., Pardalos, P. (eds.) Frontiers in Global Optimization. Nonconvex Optimization and Its Applications, vol. 74, pp. 109–121. Springer, Boston (2004). https://doi.org/10.1007/978-1-4613-0251-3_6
23. Garg, J., Mehta, R., Vazirani, V.V., Yazdanbod, S.: ETR-completeness for decision versions of multi-player (symmetric) Nash equilibria. ACM Transactions on Economics and Computation (TEAC) **6**(1), 1 (2018)
24. Gilboa, I., Zemel, E.: Nash and correlated equilibria: some complexity considerations. Games and Economic Behavior **1**(1), 80–93 (1989)
25. Hansen, K. A., Koucký, M., Lauritzen, N., Miltersen, P. B., Tsigaridas, E. P.: Exact algorithms for solving stochastic games: extended abstract. In: Proceedings of STOC, pp. 205–214 (2011)
26. Hazan, E., Krauthgamer, R.: How hard is it to approximate the best Nash equilibrium? SIAM J. Comput. **40**(1), 79–91 (2011)
27. Hillar, C.J., Lim, L.-H.: Most tensor problems are NP-hard. Journal of the ACM (JACM) **60**(6), 45 (2013)
28. Lipton, R. J., Markakis, E., Mehta, A.: Playing large games using simple strategies. In Proceedings of EC, pp. 36–41. ACM (2003)
29. Schaefer, M., Stefankovic, D.: Fixed points, Nash equilibria, and the existential theory of the reals. Theor. Comput. Syst. **60**(2), 172–193 (2017)
30. Shapley, L.S.: Stochastic games. Proc. Nat. Acad. Sci. **39**(10), 1095–1100 (1953)

Pricing Multi-unit Markets

Tomer Ezra[1], Michal Feldman[1], Tim Roughgarden[2],
and Warut Suksompong[2(✉)]

[1] Blavatnik School of Computer Science, Tel-Aviv University, Tel Aviv, Israel
tomer.ezra@gmail.com, michal.feldman@cs.tau.ac.il
[2] Department of Computer Science, Stanford University, Stanford, USA
{tim,warut}@cs.stanford.edu

Abstract. We study the power and limitations of posted prices in multi-unit markets, where agents arrive sequentially in an arbitrary order. We prove upper and lower bounds on the largest fraction of the optimal social welfare that can be guaranteed with posted prices, under a range of assumptions about the designer's information and agents' valuations. Our results provide insights about the relative power of uniform and non-uniform prices, the relative difficulty of different valuation classes, and the implications of different informational assumptions. Among other results, we prove constant-factor guarantees for agents with (symmetric) subadditive valuations, even in an incomplete-information setting and with uniform prices.

1 Introduction

We consider the problem of allocating identical items to agents to maximize the social welfare. More formally, there are m identical items, each agent $i \in [n]$ has a valuation function $v_i : [m] \to \mathbb{R}_{\geq 0}$ describing her value for a given number of items, and the goal is to compute nonnegative and integral quantities $q_1, \ldots, q_n$, with $\sum_{i=1}^{n} q_i \leq m$, to maximize the total value $\sum_{i=1}^{n} v_i(q_i)$ to the agents.

This problem underlies the design of *multi-unit auctions*, which have played a starring role in the fields of classical and algorithmic mechanism design, and in both theory and practice. As with any welfare-maximization problem, the problem can be solved in principle using the VCG mechanism. There has been extensive work on the design and analysis of more practical multi-unit auctions. There are indirect implementations of the VCG mechanism, most famously Ausubel's ascending *clinching auction* for downward-sloping (a.k.a. submodular) valuations [1]. Work in algorithmic mechanism design has identified mechanisms that retain the dominant-strategy incentive-compatibility of the VCG mechanism while running in time polynomial in n and $\log m$ (rather than polynomial in n and m), at the cost of a bounded loss in the social welfare. Indeed, Nisan [30] argues that the field of algorithmic mechanism design can be fruitfully viewed through the lens of multi-unit auctions.

The multi-unit auction formats used in practice typically sacrifice dominant-strategy incentive-compatibility in exchange for simplicity and equitability; a

© Springer Nature Switzerland AG 2018
G. Christodoulou and T. Harks (Eds.): WINE 2018, LNCS 11316, pp. 140–153, 2018.
https://doi.org/10.1007/978-3-030-04612-5_10

canonical example is the uniform-price auctions suggested by Milton Friedman (see [22]) and used (for example) by the U.S. Treasury to sell government securities. Uniform-price auctions do not always maximize the social welfare (e.g., because of demand reduction), but they do admit good "price-of-anarchy" guarantees [29], meaning that every equilibrium results in social welfare close to the maximum possible.

A key drawback of all of the mechanisms above is that they require all agents to participate simultaneously, in order to coordinate their allocations and respect the supply constraint. For example, in a uniform-price auction, all of the agents' bids are used to compute a market-clearing price-per-unit, which then determines the allocations of all of the agents. It is evident from our daily experience that, in many different markets, buyers arrive and depart asynchronously over time, making purchasing decisions as a function of their preferences and the current prices of the goods for sale.[1] The goal of this paper is to develop theory that explains the efficacy of such *posted prices* in markets where agents arrive sequentially rather than simultaneously, and that gives guidance on how to set prices to achieve an approximately welfare-maximizing outcome.

1.1   The Model

We consider a setting where a designer must post prices in advance, before the arrival of any agents. We assume that the supply m is known. The designer is given full or incomplete information about agents' valuations, and must then set a price for each item.[2] Agents then arrive in an arbitrary (worst-case) order, with each agent taking a utility-maximizing bundle (breaking ties arbitrarily), given the set of items that remain. These prices are *static*, in that they remain fixed throughout the entire process.

Example 1.1. *Suppose $m = 3$ and there are two agents, each with the valuation $v(1) = 5$, $v(2) = 9$, and $v(3) = 11$, and suppose a designer prices every item at 4. The first agent will choose either 1 or 2 items (breaking the tie arbitrarily). If the first agent chooses 2 items, the second agent will take the only item remaining; if the first agent chooses 1 item, then the second agent will take either 1 or 2 items.*

In general, we allow different items to receive different prices (as will be the case in the VCG mechanism for this problem, for example.) With identical items, however, it is natural to focus on *uniform prices*, where every item is given the same price. Generally speaking, we are most interested in positive results for uniform prices, and negative results for non-uniform prices.

---

[1] For examples involving identical items, think about general-admission concert tickets, pizzas at Una Pizza Napoletana (which shuts down for the night when the dough runs out), or shares in an IPO (other than Google [33]).

[2] No non-trivial guarantees are possible without at least partial knowledge about agents' valuations.

The overarching goal of this paper is to characterize the largest fraction of the optimal social welfare that can be guaranteed with posted prices, under a range of assumptions about the designer's information and agents' valuations. This goal is inherently quantitative, but our results also provide qualitative insights, for example about the relative power of uniform and non-uniform prices, the relative difficulty of different valuation classes, and the implications of different informational assumptions.

Table 1. Summary of results. All results are new to this paper unless indicated otherwise. Numbers in parentheses refer to the corresponding theorem or proposition number.

	Uniform prices	Non-uniform prices
Submodular	$\frac{1}{2}$ (4.6, 4.7, 4.8)	$\frac{2}{3}$ (4.1, 4.2) [2 items]
		$\geq \frac{5}{7} - \frac{1}{m}$ (4.3), ≤ 0.802 (4.4) [m items]
XOS	$\geq \frac{1}{2}$ (8.2)	$\leq 1 - \frac{1}{e}$ (5.1)
Subadditive	$\frac{1}{3}$ (6.1, 6.4)	$\leq \frac{1}{2}$ (6.3) [even with 2 buyers]
	$\frac{2}{3}$ (6.2, 6.6) [2 identical buyers]	$\leq \frac{3}{4}$ (6.5) [even with 2 identical buyers]
General	$\frac{1}{m}$ (7.1)	$\frac{1}{m}$ (7.2)

(a) Full information

	Uniform prices	Non-uniform prices
XOS	$\frac{1}{2}$ (8.2)	$\frac{1}{2}$ [21]
Subadditive	$\geq \frac{1}{4}$ (8.4)	$\leq \frac{1}{2}$ (6.3) [even with 2 buyers]
		$\leq \frac{3}{4}$ (6.5) [even with 2 identical buyers]

(b) Incomplete information

1.2 Our Results

The majority of our results are summarized in Table 1; we highlight a subset of these next. First, consider the case of a Bayesian setting with XOS agent valuations (see Sect. 2 for definitions). That is, each agent's valuation is drawn independently from a known (possibly agent-specific) distribution over XOS valuations. Feldman et al. [21] show that, even with non-identical items, posted prices can always obtain expected welfare at least 1/2 times the maximum possible. This factor of 1/2 is tight, even for the special case of a single item and i.i.d. agents. The posted prices used by Feldman et al. [21] are non-uniform, even when the result is specialized to the case of identical items (the price of an item is based on its expected marginal contribution to an optimal allocation, which can vary across items). We prove in Theorem 8.2 that with identical items, and agents with independent (not necessarily identical) XOS valuations, uniform prices suffice to achieve the best-possible guarantee of half the optimal expected welfare. Moreover, this result extends to any class of valuations that is c-close to XOS valuations, with an additional loss of a factor of c (Theorem 8.3).

While the 1/2-approximation above is tight for an incomplete-information setting, this problem is already interesting in the full-information case where the buyers' valuations are known (with the order of arrival still worst-case). Can we improve over the approximation factor of $1/2$ under this stronger informational assumption?

We prove that uniform prices cannot achieve an approximation factor better than $1/2$, even for the more restrictive class of submodular valuations, and even with two agents (Proposition 4.7) or identical agents (Proposition 4.8). In contrast, with non-uniform prices (still for submodular valuations), we prove that an approximation of $2/3$ is possible (Theorem 4.1). This is tight for the case of two items (Proposition 4.2), but in large markets (with $m \to \infty$) we show how to obtain an approximation guarantee of $5/7$ (Theorem 4.3). In addition, if the order of arrival is known beforehand, we can extract the full optimal welfare (Theorem 4.5).

We next consider the family of subadditive valuations, which strictly generalize XOS valuations and are regarded as the most challenging class of valuations that forbid complements. For example, with non-identical items, it is not known whether or not posted prices can guarantee a constant fraction of the optimal social welfare. For identical items, we prove that this is indeed possible. In the incomplete-information setting (and identical items), we show that subadditive valuations are 2-close to XOS valuations (Sect. 3), which leads to an approximation factor of $1/4$ (Theorem 8.4). We can also do better in the full-information setting: uniform prices can guarantee a $1/3$ fraction of the optimal social welfare (Theorem 6.1), and the approximation is tight (Proposition 6.4), while even non-uniform prices cannot guarantee a factor bigger than $1/2$, even with only two agents (Proposition 6.3). In the case of two identical agents, uniform prices can guarantee a $2/3$ fraction of the optimal welfare (Theorem 6.2), and this is tight (Proposition 6.6).

With all these positive results, the reader might wonder whether constant factor guarantees can be provided for general valuations. Unfortunately, this is not the case. For general valuations, we show that even in the full-information setting and with non-uniform prices, and even when there are only two agents and the arrival order is known, posted prices can guarantee a $1/m$ fraction of the optimal social welfare, but not more (Proposition 7.1, Theorem 7.2). If the seller can control the arrival order, however, then even uniform prices can guarantee half of the optimal social welfare (Theorem 7.3). No better bound is possible, even for identical valuations and with non-uniform prices (Proposition 7.4).

1.3 Further Related Work

The design and analysis of simple mechanisms has been an active area of study in algorithmic mechanism design, particularly within the last decade. This focus is motivated in part by the observation that simple mechanisms are highly desired in practical scenarios. Examples of simple mechanisms that are used in practice are the generalized second price auctions (GSP) for online advertising [15,27,28,31,34], and simultaneous item auctions (where the agents bid separately and simultaneously on multiple items) [6,11,20,24]. These mechanisms are not truthful and are evaluated in equilibrium using the price of anarchy measure.

Posting prices is perhaps the most prevalent method for selling goods in practice. By simply publishing prices for individual items, posted price mechanisms are extremely easy to understand and participate in. It should therefore not come as a surprise that these mechanisms have been studied extensively for various objective functions (e.g., welfare, revenue, makespan), information structures of values (e.g., full-information, Bayesian, online), and valuation functions (e.g., unit-demand, submodular, XOS). For example, a long line of work has focused on sequential posted prices for revenue maximization and has shown, among other things, that a form of posted price mechanisms can achieve a constant fraction of the optimal revenue for agents with unit-demand valuations [8–10]. Revenue maximization with sequential posted prices has also been studied for a single item, both in large markets [7] and when the distributions are unknown [2], for additive valuations [4,5], and for a buyer with complements [17]. Dütting et al. [14] provides a general framework for posted price mechanisms. In several of these works, posted price mechanisms are allowed to discriminate between agents and set different prices for each of them. In contrast, in this work we do not consider discriminatory prices.

Another line of research relevant to our work considers market equilibria, for example those achieved by Walrasian prices. A result of Kelso and Crawford [25] states that for the class of gross-substitute valuations, there always exists a Walrasian equilibrium, meaning that one can assign prices to items so as to achieve the optimal social welfare. However, this result is based on the assumption that agents break ties in a particular way. As such, the existence of Walrasian prices does not carry over welfare guarantees to our setting, even for unit-demand valuations. We believe that the worst-case perspective that we take is more realistic in our setting, where we do not have control over how agents break ties.

In addition to the aforementioned works, a new line of research has considered dynamic posted prices in online settings such as for the k-server and parking problems [12]. Moreover, posted price mechanisms have been studied in the context of welfare maximization in matching markets, where prices are dynamic (i.e., can change over the course of the mechanism) but do not depend on the identity of the agents [13]. With static prices, it was recently shown that one can achieve strictly more than half of the welfare in the full information setting with binary unit-demand valuations [16].

The sequential arrival of agents considered in posted price mechanisms fits into the framework of online mechanisms, which deals with dynamic environments with multiple agents having private information [3,23,32]. Our work shows that for identical items and agents with subadditive valuations, posted prices can guarantee a constant fraction of the welfare even while setting the (uniform) prices up front.

2   Preliminaries

We consider a setting with a set M of m *identical* items, and a set N of n buyers. Each buyer has a valuation function $v_i : 2^M \to \mathbb{R}_{\geq 0}$ that indicates his value for every set of objects. Since items are identical, the valuation depends only on the number of items. We assume that valuations are monotone non-decreasing (i.e., $v_i(T) \leq v_i(S)$ for $T \subseteq S$) and normalized (i.e., $v_i(\emptyset) = 0$). We use $v_i(S|T) = v_i(S \cup T) - v_i(T)$ to denote the marginal value of bundle S *given* bundle T.

A buyer valuation profile is denoted by $\mathbf{v} = (v_1, \dots, v_n)$. An *allocation* is a vector of disjoint sets $\mathbf{x} = (x_1, \dots, x_n)$, where x_i denotes the bundle associated with buyer $i \in [n]$ (note that it is not required that all items are allocated). As with valuations, since we consider identical items, an allocation can be represented by the number of items allocated to each buyer. The *social welfare* of an allocation $\mathbf{x}$ is $\mathrm{SW}(\mathbf{x}, \mathbf{v}) = \sum_{i=1}^{n} v_i(x_i)$, and the optimal social welfare is denoted by $\mathrm{OPT}(\mathbf{v})$. When clear from the context we omit $\mathbf{v}$ and write OPT for the optimal social welfare.

For two valuation functions v, v', we say that $v \geq v'$ iff $v(S) \geq v'(S)$ for every set S. A hierarchy over complement-free valuations is given by Lehmann et al. [26].

Definition 2.1. *A valuation function v is*

- additive *if $v(S) = \sum_{i \in S} v(\{i\})$ for every set $S \subseteq M$.*
- submodular *if $v(\{i\}|S) \geq v(\{i\}|T)$ for every item $i \notin T$ and sets S, T such that $S \subseteq T \subseteq M$.*
- XOS *if there exist additive valuation functions $v^1, \dots, v^k$ such that $v(S) = \max_{j=1,\dots,k} v^j(S)$ for every set $S \subseteq M$.*
- subadditive *if $v(S) + v(T) \geq v(S \cup T)$ for any sets $S, T \subseteq M$.*

Since we assume throughout the paper that all items are identical, we only work with symmetric valuation functions.

Definition 2.2. *A valuation function v is* symmetric *if $v(S) = v(T)$ for every sets $S, T \subseteq M$ such that $|S| = |T|$. A symmetric valuation function can thus be represented by a monotone non-decreasing function $v : \{0, 1, \dots, m\} \to \mathbb{R}_{\geq 0}$, which assigns a non-negative real value to any integer in $[m]$ (recall $v(0) = 0$ as we assume normalized functions).*

In what follows we adjust the definitions of additive, submodular, XOS, and subadditive functions in Definition 2.1 to the case of symmetric valuation functions. The simplified definition for XOS functions follows from the equivalence between XOS and fractional subadditivity [19].

Definition 2.3. *A symmetric valuation function v is said to be*

- additive *if $v(i) = a \cdot i$ for every integer $0 \leq i \leq m$ for some constant a.*
- submodular *if $v(i) - v(i-1) \geq v(i+1) - v(i)$ for every integer $1 \leq i \leq m-1$.*

- XOS *if* $v(i) \geq \frac{i}{j} \cdot v(j)$ *for any integers* $1 \leq i < j \leq m$.
- subadditive *if* $v(i) + v(j) \geq v(i+j)$ *for any integers* $1 \leq i, j \leq m$ *with* $i + j \leq m$.

We assume that the agents arrive sequentially. We will for the most part set static prices for the items; each arriving agent takes a bundle from the remaining items that maximizes her utility, with ties broken arbitrarily. For some results we will assume dynamic prices, i.e., the seller can set new prices for the remaining items for each iteration (but without knowing which agent will arrive next). If prices $\mathbf{p} = (p_1, \ldots, p_m)$ are set on the m items, and an agent buys a subset S of them, then her utility is given by $v(|S|) - \sum_{i \in S} p_i$. For most of the paper we will assume that the arrival order of the agents is unknown, but we will also consider settings where we know this order or where we even have control over the order. We are interested in the social welfare that we can obtain by setting prices in comparison to the optimal social welfare with respect to the worst case arrival order.

Due to space constraints, omitted results and proofs can be found in the full version of this paper [18].

3   Properties of Symmetric Functions

In this section, we consider properties of symmetric functions. In addition to being interesting in their own right, these properties will later help us establish welfare guarantees for posted prices (Theorem 8.4).

We are interested in approximating functions with "simpler" functions. Specifically, for two classes of functions $\mathcal{V}_1 \subseteq \mathcal{V}_2$, we want to determine the smallest constant c such that for any function $v \in \mathcal{V}_2$, there exists a function $\tilde{v} \in \mathcal{V}_1$ such that $v \leq \tilde{v} \leq cv$. We answer this question for each pair from the classes of subadditive, XOS, and submodular functions and show that the best constant is $c = 2$ for all of these pairs. (Note that since all three classes are closed under scalar multiplication, the inequality $v \leq \tilde{v} \leq cv$ above can also be replaced by $v/c \leq \tilde{v} \leq v$.) The details can be found in the full version of this paper [18].

4   Submodular Valuations

In this section we consider submodular valuations and establish bounds on the approximation ratio that can be obtained using different types of pricing.

4.1   Non-uniform Pricing

We first show that we can obtain 2/3 of the optimal welfare for submodular valuations if we are allowed to set non-uniform prices, and this bound is tight.

Theorem 4.1. *For every market with symmetric submodular valuations, there exists a static item pricing* **p** *that guarantees at least 2/3 of the optimal social welfare.*

Proposition 4.2. *There exists a market with two items and two buyers with symmetric submodular valuations such that every static pricing can guarantee a social welfare of at most 2/3 of the optimal social welfare.*

The negative result in Proposition 4.2 is obtained for a market with two items. In what follows we show that the guaranteed social welfare is higher when the number of items is large.

Theorem 4.3. *For every market of m items with symmetric submodular valuations, there exists a static item pricing* **p** *that guarantees at least $5/7 - 1/m$ of the optimal social welfare.*

The guarantee in Theorem 4.3 approaches $5/7 \approx 0.714$ as the number of items grows. The next theorem shows that this bound cannot exceed 0.802 even for an arbitrarily large number of items.

Theorem 4.4. *For every constant c, there exists a market with $m > c$ items with symmetric submodular valuations such that for any static item pricing* **p**, *the social welfare guaranteed by the pricing is at most 0.802 of the optimal social welfare.*

The next result shows that if we know the order of the agents beforehand (while having no control over this order), then we can extract the full optimal welfare.

Theorem 4.5. *For every market with symmetric submodular valuations with a known order of arrival, there exists a static pricing* **p** *that guarantees the optimal social welfare.*

4.2   Uniform Pricing

We now show that if we restrict ourselves to using uniform pricing with submodular valuations, we can still guarantee 1/2 of the optimal welfare. This bound is also tight.

Theorem 4.6. *For every market with symmetric submodular valuations, there exists a static uniform pricing p that guarantees at least 1/2 of the optimal social welfare.*

Proposition 4.7. *There exists a market with m items and two buyers with symmetric submodular valuations such that every uniform static pricing yields a social welfare of at most $\frac{m}{2m-1}(\approx \frac{1}{2})$ of the optimal social welfare.*

Proposition 4.8. *There exists a market with identical buyers with symmetric submodular valuation such that every uniform static pricing yields a social welfare of at most $\frac{n+1}{2n}(\approx \frac{1}{2})$ of the optimal social welfare.*

4.3 Dynamic Pricing

If we allow dynamic pricing, the following result shows that we can extract the full optimal welfare.

Theorem 4.9. *For every market with n agents with symmetric submodular valuations over m items, there exists a dynamic item pricing that guarantees the optimal social welfare.*

5 XOS Valuations

In this section we consider XOS valuations. We give upper bounds on the approximation ratio for both static and dynamic pricing.

Theorem 5.1. *There exists a market of m items and two agents with symmetric XOS valuations for which no static pricing yields more the $1 - 1/e$ of the optimal social welfare.*

Theorem 5.2. *There exists a market of three items and two agents with symmetric XOS valuations for which no dynamic pricing yields more the 5/6 of the optimal social welfare.*

6 Subadditive Valuations

In this section we consider subadditive valuations. Our main result of this section is the existence of a uniform price that guarantees at least 1/3 of the optimal welfare.

Theorem 6.1. *For every market of m items with symmetric subadditive valuations, there exists a uniform static item pricing $\mathbf{p}$ that guarantees at least 1/3 of the optimal social welfare.*

If there are two identical agents, this bound can be improved to 2/3.

Theorem 6.2. *For every market of m items and two identical agents with symmetric subadditive valuations, there exists a uniform static item pricing $\mathbf{p}$ that guarantees at least 2/3 of the optimal social welfare.*

The next propositions show that the bound in Theorem 6.1 cannot be improved to more than 1/2, and in the case of using only uniform pricing, cannot be improved to more than 1/3. Hence, this bound is tight for uniform pricing.

Proposition 6.3. *There is a market with symmetric subadditive valuations with m items and two agents such that no static pricing $\mathbf{p}$ guarantees more than 1/2 of the optimal social welfare.*

Proposition 6.4. *There is a market with symmetric subadditive valuations with m items and three agents such that no uniform static pricing $\mathbf{p}$ guarantees more than 1/3 of the optimal social welfare.*

In the case of two identical agents, the approximation cannot be improved to more than 3/4. In this special case, we can guarantee at least half of the social welfare by applying Theorem 7.3.

Proposition 6.5. *There exists a market of m items and two identical agents with a subadditive valuation such that no static pricing guarantees more the 3/4 of the optimal social welfare.*

If we use uniform pricing, we cannot guarantee more than 2/3 of the welfare for two identical agents. This means that the bound in Theorem 6.2 is tight.

Proposition 6.6. *There is a market with symmetric subadditive valuations with m items and two identical agents such that no uniform static pricing $\mathbf{p}$ guarantees more than 2/3 of the optimal social welfare.*

7   General Valuations

In this section we consider general valuations. While the analysis assumes monotonicity, all results hold even for non-monotone valuations: simply do all calculations based on the monotone closure of the valuations.

7.1   Worst-Case Ordering

We first show that for general valuations, we cannot guarantee more than $1/m$ of the optimal welfare even if we know the order of arrival, and this is tight.

Proposition 7.1. *There is a market with symmetric valuations over m items and two agents such that no static pricing $\mathbf{p}$ guarantees more than $1/m$ of the optimal social welfare even for a known order of arrival.*

Theorem 7.2. *For every market of m items, there exists a uniform static item pricing $\mathbf{p}$ that guarantees at least $1/m$ of the optimal social welfare.*

7.2   Best-Case Ordering

Next, we show that if we can choose the order of arrival, then we can guarantee at least half of the optimal welfare. We remark that when agents are identical, the order of arrival does not matter, and therefore our result holds for the setting with identical agents as well. This bound is also tight.

Theorem 7.3. *For every market of m items, there exists a uniform static item pricing $\mathbf{p}$ along with an order of arrival that guarantees at least 1/2 of the optimal social welfare.*

Proposition 7.4. *There exists a market of m items for which no static pricing and order of arrival yields more than 1/2 of the optimal social welfare.*

8   Bayesian Setting

In this section, we consider the Bayesian setting, where the valuation function of each agent is drawn independently from a distribution which can be different for different agents.

8.1   XOS Valuations

Feldman et al. [21] showed that if agents' valuations are drawn independently from a distribution over XOS valuation functions, then there exist prices that yield expected welfare at least half of the expected optimal welfare. These posted prices are non-uniform, even when the result is specialized to the case of identical items. We first restate Feldman et al.'s result and then show that if the items are identical, then the same bound can be obtained using uniform prices.

Theorem 8.1 [21]. *Let $\mathcal{F} = F_1 \times \cdots \times F_n$ be a product distribution over XOS valuation functions. For every $\mathbf{v} = (v_1, \ldots, v_n) \in \mathcal{F}$, let $X^*(\mathbf{v}) = (X_1^*(\mathbf{v}), \ldots, X_n^*(\mathbf{v}))$ be any allocation that maximizes the social welfare. Let $\mathbf{a} = (a_1, \ldots, a_n)$ be additive functions such that $v_i(S) \geq a_i(S)$ for any subset S of items, and $v_i(X_i^*) = a_i(X_i^*)$. When the items are offered at prices $p_j = E_{\mathbf{v} \in \mathcal{F}}[a_i(j)/2$ where $j \in X_i^*(\mathbf{v})]$, the expected social welfare is at least $OPT/2$.*

Theorem 8.2. *Let $\mathcal{F} = F_1 \times \cdots \times F_n$ be a product distribution over symmetric XOS valuation functions. Let OPT be the expected optimal social welfare. When all items are offered at the uniform price $OPT/(2m)$, the expected social welfare is at least $OPT/2$.*

8.2   Subadditive and General Valuations

We now define a notion that describes how close an arbitrary valuation function is to an XOS function and derive approximation results in terms of this closeness quantity. The proof of Theorem 8.3 follows the analysis presented by Feldman et al. [21].

Definition 8.1. *We say that a (not necessarily symmetric) valuation function v is c-close to XOS if there exists an XOS function $\tilde{v}$ such that for every set of item S, it holds that $v(S)/c \leq \tilde{v}(S) \leq v(S)$.*

Theorem 8.3. *For any product distribution $\mathcal{F}$ over (not necessarily symmetric) valuation functions that are c-close to XOS, there exist anonymous prices $\mathbf{p}$ that guarantee an expected social welfare of at least $1/(2c)$ of the optimal expected welfare.*

Since any symmetric subadditive function is 2-close to XOS (see Sect. 3), Theorem 8.3 implies that we can obtain at least 1/4 of the expected optimal welfare when the agents' valuations are drawn from a product distribution over subadditive valuations. In addition, using techniques similar to those in the proof of Theorem 8.2, we can achieve this with uniform prices.

Theorem 8.4. *Let $\mathcal{F} = F_1 \times \cdots \times F_n$ be a product distribution over symmetric subadditive valuation functions. Let OPT be the expected maximal social welfare. There exists a uniform price on the items for which the expected social welfare is at least $OPT/4$.*

Our results cease to hold for general valuations, even if we can control the order of arrival.

Proposition 8.5. *There is a market with n agents and $m = n^2$ items and a distribution over symmetric valuations such that no static pricing $\mathbf{p}$ yields expected welfare more than $\Theta(1/n)$ of the optimal expected welfare, even if we can control the arrival order.*

9   Discussion

In this paper, we study the fraction of the optimal social welfare that can be achieved via posted prices in markets with identical items under various assumptions on the designer's information and agents' valuations. We show that in the Bayesian setting, uniform posted prices can guarantee $1/2$ and $1/4$ of the optimal welfare for XOS and subadditive valuations, respectively. If the designer has full information on agents' valuations, then $1/3$ of the optimal welfare can be obtained via uniform prices for subadditive valuations. For general valuations, we exhibit a tight bound of $1/m$ for both uniform and non-uniform prices; on the other hand, if the designer can control the arrival order, then $1/2$ of the optimal welfare can be guaranteed for such valuations.

Our work sheds light on the power of uniform prices for settings with identical items. For submodular valuations in the full-information setting, there is a gap between the guarantee that can be obtained by uniform and non-uniform prices, while for XOS valuations in the Bayesian setting there is no gap. It would be interesting to determine whether such a gap exists for subadditive valuations, both for the full-information and the Bayesian setting. Finally, it also remains open whether the constant approximation guarantee provided here for subadditive valuations over identical items holds also for subadditive valuations over heterogeneous items. This problem has been raised by Feldman et al. [21] who provide a logarithmic (in m) bound for this setting.

References

1. Ausubel, L.M.: An efficient ascending-bid auction for multiple objects. Am. Econ. Rev. **94**(5), 1452–1475 (2004)
2. Babaioff, M., Blumrosen, L., Dughmi, S., Singer, Y.: Posting prices with unknown distributions. In: Proceedings of the 1st Innovations in Computer Science, pp. 166–178 (2011)
3. Babaioff, M., Dughmi, S., Kleinberg, R., Slivkins, A.: Dynamic pricing with limited supply. ACM Trans. Econ. Comput. **3**(1), 4:1–4:26 (2015)

4. Babaioff, M., Immorlica, N., Lucier, B., Weinberg, S.M.: A simple and approximately optimal mechanism for an additive buyer. In: Proceedings of the 55th IEEE Annual Symposium on Foundations of Computer Science, pp. 21–30 (2014)
5. Bateni, M.H., Dehghani, S., Hajiaghayi, M.T., Seddighin, S.: Revenue maximization for selling multiple correlated items. In: Bansal, N., Finocchi, I. (eds.) ESA 2015. LNCS, vol. 9294, pp. 95–105. Springer, Heidelberg (2015). https://doi.org/10.1007/978-3-662-48350-3_9
6. Bhawalkar, K., Roughgarden, T.: Simultaneous single-item auctions. In: Goldberg, P.W. (ed.) WINE 2012. LNCS, vol. 7695, pp. 337–349. Springer, Heidelberg (2012). https://doi.org/10.1007/978-3-642-35311-6_25
7. Blumrosen, L., Holenstein, T.: Posted prices vs. negotiations: an asymptotic analysis. In: Proceedings of the 9th ACM Conference on Electronic Commerce, p. 49 (2008)
8. Chawla, S., Hartline, J.D., Kleinberg, R.D.: Algorithmic pricing via virtual valuations. In: Proceedings of the 8th ACM Conference on Electronic Commerce, pp. 243–251 (2007)
9. Chawla, S., Hartline, J.D., Malec, D.L., Sivan, B.: Multi-parameter mechanism design and sequential posted pricing. In: Proceedings of the 42nd ACM Symposium on Theory of Computing, pp. 311–320 (2010)
10. Chawla, S., Malec, D.L., Sivan, B.: The power of randomness in Bayesian optimal mechanism design. In: Proceedings of the 11th ACM Conference on Electronic Commerce, pp. 149–158 (2010)
11. Christodoulou, G., Kovács, A., Schapira, M.: Bayesian combinatorial auctions. In: Aceto, L., Damgård, I., Goldberg, L.A., Halldórsson, M.M., Ingólfsdóttir, A., Walukiewicz, I. (eds.) ICALP 2008. LNCS, vol. 5125, pp. 820–832. Springer, Heidelberg (2008). https://doi.org/10.1007/978-3-540-70575-8_67
12. Cohen, I.R., Eden, A., Fiat, A., Jez, L.: Pricing online decisions: beyond auctions. In: Proceedings of the 26th Annual ACM-SIAM Symposium on Discrete Algorithms, pp. 73–91 (2015)
13. Cohen-Addad, V., Eden, A., Feldman, M., Fiat, A.: The invisible hand of dynamic market pricing. In: Proceedings of the 2016 ACM Conference on Economics and Computation, pp. 383–400 (2016)
14. Dütting, P., Feldman, M., Kesselheim, T., Lucier, B.: Posted prices, smoothness, and combinatorial prophet inequalities. In: Proceedings of the 58th IEEE Annual Symposium on Foundations of Computer Science, pp. 540–551 (2017)
15. Edelman, B., Ostrovsky, M., Schwarz, M.: Internet advertising and the generalized second-price auction: selling billions of dollars worth of keywords. Am. Econ. Rev. 97(1), 242–259 (2007)
16. Eden, A., Feige, U., Feldman, M.: Max-min greedy matching. arXiv preprint (2018). http://arxiv.org/abs/1803.05501
17. Eden, A., Feldman, M., Friedler, O., Talgam-Cohen, I., Weinberg, S.M.: A simple and approximately optimal mechanism for a buyer with complements. In: Proceedings of the 2017 ACM Conference on Economics and Computation, p. 323 (2017)
18. Ezra, T., Feldman, M., Roughgarden, T., Suksompong, W.: Pricing multi-unit markets. arXiv preprint (2018). http://arxiv.org/abs/1705.06623
19. Feige, U.: On maximizing welfare when utility functions are subadditive. SIAM J. Comput. 39(1), 122–142 (2009)
20. Feldman, M., Fu, H., Gravin, N., Lucier, B.: Simultaneous auctions are (almost) efficient. In: Proceedings of the 45th Symposium on Theory of Computing, pp. 201–210 (2013)

21. Feldman, M., Gravin, N., Lucier, B.: Combinatorial auctions via posted prices. In: Proceedings of the 26th Annual ACM-SIAM Symposium on Discrete Algorithms, pp. 123–135 (2015)
22. Friedman, M.: How to sell government securities. Wall Street J. A8 (1991)
23. Hajiaghayi, M.T., Kleinberg, R., Parkes, D.C.: Adaptive limited-supply online auctions. In: Proceedings of the 5th ACM Conference on Electronic Commerce, pp. 71–80 (2004)
24. Hassidim, A., Kaplan, H., Mansour, Y., Nisan, N.: Non-price equilibria in markets of discrete goods. In: Proceedings of the 12th ACM Conference on Electronic Commerce, pp. 295–296 (2011)
25. Kelso Jr., A.S., Crawford, V.P.: Job matching, coalition formation, and gross substitutes. Econometrica $50(6)$, 1483–1504 (1982)
26. Lehmann, B., Lehmann, D.J., Nisan, N.: Combinatorial auctions with decreasing marginal utilities. Games Econ. Behav. $55(2)$, 270–296 (2006)
27. Lucier, B., Paes Leme, R.: GSP auctions with correlated types. In: Proceedings of the 12th ACM Conference on Electronic Commerce, pp. 71–80 (2011)
28. Lucier, B., Paes Leme, R., Tardos, É.: On revenue in the generalized second price auction. In: Proceedings of the 21st World Wide Web Conference, pp. 361–370 (2012)
29. Markakis, E., Telelis, O.: Uniform price auctions: equilibria and efficiency. Theory Comput. Syst. $57(3)$, 549–575 (2015)
30. Nisan, N.: Algorithmic mechanism design through the lens of multi-unit auctions, Chap. 9. In: Young, H.P., Zamir, S. (eds.) Handbook of Game Theory with Economic Applications, vol. 4, pp. 477–515. Elsevier (2015)
31. Paes Leme, R., Tardos, É: Pure and Bayes-Nash price of anarchy for generalized second price auction. In: Proceedings of the 51st Annual IEEE Symposium on Foundations of Computer Science, pp. 735–744 (2010)
32. Parkes, D.C.: Online mechanisms, Chap. 16. In: Nisan, N., Roughgarden, T., Tardos, É., Vazirani, V. (eds.) Algorithmic Game Theory, pp. 411–439. Cambridge University Press (2007)
33. Ritter, J.: Google's IPO, 10 years later (2014). http://www.forbes.com/sites/jayritter/2014/08/07/googles-ipo-10-years-later. Accessed 09 Feb 2017
34. Varian, H.R.: Position auctions. Int. J. Ind. Org. $25(6)$, 1163–1178 (2007)

Optimal Pricing for MHR Distributions

Yiannis Giannakopoulos[1]([✉])([iD]) and Keyu Zhu[2]

[1] TU Munich, Munich, Germany
`yiannis.giannakopoulos@tum.de`
[2] Georgia Institute of Technology, Atlanta, GA, USA
`keyu.zhu@gatech.edu`

Abstract. We study the performance of anonymous posted-price selling mechanisms for a standard Bayesian auction setting, where n bidders have i.i.d. valuations for a single item. We show that for the natural class of Monotone Hazard Rate (MHR) distributions, offering the same, take-it-or-leave-it price to all bidders can achieve an (asymptotically) optimal revenue. In particular, the approximation ratio is shown to be $1 + O(\ln \ln n / \ln n)$, matched by a tight lower bound for the case of exponential distributions. This improves upon the previously best-known upper bound of $e/(e-1) \approx 1.58$ for the slightly more general class of regular distributions. In the worst case (over n), we still show a global upper bound of 1.35. We give a simple, closed-form description of our prices which, interestingly enough, relies only on minimal knowledge of the prior distribution, namely just the expectation of its second-highest order statistic.

1 Introduction

In this paper we study a traditional Myersonian auction setting: an auctioneer has an item to sell and he is facing n potential buyers. Each buyer has a (private) valuation for the item, and these valuations are i.i.d. according to some known continuous probability distribution F. You can think of this valuation, as modelling the amount of money that the buyer is willing to spend in order to get the item. An auction is a mechanism that receives as input a bid from each buyer, and then decides if the item is going to be sold and to whom, and for what price. Our goal is to design auctions that maximize the seller's expected revenue.

We focus only on truthful auctions, that is, selling mechanisms that give no incentives to the bidders to lie about their true valuation. Such auctions are both

---

Supported by the Alexander von Humboldt Foundation with funds from the German Federal Ministry of Education and Research (BMBF). Most of this work was done while the second author was visiting the Chair of Operations Research at TU Munich. A full version of this paper can be found at [18].

G. Christodoulou and T. Harks (Eds.): WINE 2018, LNCS 11316, pp. 154–167, 2018.
https://doi.org/10.1007/978-3-030-04612-5_11

conceptually and practically convenient. This restriction is essentially without loss for our revenue maximization objective, due to the Revelation Principle[1].

In general, such an optimal auction can be rather complicated and even randomized (aka a lottery). However, in his celebrated result, Myerson [24] proved that (under some standard assumptions on the valuations' distribution) revenue maximization can be achieved by a very simple deterministic mechanism, namely a second-price auction paired with a reserve value r. In such an auction, all buyers with bids smaller than r are ignored and the item is sold to the highest bidder for a price equal to the second-highest bid (or r, if no other bidder remains). Equivalently, you can think of this as the seller himself taking part in the auction, with a bid equal to r, and simply running a standard, Vickrey second-price auction; if the auctioneer is the winning bidder, then the item stays with him, that is, it remains unsold.

No matter how simple and powerful the above optimal auction seems, it still requires explicitly soliciting bids from all buyers and using the second-highest as the "critical payment"; this is essentially a centralized solution, that asks for a certain degree of coordination. Arguably, there is an even simpler selling mechanism which, as a matter of fact, is being used extensively in practice, known as *anonymous pricing*: the seller simply decides on a selling price p, and then the item goes to any buyer that can afford it (breaking ties arbitrarily); that is, we sell the item to any bidder with a valuation greater or equal to p, for a price of exactly p.

The question we investigate in this paper, is how well can such an extremely simple selling mechanism perform when compared to an arbitrary, optimal auction. We resolve this in a very positive way proving that, under natural assumptions on the valuation distribution, as the number of buyers grows large, anonymous pricing achieves optimal revenue. More precisely, its approximation ratio is $1 + O(\ln \ln n / \ln n)$. Furthermore, we show that in order to get such a near-optimal performance, the seller does not really need to have full knowledge of the bidders' population; he just needs to know the expectation of the second-highest order statistic of the valuation distribution, that is, (a good estimate of) the *expected* second-highest bid is enough.

1.1   Related Work

The seminal reference in auction theory is the work of Myerson [24] who completely characterized the revenue-maximizing auction in single-item settings with bidder valuations drawn from independent (but not necessarily identical) distribution. Under his standard regularity condition (see Footnote 4), this optimal auction has a very simple description when the valuation distributions are identical: it is a second-price auction with a reserve. Furthermore, there is an elegant, closed-form formula that gives the reserve price (see Sect. 2).

---

[1] In this paper we will avoid discussing such subtler issues as implementability and truthfulness, since our goal is to study the performance of specific and very simple pricing mechanisms. The interested reader is pointed to [25] as a good starting point for a deeper investigation of those ideas.

One can achieve good, constant approximations to that optimal revenue by using even simpler auctions, namely *anonymous pricing* mechanisms. These mechanisms offer the same take-it-or-leave-it price to all bidders, and the item is sold to someone who can afford it (breaking ties arbitrarily). An upper bound of $e/(e-1) \approx 1.58$ on the approximation ratio of anonymous pricing can be shown from the work of Chawla et al. [10]. Blumrosen and Holenstein [6] study the asymptotic performance of pricing when the number of bidders grows large and demonstrate a lower bound on the approximation ratio of $0.88/0.65 = 1.37$ for anonymous pricing[2]. If we allow for non-continuous distributions that have point-masses, then Dütting et al. [14] provide a matching lower bound of $e/(e-1)$. Although the class of MHR distributions (see Sect. 2) is a natural restriction of Myerson's regularity, that has been extensively studied in optimal auction theory, mechanism design and complexity to derive powerful positive results (see, e.g., [3,5,8,12,13,16,17,21]), no better bounds are known for anonymous pricing in this class. This is our goal in this paper.

Although not immediately related to our model, an important line of work studies the performance of "simple" auctions, such as pricing and auctions with reserves, for the more general case where bidders' valuations may be non-identically distributed. In such settings, the elegance of Myerson's characterization is not in effect any more, and the optimal auction can be rather complicated. Nevertheless, in an influential paper, Hartline and Roughgarden [21] showed that, for regular distributions, a second-price auction with a single anonymous reserve guarantees a 4-approximation to the optimal ratio, and also provided a lower bound of 2. This upper bound was subsequently improved to $e \approx 2.72$ by Alaei et al. [2], achieved even by the simpler class of anonymous pricing mechanisms. At the same paper, they also provided a lower bound of 2.23 for the approximation ratio of anonymous pricing for non-i.i.d. bidders.[3] For bounds on the approximation ratios between different pricing and reserve mechanisms, under various assumptions on the underlying distributions and the order of the bidders' arrival, see [2,10,14,20,22].

Finally, we briefly mention that there is a very rich theory about sequential pricing that deals with dynamically arriving buyers and which is inspired by and related to secretary-like online problems and the powerful theory of prophet inequalities. See, e.g., [1,9,11,19,23,26].

---

[2] As a matter of fact, one can use the techniques of Blumrosen and Holenstein [6] to get a slightly better lower bound of (at least) 1.4: a corollary of their work is that, for any $k > 1$, if the valuations are drawn from a Pareto distribution with cdf $F(x) = 1 - 1/x^k$ and the number of bidders grows arbitrarily large, then the separation between the optimal revenue and that of anonymous pricing is $\Gamma\left(\frac{k-1}{k}\right)\left(1 - \frac{1}{k}\right) / \frac{k}{e^{\eta(k)}}\eta(k)^{1-1/k}$, where Γ is the standard gamma function and $\eta(k)$ is the unique positive solution of equation $e^x = 1 + k \cdot x$. Optimizing this ratio over $k \in (1,2)$, we can get a lower bound greater than 1.403.

[3] This lower bound was very recently improved to 2.62 by Jin et al. [22].

1.2 Our Results

In this paper we study the performance of anonymous pricing mechanisms in single-item auction settings with n bidders that have i.i.d. valuations from the same MHR distribution F. These mechanisms are extremely simple: the seller simply offers the same take-it-or-leave-it price p to all potential buyers; the item is then sold to a buyer that can meet this price, that is, has a valuation greater or equal than p; the winning bidder pays p to the seller. Our benchmark is the seller's expected revenue (with respect to his incomplete, prior knowledge of the buyers' bids via distribution F) and we compare against the maximum revenue achievable by any auction. For our particular model, this optimal auction is a second-price auction with a reserve [24].

Our main result (Sect. 5; see also Fig. 1) is an explicit, closed-form upper bound on the approximation ratio of the revenue of anonymous pricing. As the number n of buyers grows large, this ratio tends to the optimal value of 1, at a rate of $1 + O(\ln \ln n / \ln)$ (Theorem 1). Additionally, we design an upper bound that is fine-tuned to handle also small values of n (Theorem 2), and using this we provide a global, worst-case (with respect to n) upper bound of 1.35 on the approximation ratio. Previously, only an upper bound of $e/(e-1) \approx 1.58$ was known (for any value of n), holding for the slightly more general class of regular distributions.

In Sect. 7 we demonstrate how the aforementioned positive guarantee on the revenue of anonymous pricing can still be (within an exponentially decreasing *additive* constant) achieved even if the seller does not have full knowledge of the prior distribution F (see Fig. 2). In particular (Theorem 4), we give an explicit formula for such a "good" pricing rule that only depends on the expectation of the second-highest order statistic of F.

Finally, in Sect. 6 we prove that our upper bound analysis is essentially tight, by showing that the exponential distribution provides an (almost) tight gap instance between the revenue of anonymous pricing and that of the optimal auction (Theorem 3; see also Fig. 2).

Our upper bound technique differs from related previous approaches [2, 10] in that we do not use the ex-ante relaxation of the revenue-maximization objective. Instead, we deploy explicit upper bounds on the optimal revenue (Sect. 3) that depend on key parameters of the valuation distribution F, namely its order statistics and its monopoly reserve. Then, we pair these with a range of critical properties of MHR distributions that we develop in Sect. 4. We believe that some of these auxiliary results may be of independent interest, in particular the order statistics tail-bound of Lemma 3 and the reserve-quantile optimal revenue bound of Lemma 4.

Due to space constraints, all omitted proofs can be found in the full version of our paper [18].

2   Model and Notation

A seller wants to sell a single item to $n \geq 2$ bidders. The valuations of the bidders for the item are i.i.d. from a continuous probability distribution supported over an interval $D_F \subseteq [0, \infty)$, with cdf F and pdf f. Throughout this paper we will assume that F has *Monotone Hazard Rate (MHR)*, that is, $\frac{f(x)}{1-F(x)}$ is monotonically nondecreasing with respect to $x \in D_F$. Equivalently, this means that $\ln(1-F)$ is a concave function. The MHR condition is a slight refinement of Myerson's standard regularity condition[4] that is still general enough to give rise to a wide family of natural distributions, like the uniform, exponential, normal and gamma. Intuitively, MHR distributions have exponentially decreasing tails. For an in-depth treatment of MHR distributions we refer to the book of Barlow and Proschan [4, Chap. 2].

For a random variable $X \sim F$ drawn from F and $1 \leq k \leq n$, we will use $X_{k:n}$ to denote the k-th lowest order statistic out of n i.i.d. draws from F. That is, $X_{1:n} \leq X_{2:n} \leq \cdots \leq X_{n:n}$. For completeness and ease of reference, we discuss some useful properties of order statistics in [18, Appendix A]. The exponential distribution will play a significant role in some parts of our paper; we denote it by $\mathcal{E}$, and its cdf and pdf are $F_{\mathcal{E}}(x) = 1 - e^{-x}$ and $f_{\mathcal{E}} = e^{-x}$, respectively. Finally, we use H_n to denote the n-th harmonic number $H_n = \sum_{i=1}^{n} \frac{1}{i}$, and $\gamma \approx 0.577$ for the Euler-Mascheroni constant (see also [18, Lemma 8]).

A pricing mechanism that offers a take-it-or-leave-it price of $p \in D_F$ to all bidders gives to the seller an expected revenue of

$$\text{PRICE}(F, n, p) \equiv p[1 - F^n(p)],$$

since the probability of no bidder being able to afford price p is $F^n(p)$. We will refer to such a mechanism simply as *(anonymous) pricing*. Thus, the optimal (maximum) revenue achievable via pricing is

$$\text{PRICE}(F, n) \equiv \sup_{p \in D_F} \text{PRICE}(F, n, p).$$

On the other hand, as discussed in the introduction, the optimal revenue attainable by any mechanism may be higher; as a matter of fact, Myerson [24] showed that it is achieved by a second-price auction with a reserve equal to the monopoly reserve $r^* = \text{argmax}_{r \in D_F} r(1 - F(r))$ of the valuation distribution. We denote this optimum revenue by $\text{MYERSON}(F, n)$, and it can be shown that

$$\text{MYERSON}(F, n) = \mathbb{E}\left[\max\{0, \phi(X_{n:n})\}\right],$$

where $\phi(x) \equiv x - \frac{1-F(x)}{f(x)}$ is the *virtual valuation* function of F (see also Footnote 6) and $X_{n:n}$ its maximum order statistic. Keep in mind that, due to the

---

[4] Regularity a la Myerson [24] requires the virtual valuation $x - \frac{1-F(x)}{f(x)}$ to be nondecreasing. Notice that this is a (strictly) weaker condition than MHR. For example, some Pareto distributions $\propto x^{-\alpha}$ with $\alpha \geq 2$ are regular but *not* MHR.

monotonicity of ϕ and the definition of the reserve r^*, we know that $\phi(x) \geq 0$ for all $x \geq r^*$.

Sometimes it is more convenient to work in quantile space instead of the actual valuation domain. More precisely, the quantile of distribution F corresponding to a value $x \in D_F$ is $q(x) = 1 - F(x)$. Using this, we can define what is known as the *revenue curve* of distribution F, by $R(q) = F^{-1}(1-q) \cdot q$. In other words, if $p \in D_F$ is a price and q is its corresponding quantile, then $R(q)$ is the expected revenue of selling the item to a single bidder, using a price p. Thus, the monopoly reserve quantile q^* that corresponds to the monopoly reserve r^* defined above is exactly the maximizer of the revenue curve $R(q)$. So, for a single bidder ($n = 1$):

$$\text{MYERSON}(F, 1) = \text{PRICE}(F, 1) = \sup_{p \in D_F} p(1 - F(p)) = \sup_{q \in [0,1]} R(q) = R(q^*).$$

In general though for more players ($n \geq 2$) this is not the case, and our goal in this paper is exactly to study how well the optimal revenue $\text{MYERSON}(F, n)$ can be approximated by pricing $\text{PRICE}(F, n)$. That is, we want to bound the following approximation ratio:

$$\text{APX}(F, n) \equiv \frac{\text{MYERSON}(F, n)}{\text{PRICE}(F, n)}.$$

Finally, we need to define an auxiliary function that will help us with stating and proving our main results. For any positive integer n, we define the function $g_n : [0, \infty) \longrightarrow [0, \infty)$ with

$$g_n(x) = x[1 - (1 - e^{-x})^n]. \tag{1}$$

Some properties of this function, that will be very useful to us in the following, are proven in [18, Appendix C].

3   Bounds on the Optimal Revenue

In this section we collect the bounds on the optimal revenue $\text{MYERSON}(F, n)$ that we will use for our main result in Sect. 5 to bound the approximation ratio of pricing. They rely on the fact that the valuation distribution is MHR. The first one is essentially a refinement of the well-known Bulow-Klemperer bound [7], and it was proven by Fu et al. [15]:

Lemma 1 (Fu et al. [15]). *For n bidders with i.i.d. values from an MHR*[5] *distribution F,*

$$\text{MYERSON}(F, n) \leq \mathbb{E}[X_{n-1:n}] + R(q^*)(1 - q^*)^{n-1},$$

where $X \sim F$ and R is the revenue curve of F and q^* is the quantile corresponding to the monopoly reserve price r^* of F, $q^* = 1 - F(r^*)$.

---

[5] As a matter of fact, this bound holds even for the more general class of regular distributions (see also the discussion in Sect. 2.).

Our second bound on the optimal revenue is a new one, that might also be of independent interest for future work:

Lemma 2. *For every MHR distribution F with monopoly reserve price r^* and quantile $q^* = 1 - F(r^*)$, and any positive integer n,*

$$\text{MYERSON}(F, n) \leq r^* \int_0^{q^*} \frac{1 - (1 - z)^n}{z} \, dz$$

4   Properties of MHR Distributions

In this section we state some properties of MHR distributions that will play a critical role into deriving our main results in the rest of the paper. The first in particular, Lemma 3, might be of independent interest, since it is providing powerful tail-bounds on with respect to the order statistics of the distribution:

Lemma 3. *For any continuous MHR random variable X, integers $1 \leq k \leq n$ and real $c \in [0, 1]$,*

$$\Pr\left[X < c \cdot \mathbb{E}[X_{k:n}]\right] \leq 1 - e^{-c(H_n - H_{n-k})}.$$

The next lemma states some useful bounds on the monopoly reserve of an MHR distribution:

Lemma 4. *For any MHR distribution with expectation μ, monopoly reserve r^* and corresponding quantile q^*:*

1. $q^* \geq 1/e$
2. $\ln(1/q^*) \cdot \mu \leq r^* \leq \frac{\ln(1/q^*)}{1-q^*} \cdot \mu$.

Finally, the following lemma shows that the high-order statistics of MHR distributions are "well-behaved", in the sense that they cannot be away from the expectation:

Lemma 5. *For any MHR random variable X and integer $n \geq 2$,*

$$\mathbb{E}[X_{n-1:n}] \geq \left(1 - \frac{H_n - 1}{n - 1}\right) \cdot \mathbb{E}[X].$$

5   Upper Bounds

This section is dedicated to proving the main result of our paper. First (Theorem 1) we show that pricing is indeed asymptotically optimal with respect to revenue and then (Theorem 2) we also provide a more refined upper-bound on the approximation ratio that is fine-tuned to work well for a small number of bidders n. As we will see in the following Sect. 6, our upper bound analysis of this section is essentially tight (see also Fig. 2).

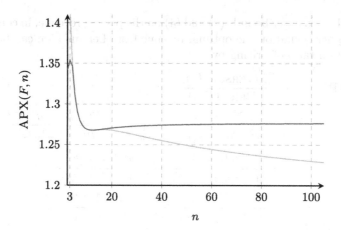

Fig. 1. The upper bounds on the approximation ratio APX(F, n) of anonymous pricing for n i.i.d. bidders with MHR valuations, given by Theorem 1 (blue) and Theorem 2 (red). The best (smallest) of the two converges to the optimal value of 1 as the number of bidders grows large, at a rate of $1 + O\left(\ln\ln n / \ln n\right)$. A single, unified plot of this can be seen in Fig. 2 (black), together with a matching lower bound (red). In the worst case ($n = 3$), our upper bound is at most 1.354. (Color figure online)

Theorem 1. *Using the same take-it-or-leave-it price, to sell an item to n buyers with i.i.d. valuations from a continuous MHR distribution F, is asymptotically optimal with respect to revenue. In particular,*

$$\mathrm{APX}(F, n) = 1 + O\left(\frac{\ln\ln n}{\ln n}\right).$$

A plot of the exact values[6] of this upper bound can be seen in Fig. 1 (blue).

Proof. First notice that by using the monopoly reserve price r^* of F as a take-it-or-leave it price to the n bidders, we get an expected revenue of

$$\mathrm{PRICE}(F, n, r^*) = r^*(1 - F(r^*)^n) = r^*[1 - (1 - q^*)^n] = R(q^*)\frac{1 - (1 - q^*)^n}{q^*},\quad (2)$$

where $q^* = 1 - F(r^*)$ is the quantile of the monopoly reserve price, for which we know that $q^* \geq \frac{1}{e}$ (Lemma 4), and R denotes the revenue curve (see Sect. 2).

Next, for simplicity denote $\nu = \mathbb{E}\left[X_{n-1:n}\right]$. For any real $c \in [0, 1]$, if we offer a price of $c \cdot \nu$ we have

$$\mathrm{PRICE}(F, n, c\nu) = c\nu[1 - F(c\nu)^n] \geq c\nu\left[1 - \left(1 - e^{-c(H_n - 1)}\right)^n\right],\quad (3)$$

the inequality holding due to Lemma 3 (for $k = n - 1$). Optimizing with respect to c we get that

$$\mathrm{PRICE}(F, n) \geq \frac{\nu}{H_n - 1} \max_{x \in [0, H_n - 1]} g_n(x).\quad (4)$$

---

[6] See (6) below.

Using the two lower bounds (2) and (4) on the pricing revenue, in conjunction with the upper bound on the optimal revenue from Lemma 1 we can bound the approximation ratio of pricing by

$$\text{APX}(F, n) = \frac{\text{MYERSON}(F, n)}{\text{PRICE}(F, n)}$$

$$\leq \frac{\nu}{\frac{\nu}{H_n - 1} \max_{x \in [0, H_n - 1]} g_n(x)} + \frac{R(q^*)(1 - q^*)^{n-1}}{R(q^*)\frac{1 - (1-q^*)^n}{q^*}}$$

$$= \frac{H_n - 1}{\max_{x \in [0, H_n - 1]} g_n(x)} + \frac{q^*(1 - q^*)^{n-1}}{1 - (1 - q^*)^n} \qquad (5)$$

$$\leq \frac{H_n - 1}{\max_{x \in [0, H_n - 1]} g_n(x)} + \frac{(e - 1)^{n-1}}{e^n - (e - 1)^n} \qquad (6)$$

$$= 1 + O\left(\frac{\ln \ln n}{\ln n}\right) + O\left(\left(\frac{e}{e - 1}\right)^{-n}\right). \qquad (7)$$

Equation (6) holds by observing that function $x \mapsto \frac{x(1-x)^{n-1}}{1-(1-x)^n}$ is decreasing over $(0, 1]$, for any $n \geq 2$, and taking into consideration that $q^* \geq 1/e$, while for (7) we make use of the asymptotics from [18, Lemma 12]. The upper bound given by (6) is plotted by the blue line in Fig. 1.

Theorem 2. *The approximation ratio of the revenue obtained by using the same take-it-or-leave-it price, to sell an item to n buyers with i.i.d. valuations from a continuous MHR distribution F, is at most*

$$\text{APX}(F, n) \leq \max_{q \in [1/e, 1]} \min \left\{ \frac{1}{1 - (1 - e^{-H_n + 1})^n} + \frac{q(1 - q)^{n-1}}{1 - (1 - q)^n}, \frac{\int_0^q \frac{1 - (1-z)^n}{z} dz}{1 - (1 - q)^n} \right\}.$$

In particular, the worst case (maximum) of this quantity is attained at $n = 3$ and is at most $\text{APX}(F, 3) \leq 1.354$. A plot of the exact values of this upper bound can be seen in Fig. 1 (red).

Proof. From (5) in the proof of Theorem 1 we can get the following upper bound on the approximation ratio, by using (possibly suboptimally) $x \leftarrow H_n - 1$ for the maximization operator:

$$\text{APX}(F, n) \leq \frac{H_n - 1}{g_n(H_n - 1)} + \frac{q^*(1 - q^*)^{n-1}}{1 - (1 - q^*)^n} = \frac{1}{1 - (1 - e^{-H_n + 1})^n} + \frac{q^*(1 - q^*)^{n-1}}{1 - (1 - q^*)^n}.$$

On the other hand, using the reserve price of F as a price and combining the guarantee of (2) with the upper bound on the optimal revenue from Lemma 2, gives us

$$\text{APX}(F, n) \leq \frac{r^* \int_0^{q^*} \frac{1 - (1-z)^n}{z} dz}{R(q^*)\frac{1 - (1-q^*)^n}{q^*}} = \frac{\int_0^{q^*} \frac{1 - (1-z)^n}{z} dz}{1 - (1 - q^*)^n},$$

since $R(q^*) = r^* q^*$. Recalling that $q^* \in [1/e, 1]$ and taking the best (i.e., minimum) of the two bounds above, finishes the proof.

6   Lower Bound

The lower bound instance of this section (Theorem 3) shows that our main positive result for the approximation ratio of pricing under MHR distributions in Theorem 1 is essentially tight (see also Fig. 2). It is achieved by an exponential distribution instance. Before proving it, we need the following auxiliary lemma about the maximizers of functions g_n that we introduced in (1).

Lemma 6. *For any positive integer n, function g_n (defined in (1)) has a unique maximizer. Furthermore, for all $n \geq 17$,*

$$\operatorname*{argmax}_{x \geq 0} g_n(x) \leq H_n - 1.$$

Theorem 3. *For $n \geq 2$ bidders with exponentially i.i.d. valuations, the approximation ratio of anonymous pricing is at least*

$$APX(\mathcal{E}, n) \geq \frac{H_n - 1}{\max_{x \geq 0} g_n(x)},$$

where function g_n is defined in (1). A plot of this lower bound can be seen in Fig. 2 (red). In particular, the upper bound derived in the proof of Theorem 1 is tight (up to an exponentially vanishing additive factor).

Proof. Let $X \sim \mathcal{E}$ be an exponential random variable. Then, we have

$$\textsc{Myerson}(\mathcal{E}, n) \geq \mathbb{E}\left[X_{n-1:n}\right] = H_n - 1$$

and

$$\textsc{Price}(\mathcal{E}, n) = \sup_{x \geq 0} x \left[1 - \left(1 - e^{-x}\right)^n\right] = \max_{x \geq 0} g_n(x).$$

Putting the above together, we get the desired lower bound on the approximation ratio.

For the tightness, we need to show that our lower bound is within an additive, exponentially decreasing factor of the upper bound given in (6). Since the second term in (6) is at most $O\left(\left(\frac{e}{e-1}\right)^{-n}\right)$, it is enough to show that, for a sufficiently large number of bidders n,

$$\max_{x \in [0, \infty)} g_n(x) = \max_{x \in [0, H_n - 1]} g_n(x).$$

This is exactly what we proved in Lemma 6, for any $n \geq 17$.

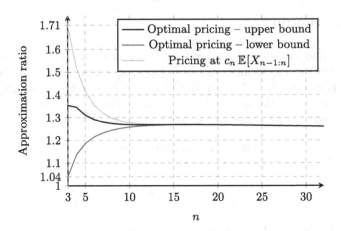

Fig. 2. Bounds on the approximation ratio of anonymous pricing for $n = 3, \ldots, 30$ i.i.d. bidders with MHR valuations: the upper bound on optimal pricing (black) derived in Sect. 5 (see also Fig. 1), the lower bound (red) given by Theorem 3, and the upper bound of pricing at the expected value of the second-highest order statistic, scaled down by parameter c_n (blue), given in Theorem 4 of Sect. 7. They are all (asymptotically) optimal, their (additive) difference decreasing exponentially fast. They all converge to the optimal value of 1 at a rate of $1 + O(\ln \ln n / \ln n)$. (Color figure online)

7   Explicit Prices – Knowledge of the Distribution

Our main result from Sect. 5 demonstrates that a seller, facing n bidders with i.i.d. valuations from and MHR distribution F, can achieve (asymptotically) optimal revenue by using just an anonymous, take-it-or-leave-it price. Taking a careful look within the proof of Theorem 1, we see that this upper bound is derived by comparing the optimal Myersonian revenue (via the bound provided by Lemma 1) to that of two different anonymous pricings; namely, first (see (2)) we use the monopoly reserve r^* of F, and then (see (3) and (4)) a multiple of the expectation $\nu = \mathbb{E}[X_{n-1:n}]$ of the second-highest order statistic of F, in particular $c_n \cdot \nu$ where

$$c_n = \frac{\mathrm{argmax}_{x \in [0, H_n - 1]} g_n(x)}{H_n - 1}. \tag{8}$$

Although the latter price requires only the knowledge of $\nu = \mathbb{E}[X_{n-1:n}]$, that is not the case for the former; determining the reserve price r^* demands, in general, a detailed knowledge of the distribution F: it is the maximizer of $r(1 - F(r))$.

As a result, we would ideally like to provide a more robust solution, that would still provide optimality but depend only in limited information about F. If we pay even closer attention to the proof of Theorem 1, and the derivation of (7) in particular, we will see that the summand of our upper bound that corresponds to the pricing using r^* is exponentially decreasing, according to $\left(\frac{e}{e-1}\right)^{-n}$. Therefore, if we could show that the expected revenue achieved by using the other price $c_n\nu$ is within a constant factor from that of using r^*, then we could deduce that using only price $c_n\nu$ has an insignificant effect on the approximation ratio of pricing. We do exactly that in the following lemma:

Lemma 7. *For $n \geq 2$ bidders with i.i.d. valuations from an MHR distribution F with monopoly reserve r^* and parameters $c_n \in [0,1]$ given by (8),*

$$\text{PRICE}(F, n, c_n \cdot \mathbb{E}[X_{n-1:n}]) \geq (1 - o(1))\frac{e-1}{e} \cdot \text{PRICE}(F, n, r^*),$$

where $X \sim F$.

Proof. For convenience, denote $\mu = \mathbb{E}[X]$ and $\nu = \mathbb{E}[X_{n-1:n}]$. By the proof of Theorem 1 (see (3) and (4)) we know that by offering an anonymous price of $c_n \cdot \nu$ gives us an expected revenue of at least

$$\text{PRICE}(F, n, c_n \cdot \nu) \geq \frac{\nu}{H_n - 1} \max_{x \in [0, H_n - 1]} g_n(x) \geq \frac{\max_{x \in [0, H_n - 1]} g_n(x)}{H_n - 1} \frac{n - H_n}{n - 1} \cdot \mu,$$

the second inequality holding due Lemma 5.

On the other hand, from (2) we know that using the reserve price r^* as an anonymous price to all bidders gives an expected revenue of at most

$$\text{PRICE}(F, n, r^*) = r^*[1 - (1 - q^*)^n] \leq \frac{\ln(1/q^*)}{1 - q^*}[1 - (1 - q^*)^n] \cdot \mu,$$

the inequality holding due to Lemma 4.

Putting everything together, we finally get that

$$\begin{aligned}
\frac{\text{PRICE}(F, n, r^*)}{\text{PRICE}(F, n, c_n\nu)} &\leq \frac{\ln(1/q^*)}{1 - q^*}[1 - (1 - q^*)^n]\frac{n - 1}{n - H_n}\frac{H_n - 1}{\max_{x \in [0, H_n - 1]} g_n(x)} \quad (9)\\
&\leq \frac{e}{e - 1}\frac{n - 1}{n - H_n}\frac{H_n - 1}{\max_{x \in [0, H_n - 1]} g_n(x)}\\
&\leq (1 + o(1))\frac{e}{e - 1}.
\end{aligned}$$

The second inequality holds because $\frac{\ln(1/q^*)}{1-q^*}[1 - (1 - q^*)^n] \leq \frac{\ln(1/q^*)}{1-q^*} \leq \frac{e}{e-1}$, since function $x \mapsto \frac{\ln(1/x)}{1-x}$ is decreasing for $x > 0$ and $q^* \geq 1/e$ (from Property 1 of Lemma 4). The last inequality is a consequence of [18, Lemma 12] and the fact that $H_n \leq \ln(n) + 1$.

As discussed before, Lemma 7 shows us that there indeed exists an anonymous price that depends on the knowledge of only the expectation of the second-order statistic of the valuation distribution and which, furthermore, guarantees an (asymptotically) optimal revenue. We can even provide a closed-form upper bound for it:

Theorem 4. *Let F be an MHR distribution and $X_{n-1:n}$ denote the second-highest, out of n i.i.d. draws from F. Then, using an anonymous price of $c_n \cdot$*

$\mathbb{E}[X_{n-1:n}]$, where c_n is given in (8), to sell an item to $n \geq 2$ bidders with i.i.d. valuations from F, guarantees a revenue with approximation ratio of at most

$$\frac{\text{MYERSON}(F,n)}{\text{PRICE}(F,n,c_n\,\mathbb{E}[X_{n-1:n}])} \leq \frac{H_n-1}{\max_{x\in[0,H_n-1]}g_n(x)}\left[1+\frac{1}{e}\frac{n-1}{n-H_n}\left(\frac{e-1}{e}\right)^{n-2}\right].$$

A plot of this upper bound can be seen in Fig. 2 (blue).

Proof. Simulating the proof of the approximation upper bound in Theorem 1, but now using (9) to approximate $\text{PRICE}(F,n,r^*)$ by $\text{PRICE}(F,n,c_n\,\mathbb{E}[X_{n-1:n}])$, the derivation in (5) gives us that

$$\frac{\text{MYERSON}(F,n)}{\text{PRICE}(F,n,c_n\,\mathbb{E}[X_{n-1:n}])} \leq \frac{H_n-1}{\max_{x\in[0,H_n-1]}g_n(x)}$$

$$+ \frac{\ln(1/q^*)}{1-q^*}[1-(1-q^*)^n]\frac{n-1}{n-H_n}\frac{H_n-1}{\max_{x\in[0,H_n-1]}g_n(x)}\cdot\frac{q^*(1-q^*)^{n-1}}{1-(1-q^*)^n}$$

$$= \frac{H_n-1}{\max_{x\in[0,H_n-1]}g_n(x)}\left[1+\frac{n-1}{n-H_n}\ln(1/q^*)q^*(1-q^*)^{n-2}\right]$$

$$\leq \frac{H_n-1}{\max_{x\in[0,H_n-1]}g_n(x)}\left[1+\frac{1}{e}\frac{n-1}{n-H_n}\left(\frac{e-1}{e}\right)^{n-2}\right],$$

the last inequality coming from [18, Lemma 11], together with the fact that $q^* \in [1/e,1]$ (see Property 1 of Lemma 4).

References

1. Alaei, S.: Bayesian combinatorial auctions: expanding single buyer mechanisms to many buyers. SIAM J. Comput. **43**(2), 930–972 (2014)
2. Alaei, S., Hartline, J., Niazadeh, R., Pountourakis, E., Yuan, Y.: Optimal auctions vs. anonymous pricing. In: Proceedings of IEEE 56th Annual Symposium on Foundations of Computer Science, FOCS, pp. 1446–1463 (2015)
3. Babaioff, M., Blumrosen, L., Dughmi, S., Singer, Y.: Posting prices with unknown distributions. ACM Trans. Econ. Comput. **5**(2), 13:1–13:20 (2017)
4. Barlow, R.E., Proschan, F.: Mathematical Theory of Reliability. Society for Industrial and Applied Mathematics, Philadelphia (1996)
5. Bhattacharya, S., Goel, G., Gollapudi, S., Munagala, K.: Budget constrained auctions with heterogeneous items. In: Proceedings of the 42nd ACM symposium on Theory of Computing, pp. 379–388 (2010)
6. Blumrosen, L., Holenstein, T.: Posted prices vs. negotiations: an asymptotic analysis. In: Proceedings of the 9th ACM Conference on Electronic Commerce, EC, p. 49 (2008)
7. Bulow, J., Klemperer, P.: Auctions versus negotiations. Am. Econ. Rev. **86**(1), 180–194 (1996)
8. Cai, Y., Daskalakis, C.: Extreme-value theorems for optimal multidimensional pricing. Proceedings of IEEE 52nd Annual Symposium on Foundations of Computer Science, FOCS, pp. 522–531 (2011)

9. Cesa-Bianchi, N., Gentile, C., Mansour, Y.: Regret minimization for reserve prices in second-price auctions. IEEE Trans. Inf. Theory **61**(1), 549–564 (2015)

10. Chawla, S., Hartline, J.D., Malec, D.L., Sivan, B.: Multi-parameter mechanism design and sequential posted pricing. In: Proceedings of the 42nd ACM symposium on Theory of Computing, STOC, pp. 311–320 (2010)

11. Correa, J., Foncea, P., Hoeksma, R., Oosterwijk, T., Vredeveld, T.: Posted price mechanisms for a random stream of customers. In: Proceedings of the 18th ACM Conference on Economics and Computation, EC, pp. 169–186 (2017)

12. Daskalakis, C., Weinberg, S.M.: Symmetries and optimal multi-dimensional mechanism design. In: Proceedings of the 13th ACM Conference on Electronic Commerce, EC, pp. 370–387 (2012)

13. Dhangwatnotai, P., Roughgarden, T., Yan, Q.: Revenue maximization with a single sample. Games Econ. Behav. **91**(C), 318–333 (2014)

14. Dütting, P., Fischer, F.A., Klimm, M.: Revenue gaps for discriminatory and anonymous sequential posted pricing. CoRR abs/1607.07105 (2016). http://arxiv.org/abs/1607.07105v1

15. Fu, H., Immorlica, N., Lucier, B., Strack, P.: Randomization beats second price as a prior-independent auction. In: Proceedings of the 16th ACM Conference on Economics and Computation, EC, p. 323 (2015)

16. Giannakopoulos, Y., Koutsoupias, E., Lazos, P.: Online market intermediation. In: Chatzigiannakis, I., Indyk, P., Kuhn, F., Muscholl, A. (eds.) 44th International Colloquium on Automata, Languages, and Programming, ICALP. LIPIcs, vol. 80, pp. 47:1–47:14 (2017)

17. Giannakopoulos, Y., Kyropoulou, M.: The VCG mechanism for Bayesian scheduling. ACM Trans. Econ. Comput. **5**(4), 19:1–19:16 (2017)

18. Giannakopoulos, Y., Zhu, K.: Optimal pricing for MHR distributions. CoRR abs/1810.00800, October 2018. http://arxiv.org/pdf/1810.00800.pdf

19. Hajiaghayi, M.T., Kleinberg, R.D., Sandholm, T.: Automated online mechanism design and prophet inequalities. In: Proceedings of the 22nd Conference on Artificial Intelligence, AAAI (2007)

20. Hartline, J.D.: Mechanism design and approximation, Chap. 4, Manuscript (2017). http://jasonhartline.com/MDnA/

21. Hartline, J.D., Roughgarden, T.: Simple versus optimal mechanisms. In: Proceedings of the 10th ACM Conference on Electronic Commerce, EC, pp. 225–234 (2009)

22. Jin, Y., Lu, P., Tang, Z.G., Xiao, T.: Tight revenue gaps among simple mechanisms. CoRR abs/1804.00480, April 2018. http://arxiv.org/abs/1804.00480

23. Kleinberg, R., Weinberg, S.M.: Matroid prophet inequalities. In: Proceedings of the 44th Annual ACM Symposium on Theory of Computing, STOC, pp. 123–136 (2012)

24. Myerson, R.B.: Optimal auction design. Math. Oper. Res. **6**(1), 58–73 (1981)

25. Nisan, N.: Introduction to mechanism design (for computer scientists), Chap. 9. In: Nisan, N., Roughgarden, T., Tardos, É., Vazirani, V. (eds.) Algorithmic Game Theory. Cambridge University Press, Cambridge (2007)

26. Yan, Q.: Mechanism design via correlation gap. In: Proceedings of the 22nd Annual ACM-SIAM Symposium on Discrete Algorithms, SODA, pp. 710–719 (2011)

Learning Convex Partitions and Computing Game-Theoretic Equilibria from Best Response Queries

Paul W. Goldberg[(✉)] and Francisco J. Marmolejo-Cossío

University of Oxford, Oxford, UK
{paul.goldberg,francisco.marmolejo}@cs.ox.ac.uk

Abstract. Suppose that an m-simplex is partitioned into n convex regions having disjoint interiors and distinct labels, and we may learn the label of any point by querying it. The learning objective is to know, for any point in the simplex, a label that occurs within some distance ε from that point. We present two algorithms for this task: Constant-Dimension Generalised Binary Search (CD-GBS), which for constant m uses $poly(n, \log\left(\frac{1}{\varepsilon}\right))$ queries, and Constant-Region Generalised Binary Search (CR-GBS), which uses CD-GBS as a subroutine and for constant n uses $poly(m, \log\left(\frac{1}{\varepsilon}\right))$ queries. We show via Kakutani's fixed-point theorem that these algorithms provide bounds on the best-response query complexity of computing approximate well-supported equilibria of bimatrix games in which one of the players has a constant number of pure strategies.

Keywords: Query protocol · Equilibrium computation Revealed preferences

1 Introduction

The computation of game-theoretic equilibria is a topic of long-standing interest in the algorithmic and AI communities. This includes computation in the "classical" setting of complete information about a game, as well as settings of partial information, communication-bounded settings, and distributed algorithms (for example, best-response dynamics). A recent line of research has studied computation of equilibria based on query access to players' payoff functions. That work, along with the notion of revealed preferences in economics, inspires the new setting we study here.

We study algorithms that have query access to the players' best-response behaviour: an algorithm may query a mixed-strategy profile (i.e. probability distributions constructed by the algorithm, over each player's pure strategies) and

Full Online Version of Paper: https://arxiv.org/abs/1807.06170.

F. J. Marmolejo Cossío—Supported by the Mexican National Council of Science and Technology (CONACyT).

© Springer Nature Switzerland AG 2018
G. Christodoulou and T. Harks (Eds.): WINE 2018, LNCS 11316, pp. 168–187, 2018.
https://doi.org/10.1007/978-3-030-04612-5_12

learn the players' best responses. Our focus is on standard bimatrix games, which is arguably the most natural starting-point for an investigation of this new query model. The solution concept of interest is ε-approximate Nash equilibria (exact equilibria are typically impossible to find using finitely many such queries). A basic challenge is to identify algorithms that achieve this goal with good bounds on their query complexity (and also, ideally, their runtime complexity).

In more detail, we assume an $m \times n$ game G: a row player has m pure strategies and a column player has n pure strategies. G has two unknown $m \times n$ payoff matrices that represent payoffs to the players for all combinations of pure strategy choices they may make. A query consists of a probability distribution over the pure strategies of one of the players, and elicits an answer consisting of a best response for the other player (i.e. a pure strategy that maximises that player's expected payoff). We seek an ε-well-supported Nash equilibrium (ε-WSNE): a pair of probability distributions over their pure strategies with the property that any strategy of player p whose expected payoff is more than ε below the value of p's best response, gets probability zero. The general question of interest is: how many queries are needed, as a function of m, n, ε.

Using Kakutani's fixed point theorem, we reduce this question to a novel and more geometrical challenge in the design of query protocols. Suppose that the m-simplex Δ^m is partitioned into n convex regions having labels in $[n] = \{1, \ldots, n\}$. When we query a point $x \in \Delta^m$ we are told the label of x. How many queries (in terms of m, n, ε) are needed in order to ensure that all points in Δ^m are within ε of a point whose label we know? We show how to achieve this using time and queries polynomial in $\log \varepsilon$ and $\max(m, n)$ provided that $\min(m, n)$ is constant. This leads to a polynomial query complexity algorithm for 2-player games, provided that one of the players has a constant number of strategies.

1.1   Further Details

In essence, we consider partitions of the unit m-simplex Δ^m into n convex polytopes, $P_1, \ldots, P_n$, with disjoint interiors, and aim to approximately learn the partition with access to a membership oracle that for a given $x \in \Delta^m$, returns a polytope to which x belongs. The notion of approximation we study is that of ε-*close labellings*, a collection of empirical polytopes, $\{\widehat{P}_i\}_{i=1}^{n}$, such that $\widehat{P}_i \subseteq P_i$ for $i = 1, \ldots, n$ and $\cup_{i=1}^{n} \widehat{P}_i$ is an ε-net of $\Delta^m \subset \mathbb{R}^m$ in the ℓ_2 norm.

Note that in one dimension ($m = 1$) we can use binary search to solve this problem using $n \log(1/\varepsilon)$ queries. We generalise to higher dimension, exploiting convexity of the regions to reduce query usage in computing ε-close labellings. We present two algorithms for this task: Constant-Dimension Generalised Binary Search (CD-GBS), which for constant m uses $poly(n, \log\left(\frac{1}{\varepsilon}\right))$ queries, and Constant-Region Generalised Binary Search (CR-GBS), which uses CD-GBS as a subroutine and for constant n uses $poly(m, \log\left(\frac{1}{\varepsilon}\right))$ queries.

This problem derives from the question of how to compute approximate (well-supported) Nash equilibra (ε-WSNE) using only best response information, obtained via queries in which the algorithm selects a mixed strategy profile and a player, and receives a best response for that player to the mixed profile. Via

Kakutani's fixed-point theorem [18] we reduce this variant of equilibrium computation to finding ε-close labellings of polytope partitions. For $m \times n$ games where m is constant (or n equivalently, by symmetry), we show that an ε-WSNE can be computed using $poly(n, \log\left(\frac{1}{\varepsilon}\right))$ best response queries.

1.2   Related Work

Earlier work in computational learning theory has studied exact learning of geometrical regions over a discretised domain, where algorithms are sought with query complexity logarithmic in a resolution parameter and binary search is repeatedly applied in a systematic way [6]. Goldberg and Kwek [12] specifically study the learnability of polytopes in this context, deriving query efficient algorithms, and precisely classifying polytopes learnable in this setting. These algorithms can be adapted to approximately learn a single polytope with membership queries, but the obtained notion of approximation is not directly applicable to computing ε-close labellings.

The Nash equilibrium (NE) is a fundamental concept in game theory [21]. They are guaranteed to exist in finite games, yet computational challenges in finding one abound, most notably, the PPAD-completeness of computing an exact equilibrium even for two-player normal form games [7,9]. For this reason, query complexity has been extensively used as a tool to differentiate hardness of equilibrium concepts in games. For payoff queries, some notable examples include: exponential lower bounds for randomised computation of exact Nash equilibria and exact correlated equilibria via communication complexity lower bounds in multiplayer games [16,17]; exponential lower bounds for randomised computation of approximate well-supported equilibria and general approximate equilibria for a small enough approximation factor in multiplayer games [1]; upper and lower bounds for equilibrium computation in bimatrix games, congestion games [10] and anonymous games [15]; upper and lower bounds for randomised algorithms computing approximate correlated equilibria [13]. Babichenko et al. have also proved lower bounds in communication complexity for computing ε-WSNE for small enough ε in both bimatrix and multiplayer games [2].

Best response queries are a weaker but natural query model which is powerful enough to implement fictitious play, a dynamic first proposed by Brown [5], and proven to converge by Robinson [22] in two-player zero-sum games to an approximate NE. Fictitious play does not always converge for general games where both players have more than two strategies [11]. Furthermore, Daskalakis and Pan have proven that the rate of convergence of the dynamic is quite slow in the worst case (with arbitrary tie-breaking) [8]. Also, beyond non-convergence, the dynamic can have a poor approximation value for general games [14]. In addition, the relationship between best responses and convex partitions of simplices has been studied by Von Stengel [23] in the context of sequential games where one player has to commit to and announce a strategy before playing.

For a bimatrix game, simple ε-close labellings can be constructed by querying best responses at mixed strategies arising as uniform distributions over sufficiently large multisets of pure strategies. As a consequence of our main theorem, best responses to these multiset distributions contain enough information to compute approximate WSNE. This result is in the spirit of [3] and [20], who aim to quantify specific k such that some approximate equilibrium arises as a uniform mixture over multisets of size k. We note in our scenario that there is also a guaranteed existence of an approximate equilibrium using sufficiently large multisets, however *verifying* that a *specific* pair of mixed strategies is an approximate WSNE is not straightforward using only best response queries. This is in contrast to the verification of approximate equilibria via utility queries as studied in [3].

Separately, we note that the present paper is possibly relevant to the search for a price equilibrium in certain markets. Baldwin and Klemperer study markets consisting of *strong-substitutes* demand functions for N different goods available in multiple discrete units [4]. These markets are a generalisation of the *product-mix auction* of [19]; a basic task is to identify prices at which some desired bundle of the goods is demanded. Consider the space $(\mathbb{R}^+)^N$ of all price vectors. As analysed in [4], a strong-substitutes demand function partitions this price space into convex polytopes, each of which comprises the prices at which some particular bundle of goods is demanded. So, the present paper relates to a setting where price vectors may be queried, and responses consist of demand bundles. The connection is imperfect, since the main objective in the context of [4] would be to learn a price at which some target bundle is demanded, rather than the entire demand function. The ideas here may be useful for learning the values that the market has for various bundles.

2 Preliminaries and Notation

Our main object of study will be families of polytopes that precisely cover the unit simplex, with the property that any two distinct polytopes from the family are either disjoint, meet at their boundary, or entirely coincide. Throughout, the polytopes we work with are convex.

Definition 1 ((m, n)-Polytope Partition). *A (m, n)-polytope partition is a set of n convex polytopes, $\mathscr{P} = \{P_1, ..., P_n\}$ such that $\bigcup P_i = \Delta^m = \{x \in \mathbb{R}^m \mid \forall i, \ x_i \geq 0, \ \sum_i x_i \leq 1\}$ and for each $i \neq j$, either $relint(P_i) \cap relint(P_j) = \emptyset$ or $P_i = P_j$ (relint(H) being the relative interior of H).*

Definition 2 (Cross-sections and Slices). *Let $P \subset \mathbb{R}^m$ be a polytope and $\pi : \mathbb{R}^m \to \mathbb{R}$ the projection function into the first coordinate. For $x \in \mathbb{R}$, we define the x-cross-section of P as $P^x = \pi^{-1}(x) \cap P$. For any $I = [x, y] \subset \mathbb{R}$ we define the $[x, y]$-slice of P as $P^I = P^{x,y} = \pi^{-1}([x, y]) \cap P$. Suppose that $\mathscr{P} = \{P_i\}_i$ is an (m, n)-polytope partition. The definitions of cross-sections and slices extend to $\mathscr{P}^x = \{P_i^x\}_i$ and $\mathscr{P}^I = \mathscr{P}^{x,y} = \{P_i^{x,y}\}_i$.*

Figure 1 gives a visualisation of these two definitions. Notice that in the same figure, $\mathscr{P}^x$ is essentially a lower-dimensional polytope partition linearly scaled

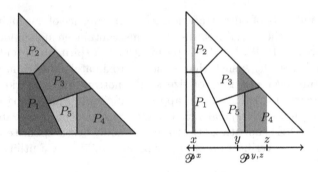

Fig. 1. Polytope partition, cross-section and slices.

by a factor of $(1 - x)$. This however, is not the case in general. We distinguish between these two scenarios with the following formal definition:

Definition 3 (Non-Degenerate and Degenerate cross-sections). *Let $\mathscr{P}$ be an (m, n)-polytope partition. For $x \in [0, 1)$ let $f_x : \mathscr{P}^x \to \Delta^{m-1}$ be defined as $f_x(v_1, ..., v_m) = \frac{1}{1-x}(v_2, ..., v_m)$. If $f_x(\mathscr{P}^x)$ is an $(m - 1, n)$-polytope partition, we say that $\mathscr{P}^x$ is a non-degenerate cross-section. Otherwise we say that $\mathscr{P}^x$ is a degenerate cross-section.*

The recursive structure of polytope partitions on non-degenerate cross-sections will be crucial to our constructions. Luckily, Lemma 1 (the full proof can be found in the online version of the paper) shows that for any polytope partition, there are only a finite number of points $x \in [0, 1)$ that give rise to degenerate cross-sections. Before stating the lemma, we define an important discrete subset of $[0, 1]$ given by the projections of vertices of polytopes under π.

Definition 4 (Vertex Critical Coordinates). *For a given polytope $P \subset \mathbb{R}^m$ let $V_P \subset \mathbb{R}^m$ be the vertex set of P. Define the set of vertex critical coordinates as $C_P = \pi(V_P) \subset \mathbb{R}$. If $\mathscr{P} = \{P_i\}_{i=1}^n$ is an (m, n)-polytope partition, then we extend our definition to include $C_{\mathscr{P}} = \bigcup_{i=1}^n C_{P_i} \subset [0, 1]$ as the vertex critical coordinates of $\mathscr{P}$.*

Lemma 1. *Let $\mathscr{P}$ be an (m, n)-polytope partition and $x \in [0, 1) \setminus C_{\mathscr{P}}$. Then $\mathscr{P}^x$ is non-degenerate.*

2.1 Membership Query Oracles and ε-Close Labellings

We study two natural query oracle models for polytope membership in any $\mathscr{P}$.

Definition 5 (Membership Query Oracles for Polytope Partitions). *Any (m, n)-polytope partition, $\mathscr{P}$ has the following membership query oracles:*

- *Lexicographic query oracle: $Q_\ell : \Delta^m \to [n]$, which for a given y returns the smallest index of polytope to which y belongs, namely $Q_\ell(y) = \min\{i \in [n] \mid y \in P_i\}$.*

– *Adversarial query oracle(s):* $Q_A : \Delta^m \to [n]$, *which can return any polytope to which y belongs. Namely, Q_A is any function such that $Q_A(y) \in \{i \in [n] \mid y \in P_i\}$ for all $y \in \Delta^m$.*

When we wish to refer to an arbitrary oracle from the above models, we use the notation Q. Before continuing, we also clarify that for $A, B \subseteq \mathbb{R}^n$, we denote $Conv(A, B) \subseteq \mathbb{R}^m$ as the convex combination of the two sets. In addition, if $A_i \subseteq \mathbb{R}^m$ is an indexed family of sets with $i = 1, ..., r$, we denote $Conv(A_i \mid i = 1, ..., r) \subseteq \mathbb{R}^n$ as the convex combination of all A_i. Upon making queries to Q, we can infer labels of $x \in \Delta^m$ by taking convex combinations. We abstract this notion in the following definition.

Definition 6 (Empirical Polytopes and Labellings). *Suppose that $\mathcal{P}$ is an (m, n)-polytope partition and $S \subset \Delta^m$ is a finite set for which queries to Q have been made. Let $\widehat{P}_i = Conv(\{x \in S \mid Q(x) = i\}) \subset P_i$. We say each $\widehat{P}_i$ is an empirical polytope of P_i and that $\widehat{\mathcal{P}} = \{\widehat{P}_i\}$ is an empirical labelling of $\mathcal{P}$. Furthermore, we use the notation $\widehat{P}_\perp = \Delta^m \setminus \cup_{i=1}^n \widehat{P}_i$. to refer to points in Δ^m unlabelled under $\widehat{\mathcal{P}}$.*

An ε-net in the ℓ_2 norm for $\Delta^m \subset \mathbb{R}^m$ is a set $N_\varepsilon^m \subseteq \Delta^m$ with the property that for all $x \in \Delta^m$, there exists a $y \in N_\varepsilon^m$ such that $\|x - y\|_2 \leq \varepsilon$. Our learning goal is to use query access to an oracle, Q, to compute an empirical labelling $\widehat{\mathcal{P}}$ such that $\cup_{i=1}^n \widehat{P}_i$ is an ε-net of Δ^m.

Definition 7 (ε-close Labelling). *Suppose that $\varepsilon \geq 0$ and that $\widehat{\mathcal{P}}$ is an empirical labelling for $\mathcal{P}$. If $\cup_{i=1}^n \widehat{P}_i$ is an ε-net of $\Delta^m \subset \mathbb{R}^m$ in the ℓ_2 norm, we say that $\widehat{\mathcal{P}}$ is an ε-close labelling of $\mathcal{P}$.*

Although ε-close labellings are defined for polytope partitions, we extend our terminology to also encompass slices of polytope partitions. As such, when we mention computing an ε-close labelling of $\mathcal{P}^{x,y}$, we mean an empirical labelling of $\mathcal{P}^{x,y}$ (in the same vein as Definition 6) with the property that the union of its empirical polytopes forms an ε-net of $(\Delta^m)^{x,y}$.

2.2   Learning in Thickness to Learning in Distance

Definition 8 (Thickness of Sets). *Suppose that $Z \subseteq \mathbb{R}^m$ is a set. We define the thickness of Z as the radius of the largest ℓ_2 ball fully contained in Z and we denote it by $\tau(Z) = \sup\{\delta \geq 0 \mid \exists x \in Z \text{ with } B_\delta(x) \subseteq Z\}$ where $B_\delta(x) = \{y \in \mathbb{R}^m \mid \|x - y\|_2 \leq \delta\}$. In the language of convex geometry, $\tau(Z)$ is the depth of the Chebyshev centre of Z.*

For a polytope partition $\mathcal{P}$, if $\widehat{\mathcal{P}}$ is an ε-close labelling, then $\tau(\widehat{P}_\perp) \leq \varepsilon$, but the converse does not hold in general. Even though $\widehat{P}_\perp$ may be of small thickness, if it contains vertices of Δ^m, these vertices may be far from labelled points. The following results (with full proofs in the online version of the paper) lead up to Lemma 4, a slightly weaker version of the aforementioned converse. Lemma 4 shows that if we are able to learn an empirical labelling where the set of unlabelled points is of small enough thickness, then we will in fact have succeeded in learning an ε-close labelling, where any unlabelled point is close in distance to a labelled point.

Lemma 2. *Let $P \subset \mathbb{R}^m$ be a full-dimensional polytope with $Diam(P) = \sup_{x,y \in P} \|x - y\|_2$.*

- *If $A \subsetneq P$ and $\gamma > \left(\frac{Diam(P)}{\tau(P)}\right)\tau(A)$, then $B_\gamma(x) \cap (P \setminus A) \neq \emptyset$ for all $x \in A$.*
- *If $A \subseteq P$ is such that $int(P) \setminus A \neq \emptyset$ (int(P) refers to the interior of P) and $\gamma > \left(\frac{Diam(P)}{\tau(P)}\right)\tau(A)$, then $B_\gamma(x) \cap (int(P) \setminus A) \neq \emptyset$ for all $x \in A$.*

Lemma 3. $Diam(\Delta^m) = \sqrt{2}$ *and* $\tau(\Delta^m) \geq \frac{1}{m+\sqrt{m}}$.

Lemma 4. *Suppose that $\mathscr{P}$ is an (m,n)-polytope partition. Furthermore suppose that $\widehat{\mathscr{P}}$ is an empirical labelling with $\tau(\widehat{P}_\perp) < \varepsilon$. For any $\gamma > \sqrt{2}(m+\sqrt{m})\varepsilon$, it follows that $\widehat{\mathscr{P}}$ is a γ-close labelling. In particular, if $\gamma > 4m\varepsilon$, the claim also holds.*

Proof. From Lemma 3, we know that $\tau(\Delta^m) \geq \frac{1}{m+\sqrt{m}}$ and $Diam(\Delta^m) = \sqrt{2}$. Suppose that $x \in \widehat{P}_\perp$. From Lemma 2 our choice of γ implies $B_\gamma(x) \cap (\Delta_m \setminus \widehat{P}_\perp) \neq \emptyset$. This in turn means that $\widehat{\mathscr{P}}$ is a γ-close labelling. As for the final claim, this holds since $m \geq 1$. □

3   Constant-Dimension Generalised Binary Search for Q_ℓ

We set up important groundwork by focusing on arbitrary polytopes $P \subset \mathbb{R}^m$. We let $\pi : \mathbb{R}^m \to \mathbb{R}$ be the projection function introduced in Definition 2, and we recall Definition 4 regarding the vertex critical coordinates of P denoted by C_P.

Lemma 5. *Suppose that $x, y \in \mathbb{R}$ are such that $[x, y] \bigcap C_P = \emptyset$. Then taking convex hulls of cross-sections we get $Conv(P^x, P^y) = P^{x,y}$.*

Proof. $[x, y] \cap C_P = \emptyset$ implies the vertices of the polytope $P^{x,y}$ lie in $\mathscr{P}^x$ and $\mathscr{P}^y$. Since the convex hull of the set of all vertices of a bounded polytope is the polytope itself, the claim follows. □

This property of polytopes whereby convex combinations give rise to complete information except when traversing a discrete set of critical points (visualised in Fig. 2) is critical to CD-GBS. With query access to polytopes however, we no longer fully recover P^x perfectly, but instead an approximation given by an ε-close labelling, $\widehat{P}^x$. It becomes more subtle to show that by taking convex hulls of $\widehat{P}^x$ and $\widehat{P}^y$, we recover the desired information along $[x, y]$.

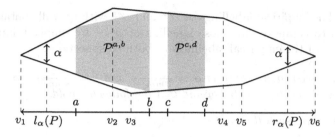

Fig. 2. $Conv(P^a, P^b) \neq P^{a,b}$ and $Conv(P^c, P^d) = P^{c,d}$

3.1   Necessary Machinery

We delve into the specifics of CD-GBS by defining important machinery. All proofs for the lemmas of this section can be found in the full online version of the paper. To begin, we recall thickness from Definition 8, and see that it satisfies a sub-additivity property when the sets being considered are convex polytopes:

Lemma 6. *Let* $P_1, .., P_k \subseteq \mathbb{R}^m$ *be convex polytopes. Then* $\tau(\cup_i P_i) < \frac{10}{3}(\sum_i \tau(P_i))(m+1)^{3/2}.$

For a given polytope partition $\mathscr{P} = \{P_i\}_i$, it will be important to establish thickness bounds on P_i at specific cross-sections.

Definition 9 (α-Critical Coordinates). *Let* $P \subset \mathbb{R}^m$ *be a polytope. For* $\alpha > 0$, *we define* $l_\alpha(P) = \inf\{x \in \mathbb{R} \mid \tau(P^x) \geq \alpha\}$ *and* $r_\alpha(P) = \sup\{x \in \mathbb{R} \mid \tau(P^x) \geq \alpha\}$ *so that* $\forall z \in \mathbb{R}, \tau(P^z) \geq \alpha$ *if and only if* $z \in [l_\alpha(P), r_\alpha(P)]$ *(Here thickness is with respect to the natural embedding of* P^x *in* $\mathbb{R}^{m-1}$*). These are called* α-critical coordinates *for* P.

The previous definition allows us to associate to each polytope P_i a segment of $[0,1]$ within which cross-sections of P_i are thick above a threshold. By combining this with Definition 4 we get the correct notion of critical coordinates.

Definition 10 (Critical Coordinates of a (m,n)-Polytope Partition). *Suppose that* $\mathscr{P} = \{P_1, ..., P_n\}$ *is an* (m,n)-polytope partition. For $\alpha > 0$, we *let* $C_{\mathscr{P}}^\alpha$ *be the union of the sets of all vertex critical coordinates of all* P_i *as defined in Definition 4, and the set of all* α-critical coordinates for all P_i as in *Definition 9. Specifically,* $C_{\mathscr{P}}^\alpha = (\cup_i C_{P_i}) \bigcup (\cup_i \{l_\alpha(P_i), r_\alpha(P_i)\}).$

CD-GBS clusters queries around critical coordinates (up to a desired toler-ance). For this reason it is important to bound the number of critical coordinates in a given (m,n)-Polytope partition.

Lemma 7. *If* $\mathscr{P}$ *is a* (m,n)-polytope partition $|C_{\mathscr{P}}^\alpha| \leq \binom{n+m}{m} + 2n.$

With this machinery in hand, we are in a position to prove the main auxi-lary lemma necessary to demonstrate correctness of CD-GBS. We show that if $x, y \in [0,1]$ are such that $[x,y]$ contains no critical coordinates, then computing sufficiently fine empirical labellings of $\mathscr{P}^x$ and $\mathscr{P}^y$ with Q_ℓ will contain enough information to compute an ε-close labelling of $\mathscr{P}^{x,y}$ by simply taking convex combinations of the empirical labellings at both cross-sections.

Lemma 8. *Given* $m, n, \varepsilon > 0$ *let* $\alpha = \frac{\varepsilon}{20nm^{5/2}}$ *and* $\beta = \frac{\varepsilon^2}{85nm^{5/2}}$. *Suppose that* $\mathscr{P}$ *is an* (m, n)-*polytope partition and that the following hold:*

- $x, y \in [0, 1]$ *are such that* $x < y \leq 1 - \frac{\varepsilon}{3}$.
- $[x, y] \cap C_{\mathscr{P}}^\alpha = \emptyset$.
- $\widehat{\mathscr{P}}^x$ *and* $\widehat{\mathscr{P}}^y$ *are empirical labellings of* $\mathscr{P}^x$ *and* $\mathscr{P}^y$ *computed via* Q_ℓ, *such that* $\cup_i \widehat{P}_i^x$ *and* $\cup_j \widehat{P}_j^y$ *are* β-*nets for* $(\Delta^m)^x$ *and* $(\Delta^m)^y$ *respectively.*

Then $\bigcup_i Conv(\widehat{P}_i^x, \widehat{P}_i^y)$ *is an* ε-*net of* $(\Delta^m)^{x,y}$.

For the following corollary, suppose that $\mathscr{P}$ is an (m, n) polytope partition and that $0 = t_0 < t_1, ..., < t_k = 1$ are points in $[0, 1]$. Furthermore suppose that $\beta = \frac{\varepsilon^2}{85nm^{5/2}}$ as in Lemma 8. For each t_i, if $t_i \notin C_{\mathscr{P}}^\alpha$, let $\widehat{\mathscr{P}}^{t_i}$ be a β-close labelling of $\mathscr{P}^{t_i}$, otherwise let $\widehat{\mathscr{P}}^{t_i} = \emptyset$. Let $\widehat{\mathscr{P}} = Conv_i(\widehat{\mathscr{P}}^{t_i})$ and for $i = 1, .., k$, let $I_j = [t_{j-1}, t_j]$. If $\widehat{\mathscr{P}}^{t_{i-1}, t_i}$ is an ε-close labelling of $\mathscr{P}^{t_{i-1}, t_i}$, we say that I_j is covered, otherwise we say I_j is uncovered.

Corollary 1. *For any collection of* $\{t_i\}_{i=1}^k$, *there are no more than* $2C_{\mathscr{P}}^\alpha$ *inter-vals* I_j *that are uncovered.*

3.2   Specification of CD-GBS and Query Usage

Terms and Notation: The details of CD-GBS are presented in Algorithm 1. We recall our notation from Definition 3 where for $x \in [0, 1)$ we defined $f_x : (\Delta^m)^x \rightarrow \Delta^{m-1}$ given by $f_x(x, ..., v_m) = \frac{1}{1-x}(v_2, ..., v_m)$. We note that this is a bijection between both polytopes, hence it is well-defined to use f_x^{-1}. In addition, we let $D^k = \{\frac{i}{2^k} \mid 1 \leq i \leq 2^k\}$ be the dyadic fractions of k-th power in the unit interval (excluding 0). For every $x \in D^k$ we can associate the interval $I_x^k = [x - \frac{1}{2^k}, x]$. For each of these intervals $midpoint(I_x^k)$ denotes its midpoint. We also use the same language as Corollary 3.1 when we talk about whether I_x^k is covered or not (with respect to the current empirical labelling, $\widehat{\mathscr{P}}$, obtained from taking convex hulls of labels in Δ^m). We note that in order to have a well-defined base case of CD-GBS (which is equivalent to binary search), we let $\Delta^0 = \mathbb{R}^0 = \{0\}$. Finally, we say that a point $x \in [0, 1]$ is an uncovered critical point if $\widehat{\mathscr{P}}^x$ is computed via a recursive call to CD-GBS and for $(a, b) = B_{\varepsilon/2}(x) \cap [0, 1]$, it holds that $\widehat{\mathscr{P}}^{a,b}$ is not an ε-close labelling of $\mathscr{P}^{a,b}$.

Theorem 1. *If   CD-GBS   is   given   access   to Q_ℓ for   a (m, n)-polytope partition, it computes an ε-close labelling of $\mathscr{P}$ using at most*

$(\prod_{i=1}^{m} \left(\binom{n+i}{i} + 2n \right)) 2^{2m^2} \log^m \left(\frac{170nm^{5/2}}{\varepsilon} \right)$ *membership queries. For constant* m
this constitutes $O(n^{m^2} \log^m \left(\frac{n}{\varepsilon} \right)) = poly(n, \log \left(\frac{1}{\varepsilon} \right))$ *queries* [1].

Proof. We first prove that CD-GBS indeed computes an ε-close labelling when given access to a valid Q_ℓ by inducting on m. It is straightforward to see that in the case $m = 1$, if CD-GBS is given access to a valid Q_ℓ for a $(1, n)$ polytope partition (a partition of the unit interval into conected subintervals), then it simply performs binary search on the interval $[0, 1] \cong \Delta^1$.

As for the inductive step, for $k = \lceil \log(2/\varepsilon) \rceil$, any two contiguous points of $\mathcal{D}^k$ are less than $\varepsilon/2$ away from each other. For now suppose that every recursive call to CD-GBS was along a non-degenerate cross section $\mathscr{P}^t$. From the inductive assumption, this means that CD-GBS computes an $\varepsilon/2$-close labellings of those cross-sections, using the triangle inequality, we know that $\widehat{\mathscr{P}}$ is an ε-close labelling of $\mathscr{P}$.

We note however that there is no guarantee for what a recursive call to CD-GBS does on a degenerate cross section $\widehat{\mathscr{P}^t}$. For this reason, it could be the case that at the end of the loop over $\mathcal{D}^i$, $\widehat{\mathscr{P}}$ is not an ε-close labelling. This can only happen if there is some $t \in C_{\mathscr{P}}^\alpha \cap \mathcal{D}^k$ which is an uncovered critical coordinate.

If t is an uncovered critical coordinate we can rectify the situation. If we find a $z \in B_{\varepsilon/2}$ that is not a critical coordinate, then $\mathscr{P}^z$ is non-degenerate and computing CD-GBS along the cross-section gives us an $\frac{\varepsilon}{2}$-close labelling of $\mathscr{P}^z$. Using the triangle inequality, we see that this in turn removes t from the set of uncovered critical coordinates, and we say that t is "fixed". Thus the final while loop of the algorithm eliminates the set of uncovered critical coordinates so that $\widehat{\mathscr{P}}$ is indeed an ε-close labelling.

It thus remains to show that the final while loop terminates. However, there are at most $|C_{\mathscr{P}}^\alpha|$ uncovered critical coordinates, and over the course of fixing all uncovered critical coordinates, there are at most $|C_{\mathscr{P}}^\alpha|$ bad guesses for $z \in B_{\varepsilon/2}(x)$ where $\mathscr{P}^z$ is degenerate. Therefore the final while loop makes at most $2|C_{\mathscr{P}}^\alpha|$ invocations to CD-GBS along cross-sections. This concludes the proof of correctness for CD-GBS.

Let us bound the total query usage of CD-GBS. For all values of k in the first for loop, we know from Corollary 3.1 that since Q_ℓ is a valid lexicographic oracle for $\mathscr{P}$, that the number of uncovered I_x^k will not exceed $2\left(\binom{n+m}{m} + 2n\right)$, and since CD-GBS is called once per uncovered interval, it follows that for each k there at most $2\left(\binom{n+m}{m} + 2n\right)$ recursive calls to CD-GBS. Furthermore, since Q_ℓ is a valid lexicographic oracle for $\mathscr{P}$, it will also never be the case that $\exists i, j \in [n]$, $z \in \Delta^m$ such that $dim(\widehat{P}_i) = m$ and $z \in int(\widehat{P}_i)$.

---

[1] CD-GBS runs in polynomial time for constant m. The time-intensive operation consists of identifying uncovered intervals, but since the dimension of the ambient simplex is constant, each empirical polytope $\widehat{P}_i$ has at most a constant number of bounding hyperplanes. These hyperplanes can each be extruded by ε, and checking whether there exists a point outside all these extrusions can be done in time polynomial in n via brute force. In fact, all other algorithms in this paper have efficient runtimes (in their relevant parameters) due to similar reasoning.

In the worst case, k loops from 1 to $\lceil \log(2/\varepsilon) \rceil$ and makes an extra $2|C_{\mathscr{P}}^{\alpha}|$ recursive calls to CD-GBS to fix all uncovered critical coordinates. In total if we let $T(m, n, \varepsilon)$ denote the query cost of running CD-GBS on a valid lexicographic oracle, we get the following recursion:

$$T(m, n, \varepsilon) \leq 2|C_{\mathscr{P}}^{\alpha}| \log \left(\frac{2}{\varepsilon} \right) T \left(m - 1, n, \frac{\varepsilon^2}{85nm^{5/2}} \right) + 2|C_{\mathscr{P}}^{\alpha}|.$$

In order to make this more amenable, we define $f(m) = \left(\binom{n+m}{m} + 2n \right)$ and use Lemma 7 to bound this expression by $T(m, n, \varepsilon) \leq 3f(m)$ $\log \left(\frac{2}{\varepsilon} \right) T \left(m - 1, n, \frac{\varepsilon^2}{85nm^{5/2}} \right)$. Furthermore, from the fact that the base case is binary search, we know $T(1, n, \varepsilon) \leq n \log \left(\frac{2}{\varepsilon} \right)$.

To unpack the recursion. Let us define $\varepsilon_0 = \varepsilon$ and $\varepsilon_{k+1} = \frac{\varepsilon_k^2}{85n(m-k)^{5/2}}$ for $k = 1, ..., m - 1$. With this in hand, we can unroll the recursion to obtain $T(m, n, \varepsilon) \leq \left(3^{m-1} \prod_{i=1}^{m-1} f(i) \right) \left(\prod_{k=1}^{m-1} \log \left(\frac{2}{\varepsilon_k} \right) \right)$. Since each $\varepsilon_{k+1} < \varepsilon_k$, we can upper bound the right-hand product by bounding each term with ε_{m-1}. If we first solve for this value, we obtain: $\varepsilon_{m-1} = \frac{\varepsilon^{2^{m-1}}}{\prod_{j=1}^{m-1}(85nj^{5/2})^{2j}} \geq \frac{\varepsilon^{2^{m-1}}}{\prod_{j=1}^{m-1}(85nm^{5/2})^{2j}} \geq$ $\left(\frac{\varepsilon}{85nm^{5/2}} \right)^{2^m}$. In the first inequality we bounded the denominator product in the base by $j \leq m$, as for the second inequality, we evaluated the geometric series in 2 for the exponent to bound the exponent by 2^m. With this in hand we obtain the desired bounds:

$$T(m, n, \varepsilon) \leq 3^m 2^{m^2} \prod_{i=1}^{m} f(i) \log^m \left(\frac{170nm^{5/2}}{\varepsilon} \right) \leq \left(\prod_{i=1}^{m} \left(\binom{n+i}{i} + 2n \right) \right) 2^{2m^2} \log^m \left(\frac{170nm^{5/2}}{\varepsilon} \right).$$

For large enough n, every term in $\prod_{i=1}^{m} \left(\binom{n+i}{i} + 2n \right)$ is bounded by $(n + m)^m + 2n$. It follows that this product is $O(n^{m^2})$. For constant m, this constitutes $O(n^{m^2} \log^m \left(\frac{n}{\varepsilon} \right)) = poly(n, \log \left(\frac{1}{\varepsilon} \right))$ queries.   □

The previous results show that for constant dimension, m, CD-GBS is query efficient in n and $\frac{1}{\varepsilon}$. In the following section we use this algorithm as a building block to construct a method for computing efficient ε-close labellings when the number of regions, n, is held constant instead.

4   Constant-Region Generalised Binary Search

The intuition behind the algorithm lies in the fact that if m is much greater than n, then any vertex of a given P_i cannot lie in the interior of the ambient simplex Δ^m. This is since a vertex in Δ^m must consist of the intersections of at least m half-spaces, all of which cannot arise from adjacencies between different P_i. Not only do all vertices lie on the boundary of Δ^m, but one can show that they are all contained in faces of the boundary of Δ^m that have dimension $O(n^2)$ which is presumed to be constant. The number of such faces in the boundary of Δ^m is thus polynomial in m, and if we could compute 0-close labellings of

---

Algorithm 1. CD-GBS(m, n, ε, Q)

Input: $m \geq 0$, $n, \varepsilon > 0$, query access to function $Q : \Delta^m \to [n]$.
Output: $\widehat{\mathscr{P}}$: an ε-close labelling of $\mathscr{P}$.
  if $m = 0$ **then**
    Query $Q(0)$
  else
    $\widehat{\mathscr{P}}^0 \leftarrow f_0^{-1} \left(\text{CD-GBS} \left(m - 1, n, \frac{\varepsilon^2}{85nm^{5/2}}, Q \circ f_0^{-1} \right) \right)$, $\widehat{\mathscr{P}}^1 \leftarrow Q(e_1)$.
    for $k = 1$ to $\lceil \log(2/\varepsilon) \rceil$ **do**
      if Number of uncovered I_x^k exceeds $2\left(\binom{n+m}{m} + 2n \right)$ **then**
        Halt
      for $x \in \mathscr{D}^k$ **do**
        if I_x^k is uncovered **then**
          $t \leftarrow midpoint(I_x)$
          $\widehat{\mathscr{P}}^t \leftarrow f_t^{-1} \left(\text{CD-GBS} \left(m - 1, n, \frac{\varepsilon^2}{85(1-t)nm^{5/2}}, Q \circ f_t^{-1} \right) \right)$
          Recompute $\widehat{\mathscr{P}}$ by taking convex hulls of labels
          if $\exists i, j \in [n]$ such that $int(\widehat{P}_i) \cap \widehat{P}_j \neq \emptyset$ or $\widehat{\mathscr{P}}$ is an ε-close labelling **then**
            Halt
    while $\exists x \in [0, 1]$ an uncovered critical point **do**
      $t \leftarrow z$ for arbitrary $z \in B_{\varepsilon/2}(x)$
      $\widehat{\mathscr{P}}^t \leftarrow f_t^{-1} \left(\text{CD-GBS} \left(m - 1, n, \frac{\varepsilon^2}{85(1-t)nm^{5/2}}, Q \circ f_t^{-1} \right) \right)$
      Recompute $\widehat{\mathscr{P}}$ by taking convex hulls of labels
  return $\widehat{\mathscr{P}}$

---

these faces we could take convex combinations and recover a 0-close labelling of the entire polytope partition. We will demonstrate that for an appropriate value of ε', if we compute ε'-close labellings of such faces in the boundary, we can recover an ε-close labelling of the entire polytope partition over all of Δ^m by taking convex combinations. CR-GBS computes the necessary ε'-close labellings of lower dimensional faces by using CD-GBS as a subroutine.

Necessary Machinery for CR-GBS: Suppose that $\mathscr{P}$ is an (m, n)-polytope partition with $m > \binom{n}{2}$. Furthermore, let $k = \binom{n}{2}$ and let $\partial_k(\Delta^m)$ denote all k-dimensional faces of Δ^m. For each face F, let $\mathscr{P}_F$ be the restriction of $\mathscr{P}$ to F. If F contains the origin, then it is an isometric embedding of Δ^k in Δ^m, so we let ϕ_F be a canonical isomorphism from F to Δ^k. If F does not contain the origin, then let $v_F \in F$ be the vertex of lowest index, i.e. $v_F = e_i$ where $i = \operatorname{argmin}\{j \mid e_j \in F\}$. In this case, we let ϕ_F be a canonical linear isomorphism from F to Δ^k that maps v_F to the origin and other vertices of F to other vertices of Δ^k.

As mentioned previously, computing empirical labellings of every face in $\partial_k(\Delta^m)$ via CD-GBS will be enough to compute an empirical labelling for $\mathscr{P}$. The only issue however, is that CD-GBS is only guaranteed to return an ε-close labelling if given access to a valid lexicographic membership oracle for a polytope partition. For an arbitrary polytope partition, $\phi_F(\mathscr{P}_F)$ is not necessarily a

(k, n)-polytope partition for all $F \in \partial_k(\Delta^m)$. For this reason, we slightly refine our notion of polytope partition.

Definition 11. *Suppose that $\mathscr{P}$ is an (m, n)-polytope partition such that for all $0 \leq k \leq m$ and $F \in \partial_k(\Delta^m)$, $\phi_F(\mathscr{P}_F)$ is a (k, n)-polytope partition. Then we say $\mathscr{P}$ is a proper polytope partition.*

Specification of CR-GBS and Query Usage: For $F \in \partial_k(\Delta^m)$, we let ϕ_F denote the canonical isomorphism from F to Δ^k. This isomorphism is different depending on whether F is axis-aligned or not. Furthermore, we let $\widehat{\mathscr{P}}_F$ empirical labelling returned by CD-GBS on $F \in \partial_k(\Delta^m)$. The full proof of correctness and query usage of Algorithm 2 can be found in the online version of the paper.

Algorithm 2. CR-GBS(m, n, ε, Q)

Input: $m, n, \varepsilon > 0$, query access to membership oracle Q for (m, n)-polytope partition $\mathscr{P}$.
Output: $\widehat{\mathscr{P}}$: an ε-close labelling of $\mathscr{P}$.

$k \leftarrow \binom{n}{2}$
for $F \in \partial_k(\Delta^m)$ **do**
$\quad \widehat{\mathscr{P}}^F \leftarrow \phi_F^{-1}\left(\text{CD-GBS}\left(k, n, \frac{3\varepsilon}{100n^2\sqrt{k+1}(m+1)^{5/2}}, Q \circ \phi_F^{-1}\right)\right).$
$\widehat{\mathscr{P}} \leftarrow \text{Conv}_F(\widehat{\mathscr{P}}_F)$
return $\widehat{\mathscr{P}}$

Theorem 2. *Let $\mathscr{P}$ be a proper (m, n)-polytope partition where n is constant and $m > k = \binom{n}{2}$. CR-GBS computes an ε-close labelling of $\mathscr{P}$ and uses $O\left(m^k \log^k\left(\frac{m}{\varepsilon}\right)\right) = poly(m, \log\left(\frac{1}{\varepsilon}\right))$ queries.*

5   Upper Envelope Polytope Partitions

We have focused completely on the lexicographic query oracle Q_ℓ, creating algorithms CD-GBS and CR-GBS that compute ε-close labellings of (m, n)-polytope partitions when given access to Q_ℓ. If these algorithms are given access to an adversarial oracle Q_A however, they may fail.

To see why CD-GBS may fail under Q_A we recall that the algorithm recursively computes ε-close labellings of cross-sections $\mathscr{P}^t$ for different values of $t \in [0, 1]$. If ever CD-GBS is called on a degenerate cross-section $\mathscr{P}^t$, it has conditions to either tell that it is being called on a degenerate cross-section (when it notices that there exist $i, j \in [n]$ and $z \in \Delta^m$ such that $z \in int(\widehat{P}_i) \cap \widehat{P}_j$), or in the worst case, prevent it from exceeding its query balance. In both cases however, the algorithm returns a valid empirical labelling, i.e., $\widehat{P} = \{\widehat{P}_i\}_{i=1}^n$ such that $\widehat{P}_i \subseteq P_i$.

When an adversarial oracle is used however, we may see $i, j \in [n]$ and $z \in \Delta^m$ such that $z \in int(\widehat{P}_i) \cap \widehat{P}_j$. Indeed this can occur if $P_i = P_j$ and both are full-dimensional. The natural solution seems to merge P_i and P_j (since the second condition of the definition of polytope partitions tells us that $P_i = P_j$ in this case). The main problem however, is that there is no way of telling when the condition above is an artifice of the adversarial oracle, or simply due to the fact that $\mathscr{P}^t$ is degenerate. If we blindly merge labels, we may in fact be performing an incorrect merge on a degenerate cross-section! This of course may return inconsistent polytope partitions.

Since the key problem is the existence of degenerate cross-sections, we consider a slightly stronger variant of polytope partitions with the key property that cross-sections are never degenerate. Furthermore, this special type of polytope partition is expressive enough for our game theoretic applications, and best of all, it allows us to prove results in the adversarial query oracle model.

Definition 12 (Upper Envelope Polytope Partition). *Suppose that $A \in \mathbb{R}^{n \times m}$ is an $n \times m$ real-valued matrix and that $b \in \mathbb{R}^n$. Let $P_i = \{y \in \Delta^m$ such that $(Ay + b)_i \geq (Ay + b)_j$ for all $j \neq i\}$. We denote the collection $\mathscr{P}(A, b) = \{P_1, \ldots, P_n\}$, as the upper envelope polytope partition (UEPP) arising from (A, b).*

It is straightforward to see that for any (A, b), $\mathscr{P}(A, b)$ is itself an (m, n)-polytope partition. Crucially however, it satisfies more properties than the previous definition of polytope partitions. The full proof of the following lemma can be found in the online version of the paper.

Lemma 9. *Suppose that A is an $n \times m$ real valued matrix and that $b \in \mathbb{R}^n$. Then $\mathscr{P}(A, b) = \{P_1, \ldots, P_n\}$ has the following properties:*

- *For any $x \in [0, 1)$ let f_x be the canonical affine transformation that maps $(\Delta^m)^x$ to Δ^{m-1}. There exists an $n \times (m-1)$ real matrix A^x and $b^x \in \mathbb{R}^n$ such that $\mathscr{P}(A^x, b^x) = f_x(\mathscr{P}(A, b)^x)$.*
- *$\mathscr{P}(A, b)$ is a proper polytope partition (Definition 11).*
- *If $A_{i,\bullet} = A_{j,\bullet}$ and $b_i = b_j$ then $P_i = P_j$. Conversely if P_i is of full affine dimension and $relint(P_i) \cap P_j \neq \emptyset$, then $A_{i,\bullet} = A_{j,\bullet}$ and $b_i = b_j$; consequently, $P_i = P_j$.*
- *Let $a_1, \ldots, a_k \in \mathbb{R}$ be such that $\sum_{i=1}^k a_i < 1$ with $k < m$. Let $H = \{(z_1, \ldots, z_m) \in \Delta^m \mid z_i = a_i, i = 1, \ldots, k\}$ where H has codimension k. If $x_1, \ldots, x_{m-k} \in \Delta^m$ are affinely independent points of $P_i \cap H$ and $y \in Conv(x_1, \ldots, x_{m-k})$ belongs to P_j, then P_i and P_j coincide in H.*

5.1   Adversarial CD-GBS

Suppose that $\mathscr{P}$ is an UEPP. Since it is also a proper (m, n)-polytope partition, it inherits all the properties from before. Along with Lemma 9 we have the necessary tools to show that Algorithm 3 is a query efficient way of computing ε-close labellings of $\mathscr{P}$ with an adversarial query oracle. In the specification of CD-GBS, we use identical terms and notation from Algorithm 1. The full proof of correctness and query usage of Algorithm 3 can be found in the online version of the paper.

Algorithm 3. Adversarial CD-GBS(m, n, ε, Q_A)

Input: $m \geq 0$, $n, \varepsilon > 0$, query access to oracle $Q_A : \Delta^m \to [n]$.
Require: Recursive calls to CD-GBS$\left(m - 1, n, \frac{\varepsilon^2}{85(1-x)nm^{5/2}}, Q_A \circ f_x^{-1}\right)$.
Output: ε-close labelling of $\mathscr{P}$.
   if $m = 0$ **then**
      Query $Q_A(0)$
   else
      $\widehat{\mathscr{P}}^0 \leftarrow f_0^{-1}\left(\text{CD-GBS}\left(m - 1, n, \frac{\varepsilon^2}{85nm^{5/2}}, Q_A \circ f_0^{-1}\right)\right)$
      $\widehat{\mathscr{P}}^1 \leftarrow Q(e_1)$.
      for $k = 1$ to $\lceil \log(2/\varepsilon) \rceil$ **do**
         for $x \in \mathscr{P}^k$ **do**
            if I_x^k is uncovered **then**
               $t \leftarrow midpoint(I_x)$
               $\widehat{\mathscr{P}}^t \leftarrow f_t^{-1}\left(\text{CD-GBS}\left(m - 1, n, \frac{\varepsilon^2}{85(1-t)nm^{5/2}}, Q_A \circ f_t^{-1}\right)\right)$
         Recompute $\widehat{\mathscr{P}}$ by taking convex hulls of labels
         while $\exists i, j \in [n]$, $z \in \Delta^m$ such that $dim(\widehat{P}_i) = m$ and $z \in int(\widehat{P}_i)$ **do**
            Merge label i with label j
            Recompute $\widehat{\mathscr{P}}$ by taking convex hulls of labels
         if $\widehat{\mathscr{P}}$ is an ε-close labelling **then**
            Break
   return $\widehat{\mathscr{P}}$

Theorem 3. *If Adversarial CD-GBS is given access to an adversarial query oracle Q_A of an (m, n)-polytope partition based on a UEPP, it computes an ε-close labelling of $\mathscr{P}$ using at most $\left(\prod_{i=1}^{m} \left(\binom{n+i}{i} + 2n\right)\right) 2^{2m^2} \log^m \left(\frac{170nm^{5/2}}{\varepsilon}\right)$ membership queries. For constant m this constitutes $O(n^{m^2} \log^m \left(\frac{n}{\varepsilon}\right)) = poly(n, \log \left(\frac{1}{\varepsilon}\right))$ queries.*

5.2 Adversarial CR-GBS

In this section we formalize an adversarial variant of CR-GBS. We note that most of the notation is identical to lexicographic CR-GBS. As before, the full proof of correctness and query usage of Algorithm 4 can be found in the online version of the paper.

Theorem 4. *Let $\mathscr{P}$ be an (m, n)-polytope partition where n is constant. Furthermore, let $k = \binom{n}{2}$. Adversarial CR-GBS computes an ε-close labelling of $\mathscr{P}$ and uses $O\left(m^k \log^k \left(\frac{m}{\varepsilon}\right)\right) = poly(m, \log \left(\frac{1}{\varepsilon}\right))$ queries.*

6 Applications to Game Theory

Let $G = (A, B)$ be an $m \times n$ bi-matrix game where $A, B \in [0, 1]^{m \times n}$ are row and column player payoff matrices respectively with payoffs normalised to $[0, 1]$. The set of row player pure strategies is $[m] = \{1, \ldots, m\}$ and that of the column player pure strategies is $[n] = \{1, \ldots, n\}$. The set of all row player mixed strategies can

Algorithm 4. Adversarial CR-GBS($m, n, \varepsilon, \mathscr{P}$)

Input: $m, n, \varepsilon > 0$, query access to Q_A for (m, n)-polytope partition $\mathscr{P}$.
Output: ε-close labelling of $\mathscr{P}$.

$\quad k \leftarrow \binom{n}{2}$
$\quad$ **for** $F \in \partial(\Delta^m)^k$ **do**
$\quad\quad \widehat{\mathscr{P}^F} \leftarrow \phi_F^{-1}\left(\text{CD-GBS}\left(k, n, \frac{3\varepsilon}{100n^2\sqrt{k+1}(m+1)^{5/2}}, Q \circ \phi_F^{-1}\right)\right).$
$\quad\quad \widehat{\mathscr{P}} \leftarrow Conv_F(\widehat{\mathscr{P}_F})$
$\quad$ **while** $\exists i, j \in [n]$, $z \in \Delta^m$ such that $dim(\widehat{P_i}) = m$ and $z \in int(\widehat{P_i})$ **do**
$\quad\quad$ Merge label i with label j
$\quad\quad$ Recompute convex hulls of labels
$\quad$ **return** $\widehat{Q}$

be identified with $\Delta^{m-1} = \{x \in \mathbb{R}^{m-1} | \sum_{i=1}^{m-1} x_i \leq 1 \text{ and } x_i \geq 0\}$. Similarly, column player mixed strategies are identified with Δ^{n-1}.

Definition 13 (Utility Functions). *Let $u \in \Delta^{m-1}$, and $v \in \Delta^{n-1}$ be row and column player mixed strategies and let $u' = (1 - \sum u_i, u_1, ..., u_{n-1})$ and $v' = (1 - \sum v_i, v_1, ..., v_{n-1})$. For strategy profile (u, v), row player utility is $U_r(u, v) = u'^T A v'$ and column player utility is $U_c(u, v) = u'^T B v'$.*

It will be useful to have shorthand for the following functions: $U_r^i(y) = U_r(e_i, y)$ as the row player utility for playing pure strategy i, and $E_R(y) = \max_{i \in [m]} U_r^i(y)$ as the maximal utility the row player can achieve against mixed strategies. In an identical fashion we can define U_c^j and E_C as the column player utility in playing strategy j and the maximal column player utility. With this notation in hand, we can define the best response oracles our algorithms will have access to.

Definition 14 (Best Response Query Oracles). *Any bimatrix game has the following best response query oracles:*

- *Strong query oracles: for the column player, $BR_s^C(u) = \{j \in [n] \mid U_c^j(u) = E_C(u)\}$ and for the row player, $BR_s^R(v) = \{i \in [m] \mid U_r^i(v) = E_R(v)\}$*
- *Lexicographic query oracles: for the column player, $BR_\ell^C(u) = \min BR_s^C(u)$ and for the row player, $BR_\ell^R(v) = \min BR_s^R(v)$*
- *Adversarial query oracles: for the column player, any function BR_A^C such that $BR_A^C(u) \in BR_s^C(u)$ and for the row player, any function such that $BR_A^R(v) \in BR_s^R(v)$*

Definition 15 (Nash Equilibrium). *Suppose that u and v are row and column player strategies respectively. We say that the pair (u, v) is a Nash Equilibrium (NE) if for all $u' \in \Delta^{m-1}$ and $v' \in \Delta^{n-1}$: $U_r(u, v) \geq U_r(u', v)$ and $U_c(u, v) \geq U_c(u, v')$.*

With utility queries the complexity of an exact Nash equilibrium is finite: we can exhaustively query the game. This is not the case for best response queries. Therefore, the relaxation we study is that of approximate well-supported

Nash equilibria. Before proceeding, we say that a row player mixed strategy $u \in \Delta^m$ is an ε best response against a column player mixed strategy $v \in \Delta^n$ if $U_r(u, v) \geq U_r(u', v) - \varepsilon$ for all $u' \in \Delta^m$. An identical notion holds for the column player.

Definition 16 (ε-Well-Supported Nash Equilibrium). *Suppose that u and v are row and column player strategies respectively. We say that the pair (u, v) is an ε-well-supported Nash equilibrium (ε-WSNE) if and only if u is supported by ε-best responses to v and vice versa.*

Definition 16 mentions approximate best responses, yet we only have access to the best response oracle in our model. To resolve this, we bound how much utilities can deviate between "close" mixed strategy profiles. The full proof of the following lemma can be found in the online version of the paper.

Lemma 10. *Fix $\varepsilon > 0$ and let $\delta_C = \frac{\varepsilon}{2\sqrt{m-1}}$. Suppose that $u \in \Delta^{m-1}$ is a row player mixed strategy with $c_j \in BR_s^C(u)$. For any u' such that $\|u - u'\|_2 \leq \delta_C$, if $c_i \in BR_s^C(u')$, then $|U_c^i(u) - U_c^j(u)| \leq \varepsilon$. In other words, c_i is an ε-best response to u. Similarly, let $\delta_R = \frac{\varepsilon}{2\sqrt{n-1}}$. Suppose that $v \in \Delta^{n-1}$ is a column player mixed strategy with $r_j \in BR_s^R(u)$. For any v' such that $\|v - v'\|_2 \leq \delta_R$, if $r_i \in BR_s^R(v')$, then $|U_r^i(v) - U_r^j(v)| \leq \varepsilon$. In other words, r_i is an ε-best response to v.*

We now prove the connection between computing ε-close labellings of polytope partitions and computing ε-WSNE for bimatrix games using best response queries.

Definition 17 (Best Response Sets). *Let $G = (A, B)$ be a bimatrix game. We define column best response sets as the collection of $C_i = \{x \in \Delta^{m-1} \mid BR_s^C(x) \ni c_i\}$. Similarly we define row player best response sets as the collection of $R_j = \{y \in \Delta^{n-1} \mid BR_s^R(y) \ni r_j\}$. We denote the collections by $\mathscr{C} = \{C_i\}_{i=1}^n$ and $\mathscr{R} = \{R_j\}_{j=1}^m$.*

Since utilities are affine functions, it is immediately clear that $\mathscr{C}$ and $\mathscr{R}$ are upper envelope polytope partitions. Best response oracles play the same role as membership oracles, Q, from before. Since adversarial oracles are a valid lexicographic oracle, we focus on using adversarial best response oracles. With our language of empirical labellings we can now define the following important concept, but first we clarify some notation: $d(x, S)$ denotes the infimum distance of a point, x to a set S.

Definition 18 (Voronoi Best Response Sets). *Suppose that $\widehat{\mathscr{C}} = \{\widehat{C}_i\}$ and $\widehat{\mathscr{R}} = \{\widehat{R}_j\}$ are empirical labellings of $\mathscr{C}$ and $\mathscr{R}$ as in Definition 6. The Voronoi Best Response Sets of the row and column player are $VR_j = \{y \in \Delta^{n-1} \mid argmin_j d(y, \widehat{R}_j) = r_j\}$ and $VC_i = \{x \in \Delta^{m-1} \mid argmin_i d(x, \widehat{C}_i) = c_i\}$, defined for any $j \in [m]$ and $i \in [n]$. Furthermore, we let $V^R(v) = \{i \mid VR_i \ni v\}$ and $V^C(u) = \{j \mid VC_j \ni u\}$ be the row and column player Voronoi Best Responses.*

Lemma 11. *Suppose that $\widehat{\mathscr{C}}$ is a $\frac{\varepsilon}{2\sqrt{m-1}}$-close labelling and $\widehat{\mathscr{R}}$ is a $\frac{\varepsilon}{2\sqrt{n-1}}$-close labelling. Then Voronoi best responses are ε best-responses in G.*

Lemma 11 (with a full proof in the online version of the paper), tells us precisely that Voronoi best response sets allow us to extend partial information from empirical labellings to approximate best response information *across the entire domains Δ^m and Δ^n.* This hints at the fact that Voronoi best response sets hold enough information to compute ε-WSNE. In fact we can prove this in the same way as Nash's theorem: via Kakutani's fixed point theorem. In order to do so, we define a Voronoi best response correspondence (which as we have shown before is an approximate best response correspondence), and show that it satisfies the properties of Kakutani's fixed point theorem. The guaranteed fixed point of this correspondence will in turn be an ε-WSNE.

Definition 19 (Voronoi Approximate Best Response Correspondence). *For a given mixed strategy profile of both the row and column player, $(u, v) \in \Delta^{m-1} \times \Delta^{n-1}$, we define $B^*(u, v)$ to be the set of all possible mixtures over Voronoi best response profiles both players may have to the other player's strategy. $B^* : \Delta^{m-1} \times \Delta^{n-1} \to \mathcal{P}(\Delta^{m-1} \times \Delta^{n-1})$ is defined as follows: $B^*(u, v) = \left(conv(V^R(v)), conv(V^C(u))\right) \subseteq \Delta^{m-1} \times \Delta^{n-1}$.*

Theorem 5 (Kakutani's Fixed Point Theorem *[18]*). *Let A be a non-empty, compact and convex subset of $\mathbb{R}^n$. Let $f : A \to \mathcal{P}(A)$ be a set-valued function on A with a closed graph and the property that $f(x)$ is non-empty and convex for all $x \in A$. Then f has a fixed point.*

Theorem 6. *B^* satisfies all the conditions of Kakutani's fixed point Theorem, and hence there exists a strategy profile (u^*, v^*) such that $(u^*, v^*) \in B^*(u^*, v^*)$. In particular, if the Voronoi best responses for B^* arise from $\widehat{\mathscr{C}}$, a $\frac{\varepsilon}{2\sqrt{m-1}}$-close labelling and $\widehat{\mathscr{R}}$, a $\frac{\varepsilon}{2\sqrt{n-1}}$-close labelling, then this in turn implies that (u^*, v^*) is an ε-WSNE of G.*

The full proof of Theorem 6 can be found in the online version of the paper. With this in hand and our algorithms for constructing ε-close labellings, we can bound the query complexity of computing an ε-WSNE in general bimatrix games.

Theorem 7. *Suppose that G is an $m \times n$ bimatrix game and let n be constant. We can compute an ε-WSNE using $O(mn^2 \log^{n^2}\left(\frac{m}{\varepsilon}\right)) = poly(m, \log\left(\frac{1}{\varepsilon}\right))$ adversarial best response queries.*

Proof. Suppose that $\mathscr{C}$ and $\mathscr{R}$ are UEPP arising from best-response sets in G. This means that $\mathscr{C}$ is a $(m-1, n)$-polytope partition and $\mathscr{R}$ is a $(n-1, m)$-polytope partition. Let $\varepsilon_C = \frac{\varepsilon}{2\sqrt{m-1}}$ and $\varepsilon_R = \frac{\varepsilon}{2\sqrt{n-1}}$. From Theorem 6, we know that computing an ε_C-close labelling of $\mathscr{C}$ and a ε_R-close labelling of $\mathscr{R}$ suffice to compute an ε-WSNE of G. We use adversarial CR-GBS on $\mathscr{C}$ and adversarial CD-GBS on $\mathscr{R}$.

n is the number of polytopes in the partition $\mathscr{C}$, which is assumed to be constant. Consequently, Theorem 4 states that computing an ε_C-close labelling of $\mathscr{C}$ using CR-GBS uses $O((m-1)^k \log^k \left(\frac{m-1}{\varepsilon_C} \right))$ adversarial queries, where $k = \binom{n}{2}$. Since $k \leq n^2$, this is bounded by $O(m^{n^2} \log^{n^2} \left(\frac{m}{\varepsilon} \right))$.

$n-1$ is the dimension of the ambient simplex in the partition $\mathscr{R}$, which is assumed to be finite. Consequently, Theorem 3 states that computing an ε_R-close labelling of $\mathscr{R}$ using CD-GBS uses $O(m^{(n-1)^2} \log^{n-1} \left(\frac{1}{\varepsilon} \right))$ queries. We trivially upper bound this quantity by $O(m^{n^2} \log^{n^2} \left(\frac{m}{\varepsilon} \right))$.

The total query usage is thus $O(m^{n^2} \log^{n^2} \left(\frac{m}{\varepsilon} \right)) = poly(m, \log \left(\frac{1}{\varepsilon} \right))$ as desired.   □

7   Conclusion and Future Directions

In this paper we introduced the concept of learning ε-close labellings of (m,n)-polytope partitions with membership queries, and derived query efficient algorithms for when either the dimension of the ambient simplex in the polytope partition, m, is held constant, or when the number of polytopes in the partition, n, is held constant. Most importantly, we introduced a novel reduction from computing ε-WSNE with best response queries to this geometric problem, thus allowing us to show that in the best response query model, computing ε-WSNE of a bimatrix game has a finite query complexity. More specifically, for $m \times n$ games with one parameter, say n, constant, the query complexity is polynomial in m and $\log \left(\frac{1}{\varepsilon} \right)$. Furthermore, in the full online version of the paper, we partially extend our results from bimatrix games to n-player games. Although the underlying geometry in n-player games prevents us from using our results from learning polytope partitions, we are still able to show that querying a fine enough ε-net of the mixed strategy space of all players suffices to compute an ε-WSNE.

As mentioned in the introduction, this geometric framework could be of use in other areas where Lipschitz continuous structures appear over domains with convex partitions. Upon further inspection, it is not difficult to see that polytope partitions do not need to be contained in Δ^m, and in fact our algorithms extend to arbitrary ambient polytopes. Furthermore, it would be of great interest to create algorithms with a better query cost, prove lower bounds with regards to computing ε-close labellings, or simply explore weaker query paradigms, such as noisy membership oracles.

References

1. Babichenko, Y.: Query complexity of approximate Nash equilibria. J. ACM **63**(4), 36:1–36:24 (2016)
2. Babichenko, Y., Rubinstein, A.: Communication complexity of approximate Nash equilibria. In: Proceedings of the 49th STOC, pp. 878–889. ACM (2017)
3. Babichenko, Y., Barman, S., Peretz, R.: Empirical distribution of equilibrium play and its testing application. Math. Oper. Res. **42**(1), 15–29 (2017)

4. Baldwin, E., Klemperer, P.: Understanding preferences: "demand types", and the existence of equilibrium with indivisibilities. Technical report, LSE, October 2016
5. Brown, G.W.: Some notes on computation of game solutions. RAND corporation report, p. 78, April 1949
6. Bshouty, N.H., Goldberg, P.W., Goldman, S.A., Mathias, H.D.: Exact learning of discretized geometric concepts. SIAM J. Comput. **28**(2), 674–699 (1998)
7. Chen, X., Deng, X., Teng, S.-H.: Settling the complexity of computing two-player Nash equilibria. J. ACM **56**(3), 14:1–14:57 (2009)
8. Daskalakis, C., Pan, Q.: A counter-example to Karlin's strong conjecture for fictitious play. In: Proceedings of 55th FOCS, pp. 11–20 (2014)
9. Daskalakis, C., Goldberg, P.W., Papadimitriou, C.H.: The complexity of computing a Nash equilibrium. SIAM J. Comput. **39**(1), 195–259 (2009)
10. Fearnley, J., Gairing, M., Goldberg, P.W., Savani, R.: Learning equilibria of games via payoff queries. J. Mach. Learn. Res. **16**, 1305–1344 (2015)
11. Fudenberg, D., Levine, D.K.: The Theory of Learning in Games. MIT Press, Cambridge (1998)
12. Goldberg, P.W., Kwek, S.: The precision of query points as a resource for learning convex polytopes with membership queries. In: Proceedings of the 13th COLT, pp. 225–235 (2000)
13. Goldberg, P.W., Roth, A.: Bounds for the query complexity of approximate equilibria. ACM Trans. Econ. Comput. **4**(4), 24:1–24:25 (2016)
14. Goldberg, P.W., Savani, R., Sørensen, T.B., Ventre, C.: On the approximation performance of fictitious play in finite games. Int. J. Game Theory **42**(4), 1059–1083 (2013)
15. Goldberg, P.W., Turchetta, S.: Query complexity of approximate equilibria in anonymous games. J. Comput. Syst. Sci. **90**, 80–98 (2017)
16. Hart, S., Mansour, Y.: How long to equilibrium? The communication complexity of uncoupled equilibrium procedures. Games Econ. Behav. **69**(1), 107–126 (2010)
17. Hart, S., Nisan, N.: The query complexity of correlated equilibria. Games and Economic Behavior (2016). ISSN 0899–8256
18. Kakutani, S.: A generalization of Brouwer's fixed point theorem. Duke Math. J. **8**(3), 457–459 (1941)
19. Klemperer, P.: The product-mix auction: a new auction design for differentiated goods. J. Eur. Econ. Assoc. **8**(2/3), 526–536 (2010)
20. Lipton, R.J., Markakis, E., Mehta, A.: Playing large games using simple strategies. In: Proceedings of the 4th ACM-EC, EC 2003, pp. 36–41. ACM, New York (2003)
21. Nash, J.: Non-cooperative games. Ann. Math. **54**(2), 286–295 (1951)
22. Robinson, J.: An iterative method of solving a game. Ann. Math. **54**(2), 296–301 (1951)
23. Von Stengel, B.: Leadership with commitment to mixed strategies. CDAM Research report (2004)

The Fluid Mechanics of Liquid Democracy

Paul Gölz$^{(\boxtimes)}$ ⓘ, Anson Kahng ⓘ, Simon Mackenzie, and Ariel D. Procaccia ⓘ

Computer Science Department, Carnegie Mellon University, Pittsburgh, USA
{pgoelz,akahng,simonm,arielpro}@cs.cmu.edu

Abstract. *Liquid democracy* is the principle of making collective decisions by letting agents transitively delegate their votes. Despite its significant appeal, it has become apparent that a weakness of liquid democracy is that a small subset of agents may gain massive influence. To address this, we propose to change the current practice by allowing agents to specify multiple delegation options instead of just one. Much like in nature, where—fluid mechanics teaches us—liquid maintains an equal level in connected vessels, so do we seek to control the flow of votes in a way that balances influence as much as possible. Specifically, we analyze the problem of choosing delegations to approximately minimize the maximum number of votes entrusted to any agent, by drawing connections to the literature on confluent flow. We also introduce a random graph model for liquid democracy, and use it to demonstrate the benefits of our approach both theoretically and empirically.

1 Introduction

Liquid democracy is a potentially disruptive approach to democratic decision making. As in direct democracy, agents can vote on every issue by themselves. Alternatively, however, agents may delegate their vote, i.e., entrust it to any other agent who then votes on their behalf. Delegations are transitive; for example, if agents 2 and 3 delegate their votes to 1, and agent 4 delegates her vote to 3, then agent 1 would vote with the weight of all four agents, including herself. Just like representative democracy, this system allows for separation of labor, but provides for stronger accountability: Each delegator is connected to her transitive delegate by a path of personal trust relationships, and each delegator on this path can withdraw her delegation at any time if she disagrees with her delegate's choices.

Although the roots of liquid democracy can be traced back to the work of Miller [15], it is only in recent years that it has gained recognition among practitioners. Most prominently, the German Pirate Party adopted the platform *LiquidFeedback* for internal decision-making in 2010. At the highest point, their installation counted more than 10 000 active users [12]. More recently, two parties—the Net Party in Argentina, and Flux in Australia—have run in national elections on the promise that their elected representatives would vote according

© Springer Nature Switzerland AG 2018
G. Christodoulou and T. Harks (Eds.): WINE 2018, LNCS 11316, pp. 188–202, 2018.
https://doi.org/10.1007/978-3-030-04612-5_13

to decisions made via their respective liquid-democracy-based systems. Although neither party was able to win any seats in parliament, their bids enhanced the promise and appeal of liquid democracy.

However, these real-world implementations also exposed a weakness in the liquid democracy approach: Certain individuals, the so-called *super-voters*, seem to amass enormous weight, whereas most agents do not receive any delegations. In the case of the Pirate Party, this phenomenon is illustrated by an article in Der Spiegel, according to which one particular super-voter's "vote was like a decree," even though he held no office in the party. As Kling et al. [12] describe, super-voters were so controversial that "the democratic nature of the system was questioned, and many users became inactive." Besides the negative impact of super-voters on perceived legitimacy, super-voters might also be more exposed to bribing. Although delegators can retract their delegations as soon as they become aware of suspicious voting behavior, serious damage might be done in the meantime. Furthermore, if super-voters jointly have sufficient power, they might find it more efficient to organize majorities through deals between super-voters behind closed doors, rather than to try to win a broad majority through public discourse. Finally, recent work by Kahng et al. [11] indicates that, even if delegations go only to more competent agents, a high concentration of power might still be harmful for social welfare, by neutralizing benefits corresponding to the Condorcet Jury Theorem.

While all these concerns suggest that the weight of super-voters should be limited, the exact metric to optimize for varies between them and is often not even clearly defined. For the purposes of this paper, we choose to minimize the weight of the heaviest voter. As is evident in the Spiegel article, the weight of individual voters plays a direct role in the perception of super-voters. But even beyond that, we are confident that minimizing this measure will lead to substantial improvements across all presented concerns.

Just how can the maximum weight be reduced? One approach might be to restrict the power of delegation by imposing caps on the weight. However, as argued by Behrens et al. [3], delegation is always possible by coordinating outside of the system and copying the desired delegate's ballot. Pushing delegations outside of the system would not alleviate the problem of super-voters, just reduce transparency. Therefore, we instead adopt a voluntary approach: If agents are considering multiple potential delegates, all of whom they trust, they are encouraged to leave the decision for one of them to a centralized mechanism. With the goal of avoiding high-weight agents in mind, our research challenge is twofold: *First, investigate the algorithmic problem of selecting delegations to minimize the maximum weight of any agent, and, second, show that allowing multiple delegation options does indeed provide a significant reduction in the maximum weight compared to the status quo.*

Put another (more whimsical) way, we wish to design liquid democracy systems that emulate the *law of communicating vessels*, which asserts that liquid will find an equal level in connected containers.

1.1   Our Approach and Results

We formally define our problem in Sect. 2. In general, our problem is closely related to minimizing congestion for confluent flow as studied by Chen et al. [5]. Not only does this connection suggest an optimal algorithm based on mixed integer linear programming, but we also get a polynomial-time $(1 + \log |V|)$-approximation algorithm, where V is the set of direct voters.[1] In addition, we show that approximating our problem to within a factor of $\frac{1}{2} \log_2 |V|$ is NP-hard.

In Sect. 3, to evaluate the benefits of allowing multiple delegations, we propose a probabilistic model for delegation behavior—inspired by the well-known *preferential attachment* model [2]. In a certain class of parameter settings, moving from single delegations to two delegation options per agent decreases the maximum weight doubly exponentially. Our analysis draws on a phenomenon called the *power of choice* that can be observed in many different load balancing models.

In Sect. 4, we show through simulations that our approach continues to outperform classical preferential attachment in more general parameter settings. These improvements in terms of maximum weight persist even if just some fraction of delegators gives two options while the others specify a single delegate. Finally, we find that the approximation algorithm and even a greedy heuristic lead to close-to-optimal maximum weights in our model.

1.2   Related Work

Kling et al. [12] conduct an empirical investigation of the existence and influence of super-voters. The analysis is based on daily data dumps, from 2010 until 2013, of the German Pirate Party installation of LiquidFeedback. As noted above, Kling et al. find that super-voters exist, and have considerable power. The results do suggest that super-voters behave responsibly, as they "do not fully act on their power to change the outcome of votes, and they vote in favour of proposals with the majority of voters in many cases." Of course, this does not contradict the idea that a balanced distribution of power would be desirable.

There are only a few papers that provide theoretical analyses of liquid democracy [7,9,11]. We would like to stress the differences between our approach and the one adopted by Kahng et al. [11]. They consider binary issues in a setting with an objective ground truth, i.e., there is one "correct" outcome and one "incorrect" outcome. In this setting, voters are modeled as biased coins that each choose the correct outcome with an individually assigned probability, or *competence level*. The authors examine whether liquid democracy can increase the probability of making the right decision over direct democracy by having less competent agents delegate to more competent ones. By contrast, our work is completely independent of the (strong) assumptions underlying the results of Kahng et al. In particular, our approach is agnostic to the final outcome of the voting process, does not assume access to information that would be inaccessible in practice, and is compatible with any number of alternatives and choice of

---

[1] Throughout this paper, let log denote the natural logarithm.

voting rule used to aggregate votes. In other words, the goal is not to use liquid democracy to promote a particular outcome, but rather to adapt the process of liquid democracy such that more voices will be heard.

2 Algorithmic Model and Results

Let us consider a delegative voting process where agents may specify multiple potential delegations. This gives rise to a directed graph, whose nodes represent agents and whose edges represent potential delegations. A distinguished subset of nodes corresponds to agents who have voted directly, the *voters*. Since voters forfeit the right to delegate, the voters are a subset of the sinks of the graph. We call all non-voter agents *delegators*.

Each agent has an inherent voting weight of 1. When the delegations will have been resolved, the weight of every agent will be the sum of weights of her delegators plus her inherent weight. We aim to choose a delegation for every delegator in such a way that the maximum weight of any voter is minimized.

This task closely mirrors the problem of congestion minimization for confluent flow (with infinite edge capacity): There, a flow network is also a finite directed graph with a distinguished set of graph sinks, the *flow sinks*. Every node has a non-negative *demand*. If we assume unit demand, this demand is 1 for every node. Since the flow is *confluent*, for every non-sink node, the algorithm must pick exactly one outgoing edge, along which the flow is sent. Then, the *congestion* at a node n is the sum of congestions at all nodes who direct their flow to n plus the demand of n. The goal in congestion minimization is to minimize the maximum congestion at any flow sink.

In spite of the similarity between confluent flow and resolving potential delegations, the two problems differ when a node has no path to a voter/flow sink. In confluent flow, the result would simply be that no flow exists. In our setting however, this situation can hardly be avoided. If, for example, several friends assign all of their potential delegations to each other, and if all of them rely on the others to vote, their weight cannot be delegated to any voter. Our mechanism cannot simply report failure as soon as a small group of voters behaves in an unexpected way. Thus, it must be allowed to leave these votes unused. At the same time, of course, our algorithm should not exploit this power to decrease the maximum weight, but must primarily maximize the number of utilized votes. We formalize these issues in the following section.

2.1 Problem Statement

All graphs $G = (N, E)$ mentioned in this section will be finite and directed. Furthermore, they will be equipped with a set of voters $V \subseteq sinks(G)$.

Some of these graphs represent situations in which all delegations have already been resolved and in which each vote reaches a voter: We call a graph (N, E) with voters V a *delegation graph* if it is acyclic, its sinks are exactly the set V, and every other vertex has outdegree one. In such a graph, define the *weight* $w(n)$ of a node $n \in N$ as

$$w(n) := 1 + \sum_{(m,n) \in E} w(m).$$

This is well-defined because E is a well-founded relation on N.

Resolving the delegations of a graph G with voters V can now be described as the MINMAXWEIGHT problem: Among all delegation subgraphs (N', E') of G with voting vertices V of maximum $|N'|$, find one that minimizes the maximum weight of the voting vertices.

2.2   Connections to Confluent Flow

We recall definitions from the flow literature as used by Chen et al. [5]. We slightly simplify the exposition by assuming unit demand at every node.

Given a graph (N, E) with V, a *flow* is a function $f : E \to \mathbb{R}_{\geq 0}$. For any node n, set $in(n) := \sum_{(m,n) \in E} f(m, n)$ and $out(n) := \sum_{(n,m) \in E} f(n, m)$. At every node $n \in N \setminus V$, a flow must satisfy *flow conservation*: $out(n) = 1 + in(n)$. The *congestion* at any node n is defined as $1 + in(n)$. A flow is *confluent* if every node has at most one outgoing edge with positive flow. We define MINMAXCONGESTION as the problem of finding a confluent flow on a given graph such that the maximum congestion is minimized.

In the full version of this paper [8], we give translations between instances of MINMAXWEIGHT and MINMAXCONGESTION that preserve the optimization objective value.

2.3   Algorithms

These translations allow us to apply algorithms—even approximation algorithms—for MINMAXCONGESTION to our MINMAXWEIGHT problem, that is, we can reduce the latter problem to the former.

Theorem 1. *Let $\mathcal{A}$ be an algorithm for MINMAXCONGESTION with approximation ratio $c \geq 1$. Let $\mathcal{A}'$ be an algorithm that, given (N, E) with V, runs $\mathcal{A}$ on the active subgraph, and translates the result into a delegation subgraph by eliminating all zero-flow edges. Then $\mathcal{A}'$ is a c-approximation algorithm for* MINMAXWEIGHT.

We relegate the proof to the full version [8]. Note that Theorem 1 works for $c = 1$, i.e., even for exact algorithms. Therefore, it is possible to solve MIN-MAXWEIGHT by adapting a standard mixed integer linear programming (MILP) formulation of MINMAXFLOW.

Since this algorithm is based on solving an NP-hard problem, it might be too inefficient for typical use cases of liquid democracy with many participating agents. Fortunately, it might be acceptable to settle for a slightly non-optimal maximum weight if this decreases computational cost. To our knowledge, the best polynomial approximation algorithm for MINMAXCONGESTION is due to Chen et al. [5] and achieves an approximation ratio of $1 + \log |V|$. Their algorithm starts by computing the optimal solution to the splittable-flow version of the problem, by solving a linear program. The heart of their algorithm is a non-trivial, deterministic rounding mechanism. This scheme drastically outperforms the natural, randomized rounding scheme, which leads to an approximation ratio of $\Omega(|N|^{1/4})$ with arbitrarily high probability [6].

2.4   Hardness of Approximation

In this section, we demonstrate the NP-hardness of approximating the MIN-MAXWEIGHT problem to within a factor of $\frac{1}{2} \log_2 |V|$. On the one hand, this justifies the absence of an exact polynomial-time algorithm. On the other hand, this shows that the approximation algorithm is optimal up to a multiplicative constant.

Theorem 2. *It is NP-hard to approximate the* MINMAXWEIGHT *problem to a factor of* $\frac{1}{2} \log_2 |V|$, *even when each node has outdegree at most* 2.

Again, the proof can be found in the full version [8]. Not surprisingly, we derive hardness via a reduction from MINMAXCONGESTION, i.e., a reduction in the opposite direction from the one given in Theorem 1. As shown by Chen et al. [5], approximating MINMAXCONGESTION to within a factor of $\frac{1}{2} \log_2 |V|$ is NP-hard. However, in our case, nodes have unit demands. Moreover, we are specifically interested in the case where each node has outdegree at most 2, as in practice we expect outdegrees to be very small, and this case plays a special role in the following section.

3   Probabilistic Model and Results

Our generalization of liquid democracy to multiple potential delegations aims to decrease the concentration of weight. Accordingly, the success of our approach should be measured by its effect on the maximum weight in real elections. Since, at this time, we do not know of any available datasets,[2] we instead propose a probabilistic model for delegation behavior, which can serve as a credible proxy. Our model builds on the well-known preferential attachment model, which generates graphs possessing typical properties of social networks.

The evaluation of our approach will be twofold: In Sects. 3.2 and 3.3, for a certain choice of parameters in our model, we establish a striking separation between traditional liquid democracy and our system. In the former case, the maximum weight at time t is $\Omega(t^{\beta})$ for a constant β with high probability,

---

[2] There is one relevant dataset that we know of, which was analyzed by Kling et al. [12]. However, due to stringent privacy constraints, the data privacy officer of the German Pirate Party was unable to share this dataset with us.

whereas in the latter case, it is in $\mathcal{O}(\log \log t)$ with high probability, even if each delegator only suggests two options. For other parameter settings, we empirically corroborate the benefits of our approach in Sect. 4.

3.1 The Preferential Delegation Model

Many real-world social networks have degree distributions that follow a power law [13,16]. Additionally, in their empirical study, Kling et al. [12] observed that the weight of voters in the German Pirate Party was "power law-like" and that the graph had a very unequal indegree distribution. In order to meld the previous two observations in our liquid democracy delegation graphs, we adapt a standard preferential attachment model [2] for this specific setting. On a high level, our *preferential delegation* model is characterized by three parameters: $0 < d < 1$, the probability of delegation; $k \geq 1$, the number of delegation options from each delegator; and $\gamma \geq 0$, an exponent that governs the probability of delegating to nodes based on current weight.

At time $t = 1$, we have a single node representing a single voter. In each subsequent time step, we add a node for agent i and flip a biased coin to determine her delegation behavior. With probability d, she delegates to other agents. Else, she votes independently. If i does not delegate, her node has no outgoing edges. Otherwise, add edges to k many i.i.d. selected, previously inserted nodes, where the probability of choosing node j is proportional to $(indegree(j) + 1)^\gamma$. Note that this model might generate multiple edges between the same pair of nodes, and that all sinks are voters. Figure 1 shows example graphs for different settings of γ.

(a) $k = 2, d = 0.5, \gamma = 0$ (b) $k = 2, d = 0.5, \gamma = 1$

Fig. 1. Example graphs generated by the preferential delegation model.

In the case of $\gamma = 0$, which we term *uniform delegation*, a delegator is equally likely to attach to any previously inserted node. Already in this case, a "rich-get-richer" phenomenon can be observed, i.e., voters at the end of large networks of potential delegations will likely see their network grow even more. Indeed, a larger network of delegations is more likely to attract new delegators. In traditional liquid democracy, where $k = 1$ and all potential delegations will be realized, this explains the emergence of super-voters with excessive weight observed by Kling et al. [12]. We aim to show that for $k \geq 2$, the resolution of potential delegations can strongly outweigh these effects. In this, we profit from an effect known as the "power of two choices" in load balancing described by Azar et al. [1].

For $\gamma > 0$, the "rich-get-richer" phenomenon additionally appears at the degrees of nodes. Since the number of received potential delegations is a proxy for an agent's competence and visibility, new agents are more likely to attach to agents with high indegree. In total, this is likely to further strengthen the inherent inequality between voters. For increasing γ, the graph becomes increasingly flat, as a few super-voters receive nearly all delegations. This matches observations from the LiquidFeedback dataset [12] that "the delegation network is slowly becoming less like a friendship network, and more like a bipartite networks of super-voters connected to normal voters." The special case of $\gamma = 1$ corresponds to preferential attachment as described by Barabási and Albert [2].

The most significant difference we expect to see between graphs generated by the preferential delegation model and real delegation graphs is the assumption that agents always delegate to more senior agents. In particular, this causes generated graphs to be acyclic, which need not be the case in practice. It does seem plausible that the majority of delegations goes to agents with more experience on the platform. Even if this assumption should not hold, there is a second interpretation of our process if we assume—as do Kahng et al. [11]—that agents can be ranked by competence and only delegate to more competent agents. Then, we can think of the agents as being inserted in decreasing order of competence. When a delegator chooses more competent agents to delegate to, her choice would still be biased towards agents with high indegree, which is a proxy for popularity.

In our theoretical results, we focus on the cases of $k = 1$ and $k = 2$, and assume $\gamma = 0$ to make the analysis tractable. The parameter d can be chosen freely between 0 and 1. Note that our upper bound for $k = 2$ directly translates into an upper bound for larger k, since the resolution mechanism always has the option of ignoring all outgoing edges except for the two first. Therefore, to understand the effect of multiple delegation options, we can restrict our attention to $k = 2$. This crucially relies on $\gamma = 0$, where potential delegations do not influence the probabilities of choosing future potential delegations. Based on related results by Malyshkin and Paquette [14], it seems unlikely that increasing k beyond 2 will reduce the maximum weight by more than a constant factor.

3.2   Lower Bounds for Single Delegation ($k = 1$, $\gamma = 0$)

As mentioned above, we begin with a lower bound on the maximum weight for the case of uniform delegation and a single delegation option per delegator. Here and in the following, we say that a sequence $(\mathcal{E}_t)_t$ of events happens *with high probability* if $\mathbb{P}[\mathcal{E}_t] \to 1$ for $t \to \infty$. Our lower bound, whose proof is relegated to the full version [8], is the following:

Theorem 3. *In the preferential delegation model with $k = 1$, $\gamma = 0$, and $d \in (0, 1)$, with high probability, the maximum weight of any voter at time t is in $\Omega(t^\beta)$, where $\beta > 0$ is a constant that depends only on d.*

3.3 Upper Bound for Double Delegation ($k = 2$, $\gamma = 0$)

Analyzing cases with $k > 1$ is considerably more challenging. One obstacle is that we do not expect to be able to incorporate optimal resolution of potential delegations into our analysis, because the computational problem is hard even when $k = 2$ (see Theorem 2). Therefore, we give a pessimistic estimate of optimal resolution via a greedy delegation mechanism, which we can reason about alongside the stochastic process. Clearly, if this stochastic process can guarantee an upper bound on the maximum weight with high probability, this bound must also hold if delegations are optimally resolved to minimize maximum weight.

In more detail, whenever a new delegator is inserted into the graph, the greedy mechanism immediately selects one of the delegation options. As a result, at any point during the construction of the graph, the algorithm can measure the weight of the voters. Suppose that a new delegator suggests two delegation options, to agents a and b. By following already resolved delegations, the mechanism obtains voters a^* and b^* such that a transitively delegates to a^* and b to b^*. The greedy mechanism then chooses the delegation whose voter currently has lower weight, resolving ties arbitrarily.

This situation is reminiscent of a phenomenon known as the "power of choice." In its most isolated form, it has been studied in the *balls-and-bins* model, for example by Azar et al. [1]. In this model, n balls are to be placed in n bins. In the classical setting, each ball is sequentially placed into a bin chosen uniformly at random. With high probability, the fullest bin will contain $\Theta(\log n / \log \log n)$ balls at the end of the process. In the choice setting, two bins are independently and uniformly selected for every ball, and the ball is placed into the emptier one. Surprisingly, this leads to an exponential improvement, where the fullest bin will contain at most $\Theta(\log \log n)$ balls with high probability.

We show that, at least for $\gamma = 0$ in our setting, this effect outweighs the "rich-get-richer" dynamic described earlier:

Theorem 4. *In the preferential delegation model with $k = 2$, $\gamma = 0$, and $d \in (0, 1)$, the maximum weight of any voter at time t is $\log_2 \log t + \Theta(1)$ with high probability.*

Due to space constraints, we defer the proof to the full version [8]. In our proof we build on work by Malyshkin and Paquette [14], who study the maximum *degree* in a graph generated by preferential attachment with the power of choice. In addition, we incorporate ideas by Haslegrave and Jordan [10].

4 Simulations

In this section, we present our simulation results, which support the two main messages of this paper: that allowing multiple delegation options significantly reduces the maximum weight, and that it is computationally feasible to resolve delegations in a way that is close to optimal.

Our simulations were performed on a MacBook Pro (2017) on MacOS 10.12.6 with a 3.1 GHz Intel Core i5 and 16 GB of RAM. All running times were measured with at most one process per processor core. Our simulation software is written in Python 3.6 using Gurobi 8.0.1 to solve MILPs. All of our simulation code is open-source and available at https://github.com/pgoelz/fluid.

4.1   Multiple vs. Single Delegations

For the special case of $\gamma = 0$, we have established a doubly exponential, asymptotic separation between single delegation ($k = 1$) and two delegation options per delegator ($k = 2$). While the strength of the separation suggests that some of this improvement will carry over to the real world, we still have to examine via simulation whether improvements are visible for realistic numbers of agents and other values of γ.

To this end, we empirically evaluate two different mechanisms for resolving delegations. First, we optimally resolve delegations by solving an MILP for confluent flow. Our second mechanism is the greedy "power of choice" algorithm used in the theoretical analysis and introduced in Sect. 3.3.

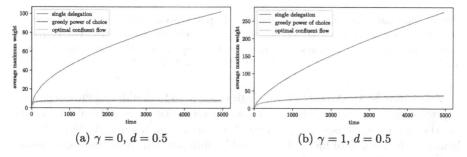

(a) $\gamma = 0$, $d = 0.5$ (b) $\gamma = 1$, $d = 0.5$

Fig. 2. Maximum weight averaged over 100 simulations of length 5 000 time steps each. Maximum weight has been computed every 50 time steps.

In Fig. 2, we compare the maximum weight produced by a single-delegation process to the optimal maximum weight in a double-delegation process, for different values of γ. Corresponding figures for different values of d and γ can be found in the full version [8].

These simulations show that our asymptotic findings translate into considerable differences even for small numbers of agents, across different values of d. Moreover, these differences remain nearly as pronounced for values of γ up to 1, which corresponds to classical preferential attachment. This suggests that our mechanism can outweigh the social tendency towards concentration of votes; however, evidence from real-world elections is needed to settle this question. Lastly, we would like to point out the similarity between the graphs for the optimal maximum weight and the result of the greedy algorithm, which indicates that a large part of the separation can be attributed to the power of choice.

If we increase γ to large values, the separation between single and double delegation disappears. As we show in the full version [8], there are even combinations of $\gamma > 1$ and d such that the curve for single delegation falls below the ones for double delegation. In these settings, since a large fraction of delegators give two identical delegation options, any resolution mechanism has virtually no leverage. In the double delegation setting, indegrees grow faster, which makes the delegations concentrate toward a single voter more quickly than in classical liquid democracy, leading to a wildly unrealistic concentration of weight. Thus, it seems that large values of γ do not actually describe our scenario of multiple delegations.

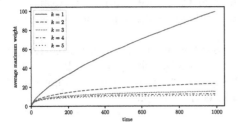

Fig. 3. Optimal maximum weight for different k averaged over 100 simulations, computed every 10 steps. $\gamma = 1$, $d = 0.5$.

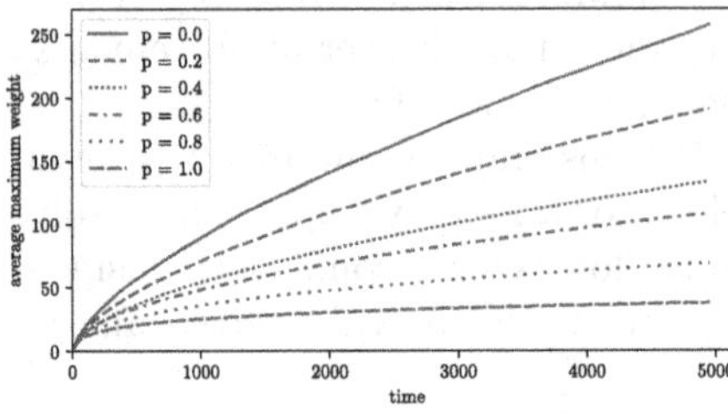

Fig. 4. Optimal maximum weight averaged over 100 simulations. Voters give two delegations with probability p; else one. $\gamma = 1$, $d = 0.5$.

As we have seen, switching from single delegation to double delegation greatly improves the maximum weight in plausible scenarios. It is natural to wonder whether increasing k beyond 2 will yield similar improvements. As Fig. 3 shows, however, the returns of increasing k quickly diminish, which is common to many incarnations of the power of choice [1].

4.2   Evaluating Mechanisms

Already the case of $k = 2$ appears to have great potential; but how easily can we tap it?

We have observed that, on average, the greedy "power of choice" mechanism comes surprisingly close to the optimal solution. However, this greedy mechanism depends on seeing the order in which our random process inserts agents and on the fact that all generated graphs are acyclic, which need not be true in practice. If the graphs were acyclic, we could simply first sort the agents topologically and then present the agents to the greedy mechanism in reverse order. On arbitrary active graphs, we instead proceed through the strongly connected components in reversed topological order, breaking cycles and performing the greedy step over the agents in the component. To avoid giving the greedy algorithm an unfair advantage, we use this generalized greedy mechanism throughout this section. Thus, we compare the generalized greedy mechanism, the optimal solution, the $(1 + \log |V|)$-approximation algorithm[3] and a random mechanism that materializes a uniformly chosen option per delegator.

On a high level, we find that both the generalized greedy algorithm and the approximation algorithm perform comparably to the optimal confluent flow solution. Across a variety of values of d and γ examined in the full paper [8], all three mechanisms seem similarly effective in exploiting the advantages of double delegation.

The similar success of these three mechanisms might indicate that our probabilistic model for $k = 2$ generates delegation networks that have low maximum weights for arbitrary resolutions. However, this is not the case: The random mechanism does quite poorly already on small instances, and the gap between random and the other mechanisms only grows further as t increases, as indicated by Fig. 5. In general, the graph for random delegations looks more similar to single delegation than to the other mechanisms on double delegation. Indeed, for $\gamma = 0$, random delegation is equivalent to the process with $k = 1$, and, for higher values of γ, it performs even slightly worse since the unused delegation options make the graph more centralized (see the full paper [8]). Thus, if simplicity is a primary desideratum, we recommend using the generalized greedy algorithm.

As Fig. 6 and additional measurements detailed in the full version [8] demonstrate, all three other mechanisms, including the optimal solution, easily scale to input sizes as large as the largest implementations of liquid democracy to date.

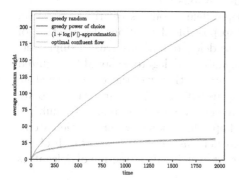

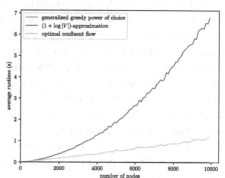

Fig. 5. Maximum weight per algorithm for $d = 0.5$, $\gamma = 1$, $k = 2$, averaged over 100 simulations.

Fig. 6. Running time of mechanisms on graphs for $d = 0.5$, $\gamma = 1$, averaged over 20 simulations.

---

[3] For one of their subprocedures, instead of directly optimizing a convex program, Chen et al. [5] reduce this problem to finding a lexicographically optimal maximum flow in $\mathcal{O}(n^5)$. We choose to directly optimize the convex problem in Gurobi, hoping that this will increase efficiency in practice.

5 Discussion

The approach we have presented and analyzed revolves around the idea of allowing agents to specify multiple delegation options, and selecting one such option per delegator. A natural variant of this approach corresponds to splittable—instead of confluent—flow. In this variant, the mechanism would not have to commit to a single outgoing edge per delegator. Instead, a delegator's weight could be split into arbitrary fractions between her potential delegates. Indeed, such a variant would be computationally less expensive, and the maximum voting weight can be no higher than in our setting. However, we view our concept of delegation as more intuitive and transparent: Whereas, in the splittable setting, a delegator's vote can disperse among a large number of agents, our mechanism assigns just one representative to each delegator. As hinted at in the introduction, this is needed to preserve the high level of accountability guaranteed by classical liquid democracy. We find that this fundamental shortcoming of splittable delegations is not counterbalanced by a marked decrease in maximum weight. Indeed, representative empirical results given in the full version [8] show that the maximum weight trace is almost identical under splittable and confluent delegations. Furthermore, note that in the preferential delegation model with $k = 1$, splittable delegations do not make a difference, so the lower bound given in Theorem 3 goes through. And, when $k \geq 2$, the upper bound of Theorem 4 directly applies to the splittable setting. Therefore, our main technical results in Sect. 3 are just as relevant to splittable delegations.

To demonstrate the benefits of multiple delegations as clearly as possible, we assumed that every agent provides two possible delegations. In practice, of course, we expect to see agents who want to delegate but only trust a single person to a sufficient degree. This does not mean that delegators should be required to specify multiple delegations. For instance, if this was the case, delegators might be incentivized to pad their delegations with very popular agents who are unlikely to receive their votes. Instead, we encourage voters to specify multiple delegations on a voluntary basis, and we hope that enough voters participate to make a significant impact. Fortunately, as demonstrated in Fig. 4, much of the benefits of multiple delegation options persist even if only a fraction of delegators specify two delegations.

Without doubt, a centralized mechanism for resolving delegations wields considerable power. Even though we only use this power for our specific goal of minimizing the maximum weight, agents unfamiliar with the employed algorithm might suspect it of favoring specific outcomes. To mitigate these concerns, we propose to divide the voting process into two stages. In the first, agents either specify their delegation options or register their intent to vote. Since the votes themselves have not yet been collected, the algorithm can resolve delegations without seeming partial. In the second stage, voters vote using the generated delegation graph, just as in classic liquid democracy, which allows for transparent decisions on an arbitrary number of issues. Additionally, we also allow delegators to change their mind and vote themselves if they are dissatisfied with how delegations were resolved. This gives each agent the final say on their share of votes, and can only further reduce the maximum weight achieved by our mechanism. We believe that this process, along with education about the mechanism's goals and design, can win enough trust for real-world deployment.

Beyond our specific extension, one can consider a variety of different approaches that push the current boundaries of liquid democracy. For example, in a recent position paper, Brill [4] raises the idea of allowing delegators to specify a ranked list of potential representatives. His proposal is made in the context of alleviating delegation cycles, whereas our focus is on avoiding excessive concentration of weight. But, on a high level, both proposals envision centralized mechanisms that have access to richer inputs from agents. Making and evaluating such proposals now is important, because, at this early stage in the evolution of liquid democracy, scientists can still play a key role in shaping this exciting paradigm.

Acknowledgments. We are grateful to Miklos Z. Racz for very helpful pointers to analyses of preferential attachment models.

References

1. Azar, Y., Broder, A.Z., Karlin, A.R., Upfal, E.: Balanced allocations. In: Proceedings of the 26th Annual ACM Symposium on Theory of Computing (STOC), pp. 593–602 (1994)
2. Barabási, A.L., Albert, R.: Emergence of scaling in random networks. Science **286**, 509–512 (1999)
3. Behrens, J., Kistner, A., Nitsche, A., Swierczek, B.: The principles of LiquidFeedback. Interaktive Demokratie (2014)
4. Brill, M.: Interactive democracy. In: Proceedings of the 17th International Conference on Autonomous Agents and Multi-agent Systems (AAMAS) (2018)
5. Chen, J., Kleinberg, R.D., Lovász, L., Rajaraman, R., Sundaram, R., Vetta, A.: (Almost) tight bounds and existence theorems for single-commodity confluent flows. J. ACM **54**(4) (2007). Article 16
6. Chen, J., Rajaraman, R., Sundaram, R.: Meet and merge: Approximation algorithms for confluent flows. J. Comput. Syst. Sci. **72**(3), 468–489 (2006)
7. Christoff, Z., Grossi, D.: Binary voting with delegable proxy: An analysis of liquid democracy. In: Proceedings of the 16th Conference on Theoretical Aspects of Rationality and Knowledge (TARK), pp. 134–150 (2017)
8. Gölz, P., Kahng, A., Mackenzie, S., Procaccia, A.D.: The fluid mechanics of liquid democracy (2018). https://arxiv.org/pdf/1808.01906.pdf
9. Green-Armytage, J.: Direct voting and proxy voting. Const. Polit. Econ. **26**(2), 190–220 (2015)
10. Haslegrave, J., Jordan, J.: Preferential attachment with choice. Random Struct. Algorithms **48**(4), 751–766 (2016)
11. Kahng, A., Mackenzie, S., Procaccia, A.D.: Liquid democracy: An algorithmic perspective. In: Proceedings of the 32nd AAAI Conference on Artificial Intelligence (AAAI) (2018)
12. Kling, C.C., Kunegis, J., Hartmann, H., Strohmaier, M., Staab, S.: Voting behaviour and power in online democracy. In: Proceedings of the 9th International AAAI Conference on Web and Social Media (ICWSM), pp. 208–217 (2015)

13. Kumar, R., Novak, J., Tomkins, A.: Structure and evolution of online social networks. In: Proceedings of the 12th International Conference on Knowledge Discovery and Data Mining (KDD), pp. 611–617 (2006)
14. Malyshkin, Y., Paquette, E.: The power of choice over preferential attachment. Lat. Am. J. Probab. Math. Stat. **12**(2), 903–915 (2015)
15. Miller, J.C.: A program for direct and proxy voting in the legislative process. Public Choice **7**(1), 107–113 (1969)
16. Newman, M.E.J.: Clustering and preferential attachment in growing networks. Phys. Rev. E **64**(2), 1–13 (2001)

Equilibrium and Inefficiency in Multi-product Cournot Games

Mohit Hota$^{(\boxtimes)}$ and Sanjiv Kapoor

Illinois Institute of Technology, Chicago, IL 60616, USA
mhota@hawk.iit.edu, kapoor@iit.edu

Abstract. We consider multi-product (m products) Cournot games played by n firms where products are substitutable goods. Such games can arise in network markets and in general is motivated by markets with differentiated goods and differing producer costs that can be arbitrary, especially due to subsidies. We provide strongly polynomial algorithms for computing the Nash equilibrium for Cournot games with quadratic utility functions. To study the inefficiency, we provide a characterization of Nash equilibrium in multi-product oligopolies with concave utilities and uniform substitutability in terms of games with quadratic utilities. We show that the *Price of Anarchy* in these games is bounded below by 2/3.

1 Introduction

We study social welfare in the context of Cournot games in a differentiated product market. One of the motivations is to investigate the impact of heterogeneous costs on the efficiency of competitive good production. Digital distribution and on-line content delivery is growing more competitive every year with increasing demand and content providers (CP) are vying for higher proportions of premium bandwidth so as to provide enhanced quality of service. Differentiation in internet markets is provided by different levels of services, which are substitutable to a certain extent. To manage costs, leading internet service providers (ISP) are providing data services to consumers that is sponsored by content providers who are vying to gain consumer subscriptions. This includes content providers like Facebook, and the Netflix- T-mobile [3] and AT&T- HBO partnerships [10].

A model of how sponsored data plans alter the payment structure to ISPs has been illustrated in [14]. The net impact of CPs providing payments to ISPs is to alter the cost structure of specific ISPs. Does this lead to loss of efficiency or unfair competition? A study of Cournot games which include producer dependent production costs provides an insight into this question. The use of sponsored data plans is one illustration of the change of costs incurred by ISPs and users. Subsidies of other types may also significantly change the cost. There has been substantial work in this area, including the use of game theoretic frameworks [2].

© Springer Nature Switzerland AG 2018
G. Christodoulou and T. Harks (Eds.): WINE 2018, LNCS 11316, pp. 203–217, 2018.
https://doi.org/10.1007/978-3-030-04612-5_14

We consider oligopolistic competition in a differentiated products market with m products. The study of markets in the context of oligopolistic competition has been modeled by Cournot and Bertrand games. Of particular interest are Cournot games where firms decide production levels in an effort to clear markets, while maximizing their profit. These games have been studied extensively by multiple authors [4,6,8,22,25,26] who have considered relationships between equilibrium prices, production costs etc. The study of social welfare in oligopolistic competition has been less extensive and is more recent [11,18]. Quantification of the loss in social welfare as measured by *Price of Anarchy* has been introduced in [19], where this term denotes the ratio between the social welfare, as defined by a specific objective function, at the worst-case Nash equilibrium and the optimum social welfare. Bounds on *Price of Anarchy* have since been extensively studied in multiple contexts, notably in transportation networks that were first studied by Roughgarden and Tardos in [24].

We use game theoretic market models to further our understanding of the impact of cost structures on pricing and profits and analyze its impact on the relative benefits accrued to the production of goods. Our model is based on a very generic Cournot model of competition with multiple goods and multiple producers. This model has been considered previously by multiple papers, most notably in [18,21], where in [21], Ledvina and Sircar consider multiple producers each producing a distinct product, the products being substitutable and the impact of substitutability being measured by a parameter γ. They analyze the number of firms that will be in the market as a function of product substitutability. In [18], Kluberg and Perakis study a general version where prices are dependent on producers and provide Price of Anarchy results when good substitutability is modeled by a matrix M, assumed to be a diagonally dominant M-matrix. In this paper, we focus on substitutability and use a model that generalizes the consumer utility model as defined in [21] and [18]. We model utilities by a concave function, generalizing the quadratic utility model used in the previous papers. In addition to incorporating multiple products, an important aspect of our model is the cross product impact or the impact of substitutability of one good on the other. Product substitutability is modeled by an interaction matrix that is positive semi-definite. When goods are similar, substitutability is modeled by an uniform interaction matrix, as defined in [21]. The provision of subsidies, either regulatory or through other market participants can alter the cost structure of the producers. Thus our model emphasizes distinct costs. We provide a constant tight bound of $\frac{2}{3}$ for the price of anarchy, using the social cost metric, in Cournot games when the games have a uniform interaction matrix. Another recent work [20] that addresses the interaction between producers and consumers in a networked market also considers substitutability in the form of parallel edges; however the price function of a good-producer pair does not include the impact of other substitutable goods.

Efficiency loss in single product games have been the attention of multiple papers, including [16] that shows a bound of $\frac{2}{3}$ for a single product game. Tsitsiklis and Xu [28,29] focus on analyzing convex inverse demand functions and its inefficiency as compared to affine price functions but do not consider multiple goods and the interaction between them via substitutability. In both papers, the model consists of a single homogeneous good; and while their inefficiency bounds (PoA) for affine inverse demand function is shown to be $\frac{2}{3}$, for convex inverse demand functions the bound can be arbitrarily low. In contrast, our bound of $\frac{2}{3}$ holds for concave inverse demand functions for games with multiple goods which are substitutable. In the case of a single good and affine inverse demand functions, this bound coincides with the bound in [28,29].

Additional research that addresses inefficiency in Cournot markets may be found in [11] which considers markets with a single homogeneous good. An analysis from the viewpoint of coalitions may be found in [13]. The study of multi-product oligopolies faces challenges, especially when the demand functions are non-linear and typically the action spaces are unbounded. In the paper [23], the existence of equilibrium in multi-product Cournot oligopolies under certain conditions has been proven. Multi-product oligopolistic competition in the context of global trade has also been considered in [7] when firms are identical. Other models that have been considered include models with user types [17]. The economics literature addresses equilibrium existence and uniqueness in the Cournot model. Equilibrium existence and uniqueness are ensured if the reciprocal of demand is convex [5].

In this paper, our first set of results provide polynomial time algorithms for computing Nash equilibrium. Recent related work includes the work in Abolhassani et al. [1] where they study multi-market Cournot model (by utilizing a bipartite graph to model consumers and products (markets)) and provide algorithms for computing Nash equilibrium. The network market defined in their paper has only one product, and the network impact is accounted for by the cost function. Since they consider a single product they do not consider substitutability of one good with another; their model thus differs substantially in this aspect. They also provide algorithms to compute equilibrium in network markets using convex programming and Linear Complementarity, and a combinatorial algorithm relying on binary search for the produced quantity in a market with one good. We use a primal-dual approach to solve the multi-dimensional problem of determining the equilibrium involving multiple goods. In [27], Harks and Timmermans consider the computation of equilibrium in multi-markets and reduce the oligopoly equilibrium to computing the equilibrium in an atomic splittable congestion game. While they use a price function that is dependent on the production of the good, they do not consider any interaction between the different goods, which is the focus of this paper.

Our Contributions. This paper

- Defines a model that considers consumer utility in a market with multiple goods and multiple producers where the consumer utility is concave and the demand of each good is based on a price function which is concave. Our model generalizes the models in [21] for multi-product oligopolistic competition with heterogeneous producers. The goods are modeled as substitutes utilizing a matrix B, termed the product interaction matrix. We provide Price of Anarchy bounds for the uniform case (modeled by an interaction matrix that has the substitutability parameter γ where the product's impact on its own demand is by a factor β). Contrast this with the analysis in [18], where the matrix M is assumed to be diagonally dominant. To understand the difference, we note that in the uniform case the property of diagonally dominant would imply that γ is asymptotically vanishing as compared to β for large number of products.

- Provides a strongly polynomial algorithm for computing Nash equilibrium in the case of quadratic utilities. The Nash equilibrium solution is determined for the most general case of n producers and m distinct products, using a primal-dual technique. To our knowledge, there are no existing comparable algorithms.
- Determines the *Price of Anarchy (PoA)* in the case of concave utilities but with uniform product interaction matrix. This model captures the impact of competition since producers can be divided into two categories, those that have cost sponsorship or subsidies and those that do not. With n producers and m product classes or goods, we show that the $PoA \geq \frac{2}{3}$ or more specifically, $\frac{2n+4}{3n+5}$, which is tight. While our bound suggests that inefficiency in social welfare is bounded, we note that the socially optimum solution favors disparity in production costs. This points us towards future research on fairness.

In Sect. 2, we define the multi-product Oligopoly model and introduce terminology. Section 3 provides the strongly polynomial algorithm for games with quadratic utility functions. In Sect. 4, we present bounds on the Price of Anarchy for games with uniform product interaction matrices. Due to space constraints, we provide proofs for our claims and results in a fuller version [12].

2   Model Formulation

We consider a general Cournot competitive model involving n producers and m goods. The game is defined as $G = (N, M, (U_i)_{i \in N}, (c_q^i)_{i \in N, q \in M})$, where $N = \{1, \ldots n\}$ is the set of producers and $M = \{1, \ldots m\}$ is the set of product classes (goods) to be produced. U_i is the utility function for producer i and c_q^i denotes the cost of producing a unit of good q incurred by producer i. We use $X^i \in \mathbb{R}_+^m$ to represent the production vector of the ith producer, i.e. $X^i = (X_1^i, \ldots, X_m^i)$ where X_q^i represents the production of the qth product. The total production schedule is represented by $X = (X^1, \ldots, X^n) \in \mathbb{R}^{n \times m}$. We also let $X_q = \sum_i X_q^i$, where $X_q \in \mathbb{R}$, represents the composite production of good q.

Given a production schedule $X = (X_q^i)_{\forall i, q}$, producer i's utility function, provided by the profit per unit of good sold, is given as $U_i(X^i, X^{-i}) = \sum_q X_q^i (p_q^i(X) - c_q^i)$. Here, $p_q^i(X)$ is the inverse demand (price) function, i.e., the price of good q offered by producer i. For simplicity, we denote $U_i(X^i, X^{-i})$ as U_i.

We consider an aggregate consumer c as a representative of all the consumers in the system, and introduce a general class of consumer utility functions, U_c, for the aggregate consumer, based on the models in [16,18,21,30]. We allow goods to be substitutable, and represent the interaction between goods by using a matrix $B \in \mathbb{R}^{nm \times nm}$, termed the *product interaction matrix*, similar to the one used in [18]. The consumer utility is defined by $U_c = L^T X - f\left(X^T B X\right)$ where $f(y)$ is a convex non-decreasing function; vector $L = \left(l_1^1, l_2^1, \cdots, l_m^1, l_1^2, \cdots, l_m^n\right)$, with nm components, represents the marginal utility of the goods for the first unit of production; and $X^T B X$ is the quadratic form that captures the effect of inter-product substitutability. We assume B to be positive semi-definite, defining a convex function. Consequently, U_c is concave w.r.t X. The optimum demand vector of the consumer generates the price function $p = L - f'\left(X^T B X\right) B X$ where $f'(X^T B X)$ is a convex function. This causes the price function p to be concave. In B, we index the row corresponding to producer i and good q as $B_{(iq)}$. Individual entries in B are indexed by the pair $(iq, i'q')$, and represented by $B_{(iq),(i'q')}$.

We also consider a meaningful restriction of the problem, i.e., when the function $f(y)$ is affine, then we can express the consumer utility as a quadratic function. For convenience, we assume consumer utility and price functions to be of the forms: $U_c = L^T X - \frac{1}{2} X^T B X$, and $p = L - BX$. This model has also been utilized in [18] and is standard in the economics literature [30]. We refer to this model as the *quadratic utility model*.

The matrix B that represents the interaction between products can, in general, be arbitrary. To model substitutability among similar products, we assume B to have a uniform structure, i.e., $\forall q \neq q' B_{(iq,iq')} = \gamma$, with γ representing a common substitutability parameter. Such uniform product interaction for producers has been considered in the context of differentiated product Cournot games in [21]. We term this model as the *uniform interaction model*. In this model, the quadratic utility function $\bar{U}_c$ yields a linear price function $p_q(X)$, described below.

$$\bar{U}_c = \sum_q \left(l_q X_q - \frac{1}{2}\left(\beta(X_q)^2 + \sum_{q'} \gamma X_q X_{q'}\right)\right) \tag{1}$$

$$p_q(X) = l_q - \beta X_q - \gamma \sum_{q' \neq q} X_{q'} \tag{2}$$

Note that $0 \leq \gamma \leq \beta$ and represents the impact of substitute products in differentiated oligopolistic markets.

Nash Equilibrium. We define that a game $G = (N, M, (U_i)_{i \in N}, (c_q^i)_{i \in N, q \in M})$ is at a Nash (Cournot-Nash) equilibrium at a feasible production vector $X = (X^i)_{i \in N}$, iff $U_i\left(X^i, X_{-i}\right) \geq U_i\left(\bar{X}^i, X_{-i}\right), \forall i$, where $\bar{X}^i$ is an alternative feasible production vector for producer i, and X_{-i} is the production schedule of the other producers at Nash equilibrium. A feasible production vector is non-negative and ensures non-negative profit per unit good.

To determine Nash equilibrium in the Cournot game, a producer $i \in N$ aims to maximize its total utility, given the production schedule of the other producers, over all the classes $q \in M$ of goods produced, by solving the following problem.

$$\arg_{X^i} \max U_i(X^i, X_{-i})$$

$$\text{s.t. } (L - BX)_{iq} \geq c_q^i, \forall q$$

$$X_q^i \geq 0, \forall q$$

$(L - BX)_{iq}$ represents the component indexed by the producer-good pair (i, q) in the vector $(L - BX)$. The inequality follows from the fact that no goods will be produced when the price of the good is below the cost of production. The natural question that arises is whether a Nash equilibrium exists in this setting. We state the following result for completion.

Claim 1. *A Nash equilibrium always exists in the m product, n producer Cournot game $G = (N, M, (U_i), (c_q^i))$.*

Social Utility. We define the *social utility* U_{social} for a game G to be the sum of utilities of all producers and the *representative consumer* involved in the game, i.e., $U_{social} = \sum_i U_i + U_c - \sum_i \sum_q X_q p(X_q)$, or equivalently, $U_{social} = \bar{L}^T X - \frac{1}{2} X^T B X$ where $\bar{L}_q^i = l_q^i - c_q^i, \forall i, q$ and vector $\bar{L} = (\bar{L}_1^1, \bar{L}_2^1, \cdots, \bar{L}_m^1, \bar{L}_1^2, \cdots, \bar{L}_m^n)$ with nm components.

Price of Anarchy. Using standard game theory terminology, the *Price of Anarchy* (PoA) for a class of games $\mathcal{G}$ is the lower bound on the ratio of the social function value at a Nash equilibrium to that of the social optimum, i.e.,

$$PoA(G) = \min_{G \in \mathcal{G}} \min_{NE \in \mathcal{N}_G} \frac{U_{social}^{NE}}{U_{social}^*} \tag{3}$$

where U_{social}^{NE} is the utility at NE, a Nash equilibrium solution in the set $\mathcal{N}_G$ of Nash solutions of G, and $U_{social}^* = \sup\{U_{social}\}$ is the optimum solution. We shall refer to $\min_{NE \in \mathcal{N}_G}\{U_{social}^{NE}\}$ as U_{social}^{NE} for simplicity in our analysis.

Preliminaries. Related to the interaction matrix B, we also define the following sub-matrices, which we use in the paper:

$$B^{ii} = \begin{bmatrix} B_{(i1),(i1)} & B_{(i1),(i2)} & \cdots & B_{(i1),(im)} \\ B_{(i2),(i1)} & B_{(i2),(i2)} & \cdots & B_{(i2),(im)} \\ \vdots & & \ddots & \vdots \\ B_{(im),(i1)} & B_{(im),(i2)} & \cdots & B_{(im),(im)} \end{bmatrix}, \text{ and } \hat{B} = \begin{bmatrix} B^{11} & 0 & \cdots & 0 \\ 0 & B^{22} & \cdots & 0 \\ \vdots & & \ddots & \vdots \\ 0 & \cdots & 0 & B^{nn} \end{bmatrix}$$

Throughout the paper, we use the following two notations for better presentation: (i) $\kappa = (n+1)(\beta - \gamma)(\beta + (m-1)\gamma)$, and (ii) $\bar{L}_q^i = l_q^i - c_q^i$

An Illustrative Example. Let us consider a game with 2 players and 2 goods. Suppose that the substitutability factor of good 2 in lieu of good 1 is γ_{12} and that of good 1 in lieu of good 2 is γ_{21}. The interaction matrix B is then represented by

$$B = \begin{bmatrix} \beta & \gamma_{12} & \beta & \gamma_{12} \\ \gamma_{21} & \beta & \gamma_{21} & \beta \\ \beta & \gamma_{12} & \beta & \gamma_{12} \\ \gamma_{21} & \beta & \gamma_{21} & \beta \end{bmatrix}$$

Vector $\bar{L}$ is given by $(\bar{L}_1^1, \bar{L}_2^1, \bar{L}_1^2, \bar{L}_2^2)$. We can now express producer utilities and the social utility respectively,

$$U_1 = X_1^1[\bar{L}_1^1 X_1^1 - \beta(X_1^1 + X_1^2) - \gamma_{12}(X_2^1 + X_2^2)] + X_2^1[\bar{L}_2^1 X_2^1 - \beta(X_2^1 + X_2^2) - \gamma_{21}(X_1^1 + X_1^2)]$$

$$U_2 = X_1^2[\bar{L}_1^2 X_1^2 - \beta(X_1^1 + X_1^2) - \gamma_{12}(X_2^1 + X_2^2)] + X_2^2[\bar{L}_2^2 X_2^2 - \beta(X_2^1 + X_2^2) - \gamma_{21}(X_1^1 + X_1^2)]$$

$$U_{social} = [(\bar{L}_1^1 X_1^1 + \bar{L}_1^2 X_1^2) - \frac{1}{2}(\beta(X_1^1 + X_1^2)^2 + \gamma_{12}((X_1^1 + X_1^2)(X_2^1 + X_2^2)))]$$

$$+ [(\bar{L}_2^1 X_2^1 + \bar{L}_2^2 X_2^2) - \frac{1}{2}(\beta(X_2^1 + X_2^2)^2 + \gamma_{21}((X_2^1 + X_2^2)(X_1^1 + X_1^2)))]$$

3   Nash Equilibrium for Quadratic Utility Cournot games

In this section, we present a polynomial time algorithm for finding Nash equilibrium in quadratic utility games. We first represent the problem of finding Nash equilibrium as an optimization problem by providing the following *composite utility function* $\hat{U}$, whose optimum solution is equivalent to the solution for Nash equilibrium. This approach has also been utilized in [15].

$\hat{U}$ is obtained as follows: Consider the utility function for a producer i, $U_i = \sum_q X_q^i \bar{L}_q^i - \sum_q X_q^i (B_{(iq)} \cdot X)$. Taking the derivative, $\frac{\partial U_i}{\partial X_q^i} = \bar{L}_q^i - B_{(iq)} \cdot X - (B_{(iq)})^i \cdot X^i = \bar{L}_q^i - (BX)_{iq} - (B^{ii}X^i)_{iq}$ where $(B_{(iq)})^i$ denotes the partial vector of $B_{(iq)}$, given by $(B_{(iq)})^i = (B_{(iq),(i1)}, B_{(iq),(i2)}, \cdots, B_{(iq),(im)})$, and X^i is the production vector of producer i. Next, we sum the partial derivatives over all producers i and goods q, $\sum_i \sum_q \frac{\partial U_i}{\partial X_q^i}$ and integrate, to get

$$\hat{U} = \sum_q \sum_i (\bar{L}_q^i X_q^i) - \frac{1}{2} X^T \hat{B} X - \frac{1}{2} X^T B X \tag{4}$$

Thus, we have the following optimization problem,

$$OP1 : \max \hat{U}$$
$$s.t. \ BX \leq L \tag{5}$$
$$X \geq 0$$

Constraints $BX \leq L$ and $X \geq 0$ ensure that all goods have positive price values, and that a producer does not produce a good with negative quantities respectively. From the construction of $\hat{U}$, we note that the derivative of the composite function $\hat{U}$ and the producer utility U_i are equal, i.e., $\frac{\partial \hat{U}}{\partial X_q^i} = \frac{\partial U_i}{\partial X_q^i}, \forall i, q$. Therefore, optimizing $\hat{U}$ w.r.t. the production value $X_q^i, \forall i, q$ is also the value that optimizes the utility function of producer i for good q. Consequently, solving $OP1$ will provide the Nash equilibrium solution to the problem.

Note that the problem $OP1$ is a convex programming problem (we use a standard result - the product of two non-decreasing and positive convex functions is convex), and although the problem of optimizing $\hat{U}$ can be solved by using the *Ellipsoid* or interior point methods, we devise a strongly polynomial algorithm.

To optimize $OP1$, consider the Lagrangian P of $\hat{U}$:

$$P = \hat{U} + \sum_i \sum_q (\mu_q^i \cdot (L - BX)_{iq}) + \sum_i \sum_q (\lambda_q^i \cdot X_q^i) \tag{6}$$

where $\mu_q^i, \forall i, q$ is the multiplier associated with the price constraint $(L - BX)_{iq} \geq 0$, whereas λ_q^i is associated with production values $X_q^i, \forall i, q$. From the Lagrangian P, we obtain the following KKT conditions:

$$
\begin{aligned}
\frac{\partial P}{\partial X_q^i} &= \frac{\partial \hat{U}}{\partial X_q^i} + \mu_q^i \cdot B_{iq,iq} + \lambda_q^i = 0 \\
(L - BX)_{iq} &> 0 \implies \mu_q^i = 0 \\
X_q^i &> 0 \implies \lambda_q^i = 0
\end{aligned}
\tag{7}
$$

If $X_q^i > 0$ then $p_q^i(X) = (L - BX)_{iq} > 0$, implying that $\mu_q^i = 0$ at optimality. We then obtain the following condition,

$$X_q^i > 0 \implies \frac{\partial P}{\partial X_q^i} = \frac{\partial \hat{U}}{\partial X_q^i} + \mu_q^i \cdot B_{iq,iq} + \lambda_q^i = \frac{\partial \hat{U}}{\partial X_q^i} = 0$$

Algorithm 1. FIND-NASH

1: $X_q^i = 0, \forall i, q$
2: $S_0 \leftarrow \{\cup_i \cup_q (i, q)\}$
3: $(\hat{i}, \hat{q}) = \arg\max_{(i,q) \in S_0} \bar{L}_q^i$
4: $S \leftarrow \{(\hat{i}, \hat{q})\}$
5: $S_0 \leftarrow \{S_0 \setminus (\hat{i}, \hat{q})\}$
6: **while** ($breakFlag == $ **false**) **do**
7:    **for each** $(i, q) \in S_0$ **do** $\lambda_{(i,q)} = $ FEASIBLE-EQ $(S, (i, q))$
8:    **if** ($\arg\max_{(i,q) \in S_0} \lambda_{(i,q)} > 0$ and $S \neq \emptyset$) **then**
9:       $(i', q') = \arg\max_{(i,q) \in S_0} \lambda_{(i,q)}$
10:      $S \leftarrow \{S \cup (i', q')\}$; $S_0 \leftarrow \{S_0 \setminus (i', q')\}$
11:   **else** $breakFlag = true$
12:   **end if**
13: **end while**
14: Assign $X_q^i = 0, \forall (i, q) \in S_0$
15: Obtain values of $X_q^i, \forall (i, q) \in S$ by solving the system of equations given by $\frac{\partial \hat{U}}{\partial X_q^i} = 0, \forall (i, q) \in S$.

We now present Algorithm 1, where we optimize the value of $\hat{U}$ while maintaining feasibility. The algorithm maintains $\mu_q^i = 0$, $\lambda_q^i = 0$ as well as $\frac{\partial L}{\partial X_q^i} = \frac{\partial \hat{U}}{\partial X_q^i}, \forall (i, q)$ where $X_q^i > 0$. In fact, the algorithm will determine a set S of producer-good pairs (i, q) that maximizes the marginal benefit to $\hat{U}$ at the given production levels X_q^i, i.e. $S = \left\{ (i,q) \mid (i,q) = \arg\sup_{(i',q')} \frac{\partial \hat{U}}{\partial X_{q'}^{i'}} \right\}$. Note that all elements in S provide the same marginal benefit to $\hat{U}$. It will also be ensured that $X_q^i > 0 \iff (i,q) \in S$. The remaining set of producer-good pairs will be in the set S_0.

At the beginning of the algorithm, there is no production and the pair with the maximum value of $\bar{L}_q^i$ will maximize the marginal utility $\frac{\partial \hat{U}}{\partial X_q^i}$. S is thus initialized with this producer-good pair and S_0 comprises of all the other (i,q) pairs. As the algorithm proceeds, (i) the value of $X_q^i, \forall (i,q) \in S$ increases and the value $\frac{\partial \hat{U}}{\partial X_q^i}, (i,q) \in S$ decreases, and (ii) the set S increases in size by addition of pairs from S_0 until the optimality condition, $\frac{\partial \hat{U}}{\partial X_q^i} = 0, \forall (i,q) \in S$ is achieved. We now present three claims that help us prove Theorem 1.

Claim 2. *Let S^* be the set of producer-good pairs (i,q) such that $X_q^i > 0$ in the optimal solution, and suppose $S \subset S^*$. There exists a pair (i',q') from S_0, where $(i',q') = \arg\max_{(i,q) \in S_0} \lambda_{(i,q)}$, such that if $\lambda_{(i',q')} > 0$ then $S \cup \{(i',q')\}$ is also part of an optimal solution.*

Claim 3. *During the course of the algorithm, the invariance $p_q^i(X) \geq 0, \forall i, q$ holds.*

Claim 4. *When breakFlag $=$ **true**, set S contains producer-good pairs (i,q) such that $\forall (i,q) \in S$, a solution to $\frac{\partial \hat{U}}{\partial X_q^i} = 0$ exists and provides the optimal solution to $\hat{U}$ and a Nash equilibrium solution.*

Given a set S and a pair $(i',q') \in S_0$, we define $\lambda_{(i',q')}$ to be the value of the gradient that satisfies the condition $\frac{\partial \hat{U}}{X_q^i} = \frac{\partial \hat{U}}{X_{q'}^{i'}}, \forall (i,q) \in S$ when $X_{q'}^{i'} = 0$. At this value of the gradient, (i',q') becomes a candidate for S. From amongst all candidates in S_0, the candidate that enters S is found using Claim 2. We next note that feasibility is maintained, by Claim 3; and when the while loop exits, the set of producer-good pairs in S determine Nash equilibrium (Claim 4). Thus, we arrive at the following theorem:

Algorithm 2. FEASIBLE-EQ$(S, (i',q'))$

1: Solve the system of simultaneous equations $\frac{\partial \hat{U}}{\partial X_q^i} = \lambda, \forall (i,q) \in S \cup (i',q')$ for X_q^i
2: Express all $X_q^i, \forall (i,q) \in S$ and λ in terms of $X_{q'}^{i'}$
3: Evaluate $\hat{X}_q^i = X_q^i, \forall (i,q) \in S$ and λ when $X_{q'}^{i'} = 0$
4: Evaluate $\hat{\lambda}$ when $X_{q'}^{i'} = 0$
5: **if** ($\hat{\lambda} > 0, \forall (i,q) \in S$) **then**
6:     return $\hat{\lambda}$
7: **else** return -1
8: **end if**

Theorem 1. *For a quadratic utility Cournot game $G = (N, M, (U_i), (c_q^i))$, Algorithm 1 correctly computes the Nash equilibrium in strongly polynomial time, $O\left((mn)^2 \cdot T_L(mn)\right)$ where $T_L(mn)$ is the time complexity of solving a system of linear equations of size $O(nm)$.*

The best known algorithm for solving a linear system of equations with nm variables provides $T_L(mn) = O\left((mn)^{2.376}\right)$ [9].

Faster Algorithm for Uniform Interaction Quadratic Utility Games. In games where the matrix B has a uniform structure, leading to a uniform interaction quadratic utility function, defined in Eq. 1, we can improve algorithm by introducing a *frontier set*, F_S of size $O(m)$. This set contains a limited number of candidates for inclusion in the set S of producer-pairs that correspond to non-zero production variables and is defined as follows:

$$X_q^i \in F_S \text{ iff } X_q^i = \arg \max_{i:(i,q)\in S_0} \bar{L}_q^i \tag{8}$$

Instead of choosing candidates from S_0, the algorithm chooses elements from F_S, supported by the following claim.

Claim 5. *Let S^* be set of producer-good pairs (i, q) such that $X_q^i > 0$ in an optimal solution, and suppose $S \subset S^*$. There exists a pair (i', q') from F_S, defined by $\arg \max_{(i,q)\in F_S} \lambda_{(i,q)}$, such that if $\lambda_{(i',q')} > 0$ then $S \cup \{(i', q')\}$ is also part of an optimal solution.*

In the algorithm, a producer-good pair is picked from the set F_S, which is then updated in $O(n)$ steps. The algorithm terminates when $F_S = \emptyset$, leading to an improved complexity algorithm, summarized as,

Theorem 2. *Nash equilibrium in the uniform interaction quadratic utility game $G = (N, M, (U_i), (c_q^i))$ can be determined in $O\left(mn \cdot m T_L(mn)\right)$ where $T_L(mn)$ is the time complexity of solving a systems of linear equations.*

4   Price of Anarchy in Concave Utility Uniform Interaction Games

In this section, we consider the Price of Anarchy in games with non-increasing concave utility functions with uniform interaction, denoted by G^U. The main result of this section is that the Price of Anarchy for the class of Cournot games G^U is bounded below by the constant $\frac{2}{3}$.

Our strategy to determine the Price of Anarchy for arbitrary concave functions will be to first show that the PoA can be determined from a class of games $\bar{G}$, having quadratic utility functions. We then determine closed form solutions for Nash equilibrium of games in the class $\bar{G}$. We also characterize the structure of optimum solutions for games in $\bar{G}$ and utilize that to determine PoA.

Establishing a Bound on PoA

We first derive a producer utility function $\bar{U}_i$, which is quadratic and with a linear price function, from the general convex function U_i, such that the Nash equilibrium solution of U_i is also a Nash equilibrium to $\bar{U}_i$, and prove that the Price of Anarchy can be obtained by considering the class of games with quadratic utility functions specified by $\bar{U}_i$, given in Claim 6.

Claim 6. *Let $G^U = (N, M, (U_i), (c_q^i))$ be a game under the uniform interaction model, where the price function is concave and non-increasing. If X^{NE} is a Nash equilibrium solution for the game G^U, then it is a Nash equilibrium solution for a game $\bar{G}$ with a quadratic utility function.*

We shall use the social utility represented as $U_{social}(X) = \sum_q U_q = \sum_q X_q(p_q - c_q^i)$, and relate the social utility function in the game G with $\bar{G}$, which has a quadratic utility function, thus arriving at Claim 7, and consequently Claim 8.

Claim 7. *Any social optimum solution for a uniform interaction model game $G^U = (N, M, (U_i), (c_q^i))$ is bounded as $U_{social}(X^*) \leq U_{social}(X^{NE}) + \bar{U}_{social}(X^*) - \bar{U}_{social}(X^{NE})$, where $\bar{U}_{social}(X)$ is the quadratic social utility in the game $\bar{G}$; X^* is the optimum solution and X^{NE} is the Nash equilibrium solution.*

Claim 8. *The PoA bound for the class of Cournot games G^U under the uniform interaction model can be obtained from the PoA bound for the class of games $\bar{G}$ with quadratic utilities.*

Nash Equilibrium Solutions

We now provide closed form solutions for Nash equilibrium for the uniform interaction model where the utility function is quadratic and prices are linear.

Theorem 3. *The Nash equilibrium solution to uniform interaction Cournot games, denoted by $\bar{G} = (N, M, (U_i), (c_q^i))$, with quadratic utilities is given by:*

$$X_q^{i\,NE} = \frac{1}{\kappa}\left((\beta + (m-2)\gamma)\left(n\bar{L}_q^i - \sum_{i' \neq i} \bar{L}_q^{i'}\right) - \gamma \sum_{q' \neq q}\left(n\bar{L}_{q'}^i - \sum_{i' \neq i} \bar{L}_{q'}^{i'}\right)\right), \text{where } \bar{L}_q^i = l_q - c_q^i$$

$$(9)$$

Using the above theorem, we express the total production of a good q at Nash equilibrium, given by $X_q^{NE} = \sum_i X_q^{i\,NE}$, where

$$\sum_i X_q^{i\,NE} = \frac{1}{\kappa}\left((\beta + (m-2)\gamma)\left(n\sum_i \bar{L}_q^i - \sum_i\sum_{i' \neq i} \bar{L}_q^{i'}\right) - \gamma \sum_{q' \neq q}\left(n\sum_i \bar{L}_{q'}^i - \sum_i\sum_{i' \neq i} \bar{L}_{q'}^{i'}\right)\right)$$

and thus represent social utility at Nash equilibrium as

$$U_{social}^{NE} = \sum_q \left(\left(\frac{1}{\kappa} ((\beta + (m-2)\gamma)(n \sum_i (\bar{L}_q^i)^2 - \sum_i \sum_{i' \neq i} \bar{L}_q^i \bar{L}_q^{i'}) - \gamma \sum_{q' \neq q} (n \sum_i \bar{L}_q^i \bar{L}_{q'}^i - \sum_i \sum_{i' \neq i} \right. \right.$$
$$\bar{L}_q^i \bar{L}_{q'}^{i'})) - \frac{1}{2} (\beta (\frac{1}{\kappa} [(\beta + (m-2)\gamma) \sum_i \bar{L}_q^i - \gamma \sum_{q' \neq q} \sum_i \bar{L}_{q'}^i])^2 + \gamma \sum_{q' \neq q} (\frac{1}{\kappa} [(\beta + (m$$
$$\left. \left. - 2)\gamma) \sum_i \bar{L}_q^i - \gamma \sum_{q' \neq q} \sum_i \bar{L}_{q'}^i]) (\frac{1}{\kappa} [(\beta + (m-2)\gamma) \sum_i \bar{L}_{q'}^i - \gamma \sum_{q'' \neq q'} \sum_i \bar{L}_{q''}^i])) \right) \right)$$

$$(10)$$

Social Optimum Solutions

Next, we determine a social optimum solution by maximizing the social utility.

$$U_{social}^* = \sup_{X_q^i} \left\{ \sum_q [(\sum_i X_q^i \bar{L}_q^i) - \frac{1}{2} [\beta (\sum_i X_q^i)^2 + \gamma \sum_{q' \neq q} (\sum_i X_q^i)(\sum_i X_{q'}^i)]] \right\}$$

$$(11)$$

We first note that the solution that optimizes U_{social} is obtained when the producer with the least cost produces to satisfy the demand of a good, and use it to provide a solution for social optimum.

Claim 9. *In a social optimum solution, any good q is produced only by the producer i_q^*, such that $\bar{L}_q^{i_q^*} = \max_i \bar{L}_q^i$.*

Theorem 4. *The socially optimum solution to the uniform interaction Cournot game, denoted by $\bar{G} = (N, M, (U_i), (c_q^i))$, with quadratic utilities is given by X^*, as*

$$X^{*i}_q = \begin{cases} \frac{(n+1)}{\kappa} \left((\beta + (m-2)\gamma)\bar{L}_q^i - \gamma \sum_{q' \neq q} \bar{L}_{q'}^i \right) & , \forall (i,q) \ s.t. \ i = \arg\max_{i'} \bar{L}_q^{i'} \\ 0 & , \forall (i,q) \ s.t. \ i \neq \arg\max_{i'} \bar{L}_q^{i'} \end{cases}$$

$$(12)$$

For simplicity, we assume that $i_q^* = 1, \forall q$ where $i_q^* = \arg\max_i \bar{L}_q^i$. Thus, at social optimum, only producer 1 will produce the goods. The social utility at optimum is then given by

$$U_{social}^* = \sum_q \left((\bar{L}_q^1 \frac{(n+1)}{\kappa} ((\beta + (m-2)\gamma)\bar{L}_q^1 - \gamma \sum_{q' \neq q} \bar{L}_{q'}^1)) - \frac{1}{2} [\beta (\frac{(n+1)}{\kappa} \right.$$
$$((\beta + (m-2)\gamma)\bar{L}_q^1 - \gamma \sum_{q' \neq q} \bar{L}_{q'}^1))^2 + \gamma \sum_{q' \neq q} (\frac{(n+1)}{\kappa} ((\beta + (m-2)\gamma)$$
$$\left. \bar{L}_q^1 - \gamma \sum_{q' \neq q} \bar{L}_{q'}^1))(\frac{(n+1)}{\kappa} ((\beta + (m-2)\gamma)\bar{L}_{q'}^1 - \gamma \sum_{q'' \neq q'} \bar{L}_{q''}^1))] \right)$$

$$(13)$$

Bound for Price of Anarchy

We now proceed to proving a bound on PoA. We first obtain conditions which yield a Nash equilibrium solution with the minimum value of social utility (Claim 10), and similarly, obtain conditions that maximize social utility (Claim 11). The approach is to determine the parameters of the game, more specifically, the values of $\bar{L}_q^i, \forall i, q$ which minimize the social cost at Nash equilibrium, noting that the solutions to Nash equilibrium have a closed form in terms of $\bar{L}_q^i$.

Claim 10. $U_{social}(X^{NE})$ *is minimized when the marginal profits for any good* q *are related as* $\bar{L}_q^i = \bar{L}_{q'}^i, \forall i, q, q'$ *and*

$$\bar{L}_q^i = -\bar{L}_q^1 \left(\frac{\alpha_3 + (m-1)\alpha_4}{(\alpha_1 + (m-1)\alpha_2 + (n-2)(\alpha_3 + (m-1)\alpha_4))} \right), \forall q, \forall i \neq 1. \quad (14)$$

$$\text{where } \alpha_1 = \frac{1}{\kappa} \frac{(2n(n+1)-1)(\beta+(m-2)\gamma)}{(n+1)}, \alpha_2 = \frac{-1}{\kappa} \frac{(2n(n+1)-1)\gamma}{(n+1)},$$

$$\alpha_3 = \frac{-1}{\kappa} \frac{(2n+3)(\beta+(m-2)\gamma)}{(n+1)}, \text{ and } \alpha_4 = \frac{1}{\kappa} \frac{(2n+3)\gamma}{(n+1)} \quad (15)$$

Claim 11. U_{social} *is maximized when* $\bar{L}_q^1 = \bar{L}_1^1, \forall q$.

Using Claim 10, we show conditions under which U_{social} is maximized. This leads us to Claim 11, which relates L_q^1 to $\bar{L}_1^1, \forall q, q'$. (Recall the assumption that $\bar{L}_q^1 = \max_i \bar{L}_q^i, \forall q$.) Further, assume w.l.o.g. that $\bar{L}_1^1 = \max_i \bar{L}_1^i$.

Theorem 5. *Let* $G^U = (N, M, (U^i), (c_q^i))$ *be a game under the uniform interaction model with concave utilities. Then,*

(a) $PoA(G^U) = \frac{2n+4}{3n+5}$. *For large number of producers, we achieve* $PoA(G^U) \geq \frac{2}{3}$.
(b) *There exists an instance of a game under the uniform interaction model* G^U *where* $PoA \leq 0.667$.

Proof. (a) Based on Claims 10 and 11, we reduce the game to one with linear prices and where the marginal profits of all producers can be expressed in terms of the maximum marginal profit achieved by a producer of one of the goods, assumed to be L_1^1. Using the definition in Eq. 3 and the expressions from Eqs. 10 and 13, we have

$$PoA \geq \frac{U_{social}^{NE}}{U_{social}^*} = \frac{m \left((\bar{L}_1^1 X_1^{1^{NE}})^2 + \bar{L}_1^2(n-1)X_1^{2^{NE}} - \frac{1}{2}(\beta(X_1^{1^{NE}} + (n-1)X_1^{2^{NE}})^2 \right.}{\left. + \gamma(m-1)((X_1^{1^{NE}} + (n-1)X_1^{2^{NE}})(X_1^{1^{NE}} + (n-1)X_1^{2^{NE}}))) \right)}{m \left(\bar{L}_1^1 X_1^{*1} - \frac{1}{2}(\beta(X_1^*)^2 + \gamma(m-1)(X_1^*)^2) \right)}$$

Expressing variables $X_q^i, \forall i, q$ and X^{*1}_1 in terms of $\bar{L}_q^i$, and simplifying, we obtain

$$PoA = \frac{2n+4}{3n+5}; \text{ and for large values of } n, \text{ we get } PoA \geq \frac{2}{3}.$$

(b) To show tightness, consider the instance of G^U with 1000 producers and 10 goods. The entries of B, have $\beta = 1$ and $\gamma = 0.6$; and $l_q = 100, \forall q$. The costs are $c_q^1 = 5, \forall q$, $c_q^i = 36.695, \forall i \neq 1, q \neq 1$.

We wrote a program written in Java to evaluate the conditions at Nash equilibrium and Social optimum, using the solutions obtained in this paper. In this instance, we get $U_{social}^{NE} = 4802.69$ and $U_{social}^* = 7199.99$, implying $PoA = 0.667 \approx \dfrac{2}{3}$.

Acknowledgement. The research was supported in part by NSF grant: CCF-1451574.

References

1. Abolhassani, M., Bateni, M.H., Hajiaghayi, M.T., Mahini, H., Sawant, A.: Network Cournot competition. In: Liu, T.-Y., Qi, Q., Ye, Y. (eds.) WINE 2014. LNCS, vol. 8877, pp. 15–29. Springer, Cham (2014). https://doi.org/10.1007/978-3-319-13129-0_2

2. Altman, E., Rojas, J., Wong, S., Hanawal, M.K., Xu, Y.: Net neutrality and quality of service. In: Jain, R., Kannan, R. (eds.) GameNets 2011. LNICST, vol. 75, pp. 137–152. Springer, Heidelberg (2012). https://doi.org/10.1007/978-3-642-30373-9_10

3. Bradley, T.: Does t-mobile Binge On violate net neutrality? (2016). https://www.forbes.com/sites/tonybradley/2016/01/08/does-t-mobile-binge-on-violate-net-neutrality/#2d501fe35312

4. Cowling, K., Waterson, M.: Price-cost margins and market structure. Economica **43**(171), 267–274 (1976)

5. Deneckere, R., Kovenock, D.: Direct demand-based Cournot existence and uniqueness conditions. Technical report, Purdue University, mimeo (1999)

6. Deneckere, R., Davidson, C.: Incentives to form coalitions with Bertrand competition. RAND J. Econ. **16**, 473–486 (1985)

7. Eckel, C., Neary, J.P.: Multi-product firms and flexible manufacturing in the global economy. Rev. Econ. Stud. **77**(1), 188–217 (2010)

8. Farrell, J., Shapiro, C.: Horizontal mergers: an equilibrium analysis. Am. Econ. Rev. **80**, 107–126 (1990)

9. Golub, G.H., Van Loan, C.F.: Matrix Computations, vol. 3. JHU Press, Baltimore (2012)

10. Granados, N.: Binge away: AT&T launches HBO bundle, going head-to-head with t-mobile-netflix plan (2017). https://www.forbes.com/sites/nelsongranados/2017/09/15/binge-away-att-responds-to-t-mobile-netflix-plan-with-unlimited-hbo-streaming/#3ace339b18fe

11. Guo, X., Yang, H.: The price of anarchy of Cournot oligopoly. In: Deng, X., Ye, Y. (eds.) WINE 2005. LNCS, vol. 3828, pp. 246–257. Springer, Heidelberg (2005). https://doi.org/10.1007/11600930_24

12. Hota, M., Kapoor, S.: Equilibrium and inefficiency in multi-product Cournot games. https://science.iit.edu/~kapoor/papers/multiproductcournot.pdf

13. Immorlica, N., Markakis, E., Piliouras, G.: Coalition formation and price of anarchy in Cournot oligopolies. In: Saberi, A. (ed.) WINE 2010. LNCS, vol. 6484, pp. 270–281. Springer, Heidelberg (2010). https://doi.org/10.1007/978-3-642-17572-5_22

14. Joe-Wong, C., Ha, S., Chiang, M.: Sponsoring mobile data: an economic analysis of the impact on users and content providers. In: IEEE Conference on Computer Communications, Infocom 2015, pp. 1499–1507. IEEE (2015)
15. Johari, R.: Efficiency loss in market mechanisms for resource allocation. Ph.D. thesis, Massachusetts Institute of Technology (2004)
16. Johari, R., Tsitsiklis, J.N.: Efficiency loss in Cournot games. Harvard University (2005)
17. Johnson, J.P., Myatt, D.P.: Multiproduct Cournot oligopoly. RAND J. Econ. **37**(3), 583–601 (2006)
18. Kluberg, J., Perakis, G.: Generalized quantity competition for multiple products and loss of efficiency. Oper. Res. **60**(2), 335–350 (2012)
19. Koutsoupias, E., Papadimitriou, C.: Worst-case equilibria. In: Meinel, C., Tison, S. (eds.) STACS 1999. LNCS, vol. 1563, pp. 404–413. Springer, Heidelberg (1999). https://doi.org/10.1007/3-540-49116-3_38
20. Kuleshov, V., Wilfong, G.: On the efficiency of the simplest pricing mechanisms in two-sided markets. In: Goldberg, P.W. (ed.) WINE 2012. LNCS, vol. 7695, pp. 284–297. Springer, Heidelberg (2012). https://doi.org/10.1007/978-3-642-35311-6_21
21. Ledvina, A., Sircar, R.: Oligopoly games under asymmetric costs and an application to energy production. Math. Financ. Econ. **6**(4), 261–293 (2012)
22. McManus, M.: Equilibrium, numbers and size in Cournot oligopoly. Bull. Econ. Res. **16**(2), 68–75 (1964)
23. Okuguchi, K., Szidarovszky, F.: On the existence and computation of equilibrium points for an oligopoly game with multi-product firms. Annales, Univ. Sci. bud. Roi. Fotvos Nom (1985)
24. Roughgarden, T., Tardos, É.: How bad is selfish routing? J. ACM (JACM) **49**(2), 236–259 (2002)
25. Salant, S.W., Switzer, S., Reynolds, R.J.: Losses from horizontal merger: the effects of an exogenous change in industry structure on Cournot-Nash equilibrium. Q. J. Econ. **98**(2), 185–199 (1983)
26. Singh, N., Vives, X.: Price and quantity competition in a differentiated duopoly. RAND J. Econ. **15**, 546–554 (1984)
27. Timmermans, V., Harks, T.: Equilibrium computation in resource allocation games. arXiv preprint arXiv:1612.00190 (2016)
28. Tsitsiklis, J.N., Xu, Y.: Profit loss in Cournot oligopolies. Oper. Res. Lett. **41**(4), 415–420 (2013)
29. Tsitsiklis, J.N., Xu, Y.: Efficiency loss in a Cournot oligopoly with convex market demand. J. Math. Econ. **53**, 46–58 (2014)
30. Vives, X.: Oligopoly Pricing: Old Ideas and New Tools. MIT Press, Cambridge (2001)

Combinatorial Assortment Optimization

Nicole Immorlica[1], Brendan Lucier[1], Jieming Mao[2(✉)], Vasilis Syrgkanis[1], and Christos Tzamos[3]

[1] Microsoft Research, Boston, USA
{nicimm,brlucier,vasy}@microsoft.com
[2] University of Pennsylvania, Philadelphia, USA
maojm517@gmail.com
[3] University of Wisconsin-Madison, Madison, USA
ctzamos@gmail.com

Abstract. Assortment optimization refers to the problem of designing a slate of products to offer potential customers, such as stocking the shelves in a convenience store. The price of each product is fixed in advance, and a probabilistic choice function describes which product a customer will choose from any given subset. We introduce the combinatorial assortment problem, where each customer may select a bundle of products. We consider a choice model in which each consumer selects a utility-maximizing bundle subject to a private valuation function, and study the complexity of the resulting optimization problem. Our main result is an exact algorithm for k-additive valuations, under a model of vertical differentiation in which customers agree on the relative value of each pair of items but differ in their absolute willingness to pay. For valuations that are vertically differentiated but not necessarily k-additive, we show how to obtain constant approximations under a "well-priced" condition, where each product's price is sufficiently high. We further show that even for a single customer with known valuation, any sub-polynomial approximation to the problem requires exponentially many demand queries when the valuation function is XOS, and that no FPTAS exists even when the valuation is succinctly representable.

1 Introduction

Imagine that you are an inventory manager, tasked with selecting which products to display on the shelves in a retail store. These products are acquired from different producers, who control the suggested retail prices. Your goal is to find a profitable assortment of items to offer, given a model of how customers choose which item(s) to ultimately purchase from the subset you display. This *assortment problem* captures a natural tradeoff. If you offer only the most expensive items, then many customers might simply leave the store without purchasing anything. On the other hand, a variety of inexpensive items might cannibalize

---

A full version of this paper can be found at https://arxiv.org/abs/1711.02601.

© Springer Nature Switzerland AG 2018
G. Christodoulou and T. Harks (Eds.): WINE 2018, LNCS 11316, pp. 218–231, 2018.
https://doi.org/10.1007/978-3-030-04612-5_15

sales from pricier goods and dilute the overall revenue. Given a collection of possible items, and a model of customer preferences, which subset of items should you display to maximize revenue?

The assortment problem is of practical importance for brick and mortar stores, but is also relevant to online shopping and travel offer aggregration platforms (e.g., Expedia) that must choose which products to display in response to a search query, and where the products' prices are set by a third party. Customers have limited patience and are more likely to select products from the first page of results, so the platform is incentivized to display a well-chosen slate of products. Since an online platform may need to choose from a vast array of potential products and offers, it is important to find computationally feasible solutions.

There is a growing literature on assortment in the field of revenue management, typically focusing on cases where each customer wants at most a single item. In such unit-demand settings, the problem is captured by a *choice function* that maps an assortment S to a probability distribution describing which good in S a customer will ultimately purchase. Commonly-studied choice functions include multinomial logit functions [16], exponential choice functions [2], and mixture models [3], among others. On the other hand, the computer science literature has mostly focused on combinatorial versions of revenue or welfare maximization when the designer controls the prices of items (see, e.g., multidimensional revenue maximization [4,6,10]) or the mode of interaction with the consumer (e.g., combinatorial auctions [9,12]). The important case of assortment optimization, where the platform designer is constrained to only design the set of available items, has been largely left untouched by the combinatorial optimization community. The goal of our work is to bridge this gap and explore the intersection of assortment and combinatorial optimization.

We introduce the *combinatorial assortment problem*, where consumers may choose to purchase bundles of goods. For example, a customer may want to buy a camera, possibly in combination with accessories, which may be either of the same brand as the camera or a cheaper off-brand variety. These items may be complementary (a camera plus an accessory), or substitutes for each other (a brand-name accessory or a generic version of the same accessory). We ask: *given the relationship between the items for sale, and possibly a cardinality constraint on the number of items that can be shown, what is a revenue-maximizing selection to offer?*

1.1   Our Results and Techniques

We explore the computational complexity of combinatorial assortment. Our goal is to characterize the limits of polynomial time computation or approximability and provide conditions under which the assortment problem is computationally feasible. In our model, each customer has a valuation function v that maps a bundle of goods T to a value $v(T) \geq 0$. This valuation is unknown to the planner choosing the assortment, but is drawn from a known distribution F. When faced with a slate of products, the customer will choose a bundle T that maximizes the value $v(T)$ less the total price of T. We will focus on the computational difficulty of the combinatorial assortment problem, parameterized by assumptions on the distribution F.

Main Result: Vertically Differentiated k-Additive Valuations. We initiate the study of combinatorial assortment by considering the problem under two assumptions. First, we focus on the case of k-additive valuations, where each buyer desires at most k items and the value of a bundle is the sum of the individual item values. This class extends unit-demand valuations to bundles of more than a single item. Second, we suppose that each valuation in the support of F agrees on the relative value of each pair of items. Equivalently, each valuation in the support of F can be expressed as $w \cdot v(T)$, where v is a fixed valuation function common to all buyers and w is a non-negative real number that describes the buyer's type. This captures settings where the relative quality and relationship between the items is unambiguous, but customers vary in their ability to extract value from the items.

It turns out that even this special case of the combinatorial assortment problem is suprisingly nontrivial. Natural heuristics like always displaying all items, or choosing items greedily by price, do not necessarily produce optimal assortments. This suggests that the vertical differentiation model is deceptively subtle. Solving the problem for this minimally combinatorial valuation class is a first step toward understanding more general cases.

We describe a dynamic programming solution to this problem that runs in time $O(n^{2k})$, where n is the total number of vouchers and k is the maximum number of items demanded by the buyer.

Theorem 1. *For vertically differentiated additive k-demand valuations, the revenue-optimal assortment can be computed in time $O(n^{2k} + n^2 \log(n))$.*

The algorithm does not impose any constraints on the price of each item, and applies whether or not there are cardinality constraints on the assortment (i.e., a bound on the number of items that can be displayed). Our solution builds an optimal assortment by first optimizing for low-type buyers and incrementally modifying the assortment to cater to higher types. An important technical challenge is that the set of customer types that choose to purchase a particular item need not be convex: it might be that a certain item is chosen by customers with low values and customers with high values, but not by intermediate types.

Further Algorithmic Results. We next consider more general settings in which valuations are still differentiated vertically, but are not necessarily k-additive. Recall our earlier intuition that offering all items might be highly suboptimal in the presence of "cheap" items that cannabilize sales from more profitable items. We show that such an issue is inherently due to items being sold at too low a price. We say that the goods are "well-priced" if, roughly speaking, the price of each bundle is at least its optimal (i.e., Myerson) reserve price, in a world where only that bundle is for sale. When goods are substitutes, this is equivalent to each individual item's price being at least its Myerson price. This may be the case if the individual product retailers are behaving like monopolists and not responding to the assortment planner, such as when the platform is driving only a small portion of the producer's overall revenue. We show that if the goods are well-priced, and the type distribution satisfies the standard regularity property, then offering all items is a 4-approximation to the optimal revenue.

Theorem 2. *For combinatorial assortment with well-priced items and vertical differentiation with a regular type distribution, the assortment that selects all items is a 4-approximation to the optimal expected revenue.*

We also show that if there is a cardinality constraint on the number of items that can be shown, then greedily accepting items to maximize marginal revenue also yields a constant approximation when the valuations satisfy the *gross substitutes* condition, which is a stronger notion of substitutability than submodularity.

Theorem 3. *For cardinality-constrained combinatorial assortment with well-priced items, a gross substitutes valuation, and vertical differentiation with a regular type distribution, the assortment that selects items greedily by revenue is a $\frac{4e}{e-1}$-approximation to the optimal expected revenue.*

Finally, for k-demand valuations that may not be additive, we show that under a certain revenue-concavity assumption on the type distribution, the optimal assortment will have size at most k. This admits an $O(n^k)$ algorithm that enumerates over all bundles of at most k items.

Negative Results. Each of the results above assume a vertical differentiation model of consumer preference, and impose constraints on either the prices or the valuations. We next show that the combinatorial assortment problem is inherently difficult even in a full-information setting. In the deterministic case, where there is a single buyer whose valuation is known to the optimizer, it is hard to approximate the revenue of the optimal assortment to a factor of $o(n^{1/2-\varepsilon})$ for any constant $\varepsilon > 0$, where n is the number of items to choose from. This is true even if there is no constraint on the number of items to be shown, and even if the valuation function is an XOS function, a subclass of subadditive functions.[1] Notably, this is a class of valuations where the welfare maximization problem can be well-approximated [9,12].

This hardness result takes the form of a communication complexity bound, independent of any computational hardness assumptions. We show that an approximation algorithm requires an exponential amount of communication with an oracle that can answer demand queries about the valuation function v.[2] Note that it is too much to hope for a lower bound in a fully general model of communication with a valuation oracle, since in particular the oracle could

---

[1] A valuation is subadditive if, for any sets of items S and T, $v(S \cup T) \leq v(S) + v(T)$. A valuation is XOS if it is the maximum of a collection of additive functions.

[2] A demand query takes as input the price of each item, and returns a utility-maximizing bundle at those prices.

simply communicate the optimal assortment, which can be described in polynomially many bits. Instead, our proof considers a communication model in which information about the valuation v is split between two oracles, and show that exponential communication between the oracles is necessary to obtain any reasonable approximation. We then show how the pair of oracles can simulate a demand query oracle. One implication of this result is that any assortment algorithm with a sub-polynomial approximation factor requires exponentially many demand queries about the valuation function v.

We next show that even for valuation functions that can be described succinctly,[3] it is still NP-hard to compute the optimal assortment. Like the communication complexity result, this holds even in the full-information case of a single buyer with known valuation.

Moving beyond the full-information case, we show that even if we impose the vertical differentiation assumption on the valuation distribution F, and even if each customer is submodular and demands at most two items, then there is no FPTAS for the combinatorial assortment problem.[4] In particular, while we provided an algorithm for solving combinatorial assortment with k-additive valuations, we cannot hope to extend this to k-demand submodular valuations. Furthermore, the natural greedy heuristic that adds items to the assortment one by one, maximizing the marginal revenue increase on each step, fails to obtain a constant approximation for submodular valuations, even in the full-information case where $\mathcal{F}$ is a point mass.

Extension: Welfare Maximization. We conclude by considering two extensions. First, we note that most of our positive results apply also to the goal of maximizing welfare, rather than maximizing revenue. The welfare maximization problem is still non-trivial, since the presence of cheap goods can result in lower-valued items being purchased. However, we show that if items are well-priced then offering all items is, in fact, the welfare-optimal assortment. Note that this is a stronger result than for revenue-maximization, where we established a 4-approximation. Under a cardinality constraint, the greedy algorithm for assortment yields a $\frac{e}{e-1}$ approximation to the optimal welfare for well-priced items and gross substitutes valuations. Finally, our dynamic programming algorithm for additive k-demand valuations applies just as well to the welfare objective, and can be used to compute a welfare-optimal assortment.

Extension: Learning. The second extension concerns a setting where $\mathcal{F}$ is not known to the seller. Rather, the seller must learn the distribution of types through demand queries: repeatedly choosing a slate of items and observing a buyer's choice. We show that the dynamic programming solution for vertically

---

[3] Formally: an XOS valuation that is the maximum of only 2 additive functions.

[4] When customers demand at most 2 items, the XOS condition is equivalent to submodularity. A valuation is submodular if, for any sets of items S and T, $v(S \cup T) + v(S \cap T) \leq v(S) + v(T)$. This is equivalent to each item having diminishing marginal value, and is more restrictive than subadditivity.

differentiated k-demand additive valuations can be implemented in this learning setting, with the loss of an $O(k\epsilon)$ additive error factor, using $\Theta(n^{k+1}\log(n)/\epsilon^2)$ queries.

1.2 Related Work

There is a growing literature on (unit-demand) assortment optimization in the management science literature. Talluri and van Ryzin [16] provide a closed-form solution when buyer choices follow the multinomial logit model. Rusmevichien-tog et al. [15] extend this solution to the case of cardinality-restricted assortment, and Davis et al. [7] show how to solve for the optimal assortment under more general nested logit models. When the choice function is described by a mixture of multinomial logit models, the assortment problem is NP-hard but various integer programming methods and approximation algorithms are known [3,8,14]. These results are incomparable with the positive results in this paper, which extend beyond unit-demand preferences but assume a model of vertical differentiation.

There has also been work studying learning in assortment, where the product slate can be adjusted to learn customer preferences. Caro and Gallein [5] consider learning in a model of assortment without substitution effects, where the demand for each product is unaffected by the other products in the assortment. Ulu et al. [17] study the dynamic learning problem when products exhibit purely horizontal differentiation, as modeled by location on a line segment. Agrawal et al. [1] consider a multi-armed bandit model of dynamic assortment, and show how to achieve near-optimal regret for multinomial logit choice models. Kleinberg et al. [11] consider a general class of comparison-based choice models, and study the complexity of learning their model from samples.

The combinatorial assortment problem can be viewed as a restricted form of mechanism design, where the design space consists only of choosing which subset of items to display. This is more restrictive than sequential posted pricing, where the designer can also choose the price at which each item can be sold (e.g., [6]).

2 The Combinatorial Assortment Optimization Problem

There is a set N of n items. Each item i has a fixed price $p_i \geq 0$. We assume items are indexed so that $p_1 \leq p_2 \leq \cdots \leq p_n$. There is an unbounded supply (i.e., number of copies) of each item.

There is a collection of buyers, each of whom wish to purchase a subset of the items. Each buyer j has a valuation function $v^j : 2^{[n]} \to \mathbb{R}_{\geq 0}$ that maps each subset of goods to a non-negative value. We assume that each valuation function is sampled independently from a distribution $\mathcal{F}$, which we refer to as the type distribution. In the full-information version where $\mathcal{F}$ is a point mass, we call the problem *noiseless*. We call the general problem *noisy*.

Given a subset of items T displayed to a buyer j, the buyer will pick $S \subseteq T$ maximizing $v^j(S) - \sum_{i \in S} p_i$ and pay $\sum_{i \in S} p_i$. Our goal as a seller is to pick an *optimal assortment*, which is a subset T of at most ℓ items that maximizes the expected revenue. Here ℓ is a parameter of the problem, which we can think of as an exogenous constraint on the number of products that fit on a shelf, the number of results that can be displayed on a search page, etc. We will focus first on the *unconstrained* case of $\ell = n$, then consider general ℓ in Sect. 5. For most of the paper we will assume that $\mathcal{F}$ is known to the seller and given as inputs to the optimization problem. In Sect. 5 we relax this assumption and suppose $\mathcal{F}$ is fixed but unknown to the seller, who must learn about it by interacting with buyers.

Valuation Classes. We focus on variants of the combinatorial assortment problem where the valuation functions in the support of $\mathcal{F}$ lie in a given class. We assume that valuations are monotone non-decreasing and normalized so that $v(\emptyset) = 0$. In this paper we will focus on the following valuation classes, which encode forms of substitutability between items.

- **additive:** there exist $v_1, \ldots, v_n \geq 0$ such that $v(S) = \sum_{i \in S} v_i$.
- **XOS:** there exist additive valuations (i.e., clauses) $v_1, \ldots, v_m$ such that $v(S) = \max_{i \in [m]} v_i(S)$.
- **submodular:** for all $S, T \subseteq [n]$, $v(S \cup T) + v(S \cap T) \leq v(S) + v(T)$.
- **gross substitutes:** for all $S, T \subseteq [n]$ and $x \in S$, one of the following is true:[5]
  1. $v(S) + v(T) \leq v(S \backslash \{x\}) + v(T \cup \{x\})$.
  2. There exists $y \in T$, $v(S) + v(T) \leq v(S \backslash \{x\} \cup \{y\}) + v(T \backslash \{y\} \cup \{x\})$.

We will also be interested in valuations that encode a constraint that a buyer does not derive benefit from receiving more than a certain number of items.

Definition 1. *Valuation v is k-demand if, for all $S \subseteq N$, $v(S) = \max_{T \subseteq S, |T| \leq k} v(T)$. That is, the buyer derives no benefit from receiving more than k items. We say that valuation v is additive (resp. XOS, submodular) k-demand if there is an additive (resp. XOS, submodular) valuation v' such that, for all $S \subseteq N$, $v(S) = \max_{T \subseteq S, |T| \leq k} v'(T)$.*

We note that these valuation classes can be ordered from most to least restrictive, as follows: *Additive k-demand $\subseteq$ gross substitutes $\subseteq$ submodular $\subseteq$ XOS.*

Vertical Differentiation. We will also be interested in the special case where the valuations in the support of $\mathcal{F}$ all agree on the relative value of bundles of items. We say that $\mathcal{F}$ satisfies *vertical differentiation* if there is a fixed valuation function v and, for all v^j in the support of $\mathcal{F}$, there exists a real number $w^j \geq 0$ such that $v^j(S) = w^j \cdot v(S)$ for all $S \subseteq [n]$. In other words, $\mathcal{F}$ is a distribution over multiplicative scalings of valuation v. For vertically differentiated settings, it is convenient to think of $\mathcal{F}$ as a distribution over scaling factors w. We will sometimes abuse notation and refer to w^j as the type of customer j.

---

[5] We use the M#-exchange characterization of gross substitutes, since it will be convenient for our proofs [13].

3   Structural and Algorithmic Results

We begin by describing instances of the combinatorial assortment problem that can be solved (or approximated) in polynomial time. All of the results in this section apply to the case where valuations are vertically differentiated, so we will assume this throughout. We will first describe an exact algorithm for k-additive valuations in Sect. 3.1. Then in Sect. 3.2 we will describe a set of structural assumptions that imply that displaying all items is approximately optimal. In Sect. 4 we will complement these algorithmic results with computational lower bounds.

3.1   Exact Combinatorial Assortment for k-demand buyers

We focus on additive k-demand valuations, which include unit-demand valuations as a special case. We will see in Sect. 4 that the combinatorial assortment problem is hard even for submodular 2-demand buyers. We show that for constant k, there is a polynomial-time algorithm that solves the combinatorial assortment problem for additive k-demand preferences.

Theorem 4. *For additive k-demand valuations, there exists an algorithm that finds the revenue-optimal assortment in time*[6] $O(n^{2k} + n^2 \log(n))$.

The full proof of Theorem 4 can be found in the full version. Importantly, this result applies even in the general noisy cases where buyer values are not fully known in advance. In general, the optimal assortment for additive k-demand valuations may include many more than k items. The algorithm we propose is a dynamic programming algorithm (DP), which incrementally builds an optimal assortment by considering how the purchasing behavior of a buyer changes with w.

3.2   Approximate Assortment for Well-Priced items

As mentioned in the introduction, it can be highly suboptimal to select all items in the combinatorial assortment problem, since the presence of a cheap but valuable item might cannibalize revenue from more expensive items. One might wonder, then, if such a situation can be made less severe if the items are all priced "reasonably." For example, suppose that each individual item is assigned the price that would maximize revenue when that item is sold by itself. Indeed, we would argue that such prices are very reasonable if the items are typically sold separately, and it is precisely the assortment platform that presents these items in combination with each other. We will show that under such an assumption, plus a regularity assumption on the type distribution, it is approximately revenue-optimal to show all of the items. Let us first define formally the assumptions needed for our result.

---

[6] Our algorithms depend on the type distribution, which may be continuous. The runtime bound assumes that the CDF of this distribution can be queried in $O(1)$ time. See the full version for a detailed discussion.

Definition 2 (Regularity). *We say that type distribution F is regular if the virtual value function $\phi(w) = w - \frac{1-F(w)}{f(w)}$ is non-decreasing, where f denotes the density function of distribution F.*

Regularity is a common assumption in the revenue maximization literature. Many natural distributions are regular, including uniform, gaussian, and exponential distributions.

Definition 3 (Revenue Curve). *The revenue curve of a type distribution F is $R(p) = p(1 - F(p))$.*

We can think of $R(p)$ as describing the revenue obtained if we were to offer a single item with value 1 and price p to a buyer whose type is drawn from distribution F. As we show the total revenue of an assortment can be expressed as a function of R.

The *optimal reserve* (or *Myerson reserve*) for F is the value r that maximizes $R(r)$ (or the supremum over such values r, if the maximum is not unique).

Definition 4 (Well-priced). *Suppose the type distribution F is regular with Myerson reserve r and non-increasing density after r.[7] Then the combinatorial assortment problem with type distribution F is well-priced if, for any item $i \in N$,*

$$p_i \geq r \cdot v_i.$$

We show that for well-priced instances of the combinatorial assortment problem, selecting all items yields a 4-approximation to the optimal revenue.

Theorem 5. *Choosing $S = N$ is a 4-approximation to the optimal revenue for well-priced combinatorial assortment when the valuation function v is subadditive.*

The idea behind Theorem 5 is to show that the revenue curve R can be well-approximated by a modified revenue curve $\hat{R}$ that is convex on the range $[r, \infty)$. We show that for convex curves, maximizing revenue reduces to the problem of maximizing utility, and hence the (modified) revenue is maximized by the assortment that maximizes buyer utility, which is to display all items. The proof complete proof can be found the full version.

Remark 1. Theorem 5 holds even for general valuations beyond subadditive, under the assumption that all subsets of items $S \subseteq N$ satisfy $\sum_{i \in S} p_i \geq r \cdot v(S)$. When valuations are subadditive this condition is equivalent to the condition on individual items given in Definition 4.

---

[7] In fact, our results for well-priced combinatorial assortment hold for distributions that satisfy a weaker condition than regularity. It is enough for the well-pricedness condition to hold for some value r (not necessarily the Myerson reserve) such that the density function f is non-increasing after r, and the revenue curve R is non-increasing after r.

Exact Assortment When Revenue is Concave. This approximation result used intuition that when revenue curves are convex, it is preferable to show as many items as possible. As it turns out, the reverse intuition holds as well: if the revenue curve is concave, then it is preferable to show fewer items. In particular, if buyers are k-demand, then the optimal assortment will consist of at most k items. The proof of Theorem 6 can be found in the full version.

Theorem 6. *Suppose that buyers are k-demand, and the revenue function R is concave over the support of the type distribution F. Then there exists an optimal assortment S with $|S| \leq k$.*

In Sect. 4, we show that in general the optimal assortment for k-demand buyers may contain far more than k items. In particular, a heuristic that simply enumerates all assortments of size at most k will not find an optimal solution in general. Theorem 6 shows that such a heuristic does find an optimal solution in cases where the revenue curve is concave.

Example 1. Suppose that buyers are k-demand with uniform type distribution over $[a, b]$. The revenue curve $R(w) = w \cdot \min \left\{ 1, \frac{b-w}{b-a} \right\}$ is concave for all $w \in [0, b]$ and thus by Theorem 6 the optimal assortment consists of at most k items.

Remark 2. If in Example 1 items are well-priced, Theorem 5 implies that even though the optimal assortment is small, showing all items yields a 4-approximation to the optimal revenue.

4 Hardness of Combinatorial Assortment

In this section we explore the hardness of the Combinatorial Assortment problem. We give a general hardness of approximation result for XOS valuations, even in the noiseless setting. We then show that even when valuations can be succinctly represented, the problem remains NP-hard. We also demonstrate that even when valuations are submodular, the natural greedy heuristic fails to obtain a good approximation. All missing proofs can be found in the full version.

Hardness of Approximation, Even Without Noise. We begin by considering the noiseless setting, where $\mathcal{F}$ is a point mass at 1 and hence the valuation of the buyer is known exactly. Our first result shows that for XOS valuations, the combinatorial nature of the problem leads to strong hardness of approximation. Indeed, it may take exponentially many demand queries to achieve better than an $O(\sqrt{n})$-approximation to the combinatorial assortment problem.

Theorem 7. *For XOS valuations, any $o(n^{1/2-\varepsilon})$-approximate algorithm for the combinatorial assortment problem requires $\Omega(\exp(n^{2\varepsilon}/24)/n)$ demand queries.*

Note that Theorem 7 is a query complexity bound, and puts no limitations on the algorithm's running time. Theorem 7 can be extended to a more general statement about communication complexity under a certain query model. See the full version for details.

Hardness for Succinct Valuations. Theorem 7's hardness is a communication bound, and relies on the fact that an XOS function may require exponentially many bits to fully describe. As we now show, the combinatorial assortment problem remains hard even for XOS valuations with succint descriptions. In particular, the problem is NP-hard, again in the noiseless setting, even if we restrict to valuations with only two clauses (i.e., the maximum of two additive functions).

Theorem 8. *For any XOS valuations with only 2 clauses, finding the optimal revenue is NP-hard in the noiseless case.*

The idea of the proof is to relate the optimal revenue of the combinatorial assortment problem to the solution to a knapsack problem, implementing the knapsack constraints by comparing values between the two clauses in the combinatorial assortment problem.

Hardness for 2-Demand Valuations in the Noisy Setting. One might also wonder if the hardness results above are driven by the large sets of goods desired by the buyers. What if we restrict attention to k-demand buyers, where k is a small constant?

One observation is that in the noiseless setting, the optimal assortment for a k-demand valuation will contain at most k items, so the problem can be solved in time $n^{O(k)}$ by evaluating the revenue for all subsets of size k. So this question is interesting only in the more general noisy setting.

Theorem 9 shows that even for submodular 2-demand valuations, and even if we assume that valuations are only vertically differentiated, there can be no FPTAS for combinatorial assortment. Therefore, we can only hope to get an efficient algorithm for k-demand valuations if we add add more restrictions, such as the restriction to k-additive valuations as in Sect. 3.1.

Theorem 9. *For vertically differentiated submodular 2-demand valuations, it is NP-hard to approximate the optimal revenue within approximation factor $1 + 1/n^c$, for some large enough constant c. In particular, there is no FPTAS in this setting unless $P = NP$.*

Greedy Assortment Fails for Submodular Valuations. We've shown that there is no FPTAS for submodular valuations in the general noisy setting. One might wonder if it's possible to obtain a constant approximation, however, by using a simple heuristic. One natural idea for submodular valuations is to use a greedy approach: repeatedly add the revenue-maximizing item to the assortment, until either no item remains or until adding any one item causes revenue to decrease. The following example shows that this heuristic can lead to approximation $\Omega(n)$, even without noise.

Example 2. There are $n = m + 1$ items, which we'll label $\{0, 1, \ldots, m\}$. The valuation v is:

$$v(S) = \begin{cases} m|S| & \text{if } 0 \notin S \\ m + (m-1)|S| & \text{otherwise} \end{cases}$$

One can verify that this valuation is indeed submodular. Suppose $p_0 = m$ and $p_j = m-2$ for all $j > 0$. The greedy algorithm selects item 0 first, as it generates revenue m which is larger than $m-2$, the revenue from any other single item. However, having selected item 0, the greedy algorithm would not add more items, since if the assortment is $\{0, i\}$ for any $i > 0$, the buyer would choose to buy only item i leading to a loss of revenue. So greedy obtains revenue m. The optimal assortment takes all items other than 0, for a revenue of $(m-2)m$.

Greedy Assortment Fails for Vertical Differentiation. There are $n = m+1$ items, which we'll label $\{0, 1, \ldots, m\}$. Each item has price $p_i = H^i$ for some large constant $H > n$. The valuation v is unit demand with $v(\{i\}) = i$ if $i > 0$ and $v(\{0\}) = n$. Buyers are vertically differentiated, where a buyer of type j has $w^j = H^j$ and appears with probability q^j proportional to H^{-j}, i.e. $q^j = \frac{H^{-j}}{\sum_{k=0}^{m} H^{-k}}$.

One can verify that every item i is only afforded by buyer types $j \geq i$. The greedy algorithm selects item 0 first as it generates revenue 1. After selecting item 0, any other item does not introduce any additional revenue as it is never bought by any type. In contrast, the optimal assortment selects all item other than 0. In this case, every buyer type $j > 0$ buys item j yielding a revenue of $\frac{m}{\sum_{k=0}^{m} H^{-k}} \to m$ as $H \to +\infty$. This implies that the optimal assortment gives revenue a factor of m worse than optimal.

5   Extensions

Constrained Assortment. To this point we focused exclusively on the case of unconstrained assortment, where $\ell = n$. For general ℓ, the lower bounds from Sect. 4 still apply. Also, the dynamic program for exact revenue-optimal assortment for additive k-demand valuations solves the constrained case; one need only track the remaining budget for additional items as part of the program. See the full version for a detailed proof.

Theorem 10. *For additive k-demand valuations and any cardinality constraint ℓ, there exists an algorithm that finds the revenue-optimal assortment of at most ℓ items in time $O(n^{2k}\ell)$.*

Theorem 5 specified conditions under which it is approximately optimal to select all items. Under a cardinality constraint, this solution may not be feasible. However, if the buyer valuations are gross substitutes, a greedy assortment algorithm is approximately optimal.

Theorem 11. *For gross substitutes valuations and well-priced items, a (6.33)-approximation to the revenue-optimal assortment of size at most ℓ can be computed in time $\tilde{O}(\ell^3 n)$.*

Welfare Maximizing Assortment. We have focused on revenue-maximization, but assortment optimization for welfare maximization is also non-trivial. The presence of cheap items in the assortment can reduce the total welfare and should be excluded. We note that the algorithm we developed for revenue-maximization under additive k-demand valuations can be easily adjusted for welfare maximization. Also, if items are well-priced, our results for revenue maximization apply to welfare maximization with even better constants. In particular, for unconstrained assortment, selecting the slate of all items is welfare-optimal if items are well-priced. See the full version.

Learning Assortments from Demand Samples. Suppose v and $\mathcal{F}$ are not known to the seller. Instead, the algorithm can learn about v and $\mathcal{F}$ via samples, taken by choosing a slate of items to sell and observing a buyer's choice. Details appear in the full version.

We show how to implement our dynamic program for k-additive valuations in this learning setting, by characterizing the algorithm's robustness to noise. We show that if the algorithm can make $\Theta(n^{k+1}\log(n)/\varepsilon^2)$ queries, then our dynamic programming solution will be within an $O(\varepsilon \cdot \max_{|S|=k} \sum_{i \in S} p_i)$ additive factor to the optimal revenue.

We also show that a variant of Theorem 6 applies to the learning setting. This requires choosing the best of a polynomial number of assortments. Since the highest revenue is bounded, standard concentration arguments imply that we can evaluate the revenue of any given assortment to within a small additive error by making polynomially many queries.

Acknowledgment. We would like to thank Aviad Rubinstein for pointing out an improvement on Theorem 7.

References

1. Agrawal, S., Avadhanula, V., Goyal, V., Zeevi, A.: A near-optimal exploration-exploitation approach for assortment selection. In: Proceedings of the 2016 ACM Conference on Economics and Computation, EC 2016, pp. 599–600. ACM, New York (2016). https://doi.org/10.1145/2940716.2940779
2. Alptekinolu, A., Semple, J.H.: The exponomial choice model: a new alternative for assortment and price optimization. Oper. Res. **64**(1), 79–93 (2016)
3. Bront, J.J.M., Méndez-Díaz, I., Vulcano, G.: A column generation algorithm for choice-based network revenue management. Oper. Res. **57**(3), 769–784 (2009). https://doi.org/10.1287/opre.1080.0567
4. Cai, Y., Daskalakis, C., Weinberg, S.M.: Optimal multi-dimensional mechanism design: reducing revenue to welfare maximization. In: Proceedings of the 2012 IEEE 53rd Annual Symposium on Foundations of Computer Science, FOCS 2012, pp. 130–139. IEEE Computer Society, Washington (2012). https://doi.org/10.1109/FOCS.2012.88
5. Caro, F., Gallien, J.: Dynamic assortment with demand learning for seasonal consumer goods. Manag. Sci. **53**, 276–292 (2007)
6. Chawla, S., Hartline, J.D., Malec, D.L., Sivan, B.: Multi-parameter mechanism design and sequential posted pricing. In: Proceedings of the Forty-second ACM Symposium on Theory of Computing, STOC 2010, pp. 311–320. ACM, New York (2010). https://doi.org/10.1145/1806689.1806733

7. Davis, J.M., Gallego, G., Topaloglu, H.: Assortment optimization under variants of the nested logit model. Oper. Res. **62**(2), 250–273 (2014). https://doi.org/10.1287/opre.2014.1256

8. Desir, A., Goyal, V.: Near-optimal algorithms for capacity constrained assortment optimization. Technical report, Department of Industrial Engineering and Operations Research, Columbia University (2015)

9. Feige, U.: On maximizing welfare when utility functions are subadditive. SIAM J. Comput. **39**(1), 122–142 (2009). https://doi.org/10.1137/070680977

10. Haghpanah, N., Hartline, J.: Reverse mechanism design. In: Proceedings of the Sixteenth ACM Conference on Economics and Computation, pp. 757–758. ACM (2015)

11. Kleinberg, J., Mullainathan, S., Ugander, J.: Comparison-based choices. In: Proceedings of the 2017 ACM Conference on Economics and Computation, EC 2017, pp. 127–144. ACM, New York (2017). https://doi.org/10.1145/3033274.3085134

12. Lehmann, B., Lehmann, D., Nisan, N.: Combinatorial auctions with decreasing marginal utilities. In: Proceedings of the 3rd ACM Conference on Electronic Commerce, EC 2001, pp. 18–28. ACM, New York (2001). https://doi.org/10.1145/501158.501161

13. Leme, R.P.: Gross substitutability: an algorithmic survey. Games Econ. Behav. **106**, 294–316 (2017). https://doi.org/10.1016/j.geb.2017.10.016. http://www.sciencedirect.com/science/article/pii/S0899825617301884

14. Méndez-Díaz, I., Miranda-Bront, J.J., Vulcano, G., Zabala, P.: A branch-and-cut algorithm for the latent-class logit assortment problem. Discrete Appl. Math. **164**, 246–263 (2014). https://doi.org/10.1016/j.dam.2012.03.003

15. Rusmevichientong, P., Shen, Z.J.M., Shmoys, D.B.: Dynamic assortment optimization with a multinomial logit choice model and capacity constraint. Oper. Res. **58**(6), 1666–1680 (2010)

16. Talluri, K., van Ryzin, G.: Revenue management under a general discrete choice model of consumer behavior. Manag. Sci. **50**(1), 15–33 (2004)

17. Ulu, C., Honhon, D., Alptekinolu, A.: Learning consumer tastes through dynamic assortments. Oper. Res. **60**(4), 833–849 (2012)

Varying the Number of Signals
in Matching Markets

Meena Jagadeesan[✉] and Alexander Wei[✉]

Harvard University, Cambridge, MA 02138, USA
{mjagadeesan,weia}@college.harvard.edu

Abstract. In large matching markets between job candidates and organizations, organizations may be unable to effectively identify interested
candidates due to a large volume of applications. The resulting congestion makes it unlikely for candidates to receive offers from their most
preferred organizations, leading to significant mismatch. We study how
signaling mechanisms can be used as a market design tool to reduce the
congestion in such markets. Specifically, we look at how the number of
signals available to market participants affects welfare and the number
of matches using a large market model. We show that for sufficiently
many signals, candidate welfare and the number of matches decrease as
a function of the number of signals, while the behavior of organization
welfare depends on the extent to which organizations value top candidates. Furthermore, we describe a class of firm utility functions for which
these limiting effects start to hold at realistic numbers of signals S.

Keywords: Signaling · Matching · Large markets

1 Introduction

In most matching markets, candidates apply to organizations and organizations
choose a subset of applicants to accept. Since the quantity of offers is typically
limited, organizations must not only determine candidates' qualities, but also
discern whether candidates are realistically attainable. However, it is often easy
for candidates to express interest in many organizations by sending out a large
number of applications. This behavior may lead to market congestion, in which
organizations become overwhelmed by the volume of applications and are unable
to select for interested applicants [1,5,7]. Candidates are also unlikely to receive
offers from their most preferred firms under such circumstances. As a result, there
can be significant mismatch between candidates and organizations, as well as
suboptimal welfare: most candidates fail to receive offers from their top choices,
and organizations waste effort on recruiting uninterested candidates.

We would like to thank Scott Kominers for his helpful guidance, as well as Alexey
Kushnir, Ran Shorrer, and Ravi Jagadeesan for feedback.

G. Christodoulou and T. Harks (Eds.): WINE 2018, LNCS 11316, pp. 232–245, 2018.
https://doi.org/10.1007/978-3-030-04612-5_16

The introduction of a signaling mechanism is one strategy to reduce the amount of congestion and mismatch in such markets. A typical signaling mechanism in this context allows candidates to signal interest to organizations, but limits the number of signals that any candidate may send. Because the signals of this mechanism are scarce, sending a signal has a tangible opportunity cost. As a result, signaling becomes more than just "cheap talk" and can serve as a means for candidates to credibly convey interest towards organizations.

An important design consideration when implementing a signaling mechanism is selecting the number of signals that each candidate may send. If candidates can signal to all or a significant fraction of the firms, then signals lose their scarcity and may devolve into cheap talk. At the other extreme, if candidates do not have enough signals, then they may not be able to communicate much information about their preferences, and the signaling mechanism may not be as effective as possible. These opposing effects show that it is not obvious how to optimally set the number of signals, e.g., so that welfare metrics such as aggregate welfare or the number of matches are maximized.

In this paper, we analyze how varying the number of signals available to candidates affects welfare in signaling markets. We consider a large market model in which a continuum of candidates is matched to a finite, discrete set of organizations. We then introduce a one-sided signaling mechanism that allows each candidate to signal to one or several organizations and examine how this mechanism affects market dynamics and statistics such as welfare and the number of matches. Our main result is a characterization of how the number of signals affects welfare metrics in the limiting case of many signals. We show that for sufficiently many signals, candidate welfare and the number of matches decrease as a function of the number of signals, while the behavior of organization welfare depends on the extent to which organizations value top candidates.

Signaling as a market design technique has been studied empirically in both the economics job market and in online matchmaking. In the economics job market, the AEA implements a mechanism through which job seekers are permitted to "signal" to up to two prospective employers. Signaling has been observed to increase the probability that a candidate lands an interview for one of their highly ranked positions and to allow for better expression of idiosyncratic preferences, e.g., over school type and over location [4]. Lee and Niederle [9] make similar empirical observations about an online matchmaking market where candidates were randomly endowed with either two or eight virtual roses. The introduction of the signal was observed to increase the total number of matches as well as increase the probability a proposal was accepted. Although these empirical studies demonstrate the benefits of a market with signals over a market without signals, there has not yet been a systematic analysis of how varying the number of available signals between different nonzero values affects market welfare.

We fill in this gap from a theoretical perspective by analyzing a large market model of signaling. In our model, a continuum of candidates gets matched to a finite, discrete set of firms, and candidates may express interest towards firms through a signaling mechanism with scarce signals. We show that a non-babbling equilibrium, i.e., an equilibrium in which firms respond to signals, always exists in this game. Furthermore, if we restrict our attention to symmetric, anonymous equilibria, a non-babbling equilibrium occurs if and only if each candidate signals to exactly their top choice firms.

Our main result is a characterization of how the number of signals affects welfare metrics once there are sufficiently many signals. In this regime, we show that a unique symmetric, anonymous, non-babbling equilibrium exists. This equilibrium is such that candidates use all of their signals and organizations only consider candidates who have signaled. For these equilibria, we show that increasing the number of signals decreases worker welfare and the number of matches, since it becomes increasingly difficult to express preferences. On the other hand, organizations may or may not prefer more signals. If organizations highly value top candidates, then they would prefer weaker signaling so that they can pursue their most preferred candidates, whereas if organizations are indifferent between strong candidates and very strong candidates, then they would prefer stronger signaling so they can pursue the candidates most interested in them.

Finally, we consider our results in the context of more explicit parameter settings. We characterize a class of firm utilities for which these limiting results start to hold when the number of signals S is at least 4. These utility functions show that although our results are for "sufficiently many" signals, they do hold for very realistic values of S. In the other direction, we describe a market with a relatively simple firm utility function that demonstrates the indeterminacies and multiple equilibria inherent for few signals. This example also concretely illustrates the potential of using many signals to achieve welfare improvements.

Past theoretical work on matching markets with a centralized signaling mechanism includes Coles et al. [5] and Kushnir [8]. Coles et al. [5] study signaling in the context of small, symmetric markets. They model signaling as a Bayesian game and characterize the set of equilibrium strategies. Furthermore, they show that under the assumption of uniform preferences, the introduction of a signal increases the utility of applicants and the expected number of matches, but has an indeterminate effect on the utility of organizations. While the uniform preferences assumption may not accurately model all real-life settings, it is crucial in making the analysis tractable: Coles et al. [5] remark that even when their model is extended to multiple blocks of firms with block-uniform preferences, the welfare comparisons across different equilibria become indeterminate. In another direction, Kushnir [8] observes that when there is heterogeneity among applicants and preferences are almost completely specified, the introduction of a signaling mechanism with one signal may improve the expression of idiosyncratic preferences, but may also hurt more "mainstream" applicants on average. These previous works show that for several market settings, signaling can allow for the better expression of preferences for various segments of the market.

We now turn to the question of how to maximize the usefulness of signals by varying their number. We borrow the game structure and uniform preferences assumptions of Coles et al. [5] in our model. However, Coles et al. [5] do not have the necessary architecture in their model to study the impact of changing the number of signals to different nonzero values, since they focus on comparing the game with signals to the game with no signal. Nonetheless, we give a generalization of their model to any nonzero number of signals, with the modification of shifting to a large market setting in order to make studying the effects of varying the number of signals tractable. Large market models have been previously considered by Azevedo and Leshno [3] in the contexts of stable matchings. In their work, the continuum matching economy serves as an analytic tool to enable comparative statistics to be taken. We similarly use the continuum assumption to allow us to analyze the impact of varying the number of signals.

The remainder of this paper is structured as follows: In Sect. 2, we present and formalize our large market model of a one-sided signaling mechanism as a Bayesian game. In Sect. 3, we state our solution concept and characterize the equilibrium strategies in our game. In Sect. 4, we investigate the signaling dynamics of our game with signals and characterize how varying the number of signals affects welfare metrics (number of matches, worker welfare, and firm welfare) in the limiting case. Proofs and useful examples are included in the online appendix.

2 Model

Our model captures the behavior of firms and workers in a large matching market with a signaling mechanism. We extend the approach of Coles et al. [5] to allow for more than one signal, making similar symmetry and uniformity assumptions on the market participants. We also shift the model to a large market setting to make our analysis of varying the number of signals tractable.

We assume the market consists of a finite number of firms and a continuum of workers. In this market, we make the standard Bayesian game assumption that the *distribution* of firm preferences over workers and worker preferences over firms is common knowledge. Furthermore, each worker knows their own preference ordering over the firms, and each firm knows the utility that it will derive from each worker, but each worker has no knowledge of how firms evaluate them relative to other workers, and each firm has no knowledge of how any given worker ranks it relative to other firms. To make our analysis tractable, we assume firms independently assign scores to workers, so that firm preferences over workers are uncorrelated, and we assume workers' preference orderings over firms are drawn uniformly at random, so that there is no common ranking of the firms.[1] Finally, we assume that firms' preference orderings over workers and workers' preference orderings over firms are independent of each other.

We model the matching market in three stages. We start with a signaling stage, during which each worker *signals* to up to a fixed number of firms through the signaling mechanism. Each signal is binary and does not transmit any further information. Then, each firm, having received the set of workers who have signaled to it, sends offers to a fixed quantity of workers. Finally, each worker

---

[1] While Coles et al. [5] also study the case of multiple blocks of firms, some of their welfare trends become indeterminate for more than one block of firms in their model. For this reason, we focus on the case of one block in our paper.

accepts their most preferred offer (if they have any offers). We assume that all workers share a common utility function and that all firms share a common utility function. Rejected offers yield 0 utility, firm utility from an accepted offer is purely determined by its score on the worker, and worker utility from an accepted offer is purely determined by their ranking of the firm.

Modeling the workers as a continuum greatly simplifies the space of possible equilibrium firm strategies. This assumption has a convexifying effect on the game [2], since no individual worker is able to influence the measure of workers who signal to a given firm. Furthermore, by a line of work on exact laws of large numbers for atomless measure spaces [10], we can assume the workers belong to an atomless measure space where such a law of large numbers holds. Stripping atomicity from the workers and assuming exact convergence to the distribution over preferences resolves complications imposed by the model of Coles et al. [5], in which the firm strategies took into account the number of signals received by the firm. The continuity of the worker types thus simplifies the equilibrium structure, allowing us to apply analytic tools and take comparative statics that would have otherwise been intractable in the discrete setting.

We describe the distribution of preferences in Sect. 2.1 and the stages of the game in Sect. 2.2.

2.1 Distribution of Preferences and Utilities

Let $\mathcal{F}$ denote the set of firms, and let $F := |\mathcal{F}|$ be the number of firms. We use $\mathcal{S}_{\mathcal{F}}$ to denote the set of permutations of $\mathcal{F}$. Let $\mathcal{W} := \mathcal{S}_{\mathcal{F}} \times [0,1]^F$ be the continuum of worker types. That is, worker rankings of firms are given by permutations drawn from $\mathcal{S}_{\mathcal{F}}$, which correspond to a total ordering of firms from highest to lowest ranked. Firm rankings of workers are given by scores in $[0,1]$, where e_f^w denotes the score of worker w given by firm f. The set of all firm scores for a given worker can thus be represented as an element of $[0,1]^F$. Finally, let $S \geq 0$ be the maximum number of signals that each worker is allowed to send.

We assume that the distribution of the worker types over $\mathcal{W}$ is given by the product of a distribution over $\mathcal{S}_{\mathcal{F}}$ of worker preferences over firms and a distribution over $[0,1]^F$ of firm preferences over workers. This product construction yields the key property that the workers' preferences over firms and the firms' preferences over workers are independent of each other. The workers themselves (and the corresponding firm preferences over workers) will then compose an atomless measure space with types drawn uniformly from $\mathcal{W}$, such that the distribution of worker types is exactly the uniform distribution over $\mathcal{W}$ [10].

By assuming that the distribution of firm preferences over workers is the uniform distribution over $[0,1]^F$, we make firm preferences over any given worker independent. Nonetheless, observe that there are some workers who have high scores from all the firms (universally "high-performing" workers), some workers who have high scores from some firms but not other firms (workers endowed with some skills but lacking other skills), and some workers who have low scores from all the firms (universally "low-performing" workers).

The uniform preferences assumption has been used throughout the signaling literature, for example in the one-signal model of Coles et al. [5] (which our model extends) to make their welfare analyses determinate. In fact, Coles et al. remark that even when their model for one signal is extended to multiple blocks of firms with block-uniform preferences, the welfare comparisons across different equilibria become indeterminate. When a single firm responds to more signals, firms in lower-ranked blocks may benefit so that there is no longer a purely negative spillover on other firms. We make the same uniform preferences assumption to make possible our analogous results on the dynamics of the signaling game and our analysis of welfare metrics in the limit.

Our welfare comparison theorems rely on the property that there is a unique equilibrium in the limit. The uniqueness comes from the fact that not receiving a signal from a worker indicates that the firm is not high on the worker's preference list. This fact, however, can fail if workers' preferences are correlated, e.g., workers may prefer to signal to lower tier firms so that their signal will carry more weight. While such strategies may be of interest in practice, it nonetheless makes the space of equilibria, which may now depend heavily on the number of signals, intractable to analyze across different numbers of signals.

We also assume that firms share a common utility function $u\colon [0,1] \to \mathbb{R}_{\geq 0}$ mapping worker score to utility. If a firm f's offer is rejected by a worker, then the firm receives 0 utility. Otherwise, if worker w accepts firm f's offer, the firm receives utility equal to $u(e_f^w)$. Naturally, the utility function u should be increasing, i.e., workers with higher score yield higher utility. For technical reasons, we also assume that u is continuously differentiable and strictly increasing. Observe that by setting u to be a quantile function, we may have $u([0,1])$ be distributed as any bounded distribution $\mathcal{D}$ on $\mathbb{R}$ whose quantile function is continuously differentiable and increasing. Using this construction, we can convert the uniform distribution over $[0,1]^F$ of worker scores to any such product distribution $\mathcal{D}^F$ over $\mathbb{R}_{\geq 0}^F$ of utilities derived from workers.

We assume that the distribution of worker preferences over firms is the uniform distribution over S_F, so that firms are equally ranked on average. Here, workers share a common utility function $v\colon \{1, \ldots, F\} \to \mathbb{R}_{\geq 0}$ mapping firm rank to utility that is a decreasing function. If a worker does not receive offers from any firm, they derive 0 utility from the game. Otherwise, a worker derives $v(i)$ utility if they accept the offer of their i-th highest ranked firm.

2.2 Stages of the Game

We model the signaling market as a game with three stages. We assume that the distribution over worker types is common knowledge, but that at the beginning of the game, firms do not have any knowledge of any given worker's preferences, and the workers do not have any knowledge of any specific firm's preferences. In other words, each firm and each worker knows only their own preferences. This assumption ensures that the only information communicated between the firms and the workers is through the presence (or lack) of the binary signal.

To define the game, we introduce a new constant $0 \leq \gamma < 1/F$, that roughly quantifies the competitiveness of the market by serving as an upper bound on the number of offers each firm is allowed to send. The constraints $0 \leq \gamma < 1/F$ ensure there are fewer positions available than workers. We require competition since when $\gamma \geq 1/F$, each worker is able to be matched to their preferred firm, leading to the existence of non-competitive equilibria in which workers only signal to their most preferred firm.

The game proceeds in the following three stages:

- During the first stage, each worker, knowing only their own preferences over firms, sends signals to up to S different firms for some fixed parameter S.
- At the start of the second stage, each firm is notified of the set of workers that signaled to it. Given the signals that it receives as well as its ranking of the workers by score, each firm then sends offers to a measurable subset of the workers of measure at most γ.
- In the third and final stage, each worker accepts the offer from the highest ranked firm from whom they receive an offer, if they receive any offers at all.

The offer structure of our game is based on the one-block, one-signal game of Coles et al. [5].[2] We allow workers to send fewer than S signals, so that workers do not have to signal if signaling lowers their probability of acceptance, thus preventing signals from becoming a negative commodity for the workers.

3   Equilibrium Strategies

Our solution concept is perfect Bayesian equilibrium restricted to strategies that are symmetric and anonymous with respect to each side of the market. The symmetry assumption comes from the fact that types on each side are drawn from the same distribution. Anonymity is necessary to rule out strategies in which there is unrealistic coordination outside of the signaling mechanism. These assumptions on strategy profiles are similar to those of related works [5,6].

We say that a worker strategy is *anonymous* if the strategy only depends on the preference profile of the worker, i.e., how a firm is treated depends only on its rank. This corresponds to the traditional definition of an anonymous strategy: a worker strategy σ_w is anonymous if for any permutation π of the firms and any preference profile θ_w, we have that $\sigma_w(\pi(\theta_w)) = \pi(\sigma_w(\theta_w))$. We say that a worker strategy profile is *symmetric* if each worker has the same response

---

[2] In Coles et al. [5], each firm is only allowed to make an offer to at most one worker, and each worker is allowed to accept at most one offer from at most one firm. In our model, we maintain that each worker can accept at most one offer. However, we instead set the maximum measure of offers that a firm is permitted to send to γ. While this offer structure does not precisely model the dating market (where candidates can "accept" multiple dates) or college admissions (where colleges can select the number of students to accept based on yield rates), the simplicity of the game, in both the setting of Coles et al. [5] and in our setting, enables the analysis to be tractable.

to a given preference profile over the firms. Next, given that a firm is facing a symmetric strategy profile from the workers, which in particular gives us a well-defined measure for the set of workers who signal to any given firm, we say that a firm strategy is an *anonymous response* if the strategy depends only on whether or not the worker signaled and the firm's own score on the worker. Finally, we say that a firm strategy profile is *symmetric* if each firm has the same response to a given preference profile and a given set of workers who signaled.

In order to determine the set of symmetric, anonymous equilibria of the signaling game, we first consider firm strategies in equilibrium. Recall that worker preferences over firms are independent from firm preferences over workers. Intuitively, given this independence assumption, the optimal firm strategy should be specified by *cutoffs*, in which the firm sends offers to all signaling workers with scores above a certain cutoff and fills the remainder of its quota with non-signaling workers with scores above another cutoff; this motivates the following definition of *cutoff strategy*:

Definition 1. *A* cutoff *strategy is a firm strategy given by cutoffs c_S and c_N such that the firm sends offers to all signaling workers whose scores is at least c_S and all non-signaling workers whose score is at least c_N. If the set of signaling workers is empty, then we take c_S to be 0; likewise, if the set of non-signaling workers is empty, then we take c_N to be 0.*

Cutoff strategies are clearly anonymous. We show that for any nonzero number of signals, firms play cutoff strategies in equilibrium. For this reason, we will only consider cutoff strategies for the remainder of our discussion.

Lemma 1. *In any symmetric, anonymous equilibrium of the game with a nonzero number of signals, firms play cutoff strategies (or measure 0 deviations from cutoff strategies) and all firms have the same cutoff c_S.*

Lemma 1 relies on the independence of worker rankings of firms from firm rankings of workers, which guarantees that the probability a worker accepts an offer is independent of the worker's score. Without this assumption, each firm may be able to strategize based on the workers' ranking of it. For example, if one firm were universally much lower ranked than the other firms, it may choose not to waste offer slots on very talented workers. As a result, our model applies to a segment of the job market in which firms are of similar caliber and workers are of similar caliber, where our independence assumptions are reasonable.

The equilibrium worker strategy depends on how c_S compares to c_N. If $c_S = c_N$, then firms effectively ignore signals, giving rise to a babbling equilibrium. Otherwise, if $c_S < c_N$, then due to the symmetry between the firms, we show that it is optimal for each worker to truthfully signal to their top S choices.

Lemma 2. *The following are the only two possibilities for symmetric, anonymous worker strategy profiles in equilibrium:*

1. *If signaling increases the probability of receiving an offer (i.e., $c_N > c_S > 0$), then workers always signal to their top S firms.*

2. *Otherwise, if firms ignore signals (i.e., $c_S = c_N$), then any symmetric, anonymous worker strategy profile leads to a babbling equilibrium.*

In particular, $c_S > c_N$ is not possible in equilibrium, since workers will never signal to a firm whose cutoff for non-signaling workers is higher than its cutoff signaling workers.

For the remainder of the paper, we focus on the equilibria where the welfare metrics change as a function of the number of signals. In the first scenario, where $c_S = c_N$, the game is in a babbling equilibrium and the welfare metrics are equivalent to the equilibrium in the game without signals. For this reason, we focus on the non-babbling equilibria.

The simplicity of worker strategies depends on the fact that the distribution of firm preferences is uniform. Without this assumption, optimal worker strategies are much less clear: the symmetry assumptions on the equilibria are no longer reasonable, and optimal strategies can no longer be parameterized by a single value. While the strategy space in our model does not capture the full range of behavior that could occur in practice, its simplicity makes our analysis of welfare metrics with varied numbers of signals tractable.

Given that the distribution of worker preferences is uniform over $\mathcal{S}_{\mathcal{F}}$, we show that there always exists at least one non-babbling equilibrium in symmetric and anonymous strategies for the signaling game.

Theorem 1. *For each firm utility function u and for any nonzero number of signals, there exists a non-babbling equilibrium in symmetric and anonymous strategies for the signaling game where workers signal to their top S firms.*

We show in the online appendix that there exist firm utility functions u that lead to multiple non-babbling equilibria where workers signal to their top S firms. In Sect. 4, we will analyze the welfare metrics (worker welfare, firm welfare, and number of matches) at these multiple equilibria.

In non-babbling equilibria in symmetric and anonymous strategies, we know by Lemma 2 that each worker signals to their top S firms. In the online appendix, we show the reverse implication: for any symmetric, anonymous equilibrium, if workers signal to their top S firms, then $c_S = c_N$, i.e., where signaling does not change the probability of receiving an offer, is never an equilibrium. For this reason, we focus for the remainder of the paper on the symmetric, anonymous equilibria that arise when each worker signals to their top S firms.

4   Impact of Signals

Having characterized the non-babbling equilibria in symmetric and anonymous strategies, we now study the welfare metrics associated to the signaling game. In Sect. 4.1, we discuss the signaling dynamics of the game with signals. In Sect. 4.2, we discuss how varying the number of signals affects the welfare metrics.

4.1   Signaling Dynamics

We now consider the dynamics of the game with signals. As we showed in Theorem 1, there is always at least one non-babbling equilibrium in the game with signals, and as we will show in the online appendix, there exist firm utility functions u that lead to multiple symmetric, anonymous, non-babbling equilibria in the signaling game. Thus, it is of interest to consider how the welfare metrics compare between the multiple equilibria that arise for a fixed number of signals.

We show that the welfare metrics of firm welfare, worker welfare, and number of matches are all monotonic in the cutoff value c_S parameterizing the equilibrium. Namely, there is an opposition of interests between firms and the workers/number of matches: at equilibria corresponding to lower cutoffs, firm welfare is lower, worker welfare is higher, and the number of matches is higher.

Theorem 2. *For any nonzero number of signals, suppose there exist symmetric, anonymous, non-babbling equilibria with cutoffs at $c_S = c_1$ and $c_S = c_2$, respectively, such that $c_1 < c_2$. Then at the equilibrium with cutoff $c_S = c_1$, firm welfare is lower, worker welfare is higher, and the number of matches is higher.*

The cutoff parameter c_S can be thought of as the extent to which firms respond to worker signals. This observation provides an explanation of the monotonicity results for firm and worker welfare: at lower cutoffs, firms consider worker preferences more, which is beneficial to workers and harmful to firms. Finally, the number of matches is intuitively decreasing in the cutoff since signaling reduces the amount of congestion in the market.

We also investigate how welfare metrics compare between equilibria in the game with signals and in the game without signals. We specifically show that the game with signals is always preferable to the game without signals with respect to worker welfare and the number of matches. The change in firm welfare is indeterminate, as we show in the online appendix.

Theorem 3. *For any nonzero number of signals, both worker welfare and the number of matches are greater at any equilibrium in the game with signals than in the game without signals.*

Since signaling forces firms to consider worker preferences, any amount of signal intuitively should improve worker welfare, which is exactly what occurs in this model. For a similar reason, the number of matches should intuitively increase with signaling. The change in firm welfare is indeterminate because the introduction of a signaling mechanism has two opposing effects. The first effect is a reduction in the market power of firms from taking into account worker preferences, which decreases firm welfare; the second is a partial resolution of the coordination problem which reduces the waste of multiple firms sending offers to the same worker.

These properties of our model for signaling markets, which hold for any nonzero number of signals, capture and reinforce many of the same structural properties as the one-block model with one signal studied by Coles et al. [5].[3] However, unlike in their model, our firm strategies are specified by one cutoff as a result of having a continuum of workers, so we can analyze the effect of varying the number of signals on these welfare metrics by taking comparative statics.

4.2   Welfare Metrics in the Limit

Our main goal is to investigate how firm welfare, worker welfare, and the number of matches change as a function of the number of signals S available to each worker. One complication that arises is the possibility of multiple equilibria in this game. In the online appendix, we present a firm utility function for which the number of equilibria and the cutoff values at the equilibria change as the number of signals varies. Furthermore, although we have shown in the previous section that welfare metrics are monotonic in the cutoff for a fixed number of signals, it is indeterminate how these values compare across different numbers of signals, as we also demonstrate through our example in the online appendix.

We resolve these complications by focusing on the relevant welfare metrics in the case where there are many signals. The regime of interest is where the signals transmit enough information to have a significant impact on firm strategy, but not enough so that signals devolve into cheap talk. Our next theorem shows that given sufficiently many signals, there exists only one symmetric, anonymous, non-babbling equilibrium, in which workers signal to their top S firms and firms only send offers to signaling workers. This phenomenon occurs when there are enough signals that *not* sending a signal becomes a strong negative indication of interest. The uniqueness of this equilibrium will allow us to take comparative statics to analyze the effect of increasing the number of signals on welfare metrics.

Theorem 4. *Suppose that* $\gamma = \frac{\alpha}{F}$ *for some constant* $\alpha < 1$. *For each utility function* u, *there exists a threshold value* $C < \infty$ *such that for any* $S \geq C$ *and* $F > S$, *there exists only one symmetric, anonymous, non-babbling equilibrium in the signaling game with* F *firms and* S *signals. This equilibrium occurs when firms only make offers to signaling workers, i.e., when firm strategies have cutoffs* $c_S = 1 - \frac{\alpha}{S}$ *and* $c_N = 1$.

When a large number of signals are available to workers, workers are able to signal to enough firms that not receiving a signal from a worker strongly indicates to a firm that it is low in the worker's preference ranking. In particular, the probability that a worker is rejected by all of its signaled-to firms and accepts

---

[3] One difference between the dynamics of our model and the dynamics of the model of Coles et al. [5] is whether $c = 1 - \gamma$ is an equilibrium if worker signal truthfully, while in our model, for any $S < F$, the cutoff $c = 1 - \gamma$ is never an equilibrium. On the other hand, in [5], the cutoff $c = 1 - \gamma$ is always an equilibrium in the game with signals. The difference arises because in our model, firms send offers to more than one candidate, and the margin now matters. One consequence is that in our model, shifts to strategies based on signaling are more likely to endogenously occur given the existence of a signaling mechanism.

the offer of a non-signaled-to firm becomes low. In contrast, in the case where there are very few signals, workers only signal to their top few firms, so it may still be worthwhile for firms to extend offers to highly valued workers who have not signaled. In this case, it is indeterminate whether $c_S = 1 - \frac{\alpha}{S}$ is an equilibrium, and there may also be other equilibria, as we show in the online appendix.

Now that we have shown that, given sufficiently many signals, the unique equilibrium is at $c_S = 1 - \frac{\gamma F}{S}$ and $c_N = 1$, our task boils down to studying how the welfare metrics change as the number of signals increases. We show that worker welfare and the number of matches decrease as the number of signals increases at this equilibrium.

Theorem 5. *Suppose that* $\gamma = \frac{\alpha}{F}$ *for some constant* $\alpha < 1$. *Let* u *be a utility function, and let* C *be a threshold value such for all* $S \geq C$, *the signaling game has only one symmetric, anonymous, non-babbling equilibrium at* $c_S = 1 - \frac{\alpha}{S}$. *(This threshold* C *is guaranteed to exist by Theorem 4.) Then, for any* $S \geq C$ *and* $F > S$, *worker welfare and the number of matches decrease as the number of the signals increases.*

The decrease in worker welfare and the number of matches occurs because firms already respond maximally to signals, and increasing the number of signals available to each worker without increasing the extent to which firms respond to signals dilutes the strength of the signal. Now, workers very interested in a firm may get their spots taken away by less interested workers. At the extreme, signaling devolves into cheap talk, and the market where the number of signals equals the number of firms is indistinguishable from the market without signals.

Now, we consider firm welfare. We show that, in the regime of many signals, the change in firm welfare as the number of signals increases is a local property of the utility function: namely, it depends solely on the rate of increase of u at the maximum worker score. We show that if this derivative is small, then firm welfare is decreasing in the number of signals in the limit, and if the derivative is large, then firm welfare is increasing in the number of signals in the limit.

Theorem 6. *Suppose that* $\gamma = \frac{\alpha}{F}$ *for some constant* $\alpha < 1$. *Let* u *be a utility function, and let* C *be such that for all* $S \geq C$, *the game with signals has only one symmetric, anonymous non-babbling equilibrium at* $c_S = 1 - \frac{\alpha}{S}$. *(This threshold value is guaranteed to exist by Theorem 4.) There exists a threshold value* $C' \geq C$ *that satisfies the following property:*

- *If* $u'(1) < \frac{\alpha u(1)}{e^\alpha - 1}$, *then for any* $S \geq C'$ *and* $F > S$, *firm welfare in the signaling game with utility function* u, S *signals, and* F *firms is a decreasing function of* S. *Furthermore, firms prefer any number of signals* $S \geq C'$ *to no signals.*
- *If* $u'(1) > \frac{\alpha u(1)}{e^\alpha - 1}$, *then for any* $S \geq C'$ *and* $F > S$, *firm welfare in the signaling game with utility function* u, S *signals, and* F *firms is an increasing function of* S. *Furthermore, firms prefer no signals to any number of signals* $S \geq C'$.

In an equilibrium where firms only send offers to signaling workers, firms face a tradeoff as the number of signals increases between being able to send offers to a stronger set of workers and getting offers rejected by very strong workers who receive multiple offers. The sign of this comparative static therefore hinges on the extent to which firms value stronger workers. If firms significantly value stronger workers, then firms would prefer more signals, i.e., to be closer to the case where signals are cheap talk. Otherwise, firms would prefer fewer signals in order to reduce the number of offers received by workers that are highly rated by many firms. In the limit, how much firms value stronger workers is captured by the derivative $u'(1)$.

When the number of signals is small, however, the effect of increasing the number of signals available has an indeterminate effect on firm utility, even if we assume that the game is in the equilibrium where firms only make offers to signaling workers. The direction of change in firm utility depends on the marginal gain of accepting a worker at the cutoff for signaling workers. If this marginal gain at the cutoff is very different from the marginal gain of accepting a top worker, then the direction of change may be indeterminate. In the online appendix, we construct an example in which the direction of change is indeterminate when S lies between the bounds C and C'. (That is, it is necessary to allow threshold C' to be greater than threshold C.)

Combining the previous two theorems yields the following result about total firm and worker welfare. If $u'(1)$ is small, then total welfare is a decreasing function of S for sufficiently large S, and if $u'(1)$ is large, then there is an opposition of interests between firms and workers. In the former case, the optimal number of signals should therefore not be too high, since the limiting behavior has negative consequences for both firms and workers. In the latter case, there exists an opposition of interests between firms and workers in the regime of many signals, and the optimal amount of signaling in this case becomes dependent on the method of welfare aggregation.

Corollary 1. *Suppose that $\gamma = \frac{\alpha}{F}$ for some constant $\alpha < 1$. Let u be a firm utility function, and assume that $u'(1) \neq \frac{\alpha u(1)}{e^\alpha - 1}$. Let C' be the threshold value such that there is a unique symmetric, anonymous, non-babbling equilibrium at $c_S = 1 - \frac{S}{S}$ and such that the change in firm welfare as the number of signals increases is monotonic. (This threshold is guaranteed to exist by Theorem 4 and Theorem 6.) Suppose that $S \geq C'$ and $F > S$. If $u'(1) < \frac{\alpha u(1)}{e^\alpha - 1}$, then total welfare is a decreasing function of S. And if $u'(1) > \frac{\alpha u(1)}{e^\alpha - 1}$, then there is an opposition of interests between firms and workers.*

Since these results are limiting results, this raises the question of how large of the thresholds C and C' needs to be for well-behaved firm utility functions. We show that for utility functions that grow sufficiently slowly between 0.75 and 1, firm welfare and total welfare are decreasing functions of S when $S \geq 4$.

Lemma 3. *Suppose that $\gamma = \frac{\alpha}{F}$ for some constant $0.8 \leq \alpha < 1$. Suppose that u satisfies $u(y) \geq u(1)(0.76 + 0.24y)$ for $1 \geq y \geq 0.75$. Suppose that $F > S \geq 4$. Then, there is only one symmetric, anonymous, non-babbling equilibrium in the signaling game with S signals and firm utility function u. Furthermore, firm and worker welfare are both decreasing functions of S for $S \geq 4$.*

This shows for certain utility functions, the optimal number of signals is between 0 and 4, demonstrating that limiting results can take effect for reasonable S.

5 Conclusion

Scarce signaling mechanisms are a useful market design tool to reduce congestion and mismatch in markets where credible communication of preferences is difficult. Such mechanisms have been implemented in practice in the economics job market and in the context of matchmaking in online dating. However, the number of signals available to each candidate varies between mechanisms. Understanding how to best set the number of signals available is therefore a key design question for the implementation of such signaling mechanisms.

In this paper, we developed a large market model for signaling markets and studied how varying the number of signals affects various welfare metrics. Assuming uniform preferences, we showed that in a symmetric, anonymous, non-babbling equilibrium, firms play cutoff strategies and workers signal to their top S firms. Our main result is that in the limit, there exists a unique equilibrium in which firms respond maximally to signals. At this equilibrium, increasing the number of signals decreases worker welfare and the number of matches, while the effect on firm welfare depends on the extent to which firms value top workers.

Future lines of research include relaxing the uniform preferences assumption and characterizing welfare behavior in the case of very small numbers of signals.

References

1. Arnosti, N., Johari, R., Kanoria, Y.: Managing congestion in decentralized matching markets. In: Proceedings of the Fifteenth ACM Conference on Economics and Computation, EC 2014, pp. 451. ACM, New York (2014)
2. Aumann, R.J.: Markets with a continuum of traders. Econometrica **32**(1/2), 39–50 (1964)
3. Azevedo, E.M., Leshno, J.D.: A supply and demand framework for two-sided matching markets. J. Polit. Econ. **124**(5), 1235–1268 (2016)
4. Coles, P., Cawley, J., Levine, P.B., Niederle, M., Roth, A.E., Siegfried, J.J.: The job market for new economists: a market design perspective. J. Econ. Perspect. **24**(4), 187–206 (2010)
5. Coles, P., Kushnir, A., Niederle, M.: Preference signaling in matching markets. Am. Econ. J.: Microecon. **5**(2), 99–134 (2013)
6. Coles, P., Shorrer, R.: Optimal truncation in matching markets. Games Econ. Beh. **87**, 591–615 (2014)
7. He, Y., Magnac, T.: A Pigouvian Approach to Congestion in Matching Markets. TSE Working Papers 17–870, Toulouse School of Economics, December 2017
8. Kushnir, A.: Harmful signaling in matching markets. Games Econ. Behav. **80**, 209–218 (2013)
9. Lee, S., Niederle, M.: Propose with a rose? Signaling in internet dating markets. Exp. Econ. **18**(4), 731–755 (2015)
10. Sun, Y., Zhang, Y.: Individual risk and Lebesgue extension without aggregate uncertainty. J. Econ. Theor. **144**(1), 432–443 (2009)

Simple and Efficient Budget Feasible Mechanisms for Monotone Submodular Valuations

Pooya Jalaly Khalilabadi[(✉)] and Éva Tardos

Department of Computer Science, Cornell University,
Gates Hall, Ithaca, NY 14853, USA
{jalaly,eva}@cs.cornell.edu

Abstract. We study the problem of a budget limited buyer who wants to buy a set of items, each from a different seller, to maximize her value. The budget feasible mechanism design problem requires the design a mechanism which incentivizes the sellers to truthfully report their cost and maximizes the buyer's value while guaranteeing that the total payment does not exceed her budget. Such budget feasible mechanisms can model a buyer in a crowdsourcing market interested in recruiting a set of workers (sellers) to accomplish a task for her.

This budget feasible mechanism design problem was introduced by Singer in 2010. We consider the general case where the buyer's valuation is a monotone submodular function. There are a number of truthful mechanisms known for this problem. We offer two general frameworks for simple mechanisms, and by combining these frameworks, we significantly improve on the best known results, while also simplifying the analysis. For example, we improve the approximation guarantee for the general monotone submodular case from 7.91 to 5; and for the case of large markets (where each individual item has negligible value) from 3 to 2.58. More generally, given an r approximation algorithm for the optimization problem (ignoring incentives), our mechanism is a $r + 1$ approximation mechanism for large markets, an improvement from $2r^2$. We also provide a mechanism without the large market assumption, where we achieve a $4r + 1$ approximation guarantee. We also show how our results can be used for the problem of a principal hiring in a Crowdsourcing Market to select a set of tasks subject to a total budget.

1 Introduction

We study *prior-free budget feasible mechanism design* problem, where a single buyer aims to buy a set of items, each from a different seller. Budget feasible

P. Jalaly Khalilabadi—Work supported in part by NSF grant CCF-1563714, ONR grant N00014-08-1-0031, and a Google Research Grant.
É. Tardos—Work supported in part by NSF grants CCF-1563714, and CCF-1422102, ONR grant N00014-08-1-0031, and a Google Research Grant.

G. Christodoulou and T. Harks (Eds.): WINE 2018, LNCS 11316, pp. 246–263, 2018.
https://doi.org/10.1007/978-3-030-04612-5_17

mechanism design focuses on maximizing the value of the buyer, while keeping the total payments bellow the budget. This problem was introduced by Singer (2010), and models problems such as the problem of a crowdsourcing platform, like a principal working on Amazon's Mechanical Turk, who wishes to procure a set of workers to accomplish a set of tasks. Each worker has a private cost for her service. We offer universally truthful mechanisms with good approximation guarantee for this problem that incentivize the workers to report their true cost and find a set of workers with close to optimal value.

We focus on simple parameterized mechanisms, assuming the buyer's valuation is a general monotone (non-decreasing) submodular function. Monotone submodular functions are widely used, with submodularity capturing the diminishing returns property of adding items. Submodular value functions are the most general class of functions where the optimization problem (without considering incentives) can be solved approximately in polynomial time using a value oracle.

We introduce new simple parameterized mechanisms for this problem, as well as parameterizing and improving the analysis of some previously known mechanisms. Our main result is to show how these simple parameterized mechanisms can be combined for the case of large markets, where each item individually does not have a significant value compared to the optimum. We also show how our results can be used for the problem of a principal hiring in a Crowdsourcing Market to select a set of tasks and agents to complete these tasks, subject to a total budget.

Our Model. We consider the problem of a single buyer with a budget B facing a set of multiple sellers A. We assume that each seller $i \in A$ has a single indivisible item, and has a private cost c_i for this item, and the buyer has no prior knowledge of the private costs. The utility of a seller for selling her item and receiving payment p_i is $p_i - c_i$. We only study universally truthful mechanisms, i.e. the mechanisms in which sellers truthfully report their costs, and do not have incentive to misreport. Since each seller $i \in A$ only has a single item, we interchangeably use i to denote the seller or her item. We assume that $v(S)$, the value of the buyer for a subset of items $S \subseteq A$, is a monotone (non-decreasing) submodular function.

The *budget feasibility constraint* requires that the total payments to the sellers may not exceed the budget. The goal of this paper is to design simple, universally truthful and budget feasible mechanisms that approximately maximize the value of the buyer. We compare the performance of our mechanism with the true optimum, without computational or incentive limitation: maximizing the value subject to keeping the total cost bellow the budget. With this comparison in mind, incentive compatible mechanisms that do not run in polynomial time are also of some interest.

We also consider a variant of the problem modeled by a bipartite graph, where one side of the graph are agents with private costs and the other side are tasks, each with a value for the principal. An edge (a, t) represents that agent a can do task t. In this model, which was introduced by Goel et al. (2014)

motivated by Crowdsourcing Markets, each agent (represented by a node) has a fixed private cost, can do a subset of the tasks, and each task has a fixed value for the principal.

Our Contribution. We offer two classes of parameterized mechanisms. The main result of our paper in Sect. 4 combines these two mechanisms in a surprising way, offering a new and improved mechanism for large markets. In Sect. 3.1, we study the class of parameterized *threshold mechanisms* that decide on adding items based on a threshold of the marginal contribution of each item over its cost (bang per buck), using a parameter γ. In Sect. 3.2, we consider another parameterized class, called the *oracle mechanisms*, which adds items in decreasing order of bang per buck, till reaching an α fraction of the true optimum, without considering the budget. In Sect. 3 we analyze these two parameterized mechanisms for general monotone submodular valuations. In Sect. 4 we combine the two mechanisms to get an improved result for large markets. See Table 1 for a summary of our results for the general problem. In Sect. 5 we focus on the application to a problem of markets with heterogeneous tasks Chen et al. (2011); Goel et al. (2014).

Table 1. The top numbers are the previously known best guarantees, $r \geq 1$ is the approximation ratio of the oracle used by the mechanism, * indicates that the mechanism has exponential running time. Rand and Det stand for randomized and deterministic mechanisms, and LM indicates the large market assumption. The $4r$ guarantee requires an additional assumption for the oracle, without the assumption the bound is $4r + 1$.

	Rand	Rand*	Det*	Det, LM	Rand, Oracle	Det, Oracle, LM
Previous work	7.91	7.91	8.34	3	–	$2r^2$
Our results	5	4	4.56	2.58	$4r$ or $4r + 1$	$1 + r$

- In Sect. 3.1 we consider threshold mechanism GREEDY-TM, and RANDOM-TM, that chooses randomly between the single item of highest value, and the output of GREEDY-TM. This framework is a direct generalization of the mechanisms presented in Singer (2010), Chen et al. (2011), Singla and Krause (2013), and some of the mechanisms of Anari et al. (2014), who used $\gamma = 0.5$. We show that for monotone submodular valuations, with the same choice of the parameter γ, the randomized threshold mechanism is universally truthful, budget feasible and can achieve a 5 approximation of the optimum. This improves on the best previous bound of 7.91 due to Chen et al. (2011).
- In Sect. 3.2, we introduce another class of parameterized mechanisms, RANDOM-OM, called *oracle mechanisms* which add items in the bang per bunk order until an α fraction of the optimum value is obtained, for a parameter α. We have included their exponential counterparts RANDOM-EOM and DETERMINISTIC-EOM in the full version of this paper (Jalaly and Tardos (2017)). The mechanisms RANDOM-EOM and DETERMINISTIC-EOM use the

true optimum value (and hence run in exponential time), while RANDOM-OM uses a polynomial time approximation instead. We show that keeping the total value of the winning set at most a fraction of the optimum guarantees that the mechanism is budget feasible. RANDOM-EOM and DETERMINISTIC-EOM use a parameterized version of the exponential time oracle mechanism of Anari et al. (2014), which we call GREEDY-EOM, as a subroutine.

For the case when the mechanism has access to an oracle computing the true optimum value, we show that with the right choice of α, our oracle mechanism RANDOM-EOM is universally truthful, budget feasible and achieves a 4 approximation of the optimum for monotone submodular values, improving the bound of 7.91 of Chen et al. (2011). For DETERMINISTIC-EOM, we use a derandomization idea, which is similar to that of Chen et al. (2011) and show it achieves 4.56 approximation of the optimum, improving the 8.34 bound of Chen et al. (2011). The main difference between our mechanism with that of Chen et al. (2011) is that their mechanism only uses the optimum value in its derandomization phase and their greedy subroutine does not take advantage of having access to the optimum value, but our mechanism uses the optimum value for both derandomization and creating a simple greedy subroutine that selects the items (called GREEDY-EOM).

The mechanism RANDOM-OM runs in polynomial time by using an r-approximation oracle as a subroutine instead of the optimum. We note that using GREEDY-EOM with a sub-optimal oracles breaks monotonicity; our RANDOM-OM mechanism guarantees monotonicity with any oracle.

We show that with the right choice of α, RANDOM-OM is universally truthful, budget feasible, and achieves a $4r + 1$ approximation of the optimum (which improves to $4r$ when the oracle used is a greedy algorithm).

– We give the main result of this paper in Sect. 4, where we combine our two parameterized mechanisms by running both and declaring the sellers in the intersection of the two sets as winners. Taking the intersection allows our CAUTIOUS-BUYER mechanism to use larger values of the parameters γ and α and keep the mechanism budget feasible. We show that for the right choice of α and γ, our mechanism is deterministic, truthful, budget feasible and has an approximation guarantee of $1 + r$, improving the bound $2r^2$ claimed by Anari et al. (2014)[1] (where r is the approximation guarantee of the oracle used). Using the greedy algorithm of Sviridenko (2004) (which was also analyzed in Khuller et al. (1999) for the special case of budgeted maximum coverage problem), the approximation guarantee of our mechanism is $1 + \frac{e}{e-1} \simeq 2.58$. In Fig. 1 we show our mechanisms and their subroutines.

– In Sect. 5, we show how our results for submodular valuations can be used for the problem of Crowdsourcing Markets with Heterogeneous Tasks of Goel et al. (2014). This implies that our large market mechanism in Sect. 4 is a

---

[1] Anari et al. (2014) achieve the claimed $2r^2$ by replacing the subroutine computing the optimal solution in their exponential mechanisms by an r-approximation algorithm. Unfortunately, this appears to break the truthfulness of the mechanism, as we point out in Sect. 3.2.

deterministic, truthful and budget feasible mechanism with $1 + \frac{e}{e-1} \approx 2.58$ approximation guarantee for this problem. The resulting deterministic mechanism matches the approximation guarantee of the randomized truthful (in expectation) mechanism of Goel et al. (2014) for this problem.

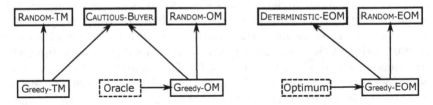

Fig. 1. Edges show mechanisms that are used as a subroutine of others. The mechanisms on left and right run polynomial and exponential time respectively. Oracle is a polynomial time algorithm that approximately solves the budgeted optimization problem for monotone submodular valuations.

Related Work. Prior free budget feasible mechanism design for buying a set of items, each from a different seller, has been introduced by Singer (2010). For monotone submodular valuations, which is the focus of our paper, Chen et al. (2011) improved the mechanism of Singer (2010) and its analysis to achieve a 7.91 approximation guarantee, and also derandomized the mechanism to get a deterministic (but exponential time) mechanism with an approximation guarantee of 8.34. Amanatidis et al. (2016) showed this mechanism can be derandomized given a linear programming (LP) relaxation of an integer program (IP) for this problem in polynomial time. Their approximation guarantee depends on some parameters of the input IP and its LP relaxation, but does not improve the 8.34 guarantee[2].

Singla and Krause (2013) considered the problem for an application in community sensing and gave a mechanism with a 4.75 approximation guarantee for large markets. Anari et al. (2014) improved the result of Singla and Krause (2013) achieving a 3 approximation guarantee for large markets with a polynomial time mechanism and a 2 approximation guarantee with an exponential time mechanism. Anari et al. (2014) also proposed a mechanism that given an r approximation oracle for the budgeted value maximization problem for monotone submodular functions, has a $2r^2$ approximation guarantee, however, their mechanism uses an optimization algorithm as an oracle, and loses monotonicity (and hence truthfulness), when using the greedy algorithm (see Sect. 3.2 for more details). We overcome this difficulty (in addition to improving the bound)

---

[2] The same authors somewhat improved this bound in Amanatidis et al. (2017) using our analysis from an earlier version of this paper, but the improved bound is at least 5.45.

by allowing winning sets of items that are no longer contiguous in the order of their marginal bang per buck.

Budget feasible mechanism design has also been considered with special valuation functions, where better bounds are known. For example, for additive valuations the best known mechanism achieves an approximation bound of $2 + \sqrt{2} \approx 3.41$ and 3 with a deterministic and randomized mechanisms respectively due to Chen et al. (2011), who also gave a $1 + \sqrt{2} \approx 2.41$ lower bound for approximation ratio of any truthful budget feasible mechanism in this setting. In large markets with additive valuations, Anari et al. (2014) improved these results and gave a budget feasible mechanism with an approximation guarantee of $\frac{e}{e-1}$ with a matching lower bound.

Singer (2010) also introduced the feasible mechanism design problem for matchings on bipartite graphs: the principal is required to select a matching of a bipartite graph, where each individual edge is an agent with a private cost and a public value. Chen et al. (2011) consider the knapsack problem with heterogeneous items, which is the special case of this problem where the bipartite graph is a set of disjoint stars. Their approximation bound mentioned above for additive item values, of $2 + \sqrt{2}$ and 3 with a deterministic and randomized mechanisms, also extend to this case. Goel et al. (2014) considered a variant of the problem motivated by Crowdsourcing Markets, as defined above, where one side of the graph are agents with private costs, and the other side are tasks, each with a value for the principal. They give a randomized truthful (in expectation) mechanism with a $1 + \frac{e}{e-1}$ approximation guarantee for this problem under the large market assumption. We consider this version of the matching problem in Sect. 5. Even though adding the hard constraint that the winning set should be a matching breaks submodularity of the valuation function (see Chen et al. (2011); Goel et al. (2014); Amanatidis et al. (2016)), we show that our mechanisms for the case of monotone submodular valuations still can be used for this case, matching the approximation guarantee of Goel et al. (2014) with a deterministic mechanism.

Prior free budget feasible mechanisms has also been studied for more general valuation functions. Monotone submodular valuations are the most general class of valuation functions for which a constant factor approximation guarantee with a polynomial time (with a value oracle), truthful and budget feasible mechanism is known. For subadditive valuations Dobzinski (2011) introduced a mechanism using a demand oracle (more powerful than the value oracle we use). The current best bound is an $O(\frac{\log n}{\log \log n})$ approximation guarantee due to Bei et al. (2012). Bei et al. (2012) also gave a randomized mechanism that achieves a constant (768) approximation guarantee for fractionally subadditive (XOS) valuations, also using a demand oracle.

Some papers consider the Bayesian setting, where cost of each agent comes from known independent distributions. Bei et al. (2012) gave a constant-competitive mechanism for subadditive valuations (with a very large constant). Balkanski and Hartline (2016) gave a $(\frac{e}{e-1})^2$-competitive posted pricing mechanism for monotone submodular valuations for large markets, using a cost ver-

sion for defining the largeness of the market. The benchmark (optimum) used
in Balkanski and Hartline (2016) is the outcome of optimal Bayesian incentive
compatible mechanism, while others (including us) have used the significantly
higher, optimum with respect to the budgeted pure optimization problem as
their benchmark. It is interesting to compare our results for large markets to
the approximation guarantee of $(\frac{e}{e-1})^2 \approx 2.5$ of the mechanism in Balkanski and
Hartline (2016). While this bound is ≈ 0.08 better than our bound, their benchmark, the optimal Bayesian incentive compatible mechanism, can be a factor of
$\frac{e}{e-1}$ lower than our benchmark of the optimum ignoring incentives Anari et al.
(2014). Even when the cost of sellers come from a uniform distributions, and the
value of each item is 1, the ratio between the two benchmarks is $\sqrt{2}$.

2   Preliminaries

We consider the problem of a single buyer with a budget B facing a set of
multiple sellers A, each selling a single item. We let n denote the number of
sellers and we assume $A = [n]$. We assume that the value $v(S)$ of the buyer for
a set of items S, is a (non-decreasing) submodular function, that is, it satisfies
$v(S) \le v(T)$ for every $S \subseteq T$, and $v(S) + v(T) \ge v(S \cup T) + v(S \cap T)$, for every
set $S, T \subseteq A$. For every $i \in A$ and $S \subseteq A$, we define $m_i(S) = v(S \cup \{i\}) - v(S)$,
i.e. the marginal value of i with respect to subset S. Note that $v(.)$ is monotone
submodular if and only if for every $S, T \subseteq A$ we have:

$$v(T) \le v(S) + \sum_{i \in T \setminus S} m_i(S).$$

The large market assumption. In Sect. 4, we consider large markets, assuming
that the value of each agent is small compared to the optimum, i.e. $v(i) \ll$
$opt(A)$ for all $i \in [n]$. For simplicity, we state our approximation bounds for
large markets in the limit[3], assuming $\theta = \max_{i \in [n]} \frac{v(i)}{opt(A)} \to 0$.

The mechanism design problem of selecting sellers maximizing the buyer's
value subject to his budget constraint, is a single parameter mechanism design
problem, in which each bidder (seller) has one private value (the cost of her
item). We design truthful, deterministic and individually rational mechanisms,
as well as universally truthful and individually rational randomized mechanisms.
A randomized mechanism is *universally truthful* if it is a randomization among
deterministic mechanisms, each of which are truthful. We use Myerson's characterization for truthful mechanisms, stating that a mechanism is truthful and
individually rational if and only if the choice of selecting each item is monotone
in its declared cost, and winners are paid threshold payments that are above
their declared cost.

---

[3] By having a θ-large market assumption instead, the approximation guarantees for
our large market mechanisms increases by a factor of $(1 - c\theta)^{-1}$, where $c \in (0, 4)$ is
a constant which is different for each mechanism. We omit stating the exact value
of c for each mechanism separately.

In order to show a mechanism is universally truthful and budget feasible, it suffices to show that the allocation is monotone and by using the threshold payments, the total payments are not more than the budget. Similar to Dobzinski (2011); Chen et al. (2011); Bei et al. (2012); Singla and Krause (2013), we assume that the payments are threshold payments and only specify the allocation rule. At the end of each section, we briefly explain how the payment rule of the mechanisms in that section can be computed. In all our mechanisms if a seller bids a cost more than B, he will not be selected in the winning set, hence will have utility 0. This combined with the fact that all our mechanisms are truthful, implies *individual rationality*, i.e. in all of our mechanisms utility of sellers are non-negative.

3   Parameterized Mechanisms for Submodular Valuations

In this section we present two simple parameterized mechanisms. We show that these parameterized mechanisms provide good approximation guarantees, and are monotone and hence can be turned into truthful mechanisms with payments defined appropriately. We analyze the approximation guarantee of these mechanisms with and without the large market assumption and give conditions that make these mechanisms budget feasible.

Let $S_0 = \emptyset$, and for each $i \in [n]$, recursively define $S_i = S_{i-1} \cup \{\arg \max_{j \in A \setminus S_{i-1}} (\frac{m_j(S_{i-1})}{c_j})\}$, adding the item with maximum marginal value to cost ratio, to S_{i-1}. To simplify notation, we will assume without loss of generality that $\{i\} = S_i \setminus S_{i-1}$. All of our mechanisms sort the items in descending order of marginal bang for buck at the beginning and consider items in this order.

3.1   The Threshold Mechanisms

Our threshold mechanism generalizes the mechanisms of Singer (2010) and Chen et al. (2011). We consider items in increasing cost-to-marginal value order, as defined above. Our *greedy threshold mechanism*, GREEDY-TM, sets a threshold for the cost to marginal value ratio of the items, compared to the ratio of the budget to the total value of the set selected. Using a parameter γ, the mechanism adds items while they are relatively cheap compared to the total so far.

The GREEDY-TM mechanism works well for large markets where each individual item has small value compared to the optimum. In the general case, we will randomly choose between just selecting the item with maximum individual value and cost below the budget, or running GREEDY-TM. We call the resulting randomized mechanism RANDOM-TM(γ, A, B).

GREEDY-TM(γ, A, B) (Greedy Threshold Mechanism) Let $k = 1$ **while** $k \leq |A|$ *and* $\frac{c_k}{m_k(S_{k-1})} \leq \gamma \frac{B}{v(S_k)}$ **do** $\quad \mid \quad k = k + 1$ **end** **return** S_{k-1}	RANDOM-TM(γ, A, B) (Random Threshold Mechanism) Let $A = \{i : c_i \leq B\}$ Let $i^* = argmax_{i \in [n]}(v(i))$ **With probability** $\frac{\gamma+1}{\gamma+2}$ **do** $\quad$ **return** GREEDY-TM(γ, A, B) $\quad$ **halt** **return** i^*

The randomized mechanisms for submodular functions in Singer (2010) is similar to RANDOM-TM with parameter $\gamma = \frac{e-1}{12e-4}$ and the improved mechanism of Chen et al. (2011) is equivalent to RANDOM-TM with $\gamma = 0.5$. In this section we offer a sketch of an improved analysis with details deferred to Jalaly and Tardos (2017).

Monotonicity of the mechanisms is easy to see: if someone is not chosen, he cannot be selected by increasing his cost (decreasing his marginal bang per buck).

Lemma 1. *For every fixed* $\gamma \in (0, 1]$, *the mechanism* GREEDY-TM(γ, A, B) *is monotone.*

We show that for every fixed $\gamma \in (0, 1]$, RANDOM-TM(γ, A, B) achieves a $1 + \frac{2}{\gamma}$ approximation of the optimum, improving the bound of Chen et al. (2011). The key difference is that we compare the output of GREEDY-TM directly with the true optimum, rather than a fractional greedy solution. Doing this not only improves the approximation factor, but also simplifies the analysis. We use this idea in the proof of the following technical lemma, which is the main ingredient for proving the approximation guarantee of our mechanism. The detailed proof is deferred to Jalaly and Tardos (2017).

Lemma 2. *For every fixed* $\gamma \in (0, 1]$, *if* $S_{k-1} =$ GREEDY-TM(γ, A, B) *then*

$$(1 + \frac{1}{\gamma})v(S_{k-1}) + \frac{1}{\gamma}v(i^*) \geq opt(A). \tag{1}$$

By using the above lemma for the performance of RANDOM-TM, we can get the following approximation bound for RANDOM-TM(γ, A, B).

Theorem 1. *For every fixed* $\gamma \in (0, 1]$, RANDOM-TM(γ, A, B) *is universally truthful, and has approximation ratio of* $1 + \frac{2}{\gamma}$.

Proof Sketch. Monotonicity of the mechanism follows from Lemma 1.

The main idea for proving the approximation ratio is calculating the expected value of the outcome of the mechanism and using inequality 1 from Lemma 2. ∎

The mechanisms GREEDY-TM(γ, A, B) and RANDOM-TM(γ, A, B) are not necessarily budget feasible for an arbitrary choice of γ. However, Chen et al. (2011) shows that RANDOM-TM$(0.5, A, B)$ (which they call RANDOM-SM) is budget feasible. We have included a simplified proof in the full version of this paper (Jalaly and Tardos (2017)).

Combining the budget feasibility proof of Chen et al. (2011) and Theorem 1 for the general case, and using inequality (1) directly, instead of Theorem 1, for the case of large market, where $v(i^*) \ll opt(A)$, we get the following theorem. The bound for large markets is matching the best approximation guarantee of Anari et al. (2014) for submodular functions with computational constraint. In Sect. 4 we improve this bound, while in Jalaly and Tardos (2017) we show that the analysis in this section for RANDOM-TM$(0.5, A, B)$ is tight.

Corollary 1. RANDOM-TM$(0.5, A, B)$ *is truthful, budget feasible and has approximation ratio of 5. For the case of large market case, where* $v(i^*) \ll opt(A)$, GREEDY-TM(γ, A, B) *is truthful, budget feasible and has approximation ratio of 3.*

Note that we need $\gamma = 0.5$ to get the above corollary, however in Sect. 4 we use this mechanism with a larger value of γ. The threshold payment of each agent i in the winning set for the threshold mechanisms in this section can be computed by increasing i's cost until he reaches the threshold that makes him not eligible to be in the winning set, while keeping the cost of other agents fixed. In order to compute this number in polynomial time, it is enough to fix other agents' costs and see where in the sorted list of marginal bang-per-bucks this agent can appear such that the inequality of GREEDY-TM(γ, A, B) still holds for her. The more detailed characterization of these threshold payments is similar to that of Singer (2010).

3.2   The Oracle Mechanisms

Here, we provide a different class of parameterized mechanisms. This class of mechanisms requires an oracle $Oracle(A, B)$, which considers the optimization problem of maximizing the value of a subset of A, subject to the total cost not exceeding the budget B, and returns a value which is close to optimum. Let $opt(A, B)$ denote the optimum value of this optimization problem. We assume that $opt(A, B) \geq Oracle(A, B)$. The oracle is an r approximation, if we also have $r \cdot Oracle(A, B) \geq opt(A, B)$. For instance, the greedy algorithm of Sviridenko (2004) can be used as an oracle with $r = \frac{e}{e-1} \approx 1.58$. Since calculating the $opt(A, B)$ is not possible in polynomial time, we call the mechanisms that use optimum value *the exponential time oracle mechanisms*. Due to lack of space, we study these mechanisms in the full version of this paper (Jalaly and Tardos (2017)). For some of our analysis in this section we offer a proof sketch here and defer the detailed proof to Jalaly and Tardos (2017).

Here we offer an oracle mechanism that uses a oracle in place of the optimum value $opt(A, B)$, as finding the optimum for monotone submodular maximization with a knapsack constraint cannot be done in polynomial time. Naively using the outcome of a sub-optimal oracle instead of optimum in our exponential time oracle mechanisms of Jalaly and Tardos (2017) can break monotonicity. To see this, note that if an item increases his cost, she cannot increase the value of $opt(A, B)$. However, if we replace $opt(A, B)$ with the outcome of a sub-optimal oracle (for instance a greedy algorithm), this is no longer true: if one increases the cost of all the items that are not in the optimum set to be more than the budget, any reasonable approximation algorithm for submodular maximization (for instance the greedy algorithm in Sviridenko (2004)) can detect and choose all the items that are in optimum set. Now if an item i increases her cost and this increases the value of the outcome of the approximation algorithm compared to $opt(A, B)$, this gets the mechanism to take more items, possibly including item i, by increasing the value of αv^* in the selection rule of the mechanism.

To make our mechanism monotone, we remove i before calling the oracle to decide if we should add i to the set S, making the items selected no longer contiguous in the order we consider them.

GREEDY-OM(α, A, B)	RANDOM-OM(α, A, B)
(Greedy Oracle Mechanism)	(Random Oracle Mechanism)
Let $S = \emptyset$	Let $A = \{i : c_i \leq B\}$
for $i = 1$ *to* n **do**	Let $i^* = argmax_{i \in [n]}(v(i))$
if	**With probability** $\frac{r}{\alpha+2r}$ **do**
$v(S_i) \leq \alpha Oracle(A \setminus \{i\}, B)$	**return**
then	GREEDY-OM(α, A, B)
$S = S \cup \{i\}$	**halt**
end	**return** i^*
end	
return S	

Next we show that GREEDY-OM(α, A, B) is monotone and provide its approximation ratio.

Lemma 3. *For every fixed* $\alpha \in (0, 1]$, *GREEDY-OM(α, A, B) is monotone. If* $S = $ *GREEDY-EOM(α, A, B), $k \in [n]$ is the biggest integer such that $S_{k-1} \subseteq S$, i^* is the item with maximum individual value, and assuming* ORACLE *is an r approximation of the optimum, then*

$$\frac{r}{\alpha}v(S_{k-1}) + (1 + \frac{r}{\alpha})v(i^*) \geq opt(A)$$

Proof. Monotonicity of the mechanism follows from the usual argument, increasing c_i does not effect ORACLE$(A \setminus \{i\}, B)$ and decreases the item's bang per buck in any step so it can only increase the value of $v(S_i)$ since the decrease in the item's bang per buck can only add more items to S_i by definition. To show the approximation factor, recall that $\{k\} = S_k \setminus S_{k-1}$. Since k was not chosen by the mechanism we have

$$v(S_{k-1}) + v(k) \geq v(S_k) > \alpha Oracle(A \setminus \{k\}) \geq \frac{\alpha}{r}opt(A \setminus \{k\})$$

$$\geq \frac{\alpha}{r}(opt(A) - v(k)) \geq \frac{\alpha}{r}(opt(A) - v(i^*))$$

$\square$

Next Lemma shows that GREEDY-OM$(0.5, A, B)$ is budget feasible. We include a sketch of the proof here with details deferred to Jalaly and Tardos (2017).

Lemma 4. *By using threshold payments,* GREEDY-OM$(0.5, A, B)$ *is budget feasible.*

Proof Sketch. We prove that if each agent i bids higher than $m_i(S_{i-1})\frac{B}{v(S)}$, where $S =$ GREEDY-OM$(0.5, A, B)$, then she will not be in the winning set. We use contradiction to prove this showing that, if an agent bids higher than this amount and is still in the winning set, then cost of optimum has to be higher than the budget which is a contradiction. ∎

In large markets $v(i^*) \ll opt(A)$ so by combining Lemmas 3 and 4 we get the following corollary.

Corollary 2. *In large markets,* GREEDY-OM$(0.5, A, B)$ *is truthful, budget feasible and given an r-approximation oracle, achieves $2r$ approximation of the optimum.*

The previously known best oracle mechanism for large markets is due to Anari et al. (2014) achieves $2r^2$. We will improve this bound for the case of large markets to $r + 1$ in Sect. 4.

By using Lemma 3, we get the following theorem, whose proof is deferred to Jalaly and Tardos (2017).

Theorem 2. RANDOM-OM(α, A, B) *is truthful and in expectation achieves $1 + \frac{2r}{\alpha}$ of the optimum, assuming the oracle used is an r-approximation.*

By combining Lemma 4 and Theorem 2 we have the following theorem.

Theorem 3. RANDOM-OM$(0.5, A, B)$ *is truthful, budget feasible and in expectation achieves $1 + 4r$ of the optimum.*

By using the greedy algorithm of Sviridenko (2004) as an oracle, we can improve the approximation ratio to $\frac{2r}{\alpha}$. To achieve this, we change GREEDY-OM(α, A, B), so that instead of using $Oracle(A \setminus \{k\}, B)$, it uses $\max_{c'_i \geq c_i} Oracle(A, (c'_i, c_{-i}))$. We also change the probability of choosing the greedy mechanism's outcome in RANDOM-OM(α, A, B) to $\frac{1}{2}$. By doing so, RANDOM-OM(α, A, B) can achieve $\frac{2r}{\alpha}$ instead of $1 + \frac{2r}{\alpha}$. By using the greedy algorithm of Sviridenko (2004), as an oracle, finding $\max_{c'_i \geq c_i} Oracle(A, (c'_i, c_{-i}))$ can be done in polynomial time, since we only have to check polynomial number of cases for c'_i. Furthermore, if i increases his cost, he cannot increase the value of $\max_{c'_i \geq c_i} Oracle(A, (c'_i, c_{-i}))$. We omit the proof of the following theorem, as it is analogous to our previous proofs.

Theorem 4. *The above modification of the* RANDOM-OM$(0.5, A, B)$ *mechanism is truthful, budget feasible, get expected value of a $4r$ fraction of the optimum. With the greedy algorithm as the oracle, it can be implemented in polynomial time, and is a $4e/(e-1)$-approximation mechanism.*

For calculating the agents' threshold payments of our oracle mechanisms in this section, it is enough to check what is the maximum cost that each agent i can declare such that she is still in the winning set. Similar to Sect. 3.1, for each agent i, this number can simply be computed by checking where in the sorted list of agents by their marginal bang-per-buck this agent can appear such that the inequality of GREEDY-EOM(α, A, B) (for the exponential time mechanisms) and the inequality of GREEDY-OM(α, A, B) (for polynomial time mechanisms) still hold. The characterization of these threshold payments is similar to the payment characterization of the oracle mechanisms of Anari et al. (2014).

4 A Simple $1 + \frac{e}{e-1}$ Approximation Mechanism for Large Markets

In this section we combine the two greedy parameterized mechanisms of Sect. 3, GREEDY-OM(α, A, B) and GREEDY-TM(γ, A, B) to improve the approximation guarantee for large markets. Given a polynomial time r approximation oracle, our simple, deterministic, truthful, and budget feasible mechanism in this section has an approximation ratio of $1 + r$ and runs in polynomial time.

At first glance, CAUTIOUS-BUYER seems worse than both of GREEDY-OM and GREEDY-TM, since its winning set is the intersection of the winning sets of these mechanisms. However, taking the intersection of these mechanisms will allow us to choose the value of α and γ to be higher than 0.5 while keeping the mechanism budget feasible. It is easy to see that the intersection of two monotone mechanisms is monotone.

> CAUTIOUS-BUYER(α, γ, A, B)
> Let $A = \{i : c_i \leq B\}$
> Let $S_\alpha = $ GREEDY-OM(α, A, B)
> Let $S_\gamma = $ GREEDY-TM(γ, A, B)
> **return** $S_\alpha \cap S_\gamma$

Proposition 1. *For two monotone mechanisms M_1 and M_2, the mechanism M that outputs the intersection of the winning set of M_1 and the winning set of M_2 is monotone.*

Next we give a parameterized approximation guarantee for CAUTIOUS-BUYER(α, γ, A, B).

Lemma 5. *Assuming the large market assumption, for every fixed value of $\alpha, \gamma \in (0, 1]$ CAUTIOUS-BUYER(α, γ, A, B) is monotone. Furthermore, with an r approximation oracle, it has a worst case approximation ratio of $\max(1 + \frac{1}{\gamma}, \frac{r}{\alpha})$.*

Proof. From Proposition 1, Lemmas 2, 3 it follows that CAUTIOUS-BUYER is monotone.

Let S be the outcome of the mechanism. Let k be the biggest integer such that $S_{k-1} \subseteq S$, i.e., $S_{k-1} \subseteq S_\alpha$ and $S_{k-1} \subseteq S_\gamma$. By definition $k \notin S$, so there are two cases

- $k \notin S_\alpha$: By Lemma 3, the large market assumption and monotonicity of $v(.)$, we have $\frac{r}{\alpha}v(S) \approx \frac{r}{\alpha}v(S) + (1 + \frac{r}{\alpha})v(i^*) \geq \frac{r}{\alpha}v(S_{k-1}) + (1 + \frac{r}{\alpha})v(i^*) \geq opt(A, B)$.

- $k \notin S_\gamma$: By Lemma 2, the large market assumption and monotonicity of $v(.)$, we have $(1 + \frac{1}{\gamma})v(S_{k-1}) \approx (1 + \frac{1}{\gamma})v(S_{k-1}) + \frac{1}{\gamma}v(i^*) \geq opt(A, B)$.

In both cases we have, $\max(1 + \frac{1}{\gamma}, \frac{r}{\alpha})v(S) \geq opt(A, B)$ assuming that $v(i^*)$ is negligible.

Now we provide a simple condition for the budget feasibility of CAUTIOUS-BUYER(α, γ, A, B).

Lemma 6. *If $\alpha \leq \frac{1}{1+\gamma}$ for any $\alpha, \gamma \geq 0$, then by using threshold payments, CAUTIOUS-BUYER(α, γ, A, B) is budget feasible.*

Proof. Let p_i be the threshold payment for agent i. Let $S =$ CAUTIOUS-BUYER(α, γ, A, B). For every $i \in S$, we show that if i deviates to bidding a cost $b_i > m_i(S_{i-1})\frac{B}{v(S)}$, he cannot be in the winning set. By proving this and by using the definition of threshold payments we get $\sum_{i \in S} p_i \leq \sum_{i \in S} m_i(S_{i-1})\frac{B}{v(S)} \leq \sum_{i \in S} m_i(S_{i-1} \cap S)\frac{B}{v(S)} = B$, so the mechanism is budget feasible.

We prove above claim by contradiction: assume that i deviates to $b_i > m_i(S_{i-1})\frac{B}{v(S)}$ and is in the winning set. Let b be the new cost vector and j be position of i in the new order of items. Let S'_z for $z \in [n]$ be defined similar to S_z but with cost vector b instead of c. So S'_j is the set of items that are in the winning set of GREEDY-TM(γ, A, B) at the end of step j once i is added. Note that S'_j is also equal to the set of all the items that has been considered by GREEDY-OM(α, A, B) at the end of its j-th step. So by using the same argument as proof of Lemma 4 we get

$$c(S^*) > B\frac{v(S^*) - v(S'_j)}{v(S)} \qquad and \qquad v(S'_j), v(S) \leq \alpha v(S^*)$$

By defining $x = \frac{v(S'_j)}{v(S)}$, we have $v(S'_j) = xv(S) \leq \alpha xv(S^*)$. So we get

$$c(S^*) > B\frac{v(S^*) - v(S'_j))}{v(S)} > B\frac{(1-\alpha)v(S^*)}{\alpha xv(S^*)} = B\frac{1-\alpha}{\alpha x}$$

so if $1 - \alpha \geq \alpha x$, or equivalently, $\alpha \leq \frac{1}{1+x}$ then we get $c(S^*) > B$ which is the desired contradiction.

Since $i \in S_\gamma$, we also have

$$\frac{b_i}{m_i(S'_j)} \leq \gamma\frac{B}{v(S'_j)} = \frac{\gamma}{x}\frac{B}{v(S)}$$

So since $m_i(S'_j) \leq m_i(S_{i-1})$, if $\gamma \leq x$, we get to a contradiction with the assumption about b_i.

The only remaining case is when $\gamma > x$ and $\alpha > \frac{1}{1+x}$. This means that $\alpha > \frac{1}{1+\gamma}$ which is a contradiction with the property in the statement of lemma, so the mechanism is budget feasible.

By using Lemmas 5 and 6, the main theorem of this section follows. We include the detailed proof in full version of this paper (Jalaly and Tardos (2017)).

Theorem 5. *By using threshold payments,* CAUTIOUS-BUYER $\left(\frac{r}{r+1}, \frac{1}{r}\right)$ *is truthful, budget feasible, and* $1+r$ *approximation of the optimum. By using the greedy algorithm with* $r = e/(e-1)$ *we get a mechanism with approximation guarantee of* ≈ 2.58.

The threshold payment of an agent in this mechanism is the minimum of the threshold payment of two mechanisms we intersected to get CAUTIOUS-BUYER(α, γ, A, B).

5   Application to Hiring in Crowdsourcing Markets

In this section, we consider an application of our mechanisms for the problem of a principal hiring in a Crowdsourcing Market. We consider the model where there is a set of agents A that can be hired and a set of tasks T that the principal would like to get done. Each agent $i \in A$ has a private cost c_i. We represent the abilities of the agents by a bipartite graph $G(A, T)$, where edge $e = (a, t)$ in the G indicates that agent a can be used for task t, where each agent hired can be used for at most one of the tasks she can do. The value of buyer for each edge e is v_e, which can be different for each edge. The principal would like to hire agents to maximize the total value of tasks done, while keeping the total payment under her budget B. The optimal solution for this problem is a maximum value matching, subject to the budget constraint on the cost of the hired agents.

This model is also known as knapsack with heterogeneous items. Knapsack with heterogeneous items and buyer with a matching constraint for budget feasible mechanism design was also studied by Singer (2010); Chen et al. (2011); Goel et al. (2014); Amanatidis et al. (2016). There are many ways for modeling the heterogeneity of items. In Chen et al. (2011), this heterogeneity has been defined by having types for items where at most one item can be chosen from each type, corresponding to a bipartite graph where agents have degree 1. Goel et al. (2014) consider our model of agents and tasks with a bipartite graph, but assume that the principal has a fixed value for each task completed, independent of the agent that took care of the task, so values of the edges entering a task node t are all equal.

In this section, we apply our technique from Sect. 4 for this problem. For the model used by Goel et al. (2014), where the principal has a value for each task independent of who completes the task, we show that a small change in our mechanism (stopping it when the marginal increase in value is 0) results in the same approximation guarantee. The small modification is needed as the value for the buyer is not a submodular function of the set of agents hired (see Amanatidis et al. (2016)).

Theorem 6. *The truthful budget feasible threshold and oracle mechanisms, as well as the large markets mechanism for submodular valuations of this paper* *without any loss in the approximation ratio can be also used for the case of heterogeneous tasks, with the constraint that each agent in the winning set should be assigned to a unique task (matching constraint).*

Before proving this theorem, we consider the general problem defied above, but similar to Singer (2010), we relax the assumption that the allocation should always assigned each agent in the winning set to a unique task, and allow instead that multiple agents get assigned to a given task, with only one contributing to the value. We define the value of the buyer for the winning set S to be the value of the maximum matching on the induced subgraph $G[S, T]$. We'll see that allowing the principal to hire extra agents makes her valuation a monotone submodular function. This can model the case where the principal asks more than one worker to accomplish a task, but only keep the best results.

General Crowdsourcing Markets. It is well-known and not hard to see that the function defined by value of maximum matching adjacent to a subset of agents S is a monotone and submodular function of S. The following proposition formalizes this statement.

Proposition 2. *For $S \subseteq A$, let $f(S)$ be the maximum value of a matching of the induced subgraph $G[S, T]$ of the bipartite graph $G(A, T)$, then $f(S)$ is a monotone submodular function.*

This proposition implies that all our truthful budget feasible mechanisms for submodular valuations can be used for this model.

Corollary 3. *Without the strict matching constraint, budgeted valuations with heterogeneous tasks (items) are a special case of monotone submodular valuations, so all the mechanisms from the previous sections can be applied to this problem.*

Hiring with Strict Matching Constraint. Consider the case where the buyer's value is defined by summation of her value for each task, i.e. for all the edges that are directly connected to the same task, the value of the buyer for those edges is the same. We argue that in this model, if we add the hard constraint each agent in the winning set should be assigned to a unique task (similar to Chen et al. (2011); Goel et al. (2014); Amanatidis et al. (2016)), then with a small change in our mechanisms, all our results still hold. This problem was considered by Goel et al. (2014) for large markets, who gave a randomized truthful (in expectation) and budget feasible mechanism with a $1 + \frac{e}{e-1}$ approximation guarantee for large markets (the main result of Goel et al. (2014)). The next lemma shows how one can use our truthful budget feasible mechanism for large markets to get the same approximation guarantee with a deterministic mechanism.

Lemma 7. *For $S \subseteq A$, let $f(S)$ be the maximum value of matching of the induced subgraph $G[S, T]$ of the bipartite graph $G(A, T)$ in which all the edges* *that connect to the same node of T have the same value. If a maximum value matching induced by $S \subseteq A$ connects all vertices in S to a vertex in T, and for $a \in A \setminus S$, $f(S \cup \{a\}) - f(S) > 0$, then there is a maximum value matching induced by $S \cup \{a\}$ which is also assigning each agent to a unique task.*

Proof. We use contradiction. Assume that there is a subset of agents $S \subseteq A$ such that there is a maximum matching M in the subgraph of G, induced by vertices of S and T that connects each agent in S to a task in T. Let $a \in A$ be an agent such that $f(S \cup \{a\}) - f(S) > 0$ and there is no maximum matching in the subgraph induced by $S' = S \cup \{a\}$ and T that connects each agent in S' to a task in T. Let M' be a maximum matching of this induced subgraph. Let G' be the union of edges in M and M' and let C and P be the set of cycles and paths that contain all the edges of G'. Since M and M' are both maximum matchings, we have $W(M \cap c) = W(M' \cap c)$ for all $c \in C$. Since the only difference between S and S' is having a, there can only be one path $p \in P$ such that $W(p \cap M') > W(p \cap M)$. Furthermore, one of the end points of p should be a and for all other paths $p' \in P$ that $p' \neq p$, $W(p' \cap M) = W(p' \cap M')$. For p there are two cases

- The edge that is connected to the other endpoint of p is in M: in this case, since the value of matching is defined by tasks, $W(p \cap M) = W(p \cap M')$. Therefore, $W(M) = W(M')$ and $F(S') - F(S) = 0$ which is a contradiction.
- The edge that is connected to the other endpoint of p is in M': In this case if we define a matching $M^* = (M \setminus p) \cup (M' \cap p)$, then M^* will connect each agent in S' to a unique task, which is a contradiction.

This means that we reach contradiction in both cases, and the proof is complete.

Proof of Theorem 6. We use our mechanism this problem, using Corollary 3, but stopping to consider items in the sorted list of marginal bang per bucks whenever the marginal bang-per-buck of the item is 0. Note that since the items are listed in decreasing order of marginal bang per buck and we know that the valuation is submodular, doing this will not have any effects on the approximation ratio (since the marginal bang-per-buck of the next items is also 0) and truthfulness (since the threshold payment of an agent whose item has 0 marginal value is 0) of our mechanisms.

By Lemma 7 in the resulting subset of agents, there is a maximum value matching assigning each agent to a unique task.   ∎

For the case of large markets, by this corollary and Theorem 5, CAUTIOUS-BUYER is a deterministic truthful and budget feasible mechanism for this problem, matching the $1 + \frac{e}{e-1}$ guarantee of the randomized truthful (in expectation) mechanism of Goel et al. (2014).

References

Amanatidis, G., Birmpas, G., Markakis, E.: Coverage, matching, and beyond: new results on budgeted mechanism design. In: Cai, Y., Vetta, A. (eds.) WINE 2016. LNCS, vol. 10123, pp. 414–428. Springer, Heidelberg (2016). https://doi.org/10.1007/978-3-662-54110-4_29

Amanatidis, G., Birmpas, G., Markakis, E.: 2017. On Budget-feasible mechanism design for symmetric submodular objectives. CoRR abs/1704.06901 (2017). http://arxiv.org/abs/1704.06901

Anari, N., Goel, G., Nikzad, A.: Mechanism design for crowdsourcing: an optimal 1–1/e competitive budget-feasible mechanism for large markets. In: 2014 IEEE 55th Annual Symposium on Foundations of Computer Science (FOCS), pp. 266–275. IEEE (2014)

Balkanski, E., Hartline, J.D.: Bayesian budget feasibility with posted pricing. In: Proceedings of the 25th International Conference on World Wide Web, International World Wide Web Conferences Steering Committee, pp. 189–203 (2016)

Bei, X., Chen, N., Gravin, N., Lu, P.: Budget feasible mechanism design: from prior-free to Bayesian. In: Proceedings of the Forty-fourth Annual ACM Symposium on Theory of Computing, pp. 449–458. ACM(2012)

Chen, N., Gravin, N., Lu, P.: On the approximability of budget feasible mechanisms. In: Proceedings of the Twenty-second Annual ACM-SIAM Symposium on Discrete Algorithms, SIAM, pp. 685–699 (2011)

Dobzinski, S., Christos, H.P., Yaron, S.: Mechanisms for complement-free procurement. In: Proceedings of the 12th ACM Conference on Electronic Commerce, pp. 273–282. ACM (2011)

Goel, G., Nikzad, A., Singla, A.: Mechanism design for crowdsourcing markets with heterogeneous tasks. In: Second AAAI Conference on Human Computation and Crowdsourcing (2014)

Jalaly, P., Tardos, É.: 2017. Simple and Efficient Budget Feasible Mechanisms for Monotone Submodular Valuations. CoRR abs/1703.10681 (2017). http://arxiv.org/abs/1703.10681

Khuller, S., Moss, A., Naor, J.S.: The budgeted maximum coverage problem. Inform. Process. Lett. **70**(1), 39–45 (1999)

Singer, Y.: Budget feasible mechanisms. In: 2010 51st Annual IEEE Symposium on Foundations of Computer Science (FOCS), pp. 765–774. IEEE (2010)

Singla, A., Krause, A.: Incentives for privacy tradeoff in community sensing. In: First AAAI Conference on Human Computation and Crowdsourcing (2013)

Sviridenko, M.: A note on maximizing a submodular set function subject to a knapsack constraint. Oper. Res. Lett. **32**(1), 41–43 (2004)

Social Welfare and Profit Maximization from Revealed Preferences

Ziwei Ji[(✉)], Ruta Mehta, and Matus Telgarsky

University of Illinois at Urbana-Champaign, Urbana, USA
{ziweiji2,rutameht,mjt}@illinois.edu

Abstract. Consider the *seller's problem* of finding optimal prices for her n (divisible) goods when faced with a set of m consumers, given that she can only observe their purchased bundles at posted prices, i.e., *revealed preferences*. We study both social welfare and profit maximization with revealed preferences. Although social welfare maximization is a seemingly non-convex optimization problem in prices, we show that (i) it can be reduced to a dual convex optimization problem in prices, and (ii) the revealed preferences can be interpreted as supergradients of the concave conjugate of valuation, with which subgradients of the dual function can be computed. We thereby obtain a simple subgradient-based algorithm for strongly concave valuations and convex cost, with query complexity $O(m^2/\epsilon^2)$, where ϵ is the additive difference between the social welfare induced by our algorithm and the optimum social welfare. We also study social welfare maximization under the online setting, specifically the random permutation model, where consumers arrive one-by-one in a *random order*. For the case where consumer valuations can be arbitrary continuous functions, we propose a price posting mechanism that achieves an expected social welfare up to an additive factor of $O(\sqrt{mn})$ from the maximum social welfare. Finally, for profit maximization (which may be non-convex in simple cases), we give nearly matching upper and lower bounds on the query complexity for separable valuations and cost (i.e., each good can be treated independently).

1 Introduction

In consumer theory, it is standard to assume that the preferences of a consumer are captured by a *valuation* function, which is often assumed to be known to the mechanism designer. However, in a real market, one can only observe what buyers buy at given prices, the *revealed preferences*. Research on revealed preferences within TCS has two primary objectives: (i) learning valuation functions from revealed preferences, with the goal of having predictive properties [6,7,10,36]; (ii) directly learning the prices that maximize social welfare or profit [4,5,8,9, 11,13,28,29,35].

The latter problem is of importance to sellers in today's online economies, where a large amount of data about consumers' buying patterns is available. For a seller, profit maximization is the primary goal in general, while she may also

© Springer Nature Switzerland AG 2018
G. Christodoulou and T. Harks (Eds.): WINE 2018, LNCS 11316, pp. 264–281, 2018.
https://doi.org/10.1007/978-3-030-04612-5_18

want to maximize social welfare in an effort to earn the goodwill of consumers, with increased market share as a byproduct.

In this paper, we consider social welfare and profit maximization using only revealed preferences.

1.1   Our Model, Results, and Techniques

Consider a market with m consumers and a producer (seller) who produces and sells a set of n *divisible* goods. In the most general case, the preferences of consumer i over bundles of goods are defined by a valuation function $v_i : C \to \mathbb{R}_+$ ($C \subset \mathbb{R}^n$ is called the feasible set), which is her private information and *unknown*. At prices $\mathbf{p}$ she demands bundle $\mathbf{x}_i(\mathbf{p})$ that maximizes her value minus payment, i.e., her *quasilinear utility*

$$\mathbf{x}_i(\mathbf{p}) \in \arg\max_{\mathbf{x} \in C} \left(v_i(\mathbf{x}) - \langle \mathbf{p}, \mathbf{x} \rangle \right).$$

Given prices $\mathbf{p}$, the *revealed preference* refers to the purchased bundle $\mathbf{x}_i(\mathbf{p})$ of each consumer in the market (*demand oracle information*), or even only $\sum_{i=1}^{m} \mathbf{x}_i(\mathbf{p})$ (*aggregate demand oracle information*). No other information of the valuations is revealed.

Producing the demanded goods incurs cost to the producer, which is represented by a convex cost function c. The producer, or the algorithm, posts prices and makes observations repeatedly, trying to maximize the *social welfare* or *profit*, as described below.

Social Welfare Maximization. The social welfare of bundles $\mathbf{x}_1, \ldots, \mathbf{x}_m \in C$ is the sum of consumers' valuations minus production cost, i.e.,

$$\mathrm{SW}(\mathbf{x}_1, \ldots, \mathbf{x}_m) = \sum_{i=1}^{m} v_i(\mathbf{x}_i) - c\left(\sum_{i=1}^{m} \mathbf{x}_i\right). \tag{1}$$

The benchmark used in this paper is the maximum social welfare SW^* and corresponding maximizing bundles $(\mathbf{x}_1^*, \ldots, \mathbf{x}_m^*)$, defined as

$$\mathrm{SW}^* = \max_{\mathbf{x}_1, \ldots, \mathbf{x}_m \in C} \mathrm{SW}(\mathbf{x}_1, \ldots, \mathbf{x}_m) \text{ and } (\mathbf{x}_1^*, \ldots, \mathbf{x}_m^*) \in \arg\max_{\mathbf{x}_1, \ldots, \mathbf{x}_m \in C} \mathrm{SW}(\mathbf{x}_1, \ldots, \mathbf{x}_m).$$
$$\tag{2}$$

In Sect. 3, offline social welfare maximization is considered. The producer tries to find good prices $\mathbf{p}$ such that $\mathrm{SW}(\mathbf{x}_1(\mathbf{p}), \ldots, \mathbf{x}_m(\mathbf{p}))$ is maximized. Although there exist many methods to maximize a concave function, the social welfare is usually a non-concave function in $\mathbf{p}$ [29]. Moreover, the producer only has access to the aggregate demand oracle; the true valuations $v_i(\mathbf{x}_i(\mathbf{p}))$ are unknown.

We first show using duality theory that the maximum social welfare SW^*, which is larger than or equal to any social welfare that can be induced by some prices, can in fact be induced by a single price vector $\mathbf{p}^*$, which is the minimizer of a convex dual function $f(\mathbf{p}) = c^*(\mathbf{p}) - \sum_{i=1}^{m} v_i^*(\mathbf{p})$, where c^* and v_i^* are respectively *convex* and *concave conjugates* [27], as reviewed in Sect. 2. Moreover, the revealed preferences are supergradients of v_i^*, with which subgradients of

f can be computed. Finally, to get a faster algorithm, we apply a smoothing technique to f and then invoke the accelerated gradient descent method. These ideas are formalized in Algorithm 1, whose guarantee is given below.

Theorem 1 (Informal statement of Theorem 4). *The additive error between the social welfare induced by Algorithm 1 and the maximum social welfare Eq. (2) is at most $O(m/\sqrt{T})$, where T is the number of queries to the aggregate demand oracle.*

In other words, to ensure an additive ϵ approximation of the maximum social welfare, Algorithm 1 needs $O(m^2/\epsilon^2)$ queries to the aggregate demand oracle.

[29] and [28] are the most relevant prior work. [29] studies profit maximization instead of social welfare maximization in a market with one consumer. However, it is assumed that the valuation function is homogeneous, under which profit maximization can be reduced to social welfare maximization. Assumptions made in [29] and this paper are basically identical. Key differences are: (i) [29] proposes a two-level algorithm, where there is an outer iterative algorithm maximizing social welfare, and for each outer iterate, the supergradient of the unknown valuation function is computed by solving a dual optimization problem. In this paper, we only need to solve a single (different) dual optimization problem. Therefore, this gives a simpler approach which may be of independent interest. (ii) The subgradient of the dual objective function in this paper can be interpreted as the excess supply (see the discussion around Eq. (8) in Sect. 3.1), which gives our algorithm a natural interpretation as a Tâtonnement process. (iii) The query complexity given in [29] can be as large as $O(1/\epsilon^6)$ to ensure an additive error of ϵ between the induced social welfare and the maximum social welfare; one reason is that they use subgradient descent, which works for non-smooth convex functions but converges slowly. In this paper, by combining a smoothing technique and accelerated gradient descent, Theorem 1 only needs $O(1/\epsilon^2)$ queries to the aggregate demand oracle. [28] assumes that the valuation is stochastic, but only considers a linear cost. It also considers unit demand consumers with indivisible goods, which is out of the scope of this paper.

Next in Sect. 4 we consider *online* social welfare maximization under the *random permutation* model. In this model, m consumers come to make purchases one by one, and correspondingly the producer is allowed to post prices dynamically, i.e., to update prices from $\mathbf{p}_i$ to $\mathbf{p}_{i+1}$ after the purchase of consumer i. Random permutation here means that those m consumers are first chosen potentially by an adversary, and then come and make purchases one by one in a uniformly random order. (The random permutation model has been extensively studied within online optimization [2,14,17], and is more general than the i.i.d. model where each valuation is an independent sample from an unknown distribution.) The objective is to maximize the expected online social welfare $\mathbb{E}[\mathrm{SW}(\mathbf{x}_1(\mathbf{p}_1), \ldots, \mathbf{x}_m(\mathbf{p}_m))]$, where the expectation is taken over random orders.

The idea to solve the online social welfare maximization problem is to run an online convex optimization algorithm on a dual problem $f_i(\mathbf{p}) = c^*(\mathbf{p})/m - \langle \mathbf{x}_i(\mathbf{p}_i), \mathbf{p} \rangle$. See Algorithm 2 for details; an introduction to online convex optimization is given in Sect. 4.

Theorem 2 (Informal statement of Theorem 5). *The expected additive error between the online social welfare achieved by Algorithm 2 and the maximum offline social welfare Eq. (2) is bounded by $O(\sqrt{mn})$, where the expectation is taken over random orders of valuations.*

For a given producer, the number of goods n can be thought of as fixed. As a result, the loss of social welfare induced by Algorithm 2 is sublinear in the number of consumers m.

The idea of Algorithm 2 comes from [2], where a general online stochastic convex programming problem is considered. It has many other advantages when applied to online social welfare maximization. First, it is enough to assume that the valuations are continuous; the consumer demand oracle may potentially need to solve some non-convex quasi-linear utility maximization problem, but our focus is on the producer side. Since f_i only depends on the revealed preference $\mathbf{x}_i(\mathbf{p}_i)$, not on v_i, it is still convex. Second, Algorithm 2 is robust, in the sense that it is not sensitive to the potential error in quasi-linear utility maximization. For details, see the discussion at the end of Sect. 4.

Profit Maximization. Next we consider profit maximization with access to the aggregate demand oracle. Given prices $\mathbf{p}$, the profit of producer is the revenue minus production cost, i.e.,

$$\text{Profit}(\mathbf{p}) = \langle \mathbf{p}, \textstyle\sum_{i=1}^{m} \mathbf{x}_i(\mathbf{p}) \rangle - c \left(\textstyle\sum_{i=1}^{m} \mathbf{x}_i(\mathbf{p}) \right). \tag{3}$$

Although it is more reasonable for the producer to maximize the profit, this problem is hard due to non-convexity. The social welfare maximization problems are solved by making a reduction to some convex optimization problem on the space of prices. However, for profit maximization, both the set of optimal bundles and the set of optimal prices may be non-convex, as shown by Example 1 in Sect. 5.

We then consider the case where both the valuations and cost are separable. A separable valuation $v_i(\mathbf{x}) = \sum_{j=1}^{n} v_{ij}(x_j)$, while similarly a separable cost $c(\mathbf{y}) = \sum_{j=1}^{n} c_j(y_j)$. Under this assumption, in Sect. 5 we give upper and lower bounds on the query complexity for profit maximization and revenue maximization (i.e., the cost is 0). These upper and lower bounds match for revenue maximization.

Theorem 3 (Informal statement of Theorems 6 and 7). *Consider a market with m consumers and n goods. If the valuations are strongly concave, and both the valuations and cost are separable and Lipschitz continuous, then Algorithm 3 maximizes the profit up to an additive ϵ error with $O(mn/\epsilon)$ queries to the aggregate demand oracle. If the cost is zero, then the strongly concave assumption on valuations can be dropped.*

On the other hand, for concave, separable and Lipschitz continuous valuations, any algorithm requires $\Omega(n/\epsilon)$ queries to the aggregate demand oracle in order to maximize the revenue up to an ϵ additive error.

1.2 Related Work

Samuelson started the theory of *revealed preferences* in 1938 [30] to facilitate
mapping observed data to valuation functions, which led to extensive work within
economics on "rationalization" or "fitting the samples" [1,15,20,22,23,26,33,34].
In TCS, there have been a lot of work on learning valuations from revealed
preferences with which predictions can be made [6,7,10,36].

Another line of research is on learning prices directly that can maximize social
welfare or profit, usually known as the dynamic pricing problem [4,5,8,9,11,13].
Some prior works assume nice properties of the demand function (oracle) itself,
such as linearity in case of large number of goods [12,21], concavity [5], Lipschitz
continuity [8,9,35]. However, these properties may not be satisfied by demands
that come from typical concave valuation functions. In [4], the valuation function
is assumed to be linear, and is first partially inferred and then used in a price
optimization step. However, if the valuation is general concave, such a learning
phase is not possible [7].

Recently, [16] studies an online linear classification problem under the
revealed preference model.

2 Preliminaries

Market Model. Our model consists of one producer (seller) who produces and sells
n divisible goods, and m consumers. Consumer i's preferences are represented
by an *unknown* valuation function $v_i : \mathcal{C} \to \mathbb{R}_+$. The feasible consumption set
$\mathcal{C} \subset \mathbb{R}_+^n$ is typically assumed to be convex and compact with non-empty interior.
It is assumed that $\mathcal{C}$ is known to the algorithm, and let $D = \max_{\mathbf{x},\mathbf{y} \in \mathcal{C}} \|\mathbf{x} - \mathbf{y}\|_2$
denote the ℓ_2 diameter of $\mathcal{C}$. Note that our algorithm can be extended to the case
where v_i's have different domains with different diameters; a common domain $\mathcal{C}$
is used here only for convenience.

Given prices $\mathbf{p} = (p_1, \ldots, p_n) \in \mathbb{R}_+^n$ of goods, the quasi-linear utility of a
bundle $\mathbf{x} \in \mathbb{R}_+^n$ is defined as

$$u_i(\mathbf{x}, \mathbf{p}) = v_i(\mathbf{x}) - \langle \mathbf{x}, \mathbf{p} \rangle.$$

Naturally, consumer i demands a bundle from $\mathcal{C}$ that maximizes her quasi-linear
utility

$$\mathbf{x}_i(\mathbf{p}) \in \arg\max_{\mathbf{x} \in \mathcal{C}} u_i(\mathbf{x}, \mathbf{p}),$$

which is known as the *revealed preference* of consumer i at prices $\mathbf{p}$. Once the
seller sets prices $\mathbf{p}$, we only get to see $\mathbf{x}_i(\mathbf{p})$ for each consumer i, where every
consumer can be thought of as a *demand oracle*, or even only $\mathbf{x}(\mathbf{p}) = \sum_{i=1}^{m} \mathbf{x}_i(\mathbf{p})$,
where the market can be seen as an *aggregate demand oracle*. v_i is always
assumed to be continuous to ensure that $\mathbf{x}_i(\mathbf{p})$ exists. This is the only assumption needed for the online social welfare maximization part of this paper; the

offline social welfare maximization part and the profit maximization part further assume that the valuations are strongly concave, which will be introduced later.

The production cost is represented by a convex, Lipschitz continuous, non-decreasing cost function $c : m\mathcal{C} \to \mathbb{R}_+$, where $m\mathcal{C} = \{m\mathbf{x} | \mathbf{x} \in \mathcal{C}\} = \{\sum_{i=1}^{m} \mathbf{x}_i | \mathbf{x}_1, \ldots, \mathbf{x}_m \in \mathcal{C}\}$ since $\mathcal{C}$ is convex. Note that the domain of c is big enough to allow production of any aggregate demand. Let λ denote the modulus of Lipschitz continuity of c with respect to the ℓ_2-norm. It is assumed that the cost function is known to the algorithm.

The producer, or the algorithm, can only post prices and observe the purchased bundles repeatedly, trying to maximize the *social welfare* Eq. (1) or *profit* Eq. (3). Note that if valuations are only continuous, $\mathbf{x}_i(\mathbf{p})$ and the induced social welfare and profit may not be uniquely defined. In this paper, the online social welfare result holds for any $\mathbf{x}_i(\mathbf{p})$, while in offline social welfare and profit maximization, $\mathbf{x}_i(\mathbf{p})$ is unique since strong concavity is assumed.

Convex and Concave Conjugates. The notion of convex and concave conjugates are crucial in our algorithms. Given a convex function $f : \mathcal{D} \to \mathbb{R}$ where $\mathcal{D} \subset \mathbb{R}^n$ is non-empty, its convex conjugate f^* is defined as:

$$f^*(\mathbf{y}) = \sup_{\mathbf{x} \in \mathcal{D}} \left(\langle \mathbf{y}, \mathbf{x} \rangle - f(\mathbf{x}) \right),$$

where the domain of f^* is given by $\mathbf{dom}\, f^* = \{\mathbf{y} \in \mathbb{R}^n | f^*(\mathbf{y}) < \infty\}$. Similarly, given a concave function $f : \mathcal{D} \to \mathbb{R}$, its concave conjugate is defined as

$$f^*(\mathbf{y}) = \inf_{\mathbf{x} \in \mathcal{D}} \left(\langle \mathbf{y}, \mathbf{x} \rangle - f(\mathbf{x}) \right),$$

where the domain of f^* is given by $\mathbf{dom}\, f^* = \{\mathbf{y} \in \mathbb{R}^n | f^*(\mathbf{y}) > -\infty\}$. Since we only compute convex conjugates of convex functions and concave conjugates of concave functions, the above notation is fine. ∂f denotes the set of subgradients of convex f or supergradients of concave f.

Note that in our case, since $\mathbf{dom}\, c = m\mathcal{C}$ is non-empty and compact and c is continuous, $\mathbf{dom}\, c^* = \mathbb{R}^n$. Similarly, for every i, $\mathbf{dom}\, v_i^* = \mathbb{R}^n$.

Lemma 1 is crucial in our algorithm: One key observation in this paper is that revealed preferences are actually supergradients of the concave conjugate of valuation, which is given by Lemma 1. Lemma 1 can be derived from [19] Corollary E.1.4.4 immediately. Although it is stated for convex functions and convex conjugates, corresponding properties hold for concave functions and concave conjugates.

Lemma 1. *Suppose f is convex continuous with non-empty domain. For every pair $(\mathbf{x}, \mathbf{y}) \in \mathbf{dom}\, f \times \mathbf{dom}\, f^*$,*

$$\mathbf{y} \in \partial f(\mathbf{x})$$
$$\Longleftrightarrow \mathbf{x} \in \partial f^*(\mathbf{y})$$
$$\Longleftrightarrow \mathbf{x} \in \arg\max_{\mathbf{x}' \in \mathbf{dom}\, f} \left(\langle \mathbf{y}, \mathbf{x}' \rangle - f(\mathbf{x}') \right)$$
$$\Longleftrightarrow \mathbf{y} \in \arg\max_{\mathbf{y}' \in \mathbf{dom}\, f^*} \left(\langle \mathbf{x}, \mathbf{y}' \rangle - f^*(\mathbf{y}') \right)$$
$$\Longleftrightarrow f(\mathbf{x}) + f^*(\mathbf{y}) = \langle \mathbf{x}, \mathbf{y} \rangle.$$

3   Offline Social Welfare Maximization

Problem Description. The goal of offline social welfare maximization is to find prices $\mathbf{p} \in \mathbb{R}_+^n$ such that the induced social welfare $\mathrm{SW}(\mathbf{x}_1(\mathbf{p}), \ldots, \mathbf{x}_m(\mathbf{p})) = \sum_{i=1}^m v_i(\mathbf{x}_i(\mathbf{p})) - c(\sum_{i=1}^m \mathbf{x}_i(\mathbf{p}))$ is maximized. As introduced below, v_i's are assumed to be strongly concave, and thus $\mathbf{x}_i(\mathbf{p})$'s are uniquely determined.

Strongly Concave Valuations. In the offline setting, the valuation functions (v_i's) are further assumed to be α-strongly concave, meaning that $v_i(x) + \alpha\|x\|_2^2/2$ is concave. Concavity is a standard assumption on valuations to capture diminishing marginal returns. Strong concavity, as its name suggests, is a strong assumption; however it is satisfied by many common valuations such as the constant elasticity of substitution functions and Cobb-Douglas functions (c.f. [29]). Furthermore, a common modulus of strong concavity α is only for convenience; the algorithm can be easily adapted to the case where different v_i have different moduli of strong concavity.

The dual notion to strong concavity (convexity) is strong smoothness. $f : \mathcal{D} \to \mathbb{R}$ is β-strongly smooth if f is differentiable and its gradient is β-Lipschitz continuous, or formally, for any $\mathbf{x}, \mathbf{y} \in \mathcal{D}$, $\|\nabla f(\mathbf{y}) - \nabla f(\mathbf{x})\|_2 \le \beta\|\mathbf{y} - \mathbf{x}\|_2$.

The following lemma can be immediately derived from [19] Theorem E.4.2.1 and E.4.2.2.

Lemma 2. *Suppose $f : \mathcal{D} \to \mathbb{R}$ is concave continuous and $\mathcal{D} \subset \mathbb{R}^n$ is non-empty. Then f is α-strongly concave if and only if f^* is $1/\alpha$-strongly smooth and concave on $\mathbb{R}^n$.*

Accelerated Gradient Descent. The accelerated gradient descent algorithm, which was first introduced in [25], gives the optimal convergence rate for smooth convex optimization problems. There have been many extensions to AGD, including [3,32]. In this paper, one variant called the AGM algorithm given in [3] will be invoked.

Lemma 3 ([3] **Theorem 4.1**). *Suppose $f : \mathcal{D} \to \mathbb{R}$ is β-strongly smooth and convex and $\mathbf{x}^* \in \arg\min_{\mathbf{x} \in \mathcal{D}} f(\mathbf{x})$. Given $\mathbf{x}_0 \in \mathcal{D}$, for any $t \ge 1$, The AGM algorithm outputs $\mathbf{x}_t \in \mathcal{D}$ such that*

$$f(\mathbf{x}_t) - f(\mathbf{x}^*) \le \frac{2\beta\|\mathbf{x}_0 - \mathbf{x}^*\|_2^2}{t^2}.$$

3.1   Algorithm and Analysis

We propose Algorithm 1 to solve the offline social welfare maximization problem.

Algorithm 1. Offline SW Maximization

$\mu \leftarrow \frac{2\lambda}{mD\sqrt{T}}$, $c_\mu(\mathbf{y}) = c(\mathbf{y}) + \frac{\mu}{2}\|\mathbf{y}\|_2^2$, $f_\mu = c_\mu^*(\mathbf{p}) - \sum_{i=1}^m v_i^*(\mathbf{p})$.

$\mathcal{P} \leftarrow \{\mathbf{p} \geq \mathbf{0} | \|\mathbf{p}\|_2 \leq \lambda\}$, $\mathbf{p}_\mu^0 = \mathbf{0}$.

Give T, the total number of rounds, and $\mathbf{p}_\mu^0$ to the AGM algorithm.

Output $\mathbf{p}_\mu^T$ returned by the AGM algorithm.

Theorem 4. *The social welfare induced by $\mathbf{p}_\mu^T$ given by Algorithm 1 is within $9\lambda mD/\sqrt{T} + 16\lambda^2 m/\alpha T$ from the maximum offline social welfare.*

The proof is based on two observations. The first one, formalized in Lemma 4, says that the optimal solution of a dual optimization problem can induce the maximum social welfare SW*.

Lemma 4. *Given concave continuous valuations $v_1, \ldots, v_m : \mathcal{C} \to \mathbb{R}$ and a convex continuous cost $c : m\mathcal{C} \to \mathbb{R}$,*

$$\text{SW}^* = \min_{\mathbf{p}\in\mathbb{R}^n}\left(c^*(\mathbf{p}) - \sum_{i=1}^m v_i^*(\mathbf{p})\right),$$

and for any dual optimal solution $\mathbf{p}^*$, $(\mathbf{x}_1(\mathbf{p}^*), \ldots, \mathbf{x}_m(\mathbf{p}^*))$ maximizes social welfare. If furthermore the cost is non-decreasing and λ-Lipschitz continuous, then

$$\text{SW}^* = \min_{\mathbf{p}\in\mathbb{R}^n}\left(c^*(\mathbf{p}) - \sum_{i=1}^m v_i^*(\mathbf{p})\right) = \min_{\mathbf{p}\geq 0, \|\mathbf{p}\|_2\leq\lambda}\left(c^*(\mathbf{p}) - \sum_{i=1}^m v_i^*(\mathbf{p})\right),$$

and for any optimal solution $\mathbf{p}^*$ of the rightmost dual problem, $(\mathbf{x}_1(\mathbf{p}^*), \ldots, \mathbf{x}_m(\mathbf{p}^*))$ maximizes social welfare.

Proof. Maximizing social welfare is equivalent to solving the following problem

$$\max_{\substack{\mathbf{x}_1,\ldots,\mathbf{x}_m\in\mathcal{C} \\ \mathbf{y}\in m\mathcal{C}}} \sum_{i=1}^m v_i(\mathbf{x}_i) - c(\mathbf{y}) \qquad (4)$$
$$\text{s.t.} \sum_{i=1}^m \mathbf{x}_i = \mathbf{y}.$$

The Lagrangian is $L(\mathbf{x}_1, \ldots, \mathbf{x}_m, \mathbf{y}, \mathbf{p}) = \sum_{i=1}^m v_i(\mathbf{x}_i) - c(\mathbf{y}) + \langle\mathbf{p}, \mathbf{y} - \sum_{i=1}^m \mathbf{x}_i\rangle$. Equation (4) equals

$$\max_{\mathbf{x}_1,\ldots,\mathbf{x}_m\in\mathcal{C}, \mathbf{y}\in m\mathcal{C}} \min_{\mathbf{p}\in\mathbb{R}^n} L(\mathbf{x}_1, \ldots, \mathbf{x}_m, \mathbf{y}, \mathbf{p})$$

$$= \min_{\mathbf{p}\in\mathbb{R}^n} \max_{\mathbf{x}_1,\ldots,\mathbf{x}_m\in\mathcal{C}, \mathbf{y}\in m\mathcal{C}} L(\mathbf{x}_1, \ldots, \mathbf{x}_m, \mathbf{y}, \mathbf{p})$$

$$= \min_{\mathbf{p}\in\mathbb{R}^n}\left(c^*(\mathbf{p}) - \sum_{i=1}^m v_i^*(\mathbf{p})\right). \qquad (5)$$

Here the second line is due to Slater's condition, and the third line is due to the definition of convex conjugate and concave conjugate.

Due to the continuity of v_i's and c and the compactness of $\mathcal{C}$, the optimal primal solution $(\mathbf{x}_1^*, \ldots, \mathbf{x}_m^*, \mathbf{y}^*)$ exists. By Slater's condition, the dual optimal solution $\mathbf{p}^*$ also exists. By the minimax property, we know that $\mathbf{y}^* = \sum_{i=1}^m \mathbf{x}_i^*$, and $\mathbf{x}_i^*$ maximizes $v_i(\mathbf{x}) - \langle \mathbf{x}, \mathbf{p}^* \rangle$, $1 \leq i \leq m$.

For the second part, due to the monotonicity of c, maximizing social welfare is equivalent to

$$
\begin{aligned}
&\max_{\substack{\mathbf{x}_1, \ldots, \mathbf{x}_m \in \mathcal{C} \\ \mathbf{y} \in m\mathcal{C}}} \quad \sum_{i=1}^m v_i(\mathbf{x}_i) - c(\mathbf{y}) \\
&\text{s.t.} \; \sum_{i=1}^m \mathbf{x}_i \leq \mathbf{y}.
\end{aligned}
\tag{6}
$$

The Lagrangian is still $L(\mathbf{x}_1, \ldots, \mathbf{x}_m, \mathbf{y}, \mathbf{p}) = \sum_{i=1}^m v_i(\mathbf{x}_i) - c(\mathbf{y}) + \langle \mathbf{p}, \mathbf{y} - \sum_{i=1}^m \mathbf{x}_i \rangle$. Equation (6) equals

$$
\begin{aligned}
&\max_{\mathbf{x}_1, \ldots, \mathbf{x}_m \in \mathcal{C}, \mathbf{y} \in m\mathcal{C}} \min_{\mathbf{p} \geq 0, \|\mathbf{p}\|_2 \leq \lambda} L(\mathbf{x}_1, \ldots, \mathbf{x}_m, \mathbf{y}, \mathbf{p}) \\
=&\min_{\mathbf{p} \geq 0, \|\mathbf{p}\|_2 \leq \lambda} \max_{\mathbf{x}_1, \ldots, \mathbf{x}_m \in \mathcal{C}, \mathbf{y} \in m\mathcal{C}} L(\mathbf{x}_1, \ldots, \mathbf{x}_m, \mathbf{y}, \mathbf{p}) \\
=&\min_{\mathbf{p} \geq 0, \|\mathbf{p}\|_2 \leq \lambda} \left(c^*(\mathbf{p}) - \sum_{i=1}^m v_i^*(\mathbf{p}) \right).
\end{aligned}
\tag{7}
$$

Here the first line is due to the Lipschitz continuity of c, the second line is due to Sion's theorem [31]. $\qquad\square$

Lemma 4 tells us that the minimizer $\mathbf{p}^*$ of $f(\mathbf{p}) = c^*(\mathbf{p}) - \sum_{i=1}^m v_i^*(\mathbf{p})$ induces SW*, and thus it is natural to try to solve this dual optimization problem. However, v_i's are unknown to us, and so are v_i^*'s and f. The second observation is that the revealed preference, $\mathbf{x}_i(\mathbf{p})$, actually gives a supergradient of v_i^* at $\mathbf{p}$. Formally, given a concave continuous valuation $v : \mathcal{C} \to \mathbb{R}$, for any $\mathbf{x} \in \mathcal{C}$, by Lemma 1,

$$
\mathbf{x} \in \arg\max_{\mathbf{x}' \in \mathcal{C}} \left(v(\mathbf{x}') - \langle \mathbf{x}', \mathbf{p} \rangle \right) \iff \mathbf{x} \in \arg\min_{\mathbf{x}' \in \mathcal{C}} \left(\langle \mathbf{x}', \mathbf{p} \rangle - v(\mathbf{x}') \right) \iff \mathbf{x} \in \partial v^*(\mathbf{p}).
\tag{8}
$$

Similarly, given a convex continuous cost $c : m\mathcal{C} \to \mathbb{R}_+$, for any $\mathbf{y} \in \mathbb{R}_+^n$, $\mathbf{y} \in \partial c^*(\mathbf{p}) \iff \mathbf{y} \in \arg\max_{\mathbf{y}'} \left(\langle \mathbf{y}', \mathbf{p} \rangle - c(\mathbf{y}') \right)$. In other words, the subgradient of c^* at $\mathbf{p}$ gives a bundle which maximizes the producer's profit, assuming everything produced can be sold.

As a result, we can run subgradient-based optimization algorithms to minimize f. (c is known to the algorithm, and so is c^*; the computation of subgradients of c^* is another problem, but does not require access to the consumer demand oracles.) Since v_i is α-strongly concave, by Lemma 2, v_i^* is $1/\alpha$-strongly smooth and concave. However, there is no guarantee on c^*, and thus in general, f is not strongly smooth. In this case the standard optimization algorithm is subgradient descent. However, for strongly smooth and convex functions, accelerated gradient descent converges much faster than subgradient descent. To invoke accelerated gradient descent, the *smoothing technique* given in [24] is used. We minimize f_μ as given in Algorithm 1 and tune the parameter μ, which finally gives Theorem 4. The detailed proof is given in the full version of this paper at https://arxiv.org/abs/1711.02211.

4   Online Social Welfare Maximization

Problem Description. In online social welfare maximization, m consumers come one by one and the producer/algorithm can post prices dynamically. Specifically, at step i, prices $\mathbf{p}_i$ are posted, and then consumer i comes and makes a purchase $\mathbf{x}_i(\mathbf{p}_i)$. Then the algorithm updates $\mathbf{p}_i$ to $\mathbf{p}_{i+1}$, based on past information. The goal is to maximize the online social welfare $\mathrm{SW}(\mathbf{x}_1(\mathbf{p}_1), \ldots, \mathbf{x}_m(\mathbf{p}_m)) = \sum_{i=1}^{m} v_i(\mathbf{x}_i(\mathbf{p}_i)) - c(\sum_{i=1}^{m} \mathbf{x}_i(\mathbf{p}_i))$.

To model the randomness in the real world, it is usually assumed that valuations are sampled i.i.d. from some unknown distribution. Here we consider a slightly stronger model, called the random permutation model. In the random permutation model, an adversary chooses m valuations $\tilde{v}_1, \ldots, \tilde{v}_m$ in advance, which then come in a uniformly random order. Formally, let $\gamma = (\gamma_1, \ldots, \gamma_m)$ be a random permutation of $(1, \ldots, m)$, then at step i the consumer with valuation $v_i = \tilde{v}_{\gamma_i}$ comes and makes a purchase, after $\mathbf{p}_i$ is posted. Note that in the random permutation model, the corresponding offline problem is fixed (with valuations $\tilde{v}_1, \ldots, \tilde{v}_m$). We still let $\mathrm{SW}^* = \max_{\mathbf{x}_1, \ldots, \mathbf{x}_m \in \mathcal{C}} \left(\sum_{i=1}^{m} \tilde{v}_i(\mathbf{x}_i) - c(\sum_{i=1}^{m} \mathbf{x}_i) \right)$, and let

$$(\tilde{\mathbf{x}}_1^*, \ldots, \tilde{\mathbf{x}}_m^*) = \arg \max_{\mathbf{x}_1, \ldots, \mathbf{x}_m \in \mathcal{C}} \left(\sum_{i=1}^{m} \tilde{v}_i(\mathbf{x}_i) - c\left(\sum_{i=1}^{m} \mathbf{x}_i \right) \right).$$

Our goal is to show that the expected online social welfare $\mathbb{E}_\gamma[\mathrm{SW}(\mathbf{x}_1(\mathbf{p}_1), \ldots, \mathbf{x}_m(\mathbf{p}_m))]$ is close to SW^*, where the expectation is taken over the random permutation γ.

Note that no more assumption is made; valuations are only required to be continuous.

Online Convex Optimization. The algorithm for online social welfare maximization invokes an online convex optimization (OCO) algorithm as a subroutine. In an OCO problem, there is a feasible domain $\mathcal{D}$ and T steps. At step t, the OCO algorithm determines $\mathbf{x}_t \in \mathcal{D}$, and then a convex function $f_t : \mathcal{D} \to \mathbb{R}$ is chosen (potentially by an adversary) and a loss of $f_t(\mathbf{x}_t)$ is induced. Based on the past information (formally, $\mathbf{x}_1, \ldots, \mathbf{x}_t$ and $f_1, \ldots, f_t$), the algorithm updates $\mathbf{x}_t$ to $\mathbf{x}_{t+1}$, and tries to minimize the *regret*

$$R(T) = \sum_{t=1}^{T} f_t(\mathbf{x}_t) - \min_{\mathbf{x} \in \mathcal{D}} \sum_{t=1}^{T} f_t(\mathbf{x}).$$

The regret of an OCO algorithm $\mathcal{A}$ is denoted by $R_\mathcal{A}(T)$.

The *online (sub)gradient descent* algorithm performs the following update at step t:

$$\mathbf{x}_{t+1} = \Pi_{\mathcal{D}}[\mathbf{x}_t - \eta_t g_t(\mathbf{x}_t)],$$

where η_t is the step size, $g_t(\mathbf{x}_t) \in \partial f_t(\mathbf{x}_t)$, and $\Pi_{\mathcal{D}}$ is the ℓ_2 projection onto $\mathcal{D}$.

Lemma 5 ([18] **Theorem 3.1**). *Let* $D = \max\{\|\mathbf{x}_1 - \mathbf{x}_2\|_2 | \mathbf{x}_1, \mathbf{x}_2 \in \mathcal{D}\}$, $G = \max\{\|\partial f_t(\mathbf{x})\|_2 | 1 \le t \le T, \mathbf{x} \in \mathcal{D}\}$, *and* $\eta_t = D/G\sqrt{T}$. *Then*

$$R_{\mathrm{OGD}}(T) = \sum_{t=1}^{T} f_t(\mathbf{x}_t) - \min_{\mathbf{x} \in \mathcal{D}} \sum_{t=1}^{T} f_t(\mathbf{x}) \le DG\sqrt{T}.$$

4.1   Algorithm and Analysis

Algorithm 2 is proposed to solve the online social welfare maximization problem. The idea of Algorithm 2 comes from [2].

Algorithm 2. Online SW Maximization

$\mathcal{P} \leftarrow \{\mathbf{p} \ge 0 | \|\mathbf{p}\|_2 \le \lambda\}$.
Give $\mathcal{P}$ to an OCO algorithm $\mathcal{A}$, and let $\mathbf{p}_1 \in \mathcal{P}$ be the initial prices chosen by $\mathcal{A}$.
for $i = 1$ **to** m **do**
    Post prices $\mathbf{p}_i$.
    Observe $\mathbf{x}_i(\mathbf{p}_i)$, the choice of the buyer who shows up in the i-th step.
    Give $f_i(\mathbf{p}) = \frac{1}{m}c^*(\mathbf{p}) - \langle \mathbf{x}_i(\mathbf{p}_i), \mathbf{p} \rangle$ with domain $\mathcal{P}$ to $\mathcal{A}$, and observe an updated
    $\mathbf{p}_{i+1}$ from $\mathcal{A}$.
end for

Theorem 5. *The expected social welfare of Algorithm 2 with respect to a uniformly random permutation of continuous valuations, is within* $R_{\mathcal{A}}(m) + 2\lambda D_\infty \sqrt{nm}$ *from the offline optimum social welfare, where* $D_\infty = \max_{\mathbf{x},\mathbf{y} \in C} \|\mathbf{x} - \mathbf{y}\|_\infty$ *is the* ℓ_∞ *diameter of* C. *Specifically, for online gradient descent, the difference is bounded by* $4\lambda D_\infty \sqrt{nm}$.

Proof. For convenience, let $\mathbf{x}_i$ denote $\mathbf{x}_i(\mathbf{p}_i)$. By the regret bound of the OCO algorithm,

$$\sum_{i=1}^{m} f_i(\mathbf{p}_i) - \min_{\mathbf{p} \ge 0, \|\mathbf{p}\|_2 \le \lambda} \sum_{i=1}^{m} f_i(\mathbf{p})$$

$$= \sum_{i=1}^{m} (f_i(\mathbf{p}_i) + v_i(\mathbf{x}_i)) - \min_{\mathbf{p} \ge 0, \|\mathbf{p}\|_2 \le \lambda} \sum_{i=1}^{m} (f_i(\mathbf{p}) + v_i(\mathbf{x}_i))$$

$$\le R_{\mathcal{A}}(m). \tag{9}$$

First, we examine the second term of Eq. (9):

$$\min_{\mathbf{p}\geq 0, \|\mathbf{p}\|_2\leq\lambda} \sum_{i=1}^{m}(f_i(\mathbf{p})+v_i(\mathbf{x}_i)) = \min_{\mathbf{p}\geq 0, \|\mathbf{p}\|_2\leq\lambda}\left(c^*(\mathbf{p}) - \left\langle\sum_{i=1}^{m}\mathbf{x}_i, \mathbf{p}\right\rangle\right) + \sum_{i=1}^{m}v_i(\mathbf{x}_i)$$

$$= \sum_{i=1}^{m}v_i(\mathbf{x}_i) - c\left(\sum_{i=1}^{m}\mathbf{x}_i\right).$$

The first equality comes from the definition of f, while the second inequality is due to the definition of c^* and the monotonicity and Lipschitz continuity of c. Thus the second term of Eq. (9) always equals the social welfare achieved by Algorithm 2. In the following we show that the first term of Eq. (9) is within $O(\sqrt{mn})$ from the offline maximum social welfare SW*.

For a permutation $(\gamma_1,\ldots,\gamma_m)$ of $1,\ldots,m$, let Γ_i denote $(\gamma_1,\ldots,\gamma_i)$. Note that $\mathbf{p}_i$ is determined by Γ_{i-1} ($\Gamma_0 = \emptyset$), v_i is determined by γ_i, and $\mathbf{x}_i$ depends on $\mathbf{p}_i$ and γ_i. Fix $1 \leq i \leq m$ and Γ_{i-1}, note that the revealed preference $\mathbf{x}_i$ maximizes the quasi-linear utility given $\mathbf{p}_i$, we have

$$\mathbb{E}_{\gamma_i}[f_i(\mathbf{p}_i) + v_i(\mathbf{x}_i)|\Gamma_{i-1}] = \mathbb{E}_{\gamma_i}\left[\frac{1}{m}c^*(\mathbf{p}_i) - \langle\mathbf{x}_i, \mathbf{p}_i\rangle + v_i(\mathbf{x}_i)\Big|\Gamma_{i-1}\right]$$

$$= \frac{1}{m}c^*(\mathbf{p}_i) + \mathbb{E}_{\gamma_i}[-\langle\mathbf{x}_i, \mathbf{p}_i\rangle + \tilde{v}_{\gamma_i}(\mathbf{x}_i)|\Gamma_{i-1}]$$

$$\geq \frac{1}{m}c^*(\mathbf{p}_i) + \mathbb{E}_{\gamma_i}[-\langle\tilde{\mathbf{x}}^*_{\gamma_i}, \mathbf{p}_i\rangle + \tilde{v}_{\gamma_i}(\tilde{\mathbf{x}}^*_{\gamma_i})|\Gamma_{i-1}]$$

$$= \frac{1}{m}c^*(\mathbf{p}_i) - \langle\mathbb{E}_{\gamma_i}[\tilde{\mathbf{x}}^*_{\gamma_i}|\Gamma_{i-1}], \mathbf{p}_i\rangle + \mathbb{E}_{\gamma_i}[\tilde{v}_{\gamma_i}(\tilde{\mathbf{x}}^*_{\gamma_i})|\Gamma_{i-1}].$$

$$(10)$$

Consider the last term of Eq. (10) and take expectation with respect to Γ_{i-1}:

$$\mathbb{E}_{\Gamma_{i-1}}\left[\mathbb{E}_{\gamma_i}[\tilde{v}_{\gamma_i}(\tilde{\mathbf{x}}^*_{\gamma_i})|\Gamma_{i-1}]\right] = \frac{1}{m}\sum_{i=1}^{m}\tilde{v}_i(\tilde{\mathbf{x}}^*_i). \tag{11}$$

Then consider the first two terms of Eq. (10):

$$\frac{1}{m}c^*(\mathbf{p}_i) - \langle\mathbb{E}_{\gamma_i}[\tilde{\mathbf{x}}^*_{\gamma_i}|\Gamma_{i-1}], \mathbf{p}_i\rangle$$

$$= \frac{1}{m}c^*(\mathbf{p}_i) - \langle\mathbb{E}_{\gamma_i}[\tilde{\mathbf{x}}^*_{\gamma_i}], \mathbf{p}_i\rangle + \langle\mathbb{E}_{\gamma_i}[\tilde{\mathbf{x}}^*_{\gamma_i}], \mathbf{p}_i\rangle - \langle\mathbb{E}_{\gamma_i}[\tilde{\mathbf{x}}^*_{\gamma_i}|\Gamma_{i-1}], \mathbf{p}_i\rangle$$

$$\geq \frac{1}{m}c^*(\mathbf{p}_i) - \langle\mathbb{E}_{\gamma_i}[\tilde{\mathbf{x}}^*_{\gamma_i}], \mathbf{p}_i\rangle - \lambda\left\|\mathbb{E}_{\gamma_i}[\tilde{\mathbf{x}}^*_{\gamma_i}] - \mathbb{E}_{\gamma_i}[\tilde{\mathbf{x}}^*_{\gamma_i}|\Gamma_{i-1}]\right\|_2$$

$$= \frac{1}{m}c^*(\mathbf{p}_i) - \frac{1}{m}\left\langle\sum_{i=1}^{m}\tilde{\mathbf{x}}^*_i, \mathbf{p}_i\right\rangle - \lambda\left\|\mathbb{E}_{\gamma_i}[\tilde{\mathbf{x}}^*_{\gamma_i}] - \mathbb{E}_{\gamma_i}[\tilde{\mathbf{x}}^*_{\gamma_i}|\Gamma_{i-1}]\right\|_2$$

$$\geq -\frac{1}{m}c\left(\sum_{i=1}^{m}\tilde{\mathbf{x}}^*_i\right) - \lambda\left\|\mathbb{E}_{\gamma_i}[\tilde{\mathbf{x}}^*_{\gamma_i}] - \mathbb{E}_{\gamma_i}[\tilde{\mathbf{x}}^*_{\gamma_i}|\Gamma_{i-1}]\right\|_2.$$

$$(12)$$

Here the first inequality is due to Cauchy-Schwarz inequality, while the second inequality is given by Fenchel-Young inequality.

Equation (10), (11) and (12) give us

$$\mathbb{E}\left[\sum_{i=1}^{m} f_i(\mathbf{p}_i) + v_i(\mathbf{x}_i)\right] \geq \sum_{i=1}^{m} \tilde{v}_i(\tilde{\mathbf{x}}_i^*) - c\left(\sum_{i=1}^{m} \tilde{\mathbf{x}}_i^*\right)$$
$$- \lambda \sum_{i=1}^{m} \mathbb{E}_{\Gamma_{i-1}}\left[\left\|\mathbb{E}_{\gamma_i}[\tilde{\mathbf{x}}_{\gamma_i}^*] - \mathbb{E}_{\gamma_i}[\tilde{\mathbf{x}}_{\gamma_i}^*|\Gamma_{i-1}]\right\|_2\right]. \tag{13}$$

Furthermore, Lemma 6 shows that the last sum in Eq. (13) is bounded by $2D_\infty\sqrt{nm}$, and thus Theorem 5 is proved for general OCO algorithms. Finally, to prove the bound for online gradient descent, it is enough to use step size $\eta_i = 2\lambda/D\sqrt{m}$ (recall that D is the ℓ_2 diameter of $\mathcal{C}$) and invoke Lemma 5. □

As we can see from the proof of Theorem 5, it is enough to have continuous valuations. Furthermore, Algorithm 2 still works if consumers only maximize their quasi-linear utilities approximately. Formally, if consumer i finds a bundle $\mathbf{x}_i$ such that $v_i(\mathbf{x}_i) - \langle \mathbf{p}_i, \mathbf{x}_i \rangle \geq \max_{\mathbf{x} \in \mathcal{C}} (v_i(\mathbf{x}) - \langle \mathbf{p}_i, \mathbf{x} \rangle) - \epsilon_i$, then an additive error of ϵ_i will be introduced in Eq. (10). However, as long as the total error $\sum_{i=1}^{m} \epsilon_i$ is not large, the expected online social welfare of Algorithm 2 will still be close to the offline optimum.

Finally, we state and prove Lemma 6.

Lemma 6. *For any $1 \leq i \leq m$, we have*

$$\mathbb{E}_{\Gamma_{i-1}}\left[\left\|\mathbb{E}_{\gamma_i}[\tilde{\mathbf{x}}_{\gamma_i}^*] - \mathbb{E}_{\gamma_i}[\tilde{\mathbf{x}}_{\gamma_i}^*|\Gamma_{i-1}]\right\|_2\right] \leq D_\infty\sqrt{\frac{n}{m-i+1}}.$$

Proof. Let $\mathcal{S} = \{s_1, \ldots, s_N\}$ denote a finite population of real numbers, and $X_1, \ldots, X_n$ ($1 \leq n \leq N$) denote n samples from $\mathcal{S}$ without replacement. Furthermore, let $\mu = \frac{1}{N}\sum_{i=1}^{N} s_i$ and $\sigma^2 = \frac{1}{N}\sum_{i=1}^{N}(s_i - \mu)^2$. Then $\overline{X} = \frac{1}{n}\sum_{i=1}^{n} X_i$ has mean μ and variance $\frac{N-n}{N-1}\frac{\sigma^2}{n}$.

Now come back the the proof of Lemma 6. By Jensen's inequality, we have

$$\mathbb{E}_{\Gamma_{i-1}}\left[\left\|\mathbb{E}_{\gamma_i}[\tilde{\mathbf{x}}_{\gamma_i}^*] - \mathbb{E}_{\gamma_i}[\tilde{\mathbf{x}}_{\gamma_i}^*|\Gamma_{i-1}]\right\|_2\right] \leq \sqrt{\mathbb{E}_{\Gamma_{i-1}}\left[\left\|\mathbb{E}_{\gamma_i}[\tilde{\mathbf{x}}_{\gamma_i}^*] - \mathbb{E}_{\gamma_i}[\tilde{\mathbf{x}}_{\gamma_i}^*|\Gamma_{i-1}]\right\|_2^2\right]}.$$

Note that each coordinate of $\mathbb{E}_{\Gamma_{i-1}}\left[\|\mathbb{E}_{\gamma_i}[\tilde{\mathbf{x}}_{\gamma_i}^*] - \mathbb{E}_{\gamma_i}[\tilde{\mathbf{x}}_{\gamma_i}^*|\Gamma_{i-1}]\|_2^2\right]$ equals the variance of the average of $m - i + 1$ without-replacement samples. Thus we further have

$$\sqrt{\mathbb{E}_{\Gamma_{i-1}}\left[\left\|\mathbb{E}_{\gamma_i}[\tilde{\mathbf{x}}_{\gamma_i}^*] - \mathbb{E}_{\gamma_i}[\tilde{\mathbf{x}}_{\gamma_i}^*|\Gamma_{i-1}]\right\|_2^2\right]} \leq \sqrt{n\frac{i-1}{m-1}\frac{D_\infty^2}{m-i+1}} \leq D_\infty\sqrt{\frac{n}{m-i+1}}.$$

□

5 Profit Maximization for Separable Valuations and Cost

Previously, social welfare maximization is solved by reducing to a convex optimization problem on the price space. However, profit maximization may be non-convex on both the bundle space and the price space.

Example 1. Consider a market where there is only one consumer, one good, and zero cost. Suppose $v' : [0,2] \to \mathbb{R}_+$ is continuous and strictly decreasing, with $v'(1) = 2$, $v'(2) = 1$, and $v'(x)x < 2$ for any $x \neq 1, 2$. The integral of v' gives a non-decreasing concave valuation v. It can be shown that the maximum profit is 2, which is attained by price 2 at quantity 1 or price 1 at quantity 2. Thus the set of optimum prices and optimum bundles are both non-convex.

Here we present an algorithm of profit maximization and a nearly matching lower bound when all v_i's and c are separable. Formally, for every $\mathbf{x} \in \mathcal{C}$ and every $1 \leq i \leq m$, $v_i(\mathbf{x}) = \sum_{j=1}^{n} v_{ij}(x_j)$, and for every $\mathbf{y} \in m\mathcal{C}$, $c(\mathbf{y}) = \sum_{j=1}^{n} c_j(y_j)$. Due to the separability assumption, we restate the assumptions on the feasible set, valuation functions and cost function:

- $\mathcal{C} = [0,1]^n$.
- For every $1 \leq i \leq m$, $1 \leq j \leq n$, v_{ij} is α-strongly concave and λ-Lipschitz continuous.
- For every $1 \leq j \leq n$, c_j is λ-Lipschitz continuous.

The i-th consumer's consumption of good j is completely determined by p_j and is denoted by $x_{ij}(p_j)$. Furthermore, $x_j(p_j) = \sum_{i=1}^{m} x_{ij}(p_j)$. Our goal is thus to maximize $\text{Profit}_j(p_j) = \sum_{j=1}^{n} x_j(p_j)p_j - c_j(x_j(p_j))$, for each $1 \leq j \leq n$. Although we can set prices for different goods independently now, to keep consistency, we still consider posting new prices $\mathbf{p} \in \mathbb{R}_+^n$ as one query.

Algorithm 3. Profit Maximization Algorithm for Separable Functions

---

$r \leftarrow \left\lceil \frac{mn\lambda(\lambda+\alpha)}{\alpha\epsilon} \right\rceil$.

$\tilde{\mathbf{p}} = (0, 0, \ldots, 0)$.

for $t = 1$ **to** r **do**

    Post prices $\mathbf{p}_t = \left(\frac{t\alpha\epsilon}{mn(\lambda+\alpha)}, \ldots, \frac{t\alpha\epsilon}{mn(\lambda+\alpha)} \right)$.

    for $j = 1$ **to** n **do**

        if $\text{Profit}_j(p_{t,j}) > \text{Profit}_j(\tilde{p}_j)$ **then**

            $\tilde{p}_j = p_{t,j}$

        end if

    end for

end for

Output $\tilde{\mathbf{p}}$.

Theorem 6. *The profit given by Algorithm 3 is no less than the optimum profit minus ϵ. The number of queries is $\lceil mn\lambda(\lambda + \alpha)/\alpha\epsilon \rceil$.*

Proof. Fix $1 \leq j \leq n$. Let p_j^* denote the profit-maximizing price for good j. Suppose $\hat{p}_j = \frac{z\alpha\epsilon}{mn(\lambda+\alpha)} \leq p_j^* \leq \frac{(z+1)\alpha\epsilon}{mn(\lambda+\alpha)}$. By the definition of strong smoothness and Lemma 2, we have

$$x_j(p_j^*)p_j^* - c_j(x_j(p_j^*))$$

$$\leq x_j(\hat{p}_j)\left(\hat{p}_j + \frac{\alpha\epsilon}{mn(\lambda+\alpha)}\right) - c_j(x_j(\hat{p}_j)) + \lambda|x_j(p_j^*) - x_j(\hat{p}_j)|$$

$$= x_j(\hat{p}_j)\left(\hat{p}_j + \frac{\alpha\epsilon}{mn(\lambda+\alpha)}\right) - c_j(x_j(\hat{p}_j)) + \lambda\left|\sum_{i=1}^{m}\left((v_{ij}^*)'(p_j^*) - (v_{ij}^*)'(\hat{p}_j)\right)\right|$$

$$\leq x_j(\hat{p}_j)\left(\hat{p}_j + \frac{\alpha\epsilon}{mn(\lambda+\alpha)}\right) - c_j(x_j(\hat{p}_j)) + \lambda\frac{m}{\alpha}\frac{\alpha\epsilon}{mn(\lambda+\alpha)}$$

$$= x_j(\hat{p}_j)\hat{p}_j - c_j(x_j(\hat{p}_j)) + x_j(\hat{p}_j)\frac{\alpha\epsilon}{mn(\lambda+\alpha)} + \frac{\lambda\epsilon}{n(\lambda+\alpha)}$$

$$\leq x_j(\hat{p}_j)\hat{p}_j - c_j(x_j(\hat{p}_j)) + \frac{\alpha\epsilon}{n(\lambda+\alpha)} + \frac{\lambda\epsilon}{n(\lambda+\alpha)}$$

$$= x_j(\hat{p}_j)\hat{p}_j - c_j(x_j(\hat{p}_j)) + \frac{\epsilon}{n}.$$

$\square$

Remark 1. Note that if the cost is 0 and thus revenue maximization i considered, then we can set $r = \lceil mn\lambda/\epsilon \rceil$ in Algorithm 3, and it is enough to assume concave valuations.

Theorem 7 shows that the dependency on n and ϵ cannot be improved, even for revenue maximization.

Theorem 7. *The revenue maximization problem needs $\Omega(n/\epsilon)$ queries to get an additive error ϵ, even if the valuations are separable, concave, non-decreasing, and Lipschitz continuous.*

Proof. Let us first consider the case where there is only one consumer and one good. Given $\lambda > 1$, consider ϵ such that there exists an integer q satisfying $(1+\epsilon)^q = \lambda$, and thus $q = \frac{\ln\lambda}{\ln(1+\epsilon)} \geq \frac{\ln\lambda}{\epsilon}$.

Now we are going to define concave functions of the amount of good on $[0, 1]$. It is enough to give a non-decreasing and integrable derivative. Let $v'(x) = \lambda$ on $[0, \frac{1}{\lambda}]$, and $\frac{1}{x}$ on $[\frac{1}{\lambda}, 1]$. One can verify that v' is non-decreasing and integrable, and thus we can integrate it into a concave function v (by shifting, we can ensure $v(0) = 0$). The maximum of $v'(x)x$ on $[0, 1]$ is 1.

We claim that the algorithm has to make at least q queries to ensure an ϵ additive or multiplicative error. If it is not true, there must exist some integer $z \in [0, q-1]$ such that no $x \in \mathcal{I} = (\frac{1}{(1+\epsilon)^{z+1}}, \frac{1}{(1+\epsilon)^z}]$ is considered. We can then set $\tilde{v}'(x) = v'(x)$ outside $\mathcal{I}$, $\tilde{v}'(x) = (1+\epsilon)^{z+1}$ on $(\frac{1}{(1+\epsilon)^{z+1}}, \frac{1}{(1+\epsilon)^z})$, and $\tilde{v}'(\frac{1}{(1+\epsilon)^z})$ does not exist. $\tilde{v}'$ is still non-decreasing and integrable, but the optimum revenue is $1 + \epsilon$ now, which is not detected by the algorithm.

In other words, for λ-Lipschitz valuations, at least $\frac{\ln \lambda}{\epsilon}$ queries is required. W.l.o.g., suppose $\lambda = e$ and thus $\frac{1}{\epsilon}$ queries is needed. Now suppose there are n goods, T_j different prices are tested for the j-th good, and the profit we get from good j is within ϵ_j from the maximum profit. Then $T_j \geq \frac{1}{\epsilon_j}$. We have

$$\frac{\sum_{j=1}^{n} T_j}{n} \geq \frac{n}{\sum_{j=1}^{n} 1/T_j} \geq \frac{n}{\sum_{j=1}^{n} \epsilon_j} = \frac{n}{\epsilon}.$$

In other words, $\sum_{j=1}^{n} T_j \geq \frac{n^2}{\epsilon}$. Since each query can set new prices for n goods, we need $\Omega(\frac{n}{\epsilon})$ queries. $\square$

6   Conclusion and Open Problems

In this paper, we study social welfare and profit maximization with only revealed preferences. The social welfare maximization problem can be solved by reducing to a convex dual optimization problem in both the offline and online case, while profit maximization is essentially non-convex, for which we give nearly matching upper and lower bounds on the query complexity when valuations and cost are separable.

While social welfare maximization is interesting and important, it is still more reasonable for a producer to maximize profit. However, as shown by Example 1, this problem is in general non-convex. While we give an algorithm for the separable case, it is a very interesting open problem to design algorithms for profit maximization in a more general setting or show some hardness result.

References

1. Afriat, S.N.: The construction of utility functions from expenditure data. Int. Econ. Rev. **8**(1), 67–77 (1967)
2. Agrawal, S., Devanur, N.R.: Fast algorithms for online stochastic convex programming. In: Proceedings of the Twenty-Sixth Annual ACM-SIAM Symposium on Discrete Algorithms, pp. 1405–1424. SIAM (2015)
3. Allen-Zhu, Z., Orecchia, L.: Linear coupling: an ultimate unification of gradient and mirror descent. arXiv preprint arXiv:1407.1537 (2014)
4. Amin, K., Cummings, R., Dworkin, L., Kearns, M., Roth, A.: Online learning and profit maximization from revealed preferences. In: AAAI, pp. 770–776 (2015)
5. Babaioff, M., Dughmi, S., Kleinberg, R., Slivkins, A.: Dynamic pricing with limited supply. ACM Trans. Econ. Comput. **3**(1), 4 (2015)
6. Balcan, M.-F., Daniely, A., Mehta, R., Urner, R., Vazirani, V.V.: Learning economic parameters from revealed preferences. In: Liu, T.-Y., Qi, Q., Ye, Y. (eds.) WINE 2014. LNCS, vol. 8877, pp. 338–353. Springer, Cham (2014). https://doi.org/10.1007/978-3-319-13129-0_28
7. Beigman, E., Vohra, R.: Learning from revealed preference. In: Proceedings of the 7th ACM Conference on Electronic Commerce, pp. 36–42. ACM (2006)
8. Besbes, O., Zeevi, A.: Dynamic pricing without knowing the demand function: risk bounds and near-optimal algorithms. Oper. Res. **57**(6), 1407–1420 (2009)

9. Besbes, O., Zeevi, A.: Blind network revenue management. Oper. Res. **60**(6), 1537–1550 (2012)
10. Blum, A., Mansour, Y., Morgenstern, J.: Learning what's going on: reconstructing preferences and priorities from opaque transactions. In: Proceedings of the Sixteenth ACM Conference on Economics and Computation, pp. 601–618. ACM (2015)
11. den Boer, A.V.: Dynamic pricing and learning: historical origins, current research, and new directions. Surv. Oper. Res. Manage. Sci. **20**(1), 1–18 (2015)
12. den Boer, A.V., Zwart, B.: Simultaneously learning and optimizing using controlled variance pricing. Manage. Sci. **60**(3), 770–783 (2013)
13. Broder, J., Rusmevichientong, P.: Dynamic pricing under a general parametric choice model. Oper. Res. **60**(4), 965–980 (2012)
14. Devanur, N.R., Hayes, T.P.: The adwords problem: online keyword matching with budgeted bidders under random permutations. In: Proceedings of the 10th ACM Conference on Electronic commerce, pp. 71–78. ACM (2009)
15. Diewert, E.: Afriat and revealed preference theory. Rev. Econ. Stud. **40**, 419–426 (1973)
16. Dong, J., Roth, A., Schutzman, Z., Waggoner, B., Wu, Z.S.: Strategic classification from revealed preferences. In: Proceedings of the 2018 ACM Conference on Economics and Computation, pp. 55–70. ACM (2018)
17. Goel, G., Mehta, A.: Online budgeted matching in random input models with applications to adwords. In: Proceedings of the Nineteenth Annual ACM-SIAM Symposium on Discrete Algorithms, pp. 982–991. Society for Industrial and Applied Mathematics (2008)
18. Hazan, E., et al.: Introduction to online convex optimization. Found. Trends® Optimi. **2**(3–4), 157–325 (2016)
19. Hiriart-Urruty, J.B., Lemaréchal, C.: Fundamentals of Convex Analysis. Springer Science and Business Media, Heidelberg (2012). https://doi.org/10.1007/978-3-642-56468-0
20. Houthakker, H.S.: Revealed preference and the utility function. Economica **17**, 159–174 (1950)
21. Keskin, N.B., Zeevi, A.: Dynamic pricing with an unknown demand model: asymptotically optimal semi-myopic policies. Oper. Res. **62**(5), 1142–1167 (2014)
22. Mas-Colell, A.: The recoverability of consumers' preferences from market demand. Econometrica **45**(6), 1409–1430 (1977)
23. Mas-Colell, A.: On revealed preference analysis. Rev. Econ. Stud. **45**(1), 121–131 (1978)
24. Nesterov, Y.: Smooth minimization of non-smooth functions. Math. Program. **103**(1), 127–152 (2005)
25. Nesterov, Y.: A method of solving a convex programming problem with convergence rate $o(1/k^2)$. Soviet Math. Dokl. **27**(2), 372–376 (1983)
26. Richter, M.: Revealed preference theory. Econometrica **34**(3), 635–645 (1966)
27. Rockafellar, R.T.: Convex Analysis. Princeton University Press, Jersey (1970)
28. Roth, A., Slivkins, A., Ullman, J., Wu, Z.S.: Multidimensional dynamic pricing for welfare maximization. In: Proceedings of the 2017 ACM Conference on Economics and Computation, pp. 519–536. ACM (2017)
29. Roth, A., Ullman, J., Wu, Z.S.: Watch and learn: optimizing from revealed preferences feedback. In: Proceedings of the Forty-Eighth Annual ACM Symposium on Theory of Computing, pp. 949–962. ACM (2016)
30. Samuelson, P.A.: A note on the pure theory of consumer's behaviour. Economica **5**(17), 61–71 (1938)

31. Sion, M., et al.: On general minimax theorems. Pacific J. Math. **8**(1), 171–176 (1958)
32. Tseng, P.: On accelerated proximal gradient methods for convex-concave optimization (2008). http://www.mit.edu/~dimitrib/PTseng/papers/apgm.pdf
33. Uzawa, H.: Preference and rational choice in the theory of consumption. In: Arrow, K.J., Karlin, S., Suppes, P. (eds.) Mathematical Models in Social Science (1960)
34. Varian, H.R.: Revealed preference. In: Szenberg, M., Ramrattand, L., Gottesman, A.A. (eds.) Samuelsonian Economics and the 21st Century, pp. 99–115 (2005)
35. Wang, Z., Deng, S., Ye, Y.: Close the gaps: a learning-while-doing algorithm for single-product revenue management problems. Oper. Res. **62**(2), 318–331 (2014)
36. Zadimoghaddam, M., Roth, A.: Efficiently learning from revealed preference. In: Goldberg, P.W. (ed.) WINE 2012. LNCS, vol. 7695, pp. 114–127. Springer, Heidelberg (2012). https://doi.org/10.1007/978-3-642-35311-6_9

Opinion Dynamics with Limited Information

Dimitris Fotakis[1,2], Vardis Kandiros[2], Vasilis Kontonis[3],
and Stratis Skoulakis[2(✉)]

[1] Yahoo Research, New York, NY, USA
dfotakis@oath.com, fotakis@cs.ntua.gr
[2] National Technical University of Athens, Athens, Greece
{vkandiros,sskoul}@corelab.ntua.gr
[3] University of Wisconsin-Madison, Madison, WI, USA
kontonis@cs.wisc.edu

Abstract. We study opinion formation games based on the Friedkin-Johnsen (FJ) model. We are interested in simple and natural variants of the FJ model that use limited information exchange in each round and converge to the same stable point. As in the FJ model, we assume that each agent i has an intrinsic opinion $s_i \in [0,1]$ and maintains an expressed opinion $x_i(t) \in [0,1]$ in each round t. To model limited information exchange, we assume that each agent i meets with one random friend j at each round t and learns only $x_j(t)$. The amount of influence j imposes on i is reflected by the probability p_{ij} with which i meets j. Then, agent i suffers a disagreement cost that is a convex combination of $(x_i(t) - s_i)^2$ and $(x_i(t) - x_j(t))^2$.

An important class of dynamics in this setting are *no regret* dynamics. We show an exponential gap between the convergence rate of no regret dynamics and of more general dynamics that do not ensure no regret. We prove that no regret dynamics require roughly $\Omega(1/\varepsilon)$ rounds to be within distance ε from the stable point x^* of the FJ model. On the other hand, we provide an opinion update rule that does not ensure no regret and converges to x^* in $\tilde{O}(\log^2(1/\varepsilon))$ rounds. Finally, we show that the agents can adopt a simple opinion update rule that ensures no regret and converges to x^* in poly$(1/\varepsilon)$ rounds.

1 Introduction

The study of *Opinion Formation* has a long history [22]. Opinion Formation is a *dynamic process* in the sense that socially connected people (family, friends, colleagues) exchange information and this leads to changes in their expressed opinions over time. Today, the advent of the internet and social media makes

---

Part of this work was performed while Vasilis Kontonis was a graduate student at the National Technical University of Athens.
Stratis Skoulakis is supported by a scholarship from the Onassis Foundation.

G. Christodoulou and T. Harks (Eds.): WINE 2018, LNCS 11316, pp. 282–296, 2018.
https://doi.org/10.1007/978-3-030-04612-5_19

the study of opinion formation in large social networks even more important; realistic models of how people form their opinions are of great practical interest for prediction, advertisement etc. In an attempt to formalize the process of opinion formation, several models have been proposed (see e.g. [8,15,20]). The common assumption underlying all these models, which dates back to DeGroot [8], is that opinions evolve through a form of repeated averaging of information collected from the agents' social neighborhoods.

Our work builds on the model proposed by Friedkin and Johnsen [15]. The FJ model is a variation on the DeGroot model capturing the fact that consensus on the opinions is rarely reached. According to FJ model each person i has a public opinion $x_i \in [0, 1]$ and an internal opinion $s_i \in [0, 1]$, which is private and invariant over time. There also exists a weighted graph $G(V, E)$ representing a social network where V stands for the persons ($|V| = n$) and E their social relations. Initially, all nodes start with their internal opinion and at each round t, update their public opinion $x_i(t)$ to a weighted average of the public opinions of their neighbors and their internal opinion,

$$x_i(t) = \frac{\sum_{j \in N_i} w_{ij} x_j(t-1) + w_{ii} s_i}{\sum_{j \in N_i} w_{ij} + w_{ii}}, \tag{1}$$

where $N_i = \{j \in V : (i, j) \in E\}$ is the set of i's neighbors, the weight w_{ij} associated with the edge $(i, j) \in E$ measures the extent of the influence that j poses on i and the weight $w_{ii} > 0$ quantifies how susceptible i is in adopting opinions that differ from her internal opinion s_i.

The FJ model is one of most influential models for opinion formation. It has a very simple update rule, making it plausible for modeling natural behavior and its basic assumptions are aligned with empirical findings on the way opinions are formed [25]. At the same time, it admits a unique stable point $x^* \in [0, 1]^n$ to which it converges with a *linear* rate [16]. The FJ model has also been studied under a game theoretic viewpoint. Bindel et al. considered its update rule as the minimizer of a quadratic disagreement cost function and based on it they defined the following opinion formation game [5]. Each node i is a selfish agent whose strategy is the public opinion x_i that she expresses incurring her a disagreement cost

$$C_i(x_i, x_{-i}) = \sum_{j \in N_i} w_{ij}(x_i - x_j)^2 + w_{ii}(x_i - s_i)^2 \tag{2}$$

Note that the FJ model is the *simultaneous best response dynamics* and its stable point x^* is the unique Nash equilibrium of the above game. In [5] they quantified its inefficiency with respect to the total disagreement cost. They proved that the *Price of Anarchy* (PoA) is 9/8 in case G is undirected and $w_{ij} = w_{ji}$. They also provided PoA bounds in the case of unweighted Eulerian directed graphs. We remark that in [5] an alternative framework for studying the way opinions evolve was introduced. The opinion formation process can be described as the *dynamics* of an opinion formation game. This framework is much more comprehensive since different aspects of the opinion formation process can be easily captured by defining suitable games. Subsequent works [3,4,9] considered variants of the above game and studied the convergence properties of the *best response dynamics*.

1.1   Motivation and Our Setting

Many recent works study the Nash equilibrium x^* of the opinion formation game defined in [5] under various perspectives. In [6] they extended the bounds for PoA in more general classes of directed graphs, while recently introduced influence maximization problems [1,17], which are defined with respect to x^*. The reason for this scientific interest is evident: the equilibrium x^* is considered as an appropriate way to model the final opinions formed in a social network, since the *well established* FJ model converges to it.

Our work is motivated by the fact that there are notable cases in which the FJ model is not an appropriate model for the dynamic of the opinions, due to the large amount of information exchange that it implies. More precisely, at each round its update rule (1) requires that every agent learns all the opinions of her social neighbors. In today's large social networks where users usually have several hundreds of friends it is highly unlikely that, each day, they learn the opinions of all their social neighbors. In such environments it is far more reasonable to assume that individuals randomly meet a small subset of their acquaintances and these are the only opinions that they learn. Such information exchange constraints render the FJ model unsuitable for modeling the opinion formation process in such large networks and therefore, it is not clear whether x^* captures the limiting behavior of the opinions. In this work we ask:

Question 1. Is the equilibrium x^* an adequate way to model the final formed opinions in large social networks? Namely, are there simple variants of the FJ model that require limited information exchange and converge fast to x^*? Can they be justified as natural behavior for selfish agents under a game-theoretic solution concept?

To address these questions, one could define precise dynamical processes whose update rules require limited information exchange between the agents and study their convergence properties. Instead of doing so, we describe the opinion formation process in such large networks as *dynamics* of a suitable opinion formation game that captures these information exchange constraints. This way we can precisely define which *dynamics* are *natural* and, more importantly, to study general classes of *dynamics* (e.g. no regret dynamics) without explicitly defining their update rule. The opinion formation game that we consider is a variant of the game in [5] based on interpreting the weight w_{ij} as a measure of how frequently i meets j.

Definition 1. *For a given opinion vector $x \in [0,1]^n$, the disagreement cost of agent i is the random variable $C_i(x_i, x_{-i})$ defined as follows:*

- *Agent i meets one of her neighbors j with probability $p_{ij} = w_{ij} / \sum_{j \in N_i} w_{ij}$.*

– Agent i suffers cost $C_i(x_i, x_{-i}) = (1 - \alpha_i)(x_i - x_j)^2 + \alpha_i(x_i - s_i)^2$, where $\alpha_i = w_{ii}/(\sum_{j \in N_i} w_{ij} + w_{ii})$.

Note that the expected disagreement cost of each agent in the above game is the same as the disagreement cost in [5] (scaled by $\sum_{j \in N_i} w_{ij} + w_{ii}$). Moreover its Nash equilibrium, with respect to the expected disagreement cost, is x^*. This game provides us with a general template of all the *dynamics* examined in this paper. At round t, each agent i selects an opinion $x_i(t)$ and suffers a disagreement cost based on the opinion of the neighbor that she randomly met. At the end of round t, she is informed only about the opinion and the index of this neighbor and may use this information to update her opinion in the next round. Obviously different update rules lead to different *dynamics*, however all of these respect the information exchange constraints: at every round each agent learns the opinion of *just one* of her neighbors. Question 1 now takes the following more concrete form.

Question 2. Can the agents update their opinions according to the limited information that they receive such that the produced opinion vector $x(t)$ converges to the equilibrium x^*? How is the convergence rate affected by the limited information exchange? Are there dynamics that ensure that the cost that the agents experience is minimal?

In what follows, we are mostly concerned about the dependence of the rate of convergence on the distance ε from the equilibrium x^*. Thus, we shall suppress the dependence on other parameters such as the size of the graph, n. We remark that the dependence of our dynamics on these constants is in fact rather good (see Sect. 2), and we do this only for clarity of exposition.

Definition 2 (Informal). *We say that a dynamics converges slowly resp. fast to the equilibrium x^* if it requires* $\mathrm{poly}(1/\varepsilon)$ *resp.* $\mathrm{poly}(\log(1/\varepsilon))$ *rounds to be within error ε.*

1.2 Contribution

The major contribution of the paper is proving an exponential separation on the convergence rate of *no regret dynamics* and the convergence rate of more general dynamics produced by update rules that do not ensure no regret.

No regret dynamics are produced by update rules that ensure no regret to any agent that adopts them. In our setting such an update rule must ensure that the total disagreement cost of an agent that adopts it is close to the total disagreement cost that she would experience by selecting the best fixed opinion in hindsight. The latter must hold even if the identities and the opinions of the neighbors that the agent meets are chosen adversarially. We prove that if all the agents adopt an update rule that ensures *no regret*, then there exists an instance of the game such that the produced opinion vector $x(t)$ requires roughly $\Omega(1/\varepsilon)$ rounds to be ε-close to x^*. The reason is that by definition such update rules only depend on the opinions that the agent observes and don't take into account the weights w_{ij} of the outgoing edges (see Sect. 5). We call the update rules with the latter property, *graph oblivious*. In Sect. 5 we use a novel information theoretic argument to prove the aforementioned lower bound for this more general class.

In Sect. 6, we present a simple update rule whose resulting dynamics converges fast, i.e. the opinion vector $x(t)$ is ε-close to x^* in $O(\log^2(1/\varepsilon))$ rounds. The reason that the previous lower bound doesn't apply is that this rule is not *graph oblivious* and it does not ensure no regret to the agents that adopt it. In fact there is a very simple example with two agents, in which the first follows the rule while the second selects her opinions adversarially, where the first agent experiences regret.

We introduce an intuitive *no regret* update rule and we show that if all agents adopt it, the resulting opinion vector $x(t)$ converges to x^*. Our rule is a *Follow the Leader algorithm*, meaning that at round t, each agent updates her opinion to the minimizer of total disagreement cost that she experienced until round $t - 1$. It also has a very simple form: it is roughly the time average of the opinions that the agent observes. In Sect. 3, we bound its convergence rate and show that in order to achieve ε distance from x^*, poly$(1/\varepsilon)$ rounds are sufficient. In view of our lower bound this rate is close to best possible. In Sect. 4, we prove its *no regret* property. This can be derived by the more general results in [19]. However, we give a short and simple proof that may be of interest.

In conclusion, our results reveal that the equilibrium x^* is a robust choice for modeling the limiting behavior of the opinions of agents since, even in our limited information setting, there exist simple and natural dynamics that converge to it. The convergence rate crucially depends on whether the agents act selfishly, i.e. they are only concerned about their individual disagreement cost.

1.3   Related Work

There exists a large amount of literature concerning the FJ model. Many recent works [3,4,9] bound the inefficiency of equilibrium in variants of opinion formation game defined in [5]. In [16] they bound the convergence time of the FJ model in special graph topologies. In [4] they proved that best response converges to PNE for a variant of the opinion formation game, in which social relations depend on the expressed opinions. Convergence results in other discretized variants of the FJ model can be found in [11,28]. In [13] they provide convergence results for a limited information variant of the FJ model. Although the considered variant is very similar to ours, their convergence results are much weaker, since they concern the expected value of the opinion vector.

Other works that relate to ours concern the convergence properties of dynamics based on no regret learning algorithms. In [12,14,27] it is proved that in a finite n-person game, if each agent updates her mixed strategy according to a no regret algorithm, the resulting *time-averaged* strategy vector converges to Coarse Correlated Equilibrium. The convergence properties of no regret dynamics for games with infinite strategy spaces were considered in [10]. They proved that for a large class of games with concave utility functions (socially concave games), the time-averaged strategy vector converges to Pure Nash Equilibrium. More recent work investigates a stronger notion of convergence of no regret dynamics. In [7] they show that, in n-person finite generic games that admit unique Nash equilibrium, the strategy vector converges *locally* and fast to it. They also provide conditions for *global* convergence. Our results fit in this line of research since we show that for a game with *infinite* strategy space, the strategy vector (and not the time-averaged) converges to the Nash equilibrium x^*.

No regret dynamics in limited information settings have recently received substantial attention from the scientific community since they provide realistic models for the practical applications of game theory. Kleinberg et al. in [23] treated load-balancing in distributed systems as a repeated game and analyzed the convergence properties of no regret learning algorithms under the *full information assumption* that each agent learns the load of every machine. In a subsequent work [24], the same authors consider the same problem in a *limited information setting* ("bulletin board model"), in which each agent learns the load of just the machine that served him. Most relevant to ours are the works [7,21,26], where they examine the convergence properties of no regret learning algorithms when the agents observe their payoffs with some additive zero-mean random noise.

2 Our Results and Techniques

As previously mentioned, an instance of the game in [5] is also an instance of the game of Definition 1. Following the notation introduced earlier we have that if $j \in N_i$ then $p_{ij} = w_{ij} / \sum_{j \in N_i} w_{ij}$ and 0 otherwise. Moreover since $w_{ii} > 0$ by the definition of the game in [5], $\alpha_i = w_{ii} / (\sum_{j \in N_i} w_{ij} + w_{ii}) > 0$. If an agent i does not have outgoing edges ($N_i = \emptyset$) then $p_{ij} = 0$ for all j. Therefore $\sum_{j=1}^{n} p_{ij} = 0$, $\alpha_i = 1$ if $N_i = \emptyset$ and $\sum_{j=1}^{n} p_{ij} = 1$, $\alpha_i \in (0,1)$ otherwise. For simplicity we adopt the following notation for an instance of the game of Definition 1.

Definition 3. *We denote an instance of the opinion formation game of Definition 1 as $I = (P, s, \alpha)$, where P is a $n \times n$ matrix with non-negative elements p_{ij}, with $p_{ii} = 0$ and $\sum_{j=1}^{n} p_{ij}$ is either 0 or 1, $s \in [0,1]^n$ is the internal opinion vector, $\alpha \in (0,1]^n$ the self confidence vector.*

An instance $I = (P, s, \alpha)$ is also an instance of the FJ model, since by the update rule (1) $x_i(t) = (1 - \alpha_i) \sum_{j \in N_i} p_{ij} x_j(t-1) + \alpha_i s_i$. It also defines the opinion vector $x^* \in [0,1]^n$ which is the stable point of the FJ model and the Nash equilibrium of the game in [5].

Definition 4. *For a given instance $I = (P, s, \alpha)$ the equilibrium $x^* \in [0,1]^n$ is the unique solution of the following linear system, for every $i \in V$, $x_i^* = (1 - \alpha_i) \sum_{j \in N_i} p_{ij} x_j^* + \alpha_i s_i$.*

It is easy to check that the above definition of x^* is equivalent to defining it as the PNE of the game in [5] or as the stable point of the FJ model. The fact that the above linear system always admits a solution follows by standard matrix norm properties.

Throughout the paper we study *dynamics* of the game of Definition 1. We denote as W_i^t the neighbor that agent i met at round t, which is a random variable whose probability distribution is determined by the instance $I = (P, s, \alpha)$ of the game, $\mathbf{P}\left[W_i^t = j\right] = p_{ij}$. Another parameter of an instance I that we often use is $\rho = \min_{i \in V} \alpha_i$.

In Sect. 3, we examine the convergence properties of the opinion vector $x(t)$ when all agents update their opinions according to the *Follow the Leader* principle. Since each agent i must select $x_i(t)$, before knowing which of her neighbors she will meet and what opinion her neighbor will express, this update rule says *"play the best according to what you have observed"*. For a given instance (P, s, a) of the game dynamics $x(t)$ is defined in Dynamics 1 and Theorem 1 shows its convergence rate to x^*.

Dynamics 1 Follow the Leader dynamics

1: Initially $x_i(0) = s_i$ for all agents i.
2: At round $t \geq 0$ each agent i:
    3: Meets neighbor with index W_i^t, $\mathbf{P}\left[W_i^t = j\right] = p_{ij}$.
    4: Suffers cost $(1 - \alpha_i)(x_i(t) - x_{W_i^t}(t))^2 + \alpha_i(x_i(t) - s_i)^2$ and learns $x_{W_i^t}(t)$.
    5: Updates her opinion as follows:

$$x_i(t+1) = \operatorname*{argmin}_{x \in [0,1]} \sum_{\tau=0}^{t} [(1 - \alpha_i)(x - x_{W_i^\tau}(\tau))^2 + \alpha_i(x - s_i)^2] \tag{3}$$

Theorem 1. *Let $I = (P, \alpha)$ be an instance of the opinion formation game of Definition 1 with equilibrium $x^* \in [0,1]^n$. The opinion vector $x(t) \in [0,1]^n$ produced by update rule (3) after t rounds satisfies*

$$\mathbf{E}\left[\|x(t) - x^*\|_\infty\right] \leq C\sqrt{\log n}\frac{(\log t)^{3/2}}{t^{\min(1/2, \rho)}},$$

where $\rho = \min_{i \in V} \alpha_i$ and C is a universal constant.

In Sect. 4 we argue that, apart from its simplicity, update rule (3) ensures no regret to any agent that adopts it and therefore the FTL dynamics can be considered as *natural dynamics* for selfish agents. Since each agent i selfishly wants to minimize the disagreement cost that she experiences, it is natural to assume that she selects $x_i(t)$ according to a *no regret algorithm* for the *online convex optimization problem* where the adversary chooses a function $f_t(x) = (1 - \alpha_i)(x - b_t)^2 + \alpha_i(x - s_i)^2$ at each round t. In Theorem 2 we prove that *Follow the Leader* is a no regret algorithm for the above OCO problem. We remark that this does not hold, if the adversary can pick functions from a different class (see e.g. chapter 5 in [18]).

Theorem 2. *Consider the function* $f : [0,1]^2 \mapsto [0,1]$ *with* $f(x,b) = (1-\alpha)(x-b)^2 + \alpha(x-s)^2$ *for some constants* $s, \alpha \in [0,1]$. *Let* $(b_t)_{t=0}^{\infty}$ *be an arbitrary sequence with* $b_t \in [0,1]$. *If* $x_t = \operatorname{argmin}_{x \in [0,1]} \sum_{\tau=0}^{t-1} f(x, b_\tau)$ *then for all* $t \geq 0$, $\sum_{\tau=0}^{t} f(x_\tau, b_\tau) \leq \min_{x \in [0,1]} \sum_{\tau=0}^{t} f(x, b_\tau) + O(\log t)$.

On the positive side, the FTL dynamics converges to x^* and its update rule is simple and ensures no regret to the agents. On the negative side, its convergence rate is outperformed by the rate of FJ model. For a fixed instance $I = (P, s, \alpha)$, the FTL dynamics converges with rate $\widetilde{O}(1/t^{\min(\rho, 1/2)})$ while FJ model converges with rate $O(e^{-\rho t})$ [16].

Question 3. Can the agents update their opinions according to other no regret algorithms such that the resulting dynamics converges fast to x^*?

The answer is no. In Sect. 5, we prove that fast convergence cannot be established for any opinion vector, produced by a no regret algorithm for the above online convex problem. The reason that FTL dynamics converges slowly is that rule (3) only depends on the opinions of the neighbors that agent i meets, α_i, and s_i. This is by definition true for any update rule based on a no regret algorithm (see Sects. 4 and 5). As already mentioned, we call this larger class of update rules *graph oblivious*, and we prove that fast convergence cannot be established for *graph oblivious dynamics*.

Definition 5 (graph oblivious update rule). *A graph oblivious update rule* A *is a sequence of functions* $(A_t)_{t=0}^{\infty}$ *where* $A_t : [0,1]^{t+2} \mapsto [0,1]$.

Definition 6 (graph oblivious dynamics). *Let a graph oblivious update rule* A. *For a given instance* $I = (P, s, \alpha)$ *the rule* A *produces a graph oblivious dynamics* $x_A(t)$ *defined as follows:*

- *Initially each agent i selects her opinion* $x_i^A(0) = A_0(s_i, \alpha_i)$
- *At round* $t \geq 1$, *each agent i updates her opinion as follows:*

$$x_i^A(t) = A_t(x_{W_i^0}(0), \ldots, x_{W_i^{t-1}}(t-1), s_i, \alpha_i),$$

where W_i^t *is the neighbors that i meets at round t.*

Theorem 3 states that for any graph oblivious dynamics there exists an instance $I = (P, s, \alpha)$, where roughly $\Omega(1/\varepsilon)$ rounds are required to achieve convergence within error ε.

Theorem 3. *Let A be a graph oblivious update rule, which all agents use to update their opinions. For any $c > 0$ there exists an instance $I = (P, s, a)$ such that* $\mathbf{E}[\|x_A(t) - x^*\|_\infty] = \Omega(1/t^{1+c})$, *where $x_A(t)$ denotes the opinion vector produced by A for the instance $I = (P, s, \alpha)$.*

To prove Theorem 3, we show that graph oblivious rules whose dynamics converge fast imply the existence of estimators for Bernoulli distributions with "small" sample complexity. The key part of the proof lies in Lemma 6, in which it is proven that such estimators cannot exist.

In Sect. 6, we present a simple update rule that achieves error rate $e^{-\tilde{O}(\sqrt{t})}$. This update rule is a function of the opinions and the indices of the neighbors that i met, s_i, α_i and the i-th row of the matrix P. Obviously this rule is not *graph oblivious*, due to its dependency on the i-th row and the indices. However it reveals that slow convergence is not a generic property of the limited information dynamics, but comes with the assumption that agents act selfishly.

3   Convergence Rate of FTL Dynamics

In this section we present the high level idea for proving Theorem 1. We note that since the produced opinion vector $x(t)$ is a random vector, the convergence metric used in Theorem 1 is $\mathbf{E}\left[\|x(t) - x^*\|_\infty\right]$, where the expectation is taken over the random meeting of the agents. At first notice that update rule (3) can be equivalently written as

$$x_i(t) = (1 - \alpha_i) \sum_{\tau=0}^{t-1} x_{W_i^\tau}(\tau)/t + \alpha_i s_i,$$

where $W_i(\tau)$ is the neighbor that i met at round τ. Using the fact that $x_i^* = (1 - \alpha_i) \sum_{j \in N_i} p_{ij} x_j^* + \alpha_i s_i$, one can prove that

$$|x_i(t) - x_i^*| \leq (1 - \alpha_i) \sum_{j \in N_i} \left| \frac{\sum_{\tau=0}^{t-1} \mathbf{1}[W_i^\tau = j] x_j(\tau)}{t} - p_{ij} x_j^* \right|$$

Now assume that $\left| \frac{\sum_{\tau=0}^{t-1} \mathbf{1}[W_i^\tau = j]}{t} - p_{ij} \right| = 0$ for all $t \geq 1$, then we easily get that $\|x(t) - x^*\|_\infty \leq e(t)$ where $e(t)$ satisfies the recursive equation $e(t) = (1 - \rho)\frac{\sum_{\tau=0}^{t-1} e(\tau)}{t}$ and $\rho = \min_{i \in V} \alpha_i$. It follows that $\|x(t) - x^*\|_\infty \leq 1/t^\rho$. Obviously the latter assumption does not hold, however since W_i^τ are independent random variables with $\mathbf{P}[W_i^\tau = j] = p_{ij}$, $\left| \frac{\sum_{\tau=0}^{t-1} \mathbf{1}[W_i^\tau = j]}{t} - p_{ij} \right|$ tends to 0 with probability 1. In Lemma 1 we use this fact to obtain a similar recursive equation for $e(t)$ and then in Lemma 2 we upper bound its solution.

Lemma 1. *Let $e(t)$ the solution of the recursion $e(t) = \delta(t) + (1 - \rho)\frac{\sum_{\tau=0}^{t-1} e(\tau)}{t}$ where $e(0) = \|x(0) - x^*\|_\infty$, $\delta(t) = \sqrt{\ln(\pi^2 n t^2/6p)/t}$ and $\rho = \min_{i \in V} \alpha_i$. Then,*

$$\mathbf{P}\left[\text{for all } t \geq 1, \ \|x(t) - x^*\|_\infty \leq e(t)\right] \geq 1 - p$$

Lemma 2. *Let $e(t)$ be a function satisfying the recursion $e(t) = \delta(t) + (1 - \rho)\sum_{\tau=0}^{t-1} e(\tau)/t$ and $e(0) = \|x(0) - x^*\|_\infty$ where $\delta(t) = \sqrt{\ln(Dt^{2.5})/t}$, $\delta(0) = 0$, and $D > e^{2.5}$ is a positive constant. Then $e(t) \leq \sqrt{2\ln(D)}\frac{(\ln t)^{3/2}}{t^{\min(\rho,\, 1/2)}}$.*

4 Follow the Leader Ensures No Regret

In this section we provide rigorous definitions of *no regret* algorithms and explain why update rule (3) ensures no regret to any agent that repeatedly plays the game of Definition 1. Based on the cost that the agents experience, we consider an appropriate *Online Convex Optimization* problem. This problem is a "game" played between an adversary and a player. *At round $t \geq 0$,*

1. *the player selects a value $x_t \in [0, 1]$.*
2. *the adversary observes the x_t and selects a $b_t \in [0, 1]$*
3. *the player receives cost $f(x_t, b_t) = (1 - \alpha)(x_t - b_t)^2 + \alpha(x_t - s)^2$.*

where s, α are constants in $[0, 1]$. The goal of the player is to pick x_t based on the history $(b_0, \ldots, b_{t-1})$ in a way that the total cost that she suffers during the "game" is minimized. Generally, different OCO problems can be defined by a set of functions $\mathcal{F}$ that the adversary chooses from and a feasibility set $\mathcal{K}$ from which the player picks her value (see [18] for an introduction to the OCO framework). In our case the feasibility set is $\mathcal{K} = [0, 1]$ and the set of functions is $\mathcal{F}_{s,\alpha} = \{x \mapsto (1 - \alpha)(x - b)^2 + \alpha(x - s)^2 : b \in [0, 1]\}$. As a result, each selection of the constants s, α leads to a different OCO problem.

Definition 7. *An algorithm A for the OCO problem with $\mathcal{F}_{s,\alpha}$ and $\mathcal{K} = [0, 1]$ is a sequence of functions $(A_t)_{t=0}^{\infty}$ where $A_t : [0, 1]^t \mapsto [0, 1]$.*

Definition 8. *An algorithm A is no regret for the OCO problem with $\mathcal{F}_{s,\alpha}$ and $\mathcal{K} = [0, 1]$ if and only if for all sequences $(b_t)_{t=0}^{\infty}$, if $x_t = A_t(b_0, \ldots, b_{t-1})$ then for all t, $\sum_{\tau=0}^{t} f(x_\tau, b_\tau) \leq \min_{x \in [0,1]} \sum_{\tau=0}^{t} f(x, b_\tau) + o(t)$.*

Informally speaking, if the player selects the value x_t according to a *no regret algorithm* then she does not regret not playing any fixed value no matter what the choices of the adversary are. Theorem 2 states that *Follow the Leader* i.e. $x_t = \operatorname{argmin}_{x \in [0,1]} \sum_{\tau=0}^{t-1} f(x, b_\tau)$ is a no regret algorithm for all the OCO problems with $\mathcal{F}_{s,\alpha}$.

Returning to the dynamics of the game in Definition 1, it is natural to assume that each agent i selects $x_i(t)$ by a no regret algorithm A_i for the OCO problem with $\mathcal{F}_{s_i,\alpha_i}$, since

$$\frac{1}{t} \sum_{\tau=0}^{t} f_i(x_i(\tau), x_{W_i^\tau}(\tau)) \leq \frac{1}{t} \min_{x \in [0,1]} \sum_{\tau=0}^{t} f_i(x, x_{W_i^\tau}(\tau)) + \frac{o(t)}{t}$$

The latter means that the time averaged total disagreement cost that she suffers is close to the time averaged cost by expressing the best fixed opinion and this holds regardless of the opinions of the neighbors that i meets. Meaning that even if the other agents selected their opinions maliciously, her total experienced cost would still be in a sense minimal. Under this perspective update rule (3) is a rational choice for selfish agents and as a result FTL dynamics is a *natural* limited information variant of the FJ model.

We now present the key steps for proving Theorem 2. We first prove that a similar strategy that also takes into account the value b_t admits no regret.

Lemma 3. *Let $(b_t)_{t=0}^{\infty}$ be an arbitrary sequence with $b_t \in [0, 1]$. Then for all t,*
$\sum_{\tau=0}^{t} f(y_\tau, b_\tau) \leq \min_{x \in [0,1]} \sum_{\tau=0}^{t} f(x, b_\tau)$, *where* $y_t = \mathrm{argmin}_{x \in [0,1]} \sum_{\tau=0}^{t} f(x, b_\tau)$.

Obviously the player cannot know b_t before selecting her value, however we can now understand why *Follow the Leader* admits no regret. Since the cost incurred by the sequence y_t is at most that of the best fixed value, we can compare the cost of x_t with that of y_t. Since the functions in $\mathcal{F}_{s,\alpha}$ are quadratic, the extra term $f(x, b_t)$ that y_t takes into account doesn't change dramatically the minimizer of the total sum i.e. x_t, y_t are relatively close.

Lemma 4. *For all $t \geq 0$, $f(x_t, b_t) \leq f(y_t, b_t) + 2\frac{1-\alpha}{t+1} + \frac{(1-\alpha)^2}{(t+1)^2}$.*

5   Lower Bound for Graph Oblivious Dynamics

In this section we prove that *graph oblivious dynamics* cannot converge much faster than FTL dynamics (Dynamics 1). The reason that this class is of particular interest is that it contains the dynamics produced by any no regret algorithm of the online convex optimation problem presented in the previous section.

Definition 9 (no regret dynamics). *Consider a collection of no regret algorithms such that for each $(s, \alpha) \in [0, 1]^2$ a no regret algorithm $A_{s,\alpha}$[1] for the OCO problem with $\mathcal{F}_{s,\alpha}$ and $\mathcal{K} = [0, 1]$, is selected. For a given instance $I = (P, s, \alpha)$ this selection produces the no regret dynamics $x(t)$ defined as follows:*

- *Initially each agent i selects her opinion $x_i(0) = A_0^{s_i, \alpha_i}(s_i, \alpha_i)$*
- *At round $t \geq 1$, each agent i selects her opinion,*

$$x_i(t) = A_t^{s_i, \alpha_i}(x_{W_i^0}(0), \ldots, x_{W_i^{t-1}}(t-1), s_i, \alpha_i),$$

where W_i^t is the neighbor that i meets at round t.

Such a selection of no regret algorithms can be encoded as a graph oblivious update rule. Specifically, the function $A_t : \{0, 1\}^{t+2} \mapsto [0, 1]$ is defined as $A_t(b_0, \ldots, b_{t-1}, s, \alpha) = A_{s,\alpha}^t(b_0, \ldots, b_{t-1})$. Thus, Theorem 3 applies. For example if agents use the Online Gradient Descent[2] to update their opinion i.e. $x_i(t+1) = x_i(t) - 1/\sqrt{t}(x_i(t) - (1 - \alpha_i)x_{W_i^t}(t) - \alpha_i s_i)$. Then we are ensured that fast convergence cannot be established in the respective no regret dynamics. The rest of the section is dedicated to prove Theorem 3. In Lemma 5 we show that any graph oblivious update rule A can be used as an estimator of the parameter $p \in [0, 1]$ of a Bernoulli random variable. Before proceeding we briefly introduce some definitions and notation.

---

[1]   These s, α are scalars in $[0, 1]$ and should not be confused with the internal opinion vector s and the self confidence vector α of an instance $I = (P, s, \alpha)$.

[2]   Online Gradient Descent is an influential no regret algorithm proposed by Zinkevic in [29] for the general OCO problem, where the adversary can select any convex function with bounded gradient. The latter directly implies that it also ensures no regret in our simpler OCO problem with $\mathcal{F}_{s_i, \alpha_i}$ and $\mathcal{K} = [0, 1]$.

Definition 10. *An estimator* $\theta = (\theta_t)_{t=1}^{\infty}$ *is a sequence of functions* θ_t *where* $\theta_t : \{0,1\}^t \mapsto [0,1]$.

Perhaps the first estimator that comes to one's mind is the *sample mean*, that is $\theta_t = \sum_{i=1}^{t} X_i / t$. To measure the efficiency of an estimator we define the *risk*, which corresponds to the expected error of an estimator.

Definition 11. *Let P be a Bernoulli distribution with mean p and P^t be the corresponding t-fold product distribution. The risk of an estimator* $\theta = (\theta_t)_{t=1}^{\infty}$ *is* $\mathbf{E}_{(X_1,\ldots,X_t) \sim P^t} [|\theta_t(X_1,\ldots,X_t) - p|]$ *or* $\mathbf{E}_p [|\theta_t - p|]$ *for brevity.*

Obviously since p is unknown, any meaningful estimator $\theta = (\theta_t)_{t=1}^{\infty}$ must guarantee that for all $p \in [0,1]$, $\lim_{t \to \infty} \mathbf{E}_p [|\theta_t - p|] = 0$. For example, *sample mean* has error rate $\mathbf{E}_p [|\theta_t - p|] \leq \frac{1}{2\sqrt{t}}$.

Lemma 5. *Let A a graph oblivious update rule such that for all instances $I = (P, s, \alpha)$, $\lim_{t \to \infty} t^{1+c} \mathbf{E} [\|x_A(t) - x^*\|_{\infty}] = 0$. Then there exists an estimator $\theta_A = (\theta_t^A)_{t=1}^{\infty}$ such that for all $p \in [0,1]$, $\lim_{t \to \infty} t^{1+c} \mathbf{E}_p [|\theta_t^A - p|] = 0$.*

Now in order to prove Theorem 3 we just need to prove the following claim.

Claim. For any estimator θ there exists p such that $\lim_{t \to \infty} t^{1+c} \mathbf{E}_p [|\theta_t - p|] > 0$.

The above claim states that for any estimator $\theta = (\theta_t)_{t=1}^{\infty}$, we can inspect the functions $\theta_t : \{0,1\}^t \mapsto [0,1]$ and then choose a $p \in [0,1]$ such that the function $\mathbf{E}_p [|\theta_t - p|] = \Omega(1/t^{1+c})$. The claim follows by Lemma 6, which states something significantly stronger: for *almost all* $p \in [0,1]$, any estimator θ cannot achieve rate $o(1/t^{1+c})$.

Lemma 6. *Let $\theta = (\theta_t)_{t=1}^{\infty}$ be a Bernoulli estimator with error rate $\mathbf{E}_p [|\theta_t - p|]$. For any $c > 0$, if we select p uniformly at random in $[0,1]$ then with probability 1, $\lim_{t \to \infty} t^{1+c} \mathbf{E}_p [|\theta_t - p|] > 0$.*

6   Limited Information Dynamics with Fast Convergence

In this section we provide an update rule that is not *graph oblivious* and converges exponentially fast to x^*. This rule is based on asynchronous distributed minimization algorithms [2] and depends not only on the opinions of the neighbors that an agent i meets, but also on the i-th row of matrix P.

In this case each agent stores the *most recent* opinions of the neighbors that she meets in an array and then updates her opinion to their weighted sum (each agent knows row i of P). For a given instance $I = (P, s, \alpha)$ we call the produced dynamics *Row Dependent dynamics* (Dynamics 2). Now the problem is that the opinions of the neighbors that she keeps in her array are *outdated*, i.e. a neighbor of agent i may have changed opinion since their last meeting. The good news are that as long as this outdatedness is bounded we can still achieve fast convergence (Lemma 7). By bounded outdatedness we mean that there exists a number B such that all agents have met all their neighbors at least once from $t - B$ to t.

Dynamics 2 Row Dependent dynamics

1: Initially $x_i(0) = s_i$ for all agent i.
2: Each agent i keeps an array M_i of length $|N_i|$, randomly initialized.
3: At round $t \geq 0$ each agent i:
    4: Meets neighbor with index W_i^t, $\mathbf{P}\left[W_i^t = j\right] = p_{ij}$.
    5: Suffers cost $(1 - \alpha_i)(x_i(t) - x_{W_i^t}(t))^2 + \alpha_i(x_i(t) - s_i)^2$ and learns $(x_{W_i^t}(t), W_i^t)$.

    6: Updates her array M_i and opinion:

$$M_i[W_i^t] \leftarrow x_{W_i^t}(t), \quad x_i(t+1) = (1 - \alpha_i) \sum_{j \in N_i} p_{ij} M_i[j] + \alpha_i s_i$$

Lemma 7. *Let $\rho = \min_i \alpha_i$, and $\pi_{ij}(t)$ be the most recent round before round t, that agent i met her neighbor j. If for all $t \geq B$, $t - B \leq \pi_{ij}(t)$ then, for all $t \geq kB$, $\|x(t) - x^*\|_\infty \leq (1 - \rho)^k$.*

In Dynamics 2 there does not exist a fixed B that satisfies Lemma 7. However we can select a length value such that the requirements hold with high probability. Observe that agent i simply needs to wait to meet the neighbor j with the smallest weight p_{ij}. Therefore, after $\log(1/\delta)/\min_j p_{ij}$ rounds, agent i met all her neighbors at least once with probability at least $1 - \delta$. In order to hold this for all agents, we shall roughly take $B = 1/\min_{p_{ij} > 0} p_{ij}$.

Theorem 4. *Let $I = (P, s, \alpha)$ be an instance of the opinion formation game of Definition 1 with equilibrium $x^* \in [0,1]^n$ and let $\rho = \min_{i \in V} \alpha_i$. The opinion vector $x(t) \in [0,1]^n$ produced by Row Dependent dynamics, after t rounds satisfies $\mathbf{E}\left[\|x(t) - x^*\|_\infty\right] \leq 2 \exp(-\rho \min_{ij} p_{ij} \sqrt{t}/(4 \ln(nt)))$.*

References

1. Abebe, R., Kleinberg, J., Parkes, D., Tsourakakis, C.E.: Opinion dynamics with varying susceptibility to persuasion. CoRR abs/1801.07863 (2018)
2. Bertsekas, D., Tsitsiklis, J.: Parallel and Distributed Computation: Numerical Methods. Athena Scientific, Belmont (1997)
3. Bhawalkar, K., Gollapudi, S., Munagala, K.: Coevolutionary opinion formation games. In: Symposium on Theory of Computing Conference, STOC 2013, pp. 41–50 (2013)
4. Bilò, V., Fanelli, A., Moscardelli, L.: Opinion formation games with dynamic social influences. In: Cai, Y., Vetta, A. (eds.) WINE 2016. Lecture Notes in Computer Science, vol. 10123, pp. 444–458. Springer, Berlin (2016). https://doi.org/10.1007/978-3-662-54110-4_31
5. Bindel, D., Kleinberg, J., Oren, S.: How bad is forming your own opinion? In: IEEE 52nd Annual Symposium on Foundations of Computer Science, FOCS 2011, pp. 57–66 (2011)
6. Chen, P., Chen, Y., Lu, C.: Bounds on the price of anarchy for a more general class of directed graphs in opinion formation games. Oper. Res. Lett. **44**(6), 808–811 (2016)

7. Cohen, J., Héliou, A., Mertikopoulos, P.: Hedging under uncertainty: regret minimization meets exponentially fast convergence. In: Bilò, V., Flammini, M. (eds.) SAGT 2017. LNCS, vol. 10504, pp. 252–263. Springer, Cham (2017). https://doi.org/10.1007/978-3-319-66700-3_20

8. DeGroot, M.: Reaching a consensus. J. Am. Stat. Assoc. **69**, 118–121 (1974)

9. Epitropou, M., Fotakis, D., Hoefer, M., Skoulakis, S.: Opinion formation games with aggregation and negative influence. In: Bilò, V., Flammini, M. (eds.) SAGT 2017. LNCS, vol. 10504, pp. 173–185. Springer, Cham (2017). https://doi.org/10.1007/978-3-319-66700-3_14

10. Even-Dar, E., Mansour, Y., Nadav, U.: On the convergence of regret minimization dynamics in concave games. In: Proceedings of the 41st Annual ACM Symposium on Theory of Computing, STOC 2009, pp. 523–532 (2009)

11. Ferraioli, D., Goldberg, P., Ventre, C.: Decentralized dynamics for finite opinion games. Theor. Comput. Sci. **648**(C), 96–115 (2016)

12. Foster, D., Vohra, R.: Calibrated learning and correlated equilibrium. Games Econ. Behav. **21**(1), 40–55 (1997)

13. Fotakis, D., Palyvos-Giannas, D., Skoulakis, S.: Opinion dynamics with local interactions. In: Proceedings of the Twenty-Fifth International Joint Conference on Artificial Intelligence, IJCAI 2016, pp. 279–285 (2016)

14. Freund, Y., Schapire, R.: Adaptive game playing using multiplicative weights. Games Econ. Behav. **29**(1), 79–103 (1999)

15. Friedkin, N., Johnsen, E.: Social influence and opinions. J. Math. Sociol. **15**(3–4), 193–206 (1990)

16. Ghaderi, J., Srikant, R.: Opinion dynamics in social networks with stubborn agents: equilibrium and convergence rate. Automatica **50**(12), 3209–3215 (2014)

17. Gionis, A., Terzi, E., Tsaparas, P.: Opinion maximization in social networks. In: Proceedings of the 13th SIAM International Conference on Data Mining, SDM 2013, pp. 387–395 (2013)

18. Hazan, E.: Introduction to online convex optimization. Found. Trends Optim. **2**(3–4), 157–325 (2016)

19. Hazan, E., Agarwal, A., Kale, S.: Logarithmic regret algorithms for online convex optimization. Mach. Learn. **69**(2), 169–192 (2007)

20. Hegselmann, R., Krause, U.: Opinion dynamics and bounded confidence models, analysis, and simulation. J. Artif. Soc. Soc. Simul. **5** (2002)

21. Héliou, A., Cohen, J., Mertikopoulos, P.: Learning with bandit feedback in potential games. In: Advances in Neural Information Processing Systems 30: Annual Conference on Neural Information Processing Systems 2017, NIPS 2017, pp. 6372–6381 (2017)

22. Jackson, M.: Social and Economic Networks. Princeton University Press, Princeton (2008)

23. Kleinberg, R., Piliouras, G., Tardos, É.: Multiplicative updates outperform generic no-regret learning in congestion games: extended abstract. In: Proceedings of 21st ACM Symposium on Theory of Computing (STOC 2009), pp. 533–542 (2009)

24. Kleinberg, R., Piliouras, G., Tardos, É.: Load balancing without regret in the bulletin board model. Distrib. Comput. **24**(1), 21–29 (2011)

25. Krackhardt, D.: A plunge into networks. Science **326**(5949), 47–48 (2009). http://science.sciencemag.org/content/326/5949/47

26. Mertikopoulos, P., Staudigl, M.: Convergence to nash equilibrium in continuous games with noisy first-order feedback. In: 56th IEEE Annual Conference on Decision and Control, CDC 2017, pp. 5609–5614 (2017)

27. Sergiu, S.H., Mas-Colell, A.: A simple adaptive procedure leading to correlated equilibrium. Econometrica **68**(5), 1127–1150 (2000)
28. Yildiz, M., Ozdaglar, A., Acemoglu, D., Saberi, A., Scaglione, A.: Binary opinion dynamics with stubborn agents. ACM Trans. Econ. Comput. **1**(4), 19:1–19:30 (2013)
29. Zinkevich, M.: Online convex programming and generalized infinitesimal gradient ascent. In: Proceedings of the Twentieth International Conference on International Conference on Machine Learning, ICML 2003, pp. 928–935. AAAI Press (2003)

The Fair Division of Hereditary Set Systems

Z. Li[1](✉) and A. Vetta[2]

[1] Computer Science Department, École Normale Supérieure, Paris, France
zl@zli2.com
[2] Department of Mathematics and Statistics, and School of Computer Science,
McGill University, Montreal, Canada
adrian.vetta@mcgill.ca

Abstract. We consider the fair division of indivisible items using the maximin shares measure. Recent work on the topic has focused on extending results beyond the class of additive valuation functions. In this spirit, we study the case where the items form an hereditary set system. We present a simple algorithm that allocates each agent a bundle of items whose value is at least 0.3667 times the maximin share of the agent. This improves upon the current best known guarantee of 0.2 due to Ghodsi et al. The analysis of the algorithm is almost tight; we present an instance where the algorithm provides a guarantee of at most 0.3738. We also show that the algorithm can be implemented in polynomial time given a valuation oracle for each agent.

1 Introduction

Consider the problem of dividing up m heterogenous goods amongst n agents. How can this be achieved in an equitable manner? This is the classical problem of *fair division* in economics and political science [13]. The issue that arises immediately is how to define "fairness". Two important concepts that have been widely studied are *proportionality* and *envy-freeness*. An allocation of the items to the agents is *proportional* if, for every agent, the value that the agent has for the *grand bundle* (all of the items) is at most a factor n times greater than the value it has for the bundle it receives. The allocation is *envy-free* if the value an agent has for the bundle it receives is at least as great as the value it has for the bundle of any other agent; that is, no agent is willing to exchange its allocated bundle for the bundle of another agent.[1]

Fair division has been extensively studied in the case of divisible items, typically in the general guise of *cake-cutting* [5,12]. Interestingly, for divisible items general equilibria can provide fair allocations in restricted settings. For example,

---

Z. Li—The first author thanks McGill University for hosting them while conducting this research.

[1] Observe that if the agents have sub-additive valuation functions then envy-freeness implies proportionality.

G. Christodoulou and T. Harks (Eds.): WINE 2018, LNCS 11316, pp. 297–311, 2018.
https://doi.org/10.1007/978-3-030-04612-5_20

assume the agents have linear valuation functions. If each agent is now given the same budget then equilibrium prices exist where all items are completely sold and each agent receives a most desired bundle; this concept of *competitive equilibrium from equal incomes* is due to Varian [14].

In practice, however, the fair division of indivisible items is more important than that of divisible items. This can be seen from the plethora of real-world examples, including course registration in universities, shift scheduling, draft assignment in sport, client assignment to sales-people, airport slot assignments, divorce settlements, and estate division [6,9]. But, at first glance, it is not clear anything useful can be said regarding the fair division of indivisible goods. For instance, what is a fair way to allocate a single indivisible good between two agents? An important concept used in understanding the case of indivisible goods was introduced by Budish [6], namely, maximin shares. The basic protocol is familiar to every child when cake cutting: "I cut, you choose". More generally, for n agents and m indivisible goods, one agent partitions the items into n bundles but that agent then gets the last choice of bundle. Intuitively, a risk averse agent seeks a partition that maximizes the value of its least desired bundle in the partition. The minimum value of a bundle in the optimal partition value is called the *maximin share* for the agent. Clearly, since the agents have different valuation functions, the optimal partitions and the corresponding maximin share values may differ for each agent. The first question that then arises is whether one can partition the items in such a way that **every** agent receives a bundle whose value is at least its maximin share. The answer is NO, even for additive valuation functions [9]. This negative result leads to the question of whether or not approximate solutions exist. Specifically, is there a partition that gives every agent a bundle with value at least an α-fraction of their maximin share? In groundbreaking work, for additive valuation functions, Kurokawa, Procaccia and Wang [9] showed the existence of a partition with $\alpha = \frac{2}{3}$; polynomial time algorithms with the same guarantee were subsequently given in [1] and [2]. A stronger guarantee of $\alpha = \frac{3}{4}$ was very recently obtained by Ghodsi et al. [8]. More general classes of valuation function have also been studied. Barman and Krishnamurthy [2] proved a bound of $\alpha = \frac{1}{10}$ for the class of submodular valuation functions. This was improved to $\frac{1}{3}$ by Ghodsi et al. [8], who also proved guarantees of $\frac{1}{5}$ for fractionally subadditive (XOS) valuations and $\Omega(1/\log n)$ for subadditive valuations.

Our Results. In this paper, we consider the fair division problem in an *hereditary set system* (or *downward-closed* set system). A set system $H = (J, \mathcal{F})$ consists of a collection J of items and a family $\mathcal{F}$ of feasible (independent) subsets of J. The set system satisfies the *hereditary property* if:

$$S \in \mathcal{F} \text{ and } T \subset S \Longrightarrow T \in \mathcal{F}$$

Hereditary set systems are ubiquitous in computer science and optimization. They arise naturally in the presence of packing or cost constraints, for example in scheduling problems and manufacturing processes. Furthermore, they are of fundamental theoretic importance; notable combinatorial and geometric objects that satisfy the hereditary property include matroids, simplicial complexes, and minor closed graph families such as networks embeddable on a surface.

In an hereditary set system, each agent i has a value $v_{i,j}$ for each item j but these values are additive only on feasible sets in the set system. A formal description of this model is given in Sect. 2 along with a proposed algorithm for dividing up the items amongst the agents. Our main result, given in Sect. 3, is that this algorithm provides a guarantee of at least 0.3667 for the maximin shares problem in a hereditary set system. This improves on the current best known bound of 0.2. In Sect. 4 we prove that our bound is almost tight by constructing an instance where the algorithm has a performance guarantee of at most 0.3738. Consequently, our lower and upper bounds for the performance guarantee of the algorithm are within an amount 0.007. The basic implementation of the algorithm runs in exponential time. So in Sect. 5 we show how to implement the procedure in polynomial time. Specifically, given a valuation oracle for each agent, the algorithm makes at most a polynomial in m number of queries to the oracles and performs a polynomial amount of computation given the responses of the oracles.

2   The Hereditary Maximin Share Problem

In this section, we describe the maximin share problem on an hereditary set system. We present a fair division algorithm for the problem and provide a simple performance analysis of the procedure (which we improve upon in the next section).

The Fair-Division Model. We have a set I of n agents and collection J of m items. The items belong to an hereditary set system $H = (J, \mathcal{F})$ and agents desire feasible (independent) sets in the set system. Specifically, each agent i has an additive valuation function over independent sets. That is, for any $S \in \mathcal{F}$ we have $v_i(S) = \sum_{j \in S} v_{i,j}$. The value the agent has for a set $S \notin \mathcal{F}$ is simply the maximum value it has for any feasible subset of S; that is $v_i(S) = \max_{T \in \mathcal{F}: T \subset S} \sum_{j \in T} v_{i,j}$.

Our aim is to fairly divide up the items amongst the agents. We measure the fairness of a division with respect to the maximin share of each agent. To define this, let $\mathbb{P}$ be the set of all partitions of the items into n sets. The value of the *maximin share* for a agent i is then

$$\text{MMS}(i) = \max_{\mathcal{P} \in \mathbb{P}} \min_{P \in \mathcal{P}} v_i(P).$$

That is, the maximin share is a partition that maximizes the value of the least valuable bundle in the partition. A partition $\mathcal{P}_i = \{P_i^1, P_i^2, \ldots, P_i^n\} \in \mathbb{P}$ that attains this value is called a *maximin partition for agent i* and the elements of $\mathcal{P}_i$ are called maximin parts. Observe that the maximin partition may be different for each agent.

Our objective is to find a partition of the items $\{S_1, S_2, \ldots, S_\ell\}$ where the bundle S_i allocated to agent i has value at least its maximin share. In general this is not possible, so instead we search for approximate solutions. Specifically, we desire the maximum fraction $\alpha > 0$ and an allocation $\{S_1, S_2, \ldots, S_\ell\}$ such that $v_i(S_i) \geq \alpha \cdot \text{MMS}(i)$, for every agent i. We call this the *hereditary maximin share problem*.

The hereditary maximin share problem has a constant factor approximation. This is because our valuation functions are *fractionally subadditive* (XOS). That is, the valuation function can be defined as the maximum over a collection of additive set functions. To show this, for each agent i, we define an additive function a_i^S over the items for each independent set $S \in \mathcal{F}$. Specifically, let

$$a_i^S(j) = \begin{cases} v_{i,j} & \text{if } j \in S \\ 0 & \text{if } j \notin S \end{cases}.$$

It is then easy to verify, for any set T (independent or not), that

$$v_i(T) = \max_{S \in \mathcal{F}} \sum_{j \in T} a_i^S(j).$$

Thus v_i is indeed fractionally subadditive. Using this fact, it follows from recent work of Ghodsi et al. [8] that a performance guarantee of 0.2 is obtainable.

Theorem 1. *[8] There is an algorithm for the fair division problem in hereditary set systems that allocates every agent a bundle with an approximation guarantee* $\alpha = \frac{1}{5}$. □

We remark that the valuation functions for hereditary set systems are not submodular functions.[2] To our knowledge, these valuation functions for the fair division problem have not been studied previously. The aim of this paper is to improve upon the $\alpha = 0.2$ performance guarantee.

A Fair-Division Algorithm. To obtain a better performance guarantee we apply a simple and natural procedure. To begin, without loss of generality, we may assume that the maximin share of every agent is exactly 1 by normalizing. Even stronger, we may assume that, for every agent i, there exists a maximin partition such that the agent has value exactly 1 for each part in the partition. To see this formally, let $\mathcal{P}_i$ be a maximin partition for agent i. Now define a new valuation function $\hat{v}_i$ with the property that $\hat{v}_{i,j} = \frac{v_{i,j}}{v_i(P)}$ if item j is in part $P \in \mathcal{P}_i$. Thus $\hat{v}_i(P) = 1$, for each part $P \in \mathcal{P}_i$. It is then easy to verify that an allocation that provides a value α with respect to $\hat{v}_i$ is a factor α allocation with respect to the true valuation v_i.

We are now ready to present our fair-division algorithm which begins with the normalization above. This normalization is not required in our polynomial

---

[2] For example, consider an hereditary set system $H = (J, \mathcal{F})$ with three items $J = \{a, b, c\}$ and let the maximal independent sets in $\mathcal{F}$ be $\{a\}$ and $\{b, c\}$. Suppose agent i has item values $v_{i,a} = 3, v_{i,b} = 2$ and $v_{i,c} = 2$. Thus $v_i(\{a, c\}) = 3$ and $v_i(\{a, b, c\}) = 4$. Consequently, the marginal value of adding item c to the set $\{a, b\}$ is larger than the marginal value of adding item c to the set $\{a\}$. Thus the valuation function is not submodular.

time algorithm; see the full version of the paper. Given a target value α, we search for a minimum cardinality feasible sets of at least the targeted value for some agent. If such a set is found we allocate that set (bundle) to that agent and then recurse on the remaining items and agents. This method is formalized in Procedure 1.

Algorithm 1. The Fair Division Algorithm

Input: A set I of agents, a set J of items, and a target value α.
for $\tau = 1$ to m **do**
    while there exists a set $S \subseteq J$ with $|S| = \tau$ *and an* $i \in I$ *with* $v_i(S) \geq \alpha$ **do**
        Allocate bundle S to agent i
        Set $I \leftarrow I \setminus \{i\}$
        Set $J \leftarrow J \setminus S$

We use the following notation. Let $\{S_1, S_2, \ldots, S_\ell\}$ be the bundles assigned by the procedure in order; note that $\ell \leq n$. We view the procedure as working in phases. In Phase τ the procedure searches for bundles of cardinality τ that provide utility at least α for some agent; note that $\tau \leq m$. We denote by $\mathcal{A}_\tau$ the collection of all items allocated during Phase τ.

A Simple Analysis. To begin, let's present a very simple analysis that shows this algorithm gives a factor $\alpha = \frac{1}{3}$ guarantee. In Sect. 4, we will give a more intricate and nearly tight analysis.

Theorem 2. *The procedure gives every agent a bundle of value at least* $\alpha = \frac{1}{3}$.

Proof. Clearly, if an agent is allocated a bundle by the procedure then it receives a bundle of value at least α. So it suffices to show that the procedure allocates every agent a bundle if it is run with a target value $\alpha = \frac{1}{3}$. For a contradiction, suppose the procedure terminates after allocating bundles to $\ell < n$ agents. Let i be an agent that is not allocated a bundle.

We may assume that $\mathcal{A}_1 = \emptyset$. That is, $v_{i,j} < \alpha$ for every agent i and every item j and so no items are allocated in Phase 1. The argument is standard [9]: if a set of cardinality one is allocated to an agent then this item intersects at most one of the n bundles in the maximin partition of any other agent. Thus $n - 1$ of the bundles in the partition are untouched and each still have total value 1. Consequently, $n - 1$ agents remain and they each have a partition of the items into $n - 1$ bundles each with value 1. Thus, we recurse on this smaller problem.

Therefore, we may assume the procedure only allocates items in Phases $\tau \geq 2$. Since the algorithm considers items in increasing size τ, agents receive a minimal bundle with value the sum of value of its elements (by definition of valuation). Now take any set S allocated to some agent k in Phase τ. It must be the case that $v_i(S) < \frac{\tau}{\tau-1} \cdot \alpha$. If not then there is a set $T \subset S$ with cardinality $\tau - 1$ such that $v_i(T) \geq \frac{\tau-1}{\tau} \cdot \frac{\tau}{\tau-1} \cdot \alpha = \alpha$. But, by the hereditary property, the bundle T is an independent set so should then have been allocated to agent i in Phase $\tau - 1$.

Let bundle S_k be the kth bundle allocated, where $1 \le k \le \ell$. Let U be the set of items unallocated by the procedure. Then the total value of unallocated items in some bundle of the maximin partition $\mathcal{P}_i$ of agent i is at least

$$\sum_{j \in U \cap \mathcal{P}_i} v_i(j) \ge n - \sum_{k=1}^{\ell} v_i(S_k) \ge n - \sum_{k=1}^{\ell} \frac{|S_k|}{|S_k| - 1} \cdot \alpha \ge n - \sum_{k=1}^{\ell} 2 \cdot \alpha \quad (1)$$

Here the final inequality arises as $\mathcal{A}_1 = \emptyset$ and so $|S_k| \ge 2$ for every allocated bundle. Since $\alpha = \frac{1}{3}$, we obtain from (1) that

$$\sum_{j \in U \cap \mathcal{P}_i} v_i(j) \ge n - \frac{2}{3}\ell > n - \frac{2}{3}n = \frac{n}{3}$$

Because the maximin partition $\mathcal{P}_i$ contains exactly n parts, there is a part that contains a set S of unallocated items where $v_i(S) \ge \frac{n}{3} \cdot \frac{1}{n} = \frac{1}{3}$. By the hereditary property, this contradicts the fact that the procedure terminated without allocating agent i a bundle. Thus every agent received a bundle of value at least $\frac{1}{3}$.   □

3   An Improved Lower Bound

In this section, we provide a much more detailed analysis of the fair division algorithm and prove it provides for an approximation guarantee of $\alpha = 0.3667$. This analysis is almost tight; in Sect. 4 we present an example showing that the performance of the procedure is not better than $\alpha = 0.3738$.

Theorem 3. *The procedure gives every agent a bundle of value at least* $\alpha = \frac{11}{30}$.

Before proving this theorem, let's give some intuition behind the analysis. The basic approach is the same as in Theorem 2. For an appropriately chosen target value α we run the procedure and assume for a contradiction that some agent i was not allocated a bundle. We then consider the maximin shares partition $\mathcal{P} = \{P_1, P_2, \ldots, P_n\}$ for agent i and show that some part $P \in \mathcal{P}$ contains items with total value at least α that are unallocated at the end of the procedure. By the hereditary property, this will contradict the fact the procedure terminated without allocating a bundle to agent i.

However, in order for this method to work for $\alpha = \frac{11}{30}$, we refine the analysis in four key ways. First, upon termination of the procedure, rather than considering the entire maximin shares partition $\mathcal{P} = \{P_1, P_2, \ldots, P_n\}$ for agent i, we focus on a restricted sub-partition $\hat{\mathcal{P}}$ of $\mathcal{P}$. To find $\hat{\mathcal{P}}$ we use combinatorial arguments on an auxiliary graph that is constructed with respect to the allocation decisions made in Phase 2 of the procedure. With this sub-partition $\hat{\mathcal{P}}$ more specialized accounting techniques can then be applied. We explain how to find the sub-partition $\hat{\mathcal{P}}$ in Sect. 3.1.

For the second improvement, note that upon termination we now wish to show that at least one of the parts $P \in \hat{\mathcal{P}}$ has unallocated items with total value

α. For this to not hold at the end of the procedure, we require that each part contains unallocated items of total value less than α. This simple observation is useful in that once a part contains unallocated items of value less than α, the removal of additional items from that part does not hurt us any more. In fact, the removal of additional items from such a part is actually beneficial to us in the analysis as such items do not reduce the value of parts that may still feasibly provide agent i with a bundle of value α. To quantify this, let

$$v_P = 1 - \sum_{\tau \geq 1} \sum_{j \in P \cap \mathcal{A}_\tau} v_{i,j}$$

be the total value of unallocated items in part $P \in \hat{\mathcal{P}}$ upon termination of the algorithm. We then denote by $s_P = \alpha - v_P$ the *superfluity* of part P. We may assume that $s_P > 0$, otherwise the procedure would have allocated agent i a bundle. Accounting for this superfluous damage will be the second key ingredient in the proof.

The third idea is to exploit any laxity the procedure provides before the start of the third phase. Essentially the *laxity* l_P of a part $P \in \hat{\mathcal{P}}$ is a measure of how much better the unallocated value of the part is after Phase 2 than a "perceived" worst case. The fourth key idea is to amortize our accounting process. Rather than simply focus independently on items in each part of the partition $\hat{\mathcal{P}}$, we will also redistribute values within allocated bundles that cross multiple parts in the partition. The concepts of superfluity and laxity are explained in Sect. 3.2 along with a description of the amortization process. The proof of Theorem 3 is then given in Sect. 3.3.

3.1   Finding a Sub-Partition

Now assume the procedure terminates after $\ell < n$ iterations leaving at least one agent i who does not receive a bundle. Let the maximin partition for agent i be $\mathcal{P} = \{P_1, P_2, \ldots, P_n\}$. Let $J^* \subseteq J$ be the set of items allocated to agents during the procedure. We will show

$$\max_{1 \leq k \leq n} v_i(P_k \setminus J^*) \geq \alpha \tag{2}$$

This will contradict the fact that the procedure terminated without allocating a bundle to agent i. So let's prove that inequality (2) holds. As in the proof of Theorem 2, without loss of generality, we may assume no items were allocated in Phase 1; that is, $\mathcal{A}_1 = \emptyset$. Next we construct an *auxiliary graph* $\mathcal{G}$ based upon the allocation decisions made in Phase 2. The graph contains n vertices, one vertex for each part P_k in the partition $\mathcal{P}$. The graph contains an edge connecting the two (possibly equal) parts containing the two items of the bundle, for each bundle allocated in Phase 2. Observe that a vertex in $\mathcal{G}$ may have degree greater than one since the part it represents may contain multiple items. This also implies that $\mathcal{G}$ may contain edges that are self-loops; this happens whenever a bundle allocated in Phase 2 consists of two items in the same part of the partition $\mathcal{P}$.

Let's further investigate the structure of $\mathcal{G}$. Let X be a maximal set of vertices of $\mathcal{G}$ that induce at least $|X|$ edges. Note that such a set exists because $\emptyset$ is a feasible choice for X. On the other hand, it must be the case that $X \neq V(\mathcal{G})$. This follows as $\mathcal{G}$ contains exactly n vertices but $\ell < n$ edges. But this, in turn, implies that X induces exactly $|X|$ edges. If X induced more than $|X|$ edges then we could add to it any other vertex in $V(\mathcal{G}) \setminus X$ and still maintain the desired property.

Now consider the subgraph $\mathcal{G} \setminus X$. This subgraph is a forest F. If it contained a cycle C then $X \cup V(C)$ would contradict the maximality of X. Furthermore, there are no edges between X and $\mathcal{G} \setminus X$; otherwise, the endpoint in $\mathcal{G} \setminus X$ of such an edge could have been added to X.

Let the forest F contain s components consisting of a single vertex – observe that, by the above argument, these vertices are also singleton components of $\mathcal{G}$. Let F contain c non-trivial components, that is, trees with at least one edge. Clearly, every non-trivial tree contains at least two leaves. Therefore, we may select a set Y that consists of every vertex in non-trivial trees in F except for exactly two leaves in each non-trivial tree. Finally, we set $Z = V(\mathcal{G}) \setminus (X \cup Y)$. An illustration of the auxiliary graph $\mathcal{G}$ and the sets X, Y and Z is shown in Fig. 1.

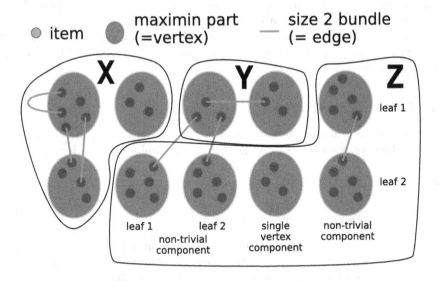

Fig. 1. The auxiliary graph

The sub-partition $\hat{\mathcal{P}}$ of $\mathcal{P}$ consisting of the vertices in $Z = V(\mathcal{G}) \setminus (X \cup Y)$ will be important to us. Let's now present a couple of combinatorial equalities that will be useful later. The first is a claim that follows trivially by definition of $\hat{\mathcal{P}}$. The second is a lemma quantifying how many agents are allocated bundles in Phase 2.

Claim. The number of parts in the sub-partition $\hat{\mathcal{P}}$ is $|Z| = 2c + s$.   □

Lemma 1. *Exactly $|X| + |Y| + c$ agents receive a bundle in Phase 2.*

Proof. Observe that, by construction, the number of agents allocated a bundle in Phase 2 is exactly $|E(\mathcal{G})|$. So we must count the number of edges in the auxiliary graph $\mathcal{G}$. We have seen that $E(\mathcal{G}) = E(X) \cup E(F)$. By the maximality of X, we have that $|E(X)| = |X|$. In addition, $|V(F)| = |Y| + 2c + s$. Thus, as F consists of exactly $c + s$ trees, $|E(F)| = (|Y| + 2c + s) - (c + s) = |Y| + c$. Putting this together gives $|E(\mathcal{G})| = |X| + |Y| + c$, as desired.   □

We will focus our counting arguments on the sub-partition $\hat{\mathcal{P}}$ in order to obtain a contradiction to the fact that agent i was not allocated a bundle. Specifically, we will show that at least one of the vertices in Z contains unallocated items that together provide value at least α to agent i.

3.2 Laxity, Superfluity and Amortization

Consider the allocated items in $\hat{\mathcal{P}}$. As $\mathcal{A}_1 = \emptyset$, every item j has $v_{i,j} < \alpha$. We now study the value (to agent i) of the items in $\hat{\mathcal{P}}$ allocated in Phase 2. To do this, recall that the vertices of $\hat{\mathcal{P}} = Z$ are of two types, vertices of degree 0 in $\mathcal{G}$ (specifically, singleton vertices in F) and vertices of degree 1 (that is, leaf vertices in F). Vertices of degree 0 contain no items that are allocated in Phase 2. Vertices of degree 1 in $\mathcal{G}$ contain exactly one item that is allocated in Phase 2. So given a part $P \in \hat{\mathcal{P}}$, we define the *laxity* of P to be $l_P = \alpha$ if P corresponds to a singleton vertex in F. Otherwise, if P corresponds to a leaf vertex in F we define $l_P = (1 - v_{i,j^*(P)}) - (1 - \alpha) = \alpha - v_{i,j^*(P)}$, where $j^*(P)$ is the unique item of P that is allocated in Phase 2. Observe that $l_P > 0$ since $v_{i,j} < \alpha$ for every item j and in particular for $j^*(P)$. We can use the laxity to quantify the total value of items in $\hat{\mathcal{P}}$ allocated in Phase 2.

Lemma 2. $\sum_{P \in \hat{\mathcal{P}}} \sum_{j \in P \cap \mathcal{A}_2} v_{i,j} = 2\alpha \cdot c - \sum_{P \in \hat{\mathcal{P}}} l_P.$

Proof. We have

$$\sum_{P \in \hat{\mathcal{P}}} \sum_{j \in P \cap \mathcal{A}_2} v_{i,j} = \sum_{P \in \hat{\mathcal{P}}} (\alpha - l_P) = 2c \cdot \alpha - \sum_{P \in \hat{\mathcal{P}}} l_P$$

Here the first equality holds since when P corresponds to singleton, the associated term in the sum is $0 = \alpha - \alpha = \alpha - l_P$, and when P corresponds to a leaf, the associated term is $v_{i,j^*(P)} = \alpha - l_P$. The final equality follows as $\hat{\mathcal{P}} = Z$ contains exactly $2 \cdot c$ vertices of degree 1.   □

Next we want to bound the value (to agent i) of items allocated in Phases 3 and beyond. First, let's count the number of bundles allocated in Phases 3 and beyond.

Lemma 3. *At most $c + s - 1$ agents receive a bundle in Phases 3 and beyond.*

We give the proof of this claim in the full version of the paper. Now let's bound the value of bundles allocated in Phases 3 and beyond. We will need two more definitions. First, recall, we defined the superfluity of a part P as $s_P = \alpha - \left(1 - \sum_{\tau \geq 1} \sum_{j \in P \cap \mathcal{A}_\tau} v_{i,j}\right)$. For the purpose of our analysis, we also define the *excess* e_P of a part $P \in \hat{\mathcal{P}}$ to be the sum of its superfluity and its laxity. Therefore

$$e_P = s_P + l_P = \left(\alpha - 1 + \sum_{\tau \geq 1} \sum_{j \in P \cap \mathcal{A}_\tau} v_{i,j}\right) + (\alpha - v_{i,j^*(P)})$$

$$= 2\alpha - 1 + \sum_{\tau \geq 3} \sum_{j \in P \cap \mathcal{A}_\tau} v_{i,j} \tag{3}$$

The final equality holds because $\mathcal{A}_1 = \emptyset$ and $P \cap \mathcal{A}_2$ is either $\{j^*(P)\}$ or $\emptyset$.

As discussed, to bound the value of bundles allocated in Phases 3 and beyond, we will amortize our accounting process. In these phases each allocated bundle has cardinality at least three. Take such a bundle, say $B = \{j_1, j_2, \ldots, j_k\}$ allocated to some agent τ. The hereditary property states that $B \setminus \{j\}$ is feasible, for every item $j \in B$. Thus, because B is a minimum cardinality feasible bundle of value at least α when it is allocated, it must be the case that $v_i(B \setminus \{j\}) < \alpha$. Observe, this applies even for the least valuable item $\hat{j} \in B$ to agent i in the bundle B. Furthermore, because B is an independent set, we have that (i) $v_i(B \setminus \{\hat{j}\}) = v_i(B) - v_{i,\hat{j}}$, and (ii) $v_{i,\hat{j}}$ is at most the average value of an item in B. Putting this all together gives

$$v_i(B) < \alpha + v_{i,\hat{j}} \leq \alpha + \frac{1}{k} \cdot v_i(B)$$

As $k \geq 3$, we obtain

$$v_i(B) < \frac{k}{k-1} \cdot \alpha \leq \frac{3}{2} \cdot \alpha \tag{4}$$

Hence, each bundle that is allocated in Phase 3 reduces the total value to agent i of items in $\hat{\mathcal{P}}$ by at most $\frac{3}{2} \cdot \alpha$. For bundles of cardinality at least 4 (allocated in Phases 4 and beyond) this is at most $\frac{4}{3} \cdot \alpha$. This bound suffices for bundles of size at least 4. We now bound bundles allocated in Phase 3 more carefully. To do so, we amortize the accounting process by defining for any item j allocated in a bundle B by the mechanism,

$$a_j = \frac{v_i(B)}{|B|}.$$

That is, a_j is the average value of items in B. Since each allocated bundle is minimal and has cardinality 3, we have, for any item j, that

$$a_j \leq \frac{1}{3}(\alpha + v_{i,j}) \tag{5}$$

We now further partition elements of $\mathcal{A}_3$ into two sets whose values we will bound independently in order to reduce the gaps in our accounting. To this end, let U consist of those items in $\mathcal{A}_3$ that belong to a part $P \in \hat{\mathcal{P}}$ which contains no other elements of $\mathcal{A}_3$. That is, $j \in U$ if item j is the only item of $\mathcal{A}_3$ in some part $P \in \hat{\mathcal{P}}$. Let $\bar{U} = \mathcal{A}_3 \setminus U$; thus, $\bar{U}$ consists of those items in $\mathcal{A}_3$ that belong to a part $P \in \hat{\mathcal{P}}$ which contains a least two items of $\mathcal{A}_3$.

Claim. If $\alpha \leq \frac{11}{30}$ then for any part P of $\hat{\mathcal{P}}$, $\sum_{j \in P \cap \bar{U}} a_j - e_P \leq \frac{1}{6} \cdot |P \cap \bar{U}|$.

Claim 3.2 is proved in the full version of the paper. Applying Claim 3.2 to every part $P \in \hat{\mathcal{P}}$, we obtain that

$$\sum_{j \in \bar{U}} a_j \leq \frac{1}{6} \cdot |\bar{U}| + \sum_{P \in \hat{\mathcal{P}}} e_P \qquad (6)$$

Moreover, because $a_j \leq \frac{1}{2} \cdot \alpha$ for all $j \in U \subseteq \mathcal{A}_3$, we also have

$$\sum_{j \in U} a_j \leq \frac{1}{2} \cdot \alpha \cdot |U| \qquad (7)$$

3.3 Proof of the Improved Bound

We are now ready to prove the stated performance guarantee of the procedure.

Theorem 3. *The procedure gives every agent a bundle of value at least* $\alpha = \frac{11}{30}$.

Proof. Let β_3 be the number of bundles of size 3 allocated in Phase 3 and let β_{4+} be number of bundles of size at least 4 allocated in Phase 4 and beyond. Applying inequality (4) for $k \geq 4$, we have

$$\sum_{\tau \geq 4} \sum_{j \in \mathcal{A}_\tau} v_{i,j} \leq \frac{4}{3} \cdot \alpha \cdot \beta_{4+} \qquad (8)$$

This gives

$$\sum_{\tau \geq 3} \sum_{j \in \mathcal{A}_\tau} v_{i,j} = \sum_{j \in \mathcal{A}_3} v_{i,j} + \sum_{\tau \geq 4} \sum_{j \in \mathcal{A}_\tau} v_{i,j} \leq \sum_{j \in \mathcal{A}_3} v_{i,j} + \frac{4\alpha}{3} \cdot \beta_{4+}$$

$$= \sum_{j \in \mathcal{A}_3} a_j + \frac{4\alpha}{3} \cdot \beta_{4+} = \sum_{j \in U} a_j + \sum_{j \in \bar{U}} a_j + \frac{4\alpha}{3} \cdot \beta_{4+}$$

where the last equality follows by definition of U and $\bar{U}$. Applying (7) and (6) then produces:

$$\sum_{\tau \geq 3} \sum_{j \in \mathcal{A}_\tau} v_{i,j} \leq \frac{\alpha}{2} \cdot |U| + \left(\frac{1}{6} \cdot |\bar{U}| + \sum_{P \in \hat{\mathcal{P}}} e_P \right) + \frac{4\alpha}{3} \cdot \beta_{4+}$$

$$= \frac{\alpha}{2} \cdot |U| + \frac{1}{6} \cdot (|\mathcal{A}_3| - |U|) + \frac{4\alpha}{3} \cdot \beta_{4+} + \sum_{P \in \hat{\mathcal{P}}} e_P$$

$$= \left(\frac{\alpha}{2} - \frac{1}{6}\right) \cdot |U| + \frac{1}{6} \cdot |\mathcal{A}_3| + \frac{4\alpha}{3} \cdot \beta_{4+} + \sum_{P \in \hat{\mathcal{P}}} e_P$$

$$\leq \left(\frac{\alpha}{2} - \frac{1}{6}\right) \cdot |U| + \frac{1}{2} \cdot \beta_3 + \frac{4\alpha}{3} \cdot \beta_{4+} + \sum_{P \in \hat{\mathcal{P}}} e_P$$

$$\leq \left(\frac{\alpha}{2} - \frac{1}{6}\right) \cdot |U| + \frac{1}{2} \cdot (\beta_3 + \beta_{4+}) + \sum_{P \in \hat{\mathcal{P}}} e_P$$

Here the second inequality is due to the fact that $|\mathcal{A}_3| = 3\beta_3$; the final inequality follows as our target value is $\alpha = \frac{11}{30} \leq \frac{3}{8}$.

Now, by definition, there is at most one element of U for each of the $2c + s$ maximin parts in $\hat{\mathcal{P}}$. So $|U| \leq 2c + s$. Furthermore, by Lemma 3, we have $\beta_3 + \beta_{4+} \leq c + s - 1$. Therefore, it follows that

$$\sum_{\tau \geq 3} \sum_{j \in \mathcal{A}_\tau} v_{i,j} \leq \left(\frac{\alpha}{2} - \frac{1}{6}\right)(2c + s) + \frac{1}{2}(c + s) + \sum_{P \in \hat{\mathcal{P}}} e_P \qquad (9)$$

We are now ready to complete the proof. The total value of non-superfluous, allocated items in $\hat{\mathcal{P}}$ at the end of the procedure is then at most

$$\sum_{\tau \geq 1} \sum_{j \in \mathcal{A}_\tau} v_{i,j} - \sum_{P \in \hat{\mathcal{P}}} s_P = \sum_{j \in \mathcal{A}_1} v_{i,j} + \sum_{j \in \mathcal{A}_2} v_{i,j} + \sum_{\tau \geq 3} \sum_{j \in \mathcal{A}_\tau} v_{i,j} - \sum_{P \in \hat{\mathcal{P}}} s_P$$

$$= 0 + \left(2\alpha c - \sum_{P \in \hat{\mathcal{P}}} l_P\right) + \sum_{\tau \geq 3} \sum_{j \in \mathcal{A}_\tau} v_{i,j} - \sum_{P \in \hat{\mathcal{P}}} s_P$$

$$\leq 2\alpha c - \sum_{P \in \hat{\mathcal{P}}} l_P + \left(\left(\frac{\alpha}{2} - \frac{1}{6}\right)(2c + s) + \frac{1}{2}(c + s) + \sum_{P \in \hat{\mathcal{P}}} e_P\right) - \sum_{P \in \hat{\mathcal{P}}} s_P$$

Here the second equality follows from the fact that $\mathcal{A}_1 = \emptyset$ and by Lemma 2. The inequality follows by (9). Simplifying now gives

$$\sum_{\tau \geq 1} \sum_{j \in \mathcal{A}_\tau} v_{i,j} - \sum_{P \in \hat{\mathcal{P}}} s_P \leq 2\alpha c + \left(\frac{\alpha}{2} - \frac{1}{6}\right)(2c + s) + \frac{1}{2}(c + s)$$

$$= 2c \cdot \left(\alpha + \frac{1}{2}\alpha - \frac{1}{6} + \frac{1}{4}\right) + s \cdot \left(\frac{1}{2}\alpha - \frac{1}{6} + \frac{1}{2}\right)$$

$$= 2c \cdot \left(\frac{1}{12} + \frac{3}{2}\alpha\right) + s \cdot \left(\frac{1}{3} + \frac{1}{2}\alpha\right)$$

By Claim 3.1, at the start of the procedure the total value of items to agent i in the sub-partition $\hat{\mathcal{P}}$ is at least $(2c + s) \cdot 1$. Thus upon termination, the total value of unallocated items (modulo superfluity) is at least

$$2c + s - \left(\sum_{\tau \geq 1} \sum_{j \in A_\tau} v_{i,j} - \sum_{P \in \hat{P}} s_P\right) \geq (2c+s) - \left(2c \cdot \left(\frac{1}{12} + \frac{3}{2}\alpha\right) + s \cdot \left(\frac{1}{3} + \frac{1}{2}\alpha\right)\right)$$

$$= 2c \cdot \left(\frac{11}{12} - \frac{3}{2}\alpha\right) + s \cdot \left(\frac{2}{3} - \frac{1}{2}\alpha\right)$$

$$\geq (2c + s) \cdot \left(\frac{11}{12} - \frac{3}{2}\alpha\right)$$

Here the final inequality holds because $\frac{11}{12} - \frac{3}{2}\alpha \leq \frac{2}{3} - \frac{1}{2}\alpha$ for $\alpha = \frac{11}{30} \geq \frac{1}{4}$. Hence the average remaining value in each part of $\hat{P}$ at the end of the procedure is at least $\frac{11}{12} - \frac{3}{2}\alpha$. This is at least α for $\alpha \leq \frac{11}{30}$. Thus agent i must receive a bundle of value at least $\alpha = \frac{11}{30} = 0.3667$. $\qquad \square$

4   An Upper Bound

We now show that the analysis in Sect. 3 is tight to within an additive amount of 0.007. Specifically, we present an example that shows the procedure cannot guarantee a performance guarantee better than $\frac{40}{107} = 0.3738$.

Theorem 4. *The procedure's performance guarantee is at most $\frac{40}{107} = 0.3738$.*

The proof of this theorem is available in the full version of the paper and is obtained from the following example.

Class	A	B	C	D	E	F
Quantity	n	$\frac{1}{3}n$	$\frac{2}{3}n$	$\frac{2}{3}n$	$\frac{1}{3}n$	$40n$
Value	$\alpha' = \frac{40}{107}$	$\frac{13}{4}\alpha' - 1 = \frac{23}{107}$	$1 - \frac{9}{4}\alpha' = \frac{17}{107}$	$\frac{1}{4}\alpha' = \frac{10}{107}$	$2 - \frac{21}{4}\alpha' = \frac{4}{107}$	$\frac{1}{40}\alpha' = \frac{1}{107}$
Capacity	2	1	2	5	11	40

Set $\alpha' = \frac{40}{107}$, $\varepsilon > 0$. The set system contains six classes of items, denoted $\{A, B, C, D, E, F\}$. The number of identical items in each class are shown in the second row of the table. Moreover these agents are identical. The value each agent has for a single item of each class is shown in the third row of the table. Finally, the feasible (independent) sets are defined by a capacity constraint for each class of items, as shown in fourth row of the table.

5   A Polynomial Time Implementation

Procedure 1 is not a polynomial time algorithm. However, it can be modified to give a polynomial time implementation given access to a valuation oracle for each agent. To do this there are two main problems. First, the use of a phase τ to search for bundles of cardinality τ is clearly exponential time if the procedure ends up searching for bundles of large cardinality. Second, the procedure, requires the maximin partition or, more specifically, the maximin share value for each agent. In the full version, we detail how to overcome both these problems using a polynomial amount of computation and a polynomial number of valuation queries.

6   Conclusion

We have presented a fair division algorithm for hereditary set systems which provides each agent with at least an $\frac{11}{30}$ fraction of its maximin share value. Several open problems remain. The first is to close the 0.007 gap between the lower and upper bounds given for the performance of the procedure given in this paper. The second is to design a new procedure with a better performance guarantee. Of course, this may be easier to do for sub-classes of hereditary set systems. One very important sub-class is that of matroids. A matroid is a hereditary set system that also satisfies the *augmentation property*: given two independent sets S and T where $|S| > |T|$, there exists an element $s \in S$ such that $T \cup \{s\}$ is independent. Because the almost-tight example presented in Sect. 4 is a (partition) matroid, the procedure presented in this paper does not have a better performance guarantee for matroids. So a third open problem would be to design a fair division algorithm that exploits the augmentation property to produce improved performance guarantees for matroids.

Acknowledgements. The authors thank Jugal Garg, Vasilis Gkatzelis and Richard Santiago for interesting discussions on fair division. We thank the anonymous reviewers for helpful suggestions.

References

1. Amanatidis, G., Markakis, E., Nikzad, A., Saberi, A.: Approximation algorithms for computing maximin share allocations. ACM Trans. Algorithms **13**(4) (2018). Article #52
2. Barman, S., Krishna Murthy, S.: Approximation algorithms for maximin fair division. In: Proceedings of the 18th Conference on Economics and Computation (EC), pp. 647–664 (2017)
3. Bouveret, S., Cechlarova, K., Elkind, E., Igarashi, A., Peters, D.: Fair division of a graph. In: Proceedings of the 26th International Joint Conference on Artificial Intelligence (IJCAI), pp. 135–141 (2017)
4. Bouveret, S., Lang, J.: Characterizing conflicts in fair division of indivisible goods using a scale of criteria. In: Proceedings of 13th Conference on Autonomous Agents and Multi-Agent Systems (AAMAS), pp. 1321–1328 (2014)
5. Brahms, S., King, D.: Fair Division: From Cake Cutting to Dispute Resolution. Cambridge University Press, Cambridge (1996)
6. Budish, E.: The combinatorial assignment problem: approximate competitive equilibrium from equal incomes. J. Polit. Econ. **119**(6), 1061–1103 (2011)
7. Caragiannis, I., Kurokawa, D., Moulin, H., Procaccia, A., Shah, N., Wang, J.: The unreasonable fairness of maximum nash welfare. In: Proceedings of 16th Conference on Economics and Computation (EC), pp. 305–322 (2016)
8. Ghodsi, M., HajiAghayi, M., Seddighin, M., Seddighin, S., Yami, H.: Fair allocation of indivisible goods: Improvement and generalization. In: Proceedings of the 19th Conference on Economics and Computation (EC), pp. 539–556 (2018)
9. Kurokawa, D., Procaccia, A., Wang, J.: Fair enough: guaranteeing approximate maximin shares. J. ACM **65**(2) (2018). Article #8

10. Moulin, H.: Uniform externalities: two axioms for fair allocation. J. Public Econ. **43**(3), 305–326 (1990)
11. Moulin, H.: Fair Division and Collective Welfare. MIT Press, Cambridge (2003)
12. Robertson, J., Webb, W.: Cake Cutting Algorithms: Be Fair If You Can. A K Peters, Natick (1998)
13. Steinhaus, H.: The problem of fair division. Econometrica **16**, 101–104 (1948)
14. Varian, H.: Equity, envy and efficiency. J. Econ. Theory **9**, 63–91 (1974)
15. Woeginger, G.: A polynomial time approximation scheme for maximizing the minimum machine completion time. Oper. Res. Lett. **20**, 149–154 (1997)

Stable Marriage with Groups of Similar Agents

Kitty Meeks[1] and Baharak Rastegari[2(✉)]

[1] School of Computing Science, University of Glasgow, Glasgow, UK
kitty.meeks@glasgow.ac.uk
[2] Electronics and Computer Science, University of Southampton, Southampton, UK
b.rastegari@soton.ac.uk

Abstract. Many important stable matching problems are known to be NP-hard, even when strong restrictions are placed on the input. In this paper we seek to identify structural properties of instances of stable matching problems which will allow us to design efficient algorithms using elementary techniques. We focus on the setting in which all agents involved in some matching problem can be partitioned into k different *types*, where the type of an agent determines his or her preferences, and agents have preferences over types (which may be refined by more detailed preferences within a single type). This situation would arise in practice if agents form preferences solely based on some small collection of agents' attributes. We also consider a generalisation in which each agent may consider some small collection of other agents to be exceptional, and rank these in a way that is not consistent with their types; this could happen in practice if agents have prior contact with a small number of candidates. We show that (for the case without exceptions), the well-known NP-hard matching problem MAX SMTI (that of finding the maximum cardinality stable matching in an instance of stable marriage with ties and incomplete lists) belongs to the parameterised complexity class FPT when parameterised by the number of different types of agents needed to describe the instance. This tractability result can be extended to the setting in which each agent promotes at most one "exceptional" candidate to the top of his/her list (when preferences within types are not refined), but the problem remains NP-hard if preference lists can contain two or more exceptions and the exceptional candidates can be placed anywhere in the preference lists.

1 Introduction

Matching problems occur in various applications and scenarios such as the assignment of children to schools, college students to dorm rooms, junior doctors to hospitals, and so on. In all the aforementioned, and similar, problems, it is understood that the participants (which we will refer to as agents) have preferences over other agents, or subsets of agents. The majority of the literature assumes that these preferences are ordinal, and that is the assumption we make in this

© Springer Nature Switzerland AG 2018
G. Christodoulou and T. Harks (Eds.): WINE 2018, LNCS 11316, pp. 312–326, 2018.
https://doi.org/10.1007/978-3-030-04612-5_21

work as well. Moreover, it is widely accepted that a "good" and "reasonable" solution to a matching problem must be *stable*, where stability is defined according to the context of the problem at hand. Intuitively speaking, a stable solution guarantees that no subset of agents find it in their best interest to leave the prescribed solution and seek an assignment amongst themselves. Unfortunately, many interesting and important stable matching problems are known to be NP-hard even for highly restricted cases.

In this paper we focus on the *Stable Marriage problem (SM)*, which is perhaps the most widely studied matching problem. In an instance of SM we have two disjoint sets of agents, men and women, each having a strict preference ordering over the individuals of the opposite sex (candidates). A solution to this problem is a *matching*, that is a mapping from men to women where each man is matched to at most one woman and vice versa. Each agent prefers being matched to remaining unmatched. A matching is *stable* if there are no two agents a and b who prefer each other to their assigned partners. If such a pair exists, we say that (a, b) is a *blocking pair*. *Stable Marriage with Incomplete lists (SMI)* is a generalisation of SM where agents are permitted to declare some candidates unacceptable. In their seminal work, Gale and Shapley [11] showed that every instance of SMI admits a stable matching that can be found in polynomial time by their proposed algorithm (GS). A simple extension of GS can be used to identify stable matchings in domains where agents are additionally allowed to express indifference between two or more candidates (*Stable Marriage with Ties and Incomplete lists (SMTI)*). However, it is known that (in contrast with SMI) an instance of SMTI might admit stable matchings of different sizes, and GS does not necessarily find the largest. In many practical applications, it is important to match as many agents as possible, but finding a matching which achieves this is much more computationally challenging: MAX SMTI, the problem of determining a maximum cardinality stable matching (i.e., a stable matching with the largest size amongst all stable matchings) in an instance of SMTI, is known to be NP-hard [3,14,16,20], even when the input is heavily restricted.

Most hardness results in the study of stable matching problems are based on the premise that agents may have arbitrary preference lists. In practice, however, agents' preferences are likely to be more structured and correlated. In this work, we consider a setting where agents can be grouped into k different "types", where the type of an agent determines (most of) the agent's preferences, and also how s/he is compared against other agents. If we allow each agent to have a different type, this setup does not place any restrictions on the instance. However, we are interested in the setting where the number of types required to describe an instance is much smaller than the total number of agents: such a situation would arise in practice if agents derive their preferences by considering some small collection of attributes of other agents (where each of these attributes has a small number of possible values). As an example, consider the hospitals-residents job market in which junior doctors or residents are to be assigned to hospital posts. It is highly plausible that agents in this market base their preferences on small collection of candidates' attributes. E.g. hospitals might rank applicants based on their exam grade, interview score, etc., and junior doctors might rank the hospitals based on the programs they offer, their reputation, their geographic location, etc. Similar observations have been made in the literature (see [2,4]) regarding stable marriage market and stable roommate market respectively, where agents form preferences based on candidates' attributes such as attractiveness, intelligence, wealth, etc. In this setting, we obtain our set of types by first partitioning agents by their profile of attributes, then further partitioning each set by the preference list over other profiles of attributes. Note that the number of possible preference lists depends only on the number of possible attribute profiles.

The notion of types is also useful if we are interested in a relaxation of stability, where agents are only willing to form a private arrangement with a partner who is distinctly superior to their current partner with respect to an important characteristic. It is reasonable to assume that in practice a certain amount of effort is required by both agents in a blocking pair to make a private arrangement outside the matching, and so agents are unlikely to make this effort for a very small improvement in their utility. Suppose that an agent is only willing to make the effort to form a private arrangement if it results in a significantly better partner, specifically one which has a significantly better value for the most important attribute. In this case we only need to consider attributes which are the most important for at least one agent, and moreover we might reasonably consider only a small number of categories of values for these attributes.

The simplest model is to assume that the agents of the same type are completely indistinguishable. That is, they have the same preference lists, and every other agent that finds their type acceptable is indifferent between them. Equivalently, we can say that each type has a preference ordering over types of the candidates, which need not be complete or strict. We also consider two generalisations of this basic model. In the first generalisation, agents no longer have to be indifferent between agents of the same type: they can refine their preference lists arbitrarily (so that agents of the same type still occur consecutively), so long as the preference lists for agents of the same type are identical. In the second generalisation, we instead enrich the basic model by allowing each agent to consider some small number of other agents "exceptional": such agents can appear anywhere in the preference list, regardless of their type. This situation with exceptions might arise in practice if, for example, an agent knows some of the candidates directly or through a third-party connection and, based on this additional information, ranks them disregarding their type, e.g. at the top or bottom of his/her preference list.

Our Contribution. We consider the parameterised complexity of MAX SMTI in all three settings. In the basic model and the extension which allows consistently refined preference lists, we show that the problem is in FPT parameterised by the number of types. In both settings the problem further becomes polynomial-time solvable if all preferences over types are strict. When exceptions are allowed in the preference lists, we demonstrate that the problem is once again in FPT, parameterised by the number of types, if each agent considers at most one agent exceptional, whom s/he promotes to the top of his/her preference list.

On the other hand, if two arbitrarily placed exceptions are allowed, MAX SMTI remains NP-hard, even if the number of types is bounded by a constant.

Due to shortage of space we have omitted or shortened the proofs. We refer the reader to the full version of the paper [18].

1.1 Definitions

Let N denote a set of n agents, which is composed of two disjoint sets. We use the term *candidates* to refer to the agents on the opposite side of the market to that of an agent under consideration. Each agent finds a subset of candidates acceptable and ranks them in order of preference. Preference orderings need not be strict, so it is possible for an agent to be indifferent between two or more candidates. We write $b \succ_a c$ to denote that agent a prefers candidate b to candidate c, and $b \simeq_a c$ to denote that a is indifferent between b and c. We write $b \succeq_a c$ to denote that a either prefers b to c or is indifferent between them. The indifference relation $\simeq_a$ implies an equivalence relation on acceptable candidates for a; each equivalence class under $\simeq_a$ is referred to as a tie.

In an instance of SMTI, a *matching* M is a pairing of men and women such that no one is paired with an unacceptable partner, each man is paired with at most one woman, and each woman is paired with at most one man. We write $(a, b) \in M$ to say that a and b are matched in M. We use $M(a)$ to denote the agent matched to a in M. We write $M(a) = \varnothing$ if agent a is unmatched in M. We assume that every agent prefers being matched to an acceptable candidate to remaining unmatched. Given an instance of SMTI, a matching M is *(weakly) stable* if there is no pair $(a, b) \notin M$ where a prefers b to his current partner in M, i.e., $b \succ_a M(a)$, and vice versa. For further background and terminology on stable matchings we refer the reader to [15].

We are concerned with the *parameterised complexity* of computational problems that are intractable in the classical sense. Parameterised complexity provides a multivariate framework for the analysis of hard problems: if a problem is known to be NP-hard, so that we expect the running-time of any algorithm to depend exponentially on some aspect of the input, we can seek to restrict this combinatorial explosion to one or more *parameters* of the problem rather than the total input size. This has the potential to provide an efficient solution to the problem if the parameter(s) in question are much smaller than the total input size. A parameterised problem with total input size n and parameter k is considered to be tractable if it can be solved by a so-called *FPT algorithm*, an algorithm whose running time is bounded by $f(k) \cdot n^{\mathcal{O}(1)}$, where f can be any computable function. Such problems are said to be *fixed parameter tractable*, and belong to the complexity class FPT. For further background on the theory of parameterised complexity, we refer the reader to [7,8,10].

1.2 Related Work

The NP-hardness of MAX SMTI has been shown for a variety of restricted settings, for example: (1) even if each man's list is strictly ordered, and each woman's list is either strictly ordered or is a tie of length 2 [16], (2) even if each mans preference list is derived from a strictly-ordered master list of women, and each woman's preference list is derived from a master list of men that contains only one tie [14], and (3) even if the SMTI instance has symmetric preferences; that is, for any acceptable (man, woman) pair (m_i, w_j), $rank(m_i, w_j) = rank(w_j, m_i)$ [20], where $rank(a, b)$ is defined to be one plus the number of candidates that a prefers to b.

There are a limited number of works addressing fixed-parameter tractability in stable matching problems. Marx and Schlotter [17] gave the first parameterised complexity results on MAX SMTI. They showed that the problem is in FPT when parameterised by the total length of the ties, but is W[1]-hard when parameterised by the number of ties in the instance, even if all the men have strictly ordered preference lists. Very recently, three different works have studied hard stable matching problems from the perspective of parameterised complexity. Mnich and Schlotter [19] obtained results on the parameterised complexity of finding a stable matching which matches a given set of distinguished agents and has as few blocking pairs as possible. Gupta et al. [13] showed that several hard stable matching problems, including MAX SMTI, are W[1]-hard when parameterised by the treewidth of the graph obtained by adding an edge between each pair of agents that find each other mutually acceptable. Gupta et al. [12] studied above guarantee parameterisations of the problem of finding a stable matching that balances between the dissatisfaction of men and women, with parameters that capture the degree of dissatisfaction.

Settings in which agents are partitioned into different types, or derive their preferences based on a set of attributes assigned to each candidate, have been considered for the problems of sampling and counting stable matchings in instances of SM or SR (Stable Roommate problem); see, e.g., [2,4,5]. Echenique et al. [9] studied the problem of characterising matchings that are rationalisable as stable matchings when agents' preferences are unobserved. They focused on a restricted setting that translates into assigning each agent a type based on several attributes, and assuming that agents of the same type are identical and have identical preferences. They remarked that empirical studies on marriage typically make such an assumption [6]. Bounded agent types have been considered by Aziz and de Keijzer [1] and Shrot et al. [22] to derive polynomial-time results for the coalition structure generation problem, an important issue in cooperative games when the goal is to partition the participants into exhaustive and disjoint coalitions in order to maximise the social welfare.

2   Our Basic Model: Agents of the Same Type are Indistinguishable

In this section we begin with a formal definition of the simplest model we consider, in which agents' preferences can be derived directly from the preferences of types over types of candidates. We then identify a necessary and sufficient condition, in terms of the type of the least desirable partner assigned to any agent of each type, for a matching to be stable in this model. We use this to show that, if there are k types, we can solve MAX SMTI by solving $k^{\mathcal{O}(k)} \cdot \log n$ instances of MAX FLOW on directed networks with $\mathcal{O}(k)$ vertices and maximum edge capacity $\mathcal{O}(n)$. This implies that MAX SMTI, parameterised by k, belongs to FPT.

2.1 Definition of Typed Instances

Assume that there are k types available for agents. Let $[k]$ denote the set $\{1, 2, \ldots, k\}$. Let N_i denote the set of agents that are of type i. Thus we have that the set of agents $N = \bigcup_{i \in [k]} N_i$. Each type i has a preference ordering over types of the candidates, which need not be complete or strict. We assume, without loss of generality, that $|N_i| > 0$ for all $i \in [k]$, and that each type finds at least one other type acceptable. We write $j \succ_i \ell$ if agents of type i strictly prefer agents of type j to agents of type ℓ. We write $j \simeq_i \ell$ to denote that agents of type i are indifferent between agents of types j and ℓ, and $j \succeq_i \ell$ if agents of type i prefer agents of type j to those of type ℓ or are indifferent between the two. We assume that given every two agents x and y of the same type:

1. x and y have identical preference lists, and
2. all other agents are indifferent between x and y.

These requirements imply that any agent either finds all agents of a given type acceptable (and is indifferent between them) or finds none of them acceptable. We say that an instance of a stable matching problem satisfying these requirements is *typed*, and refer to the standard problems with input of this form as TYPED MAX SMTI etc. Note that TYPED MAX SMTI remains NP-hard when k is considered to be part of the input: we can always create a typed instance by assigning each agent its own type.

A typed instance I of SMTI is given as input by specifying the number of types k and, for each type i, the set N_i of agents of type i as well as the preference ordering $\succ_i$ over types of the candidates. Observe that, if we are only given the preference list for each agent as input, it is straightforward to compute, in polynomial time, the coarsest partition of the agents into types that satisfies the definition of a typed instance (see [18]). Having found such a partition, the preference lists over types can also be constructed efficiently.

Example 1. Assume we have 4 types for the agents, all men are of type 1 and types 2, 3 and 4 correspond to women. Let the preference ordering of type 1 over types of women be as follows, where the preference list is ordered from left to right in decreasing order of preference, and the types in round brackets are tied: (2 3) 4. Assume that there are 7 women and w_1 and w_2 are of type 2, w_3 and w_4 are of type 3, and w_5, w_6 and w_7 are of type 4. Therefore, the preference lists of all men under the typed model are as follows: $(w_1 \ w_2 \ w_3 \ w_4) \ (w_5 \ w_6 \ w_7)$.

2.2   An FPT Algorithm for TYPED MAX SMTI

Let I be a typed instance of SMTI, and let M be a matching in I. We may assume without loss of generality that every agent is matched, by creating sufficiently many dummy agents of type $k + 1$ which are inserted at the end of each man's and woman's (possibly incomplete) preference list. We define $worst_M(i)$ to be the type of the least desirable agent with which any agent of type i is matched in M, breaking ties arbitrarily (e.g. lexicographically). Note that $worst_M(i)$ would be a dummy type if an agent of type i is unmatched (i.e. matched to a dummy agent) in M. Let $type(a)$ denote the type of a given agent a.

The key observation is that, in order to determine whether or not M is stable, it suffices to examine the values of $worst_M(i)$ for each $i \in [k]$.

Lemma 1. *Let I be a typed instance of SMTI. Then a matching M in I is stable if and only if there is no pair $(i,j) \in [k]^{(2)}$ such that $j \succ_i worst_M(i)$ and $i \succ_j worst_M(j)$.*

We say that a matching M *realises* a given function $worst : [k] \to [k+1]$ if, for each $i \in [k]$, the least desirable partner any agent of type i has in M is of type no worse than $worst(i)$. We say that a function $worst : [k] \to [k+1]$ is I-*stable* for an instance I of SMTI if there is no pair $(i,j) \in [k]^{(2)}$ such that $j \succ_i worst(i)$ and $i \succ_j worst(j)$. Given any I-stable function $worst$, we write $\max(worst)$ for the maximum cardinality of any matching in I that realises $worst$. Using Lemma 1, it is straightforward to check that, given a typed instance I of SMTI, the cardinality of a solution to MAX SMTI can be found by taking the largest value of $\max(worst)$ over all I-stable functions $worst$.

Corollary 1. *Let I be a typed instance of SMTI. Then the cardinality of the largest stable matching in I is equal to $\max\{\max(worst) : worst \text{ is } I\text{-stable}\}$.*

We next show that, given an arbitrary I-stable function $worst$, we can compute $\max(worst)$ in time polynomial in k and $\log n$. We do this by solving $\mathcal{O}(\log n)$ instances of MAX FLOW on a directed network.

Lemma 2. *Let I be a typed instance of SMTI, and fix an I-stable function $worst$. We can compute $\max(worst)$ in time $\mathcal{O}(k^3 \log^2 n)$.*

Proof (Proof sketch.). The proof is structured as follows. Suppose that in total there are n_1 women and n_2 men, so we have that $n = n_1 + n_2$. Note that $\max(worst)$ is at most $\min\{n_1, n_2\}$, which in turn is at most $\lfloor n/2 \rfloor$. Therefore, using a binary search strategy, we can determine the maximum size of a matching realising $worst$ by solving $\mathcal{O}(\log n)$ instances of the decision problem "Is $\max(worst)$ at least c?", where $c \in \{1, \ldots, \min\{n_1, n_2\}\}$. We show that we can determine whether $\max(worst) \geq c$ by solving MAX FLOW on a directed network D with $\mathcal{O}(k)$ vertices, in which the maximum capacity of any edge is $\mathcal{O}(n)$ (see Fig. 1); we can construct D from I in time $\mathcal{O}(k^2 \log n)$. MAX FLOW can be solved on D in time $\mathcal{O}(k^3 \log n)$, using an algorithm due to Orlin [21], where the $\log n$ factor is required to carry out arithmetic operations on integers of size $\mathcal{O}(n)$. Therefore, we conclude that we can compute $\max(worst)$ in time $\mathcal{O}(k^3 \log^2 n)$. $\square$

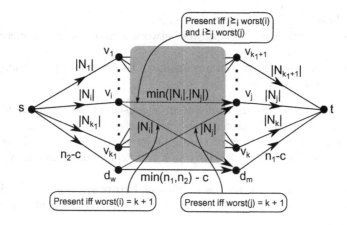

Fig. 1. Network D, constructed from an instance of SMTI in the proof of Lemma 2. Types $1, \ldots, k_1$ are types of women and types $k_1 + 1, \ldots, k$ are types of man. Each vertex v_i corresponds to type i and both vertices d_w and d_m correspond to the dummy type. In total there are n_1 women and n_2 men. We have $\max(\text{worst}) \geq c$ if and only if the maximum flow in D is equal to $n_1 + n_2 - c$.

It then follows that TYPED MAX SMTI is in FPT parameterised by the number k of different types in the instance.

Corollary 2. TYPED MAX SMTI *can be solved in time* $k^{\mathcal{O}(k)} \cdot \log^2 n + \mathcal{O}(n)$*. If we are only interested in computing the size of the maximum cardinality matching, and not the matching itself, this can be done in time* $k^{\mathcal{O}(k)} \cdot \log^2 n$*.*

3 Agents of the Same Type Refine Their Preferences in the Same Way

In this section, we generalise the model from Sect. 2 by allowing agents to refine their preferences over candidates within a particular type, so long as agents of the same type still have identical preference lists. Our key result is that refining preferences in this way can never change the size of the largest stable matching, compared with the corresponding typed instance. We also use the tools we develop to deal with this generalisation to show that MAX SMTI becomes polynomially solvable if preferences over types are strict, both in this setting and under the basic model.

3.1 Definition of Consistently-Refined-Typed Instances

Consider a generalisation of typed instances in which agents are no longer necessarily indifferent between two agents of the same type, however agents of the same type occur consecutively in preference lists. This means that for any two agents x and y of the same type i:

1. x and y have identical preference lists,
2. no agent of a different type appears between x and y in any preference list, and
3. if a tie in a preference list contains agents of two or more types, then that tie is in fact a union of types.

The third criterion allows us to define in a consistent way what it means for agents of type i to strictly prefer type j to type ℓ or to be indifferent between them. We will say that agents of type i prefer type j to type ℓ if and only if given every pair of agents x of type j and y of type ℓ all agents in N_i prefer x to y.

On the other hand, if type i is indifferent between types j and ℓ it means that, in the preference list for each agent x of type i, all agents in $N_j \cup N_\ell$ belong to a single tie.

If an instance of a stable matching problem satisfies these slightly weaker requirements, we say that the instance is *consistently-refined-typed*, and refer to the standard problems with input of this form as CONSISTENTLY-REFINED-TYPED MAX SMTI etc.

A consistently-refined-typed instance I of SMTI is given as an input by specifying the number of types k and, for each type i, the set N_i of agents of type i as well as the preference ordering $\succ_i$ over agents. Note that for typed instances $\succ_i$ specified preferences over types, whereas here the preferences are over agents. However, we can compute preferences over types from preferences over agents in time $\mathcal{O}(kn)$. Note that if we are only given the preference list for each agent as input (i.e., no information about types is given), it is straightforward to compute, in polynomial time, the coarsest partition of the agents into types that satisfies the definition of consistently-refined-typed instance (see [18]).

Example 2. Assume we have 4 types for the agents, all men are of type 1 and types 2, 3, and 4 correspond to women. Assume also that we have 3 men m_1, m_2 and m_3, and 7 women where w_1 and w_2 are of type 2, w_3 and w_4 are of type 3, and w_5, w_6 and w_7 are of type 4. Let all men have the preference ordering $(w_1\ w_2\ w_3\ w_4)\ w_6\ (w_5\ w_7)$, women of types 2 and 3 have the preference ordering $(m_1\ m_2\ m_3)$, and women of type 4 have the preference ordering $m_2\ m_1$. This setting constitutes a consistently-refined-typed instance. It is easy to compute the preferences of type 1 agents over the types of women, which is (2 3) 4, similar to that of Example 1. Allowing men to have the preference ordering $(w_1\ w_2)\ w_3\ w_4\ (w_5\ w_6\ w_7)$, while keeping everything else unchanged, also gives us a consistently-refined-typed instance. In this new instance agents of type 1 have the strict preference ordering 2 3 4 over the types of women.

3.2 An FPT Algorithm for Consistently-Refined-Typed Max SMTI

To extend the result for TYPED MAX SMTI to CONSISTENTLY-REFINED-TYPED MAX SMTI, we need the following result.

Lemma 3. *Let I be a consistently-refined-typed instance of SMTI and suppose that M is a matching in I such that there is no pair $(i,j) \in [k]^{(2)}$ where $j \succ_i$ worst$_M(i)$ and $i \succ_j$ worst$_M(j)$. Then there is a stable matching M' such that, for every $(i,j) \in [k]^{(2)}$, both M and M' contain the same number of pairs that consist of one agent of type i and another of type j. Moreover, given M, we can compute M' in time $\mathcal{O}(kn)$.*

Let I be a consistently-refined-typed instance of SMTI and let I' be a typed instance of SMTI that is obtained from I by ignoring the refined preferences within each type (i.e. every agent is indifferent between the candidates of the same type). It follows from the definition of stability that every matching that is stable in I is also stable in I'. Lemma 3 implies that for any stable matching M in I', there exists a stable matching M' in I of the same cardinality as M. Thus, in order to find a maximum cardinality matching in a consistently-refined-typed instance I of SMTI, it suffices to (1) solve the typed problem (i.e. ignore the refined preferences within each type) and then (2) use the algorithm provided in the proof of Lemma 3 (see [18]) to convert the solution to a matching of the same cardinality that is stable in the instance I. In fact, in (1) it is enough to only compute the maximum flow f (and not the matching M); the flow f, that can be computed in time $k^{\mathcal{O}(k)} \cdot \log^2 n$, provides sufficient information for the algorithm described in the proof of Lemma 3 to construct M' in time $\mathcal{O}(kn)$. Deriving a typed instance from a consistently-refined-typed instance can be done easily in time $\mathcal{O}(kn)$. It thus follows that CONSISTENTLY-REFINED-TYPED MAX SMTI is in FPT parameterised by the number k of different types in the instance.

Theorem 1. CONSISTENTLY-REFINED-TYPED MAX SMTI *can be solved in time $k^{\mathcal{O}(k)} \cdot \log^2 n + \mathcal{O}(kn)$.*

3.3   Strict Preferences over Types

Elsewhere in the paper, we assume that agents can be indifferent between agents of two or more types. It turns out that MAX SMTI becomes easier if we restrict the set of possible instances by assuming that agents have strict preferences over types. We prove this by breaking ties arbitrarily in a consistent way for each type, to obtain an instance I' of the polynomially-solvable problem SMI, and then using Lemma 3 to argue that the cardinality of the largest stable matching in I' is the same as that in our original instance. This argument is based on a private communication with David Manlove.

Theorem 2. *When preferences over types are strict, TYPED MAX SMTI and CONSISTENTLY-REFINED-TYPED MAX SMTI are polynomial-time solvable. Furthermore, all stable matchings are of the same size.*

4   Exceptions in Preference Lists

We have argued for the existence of typed instances, where $k \ll n$, based on the premise that agents' preferences are formed based on a small collection of candidates' attributes. In practice, it seems likely that an agent might have access to additional information about some small subset of the candidates, either through personal acquaintance or some third-party connection; we say that an agent considers such candidates to be *exceptional*. This additional information may alter the agent's opinion of candidates relative to that derived from the attributes alone, and so affect where these candidates are placed in his/her preference ordering. In this section we consider a generalisation of typed instances in which each agent may find some small collection of other agents to be exceptional and ranks them without regard to their types. Note that if only a small number of the agents in our instance consider one or more candidates to be exceptional, we can capture this information in a typed instance: each agent with exceptions in their preference list can be assigned their own type.

We say that an instance I of a stable matching problem is a *(c,Any)-exception-typed* instance, for a given constant c, if I is a typed instance in which each agent finds at most c number of the candidates exceptional and may rank them anywhere in his/her preference list. Two special cases are *(c, Top)-exception-typed* and *(c,Bottom)-exception-typed* instances where the exceptions are promoted to the top, or demoted to the bottom, of the preference lists, respectively. We refer to the standard problems with input of this form as (C,ANY)-EXCEPTION TYPED MAX SMTI etc.

In this section we show that (1,TOP)-EXCEPTION TYPED MAX SMTI belongs to FPT, but that (2,ANY)-EXCEPTION TYPED MAX SMTI remains NP-hard even when there are only a constant number of types. The computational complexity of (1,ANY)-EXCEPTION TYPED MAX SMTI and (2,TOP)-EXCEPTION TYPED MAX SMTI remain open.

We begin with the case of (1,TOP)-EXCEPTION TYPED MAX SMTI. For each agent a let $ex(a)$ denote the exceptional candidate from a's point of view; $ex(a) = \varnothing$ if a does not find any candidate exceptional. Formally, we say that I is a *(1, Top)-exception-typed* instance of SMTI if, given every two agents x and y of the same type:

1. x and y have identical preference lists when restricted to $N \setminus \{ex(x), ex(y)\}$, and
2. all other agents who do not find either x or y exceptional are indifferent between x and y.

Without loss of generality we can assume that there is no pair of agents who each consider the other to be exceptional in a (1,Top)-exception-typed instance of SMTI. If there are such pairs, they must be assigned to each other in any stable matching; so we can remove all such pairs to reduce to an instance that satisfies this assumption. A (1,Top)-exception-typed instance of SMTI is given as input by, in addition to the specifications needed for a typed instance (see Sect. 2.1), providing for each agent his or her exceptional candidate (if s/he has one).

Let I be a (1,Top)-exception-typed instance of SMTI, and let M be a matching in I. As in Sect. 2.2, we may assume without loss of generality that every agent is matched, by creating sufficiently many dummy agents of type $k + 1$ which are inserted at the end of each man's and woman's (possibly incomplete) preference list. In order to obtain an analogue of the stability criterion given in Lemma 1 in this setting, we need some more notation.

Recall that we write $j \simeq_i \ell$ if agents of type i are indifferent between types j and ℓ. It is straightforward to see that $\simeq_i$ defines an equivalence relation on $[k]$ for each i. Given $j \in [k]$, we write $\text{class}_i(j)$ for the equivalence class under $\simeq_i$ which contains j. For each equivalence class J under $\simeq_i$, we say that the agent x of type i has *subtype* $i[J]$ if:

1. some agent y, with $\text{type}(y) \in J$, considers x exceptional, and
2. there is no agent z, such that $\text{type}(z) \succ_i j$ for $j \in J$, who considers x exceptional.

Thus $\text{subtype}(x) = i[J]$ if the most desirable agents who consider x exceptional have types from J. If an agent x of type i is not considered exceptional by any agent, we say that x has subtype $i[\{k+1\}]$. We also introduce a second dummy type 0, which is inserted at the head of each type's preference list and corresponds to exceptional candidates. We write $N_{i[J]}$ for the set of agents of subtype $i[J]$. Observe that the sets $N_{i[J]}$ can be computed in time $\mathcal{O}(n)$: for each agent x, $\text{subtype}(x)$ can be computed in time $\mathcal{O}(n)$ with suitable data structures.

We will need a variation on the function worst_M, which we call $\text{worst}^{\text{ex}}_M$. For any non-empty set $N_{i[J]}$, $\text{worst}^{\text{ex}}_M(i[J])$ is the type of the least desirable partner received by an agent of subtype $i[J]$ who is not matched with an agent they find exceptional; if the least desirable partners assigned to agents of subtype $i[J]$ belong to two or more different types between which agents of type i are indifferent, we define $\text{worst}^{\text{ex}}_M(i[J])$ to be the lexicographically first such type. If every agent of subtype $i[J]$ is matched with a partner they find exceptional, we set $\text{worst}^{\text{ex}}_M(i[J]) = 0$. Therefore $\text{worst}_M(i)$, as defined in Sect. 2.2, is the least desirable type out of $\{\text{worst}^{\text{ex}}_M(i[J]) : N_{i[J]} \neq \emptyset\}$.

We say that a matching M in an instance I of (1,Top)-exception-typed SMTI realises the function worst^{ex}, mapping nonempty subtypes $i[J]$ to values in $\{0, 1, \ldots, k+1\}$, if $\text{worst}^{\text{ex}}_M(i[J]) \succeq_i \text{worst}^{\text{ex}}(i[J])$ whenever $N_{i[J]} \neq \emptyset$. We can now characterise stability in a (1,Top)-exception-typed instance.

Lemma 4. *Let I be a (1, Top)-exception-typed instance of SMTI. Then a matching M in I is stable if and only if there is no pair $(i, j) \in [k]^{(2)}$ such that*

1. $j \succ_i \text{worst}_M(i)$ and $i \succ_j \text{worst}_M(j)$, or
2. $i \succ_j \text{worst}^{\text{ex}}_M(j[\text{class}_j(i)])$.

We will say that a function worst^{ex} is *I-exception-stable* for a (1,Top)-exception-typed instance I of SMTI if there is no pair $(i, j) \in [k]^{(2)}$ such that either $j \succ_i \text{worst}(i)$ and $i \succ_j \text{worst}(j)$ or $i \succ_j \text{worst}^{\text{ex}}(j[\text{class}_j(i)])$. Given any I-exception-stable function worst^{ex}, we write $\max(\text{worst}^{\text{ex}})$ for the maximum cardinality of any matching in I that realises worst^{ex}. We have an analogous result to Corollary 1 in this setting.

Lemma 5. *Let I be a (1, Top)-exception-typed instance of SMTI. Then the cardinality of the largest stable matching in I is equal to*

$$\max\{\max(\text{worst}^{\text{ex}}) : \text{worst}^{\text{ex}} \text{ is } I\text{-exception-stable}\}.$$

By solving a collection of instances of MAXIMUM MATCHING in suitable undirected graphs, we are able to compute max(worst^{ex}) (and generate a stable matching of this size) in time $\mathcal{O}(n^{5/2} \log n)$, for any I-exception-stable function worst^{ex}. Our FPT result now follows.

Theorem 3. (1,TOP)-EXCEPTION TYPED MAX SMTI *can be solved in time* $\mathcal{O}\left(k^{k^2}(k^3 + n^{5/2} \log n)\right)$.

In contrast with this positive result, we show that only a small relaxation of the requirements on exceptions results in a problem that is NP-hard, even if the number of types is bounded by a constant. We show that, if we allow each agent to declare two candidates exceptional, and these two candidates can appear anywhere in the agent's preference list, then MAX SMTI remains NP-hard under sever restrictions. In fact, we give a reduction from the NP-complete problem CLIQUE to the special case COM SMTI, which involves deciding whether a given instance of SMTI admits a complete stable matching (i.e., a matching that matches all agents).

Theorem 4. (2,ANY)-EXCEPTION TYPED COM SMTI *is* NP-*complete, even if only men have exceptions in their preference lists, preferences over types are strict, and there are three types each of men and women.*

5   Discussion and Future Work

We believe that the same techniques used in this paper can be extended to prove analogous results in all three settings for the Hospitals-Residents problem (a many-one generalisation of SMTI) and the Stable Roommates problem (a non-bipartite generalisation of SMTI). For typed and consistently-refined-typed instances, a standard cloning argument gives the corresponding results for the Hospitals-Residents problem immediately.

We note that our FPT results can also be derived using Integer Linear Programming (ILP) techniques: for each function worst (or worst^{ex}), the corresponding optimisation problem can be encoded as an ILP instance. For the problems studied here, the ILP approach results in worse running times than the algorithms we have described, but this alternative approach might be helpful in tackling other stable matching problems involving types.

It would be interesting to investigate what further generalisations of our model yield FPT algorithms for NP-hard stable matching problems. In particular, the complexity of (1,BOTTOM)-EXCEPTION-TYPED MAX SMTI, (1,ANY)-EXCEPTION-TYPED MAX SMTI, and (2,TOP)-EXCEPTION-TYPED MAX SMTI remain open. Moreover, we could consider further restrictions with two or more exceptions, for example if an exceptional candidate can only be moved to the top or bottom of its type. Another intriguing question would be to understand how the complexity of MAX SMTI and other stable matching problems changes when agents on only one side of the market are associated with types.

Acknowledgements. The first author is supported by a Personal Research Fellowship from the Royal Society of Edinburgh (funded by the Scottish Government). Both authors are extremely grateful to David Manlove for his insightful comments on a preliminary version of this manuscript.

References

1. Aziz, H., de Keijzer, B.: Complexity of coalition structure generation. In: Proceedings of the 10th International Conference on Autonomous Agents and Multiagent Systems, AAMAS 2011, pp. 191–198 (2011)
2. Bhatnagar, N., Greenberg, S., Randall, D.: Sampling stable marriages: why spouse-swapping won't work. In: Proceedings of the 19th ACM/SIAM Symposium on Discrete Algorithms, SODA 2008, pp. 1223–1232. ACM-SIAM (2008)
3. Biró, P., Manlove, D., Mittal, S.: Size versus stability in the marriage problem. Theor. Comput. Sci. **411**, 1828–1841 (2010)
4. Chebolu, P., Goldberg, L.A., Martin, R.: The complexity of approximately counting stable matchings. Theor. Comput. Sci. **437**, 35–68 (2012)
5. Chebolu, P., Goldberg, L.A., Martin, R.: The complexity of approximately counting stable roommate assignments. J. Comput. Syst. Sci. **78**(5), 1579–1605 (2012)
6. Choo, E., Siow, A.: Who marries whom and why. J. Polit. Econ. **114**(1), 175–201 (2006)
7. Cygan, M., et al.: Parameterized Algorithms. Springer, Cham (2015). https://doi.org/10.1007/978-3-319-21275-3
8. Downey, R.G., Fellows, M.R.: Fundamentals of Parameterized Complexity. Springer, London (2013). https://doi.org/10.1007/978-1-4471-5559-1
9. Echenique, F., Lee, S., Shum, M., Yenmez, M.B.: The revealed preference theory of stable and extremal stable matchings. Econometrica **81**(1), 153–171 (2013)
10. Flum, J., Grohe, M.: Parameterized Complexity Theory. Springer, Heidelberg (2006). https://doi.org/10.1007/3-540-29953-X
11. Gale, D., Shapley, L.: College admissions and the stability of marriage. Am. Math. Mon. **69**, 9–15 (1962)
12. Gupta, S., Roy, S., Saurabh, S., Zehavi, M.: Balanced stable marriage: how close is close enough. Technical report 1707.09545, CoRR, Cornell University Library (2017)
13. Gupta, S., Saurabh, S., Zehavi, M.: On treewidth and stable marriage. Technical report 1707.05404, CoRR, Cornell University Library (2017)
14. Irving, R., Manlove, D., Scott, S.: The stable marriage problem with master preference lists. Discret. Appl. Math. **156**(15), 2959–2977 (2008)
15. Manlove, D.: Algorithmics of Matching Under Preferences. World Scientific, Singapore (2013)
16. Manlove, D., Irving, R., Iwama, K., Miyazaki, S., Morita, Y.: Hard variants of stable marriage. Theor. Comput. Sci. **276**(1–2), 261–279 (2002)
17. Marx, D., Schlotter, I.: Parameterized complexity and local search approaches for the stable marriage problem with ties. Algorithmica **58**(1), 170–187 (2010)
18. Meeks, K., Rastegari, B.: Solving hard stable matching problems involving groups of similar agents. Technical report 1708.04109, CoRR, Cornell University Library (2018)

19. Mnich, M., Schlotter, I.: Stable marriage with covering constraints–a complete computational trichotomy. In: Bilò, V., Flammini, M. (eds.) SAGT 2017. LNCS, vol. 10504, pp. 320–332. Springer, Cham (2017). https://doi.org/10.1007/978-3-319-66700-3_25

20. O'Malley, G.: Algorithmic aspects of stable matching problems. Ph.D. thesis, Department of Computing Science, University of Glasgow (2007)

21. Orlin, J.B.: Max flows in $\mathcal{O}(nm)$ time, or better. In: Proceedings of the 45th Annual ACM Symposium on Theory of Computing, STOC 2013, pp. 765–774. ACM (2013)

22. Shrot, T., Aumann, Y., Kraus, S.: On agent types in coalition formation problems. In: Proceedings of the 9th International Conference on Autonomous Agents and Multiagent Systems, AAMAS 2010, pp. 757–764 (2010)

Byzantine Preferential Voting

Darya Melnyk[(✉)], Yuyi Wang, and Roger Wattenhofer

ETH Zurich, Zurich, Switzerland
{dmelnyk,yuwang,wattenhofer}@ethz.ch

Abstract. In the Byzantine agreement problem, n nodes with possibly different input values aim to reach agreement on a common value in the presence of $t < n/3$ Byzantine nodes which represent arbitrary failures in the system. This paper introduces a generalization of Byzantine agreement, where the input values of the nodes are preference rankings of three or more candidates. We show that consensus on preferences, which is an important question in social choice theory, complements already known results from Byzantine agreement. In addition, preferential voting raises new questions about how to approximate consensus vectors. We propose a deterministic algorithm to solve Byzantine agreement on rankings under a generalized validity condition, which we call Pareto-Validity. These results are then extended by considering a special voting rule which chooses the Kemeny median as the consensus vector. For this rule, we derive a lower bound on the approximation ratio of the Kemeny median that can be guaranteed by any deterministic algorithm. We then provide an algorithm matching this lower bound. To our knowledge, this is the first non-trivial generalization of multi-valued Byzantine agreement to multiple dimensions which can tolerate a constant fraction of Byzantine nodes.

Keywords: Social choice · Byzantine agreement · Pareto-Validity
Distributed voting · Multivalued

1 Introduction

In distributed machine learning, different data is often collected and owned by different parties, each of which will locally train its own machine learning model. If a new data item needs to be judged, the parties could collaborate in order to make a collective decision. As an example, a hospital may be authorized to use its own collected patient data to train an image recognition model, but not to share that data with other hospitals because of patient privacy limitations. For some critical cases the hospitals would still want to collaborate and decide on the correct diagnosis together.

In order to obtain a *robust* collective decision, we need to take the following two aspects into account. On the one hand, it is possible that some of the involved parties experience hardware or software difficulties, or simply

© Springer Nature Switzerland AG 2018
G. Christodoulou and T. Harks (Eds.): WINE 2018, LNCS 11316, pp. 327–340, 2018.
https://doi.org/10.1007/978-3-030-04612-5_22

play dirty. Our decision will be robust if we can withstand even *Byzantine* parties, who are controlled by a single omnipotent adversary trying to maliciously disturb the process. On the other hand, non-Byzantine parties should use all available information to come up with the best possible decision. In standard multi-valued Byzantine agreement algorithms, each party will provide only one input, however, machine learning algorithms usually provide information about the second-best and third-best guess. For example, when doing image recognition in medicine the result can be a *ranking* of possible diagnoses: glioblastoma $\succ$ metastasis $\succ \ldots \succ$ inflammatory. Such rankings convey much more information than just the top ranked alternative (glioblastoma). While the different honest parties might completely disagree on the top alternative, the second alternative might serve as a tie breaker, and we can therefore hope to receive more meaningful results from the voting process by considering rankings.

In this paper we use social choice theory in order to investigate the most fair choice among a set of rankings to solve Byzantine agreement on rankings. In particular, we want to study how robust preferential voting is in a Byzantine environment. In Sect. 2, we first focus on some basic properties for voting rules, and see that not all of them can be satisfied if the parties should reach an agreement. This is because Byzantine voters are manipulators that modify the result to make it more favorable to themselves. In the main part of the paper (Sect. 5) we then study how well the voting result intended by the correct (non-Byzantine) voters can be approximated. For this purpose we consider the Kemeny rule which picks the most central ranking as the voting result. We will provide an algorithm that approximates the solution of the Kemeny rule in the presence of Byzantine voters and prove that this algorithm computes the best possible approximation. We believe our paper will contribute a deeper understanding of both fault-tolerant distributed systems as well as social choice theory.

2   Background and Motivation

In search of a *fair* rule to elect candidates, philosophers and mathematicians started developing various voting mechanisms and rules already in the beginning of the 18th century. In the middle of the 20th century, Kenneth Arrow [2,3] was one of the first to formalize existing voting rules and analyze possibility and impossibility results in an axiomatic fashion, thereby introducing the field of Computational Social Choice. In this section we will show how well Byzantine agreement connects to voting theory.

We start by considering the special case of n voters voting on only two candidates c_1 and c_2. In this setting, each voter (node) ranks the two candidates such that its preferred candidate (input value) is ranked first. A vote for a candidate c_1 means that the voter strictly prefers c_1 to c_2, which we here denote $c_1 \succ c_2$. A central authority then applies a *social choice function (SCF)* to a given preference profile in order to determine the winner (decision value), or set of winners in case of a tie. An SCF should typically strive to satisfy anonymity, neutrality and positive responsiveness. May's theorem [27] shows that the majority rule is the only voting rule on two candidates that satisfies all three properties.

Interestingly, most known algorithms for binary Byzantine agreement indirectly exploit the properties of May's theorem: Some of them make use of leaders who suggest their decision value to all nodes, e.g., the King and the Queen algorithms [8,9]. The leader in these algorithms temporarily plays what is known as a dictator in voting theory. Another type of algorithm, e.g., the shared coin algorithm in [39], is biased towards one of the outcomes and thus violates neutrality. In general we can say that most of the proposed algorithms try to use the majority value as the decision value if a majority exists, or an arbitrary input value otherwise, see for example [7,11]. Such settings may satisfy anonymity and neutrality, but in cases where the correct nodes are undecided, i.e., there is a tie between the two input values, Byzantine nodes have a large influence on the majority value. Thus, if a correct node decides to swap two candidates in its ranking in order to make one of the candidates win, a Byzantine node can perform an opposite swap in its own ranking and return the profile to the previous state. This shows that positive responsiveness cannot be satisfied for these algorithms in the presence of Byzantine nodes.

May's theorem does however not apply to the general case with more than two candidates. Moreover, a lot of information is lost when a single winner is sought. When it comes to preferential voting, social choice theory often wants not only the input to be rankings, but also the output. This is satisfied by *social welfare functions (SWF)* that map a preference profile to a set of consensus rankings. For an SWF, g, the following three properties are usually considered:

- g is *dictatorial* if there is one distinguished voter whose input ranking is chosen as the single consensus ranking
- g is *independent of irrelevant alternatives (IIA)* if the consensus ranking of two candidates c_i and c_j only depends on the relative preference of these candidates in each voter's ranking, and not on the ranking of some third candidate c_k
- g is *weakly Paretian* if it satisfies the weak Pareto condition [31]: for two candidates c_i and c_j which are ranked $c_i \succ c_j$ by all voters, consensus ranking has to rank $c_i \succ c_j$ as well.

Unfortunately, Arrow's impossibility theorem [2] shows that every SWF on three or more alternatives that is weakly Paretian and IIA must be dictatorial. From the viewpoint of Byzantine agreement, an SWF should not be dictatorial since one does not want a dictator to be a Byzantine node. Consequently, any reasonable Byzantine agreement protocol must either violate IIA or weak Pareto. We say that IIA or weak Pareto are satisfied in the Byzantine setting if they are satisfied with respect to the input rankings of the correct nodes only. Under this assumption, the IIA condition implies that the consensus ranking should remain the same if the input of every correct node does not change, no matter what the Byzantine nodes do. However, a Byzantine node can pretend to be a correct node but change its ranking in two executions in which the correct nodes have the same inputs. This change may lead to a different consensus ranking and thus violate IIA. For the weak Pareto condition consider the case with two candidates: if every non-Byzantine voter ranks $c_1 \succ c_2$, the consensus ranking should also rank $c_1 \succ c_2$. This corresponds to a well-known validity condition in Byzantine agreement – the *All-Same-Validity*: If all correct nodes have the same input value, all correct nodes have to decide on this value. We use the weak Pareto condition to impose a validity rule on Byzantine agreement on rankings:

Pareto Validity for any pair of candidates c_i and c_j: if all correct nodes rank $c_i \succ c_j$, then the consensus ranking should rank $c_i \succ c_j$ as well.

Given m candidates, Pareto-Validity can be viewed as All-Same-Validity applied on each of the $\binom{m}{2}$ pairs of candidates in a ranking. Note that Byzantine agreement on a ranking is at least as hard as binary Byzantine agreement: Consider a case where the nodes agree on the ranking of the candidates $c_3, \ldots c_m$ which they rank last, but not on the two first candidates c_1 and c_2. Pareto-Validity is then satisfied for every binary relation which contains at least one of the candidates $c_3, \ldots c_m$. Agreement in this case is then reduced to binary Byzantine agreement on the two candidates c_1 and c_2, under the All-Same-Validity condition.

There is no straightforward way to apply a binary Byzantine agreement protocol to solve Byzantine agreement on rankings. This is because, in contrast to binary relations on two candidates, preference profiles can contain Condorcet cycles, e.g. tree contradicting binary relations $c_i \succ c_j$, $c_j \succ c_k$ and $c_k \succ c_i$ which are each preferred by a majority of nodes. Simply agreeing on each pair of candidates can thus lead to a circular decision which does not form a ranking. In order to get rid of such cycles one could think of applying the quicksort algorithm on the candidates sorted with respect to the majority. This procedure will however violate Pareto-Validity: Consider a candidate c_i that Pareto dominates candidate c_j. Assume that the quicksort algorithm compares both candidates to some third candidate c_k first. Then c_j might win against c_k and c_i might lose, thus swapping c_i and c_j in the consensus ranking. This consideration makes the problem of finding a consensus ranking in the presence of Byzantine nodes rather an instance of multi-valued agreement, as we discuss in Sect. 4, which makes the problem both interesting and challenging.

3   Related Work

Byzantine agreement was first proposed as the Byzantine Generals problem by Pease, Shostak and Lamport [26,32]. In these papers the authors showed that three nodes cannot establish agreement in the presence of one Byzantine node even if the communication system is synchronous. Given n nodes, it was shown for the synchronous model that at least $t + 1$ rounds are required to establish agreement [20], where $t < n/3$ is the number of Byzantine nodes in the system; the corresponding upper bound was provided in [8,9]. For the asynchronous model, the FLP impossibility result [21] states that there is no deterministic agreement protocol which can tolerate even one Byzantine node. The first randomized algorithm for solving Byzantine agreement proposed in [7] had expected exponential running time for a constant fraction of Byzantine nodes. Recently, the authors of [25] claimed that it is possible to establish agreement within expected polynomial running time using spectral methods.

Byzantine agreement with more than two input values has mostly been considered in approximate agreement [17,19], where the input values of the nodes converge towards some value over rounds. More recent results seek to establish agreement on a value that makes sense for applications. In [16], the values converge towards a value at most $\sqrt{n \log n}$ positions away from the median. In [28,35] an exact algorithm to establish agreement on a value that is at most $t/2$ positions away from the median or t positions away from a minimum or a maximum was proposed. In [29,30,38], Byzantine agreement was further generalized to several dimensions. There, the nodes converge to a vector inside the convex hull of all correct input vectors. In [13,37] the authors consider voting in Byzantine systems, they do however only focus on single winners that are determined by applying the plurality rule to the top alternatives of the rankings, a setting which corresponds to standard Byzantine agreement. All previous approaches for multiple dimensions struggle to derive an algorithm which either can tolerate a constant fraction of Byzantine nodes independent on the number of dimensions, or find a solution that is not trivial.

In social choice theory, Byzantine behavior can be interpreted as manipulation of a ballot in an election, in which the manipulating party has full knowledge about all votes. Bartholdi et al. [5] defined manipulation as a preference profile where one single voter can change its ranking such that this voter's most preferred candidate wins the election. Groups of voters have also been considered in this context, but mostly from the perspective of how hard it is for a group of nodes to manipulate the voting result given a certain voting rule [10,14]. Other types of Byzantine behavior have been considered with respect to robustness of proposed voting rules. In [6], the authors investigate robustness of Borda's mean and median in the presence of outlier ballots. In [33], robustness of scoring rules is considered under arbitrary noise which is described in terms of pairwise swaps of candidates in the ranking of one voter.

In this paper we will consider the Kemeny rule which was first proposed in [22,23]. The corresponding Kemeny median satisfies additional properties to those presented in Sect. 2, but it was shown to be NP-hard to compute for an increasing number of candidates and already for four voters in [4,18]. At least three different 2-approximation algorithms for the Kemeny median have been proposed in [1] and [15]. In [1], the approximation ratio was improved to 4/3 using randomization, and later derandomized in [40]. A good overview over the Kemeny rule and an extended introduction into social choice theory can be found in [12].

4   A Deterministic Algorithm for Pareto-Validity

This section focuses on Byzantine agreement protocols for rankings that satisfy Pareto-Validity. By using a similar idea to single transferable voting [36] and a multi-valued Byzantine agreement algorithm, a ranking satisfying Pareto-Validity can be obtained in $(m-1) \cdot (t+1)$ rounds: In the first $t+1$ rounds, we let the voters apply the King algorithm [9] in order to agree on the top candidate. After this, every node removes this candidate from its ranking. In the next step, they will agree on the top candidate from the reduced rankings, and so on. While this procedure is simple, the number of rounds depends not only on the number of nodes, but also on the number of candidates.

In the following we present a deterministic algorithm which solves this problem in only $t + 1$ phases using the same number of messages. We do this by modifying the King algorithm to broadcast rankings instead of single candidates. For convenience, we assume that a broadcast operation also includes sending a message to oneself. In the proposed algorithm, we select $t + 1$ different nodes and assign each of them to one of the $t + 1$ phases of the algorithm. Such a node is called the dictator of the corresponding phase. This dictator then suggests its own, possibly adjusted, ranking to all nodes, which will always be accepted if the dictator is a correct node. This way, dictators decide on the ranking of all pairs of candidates which do not satisfy the Pareto-Validity. Algorithm 1 presents this procedure in pseudocode.

Since we are dealing with rankings, it is not trivial to see that the nodes always will be able to agree on a proper ranking at the end of the algorithm. The following lemmas state that the nodes can adjust their rankings in Step 9 of Algorithm 1 in order to guarantee Pareto-Validity and that the outcome of the algorithm thus will be a proper ranking. It is easy to see that the algorithm is correct for $t < n/4$ Byzantine nodes, since the correct nodes will not be able to propose binary relations which form a Condorcet cycle in this case. In order to show that the algorithm can tolerate $t < n/3$ Byzantine nodes a well, we need to exploit the fact that no Byzantine node can propose relations that form a Condorcet cycle at any point of the algorithm.

Lemma 1. *There is no Condorcet cycle that can be proposed by the correct nodes if $t < n/3$.*

Note that by the properties of the King algorithm, no two opposite binary relations can be proposed in Step 4 simultaneously. Lemma 1 additionally shows that a Condorcet cycle cannot be proposed in Step 4 and that all proposed pairs can form a ranking. It remains to be proven that the nodes will always be able to adjust their rankings to incorporate the proposed pairs.

Lemma 2. *In Step 9 a correct node will always be able to incorporate the proposed pairs into its own ranking.*

Proof. This is constructed based on the following strategy: Divide the candidates into two sets. The first set contains all candidates which appear in at least one

---

Algorithm 1. Byzantine agreement protocol on rankings (for $t < n/3$)

Every node v executes the following algorithm

1: **for** phase 1 to $t + 1$ **do**

   Communication Round:

2:     Broadcast own input ranking r_v

3:     **for** all pairs of candidates c_i and c_j **do**

4:         **if** c_i is ranked above c_j in at least $n - t$ rankings **then**

5:             Broadcast "propose $c_i \succ c_j$"

6:         **end if**

7:     **end for**

8:     **if** some "propose $c_k \succ c_l$" received at least $t + 1$ times **then**

9:         Adjust own ranking r_v according to Lemma 2

10:     **end if**

11:     **if** some "propose $c_k \succ c_l$" received at least $n - t$ times **then**

12:         Fix the pair $c_k \succ c_l$

13:     **end if**

    Dictator Round:

14:     Let node w be the predefined dictator of the current phase

15:     The dictator broadcasts its ranking $r_{dictator} := r_w$

    Decision Round:

16:     **if** $r_{dictator}$ agrees with r_v in all fixed pairs $c_i \succ c_j$ from Step 12 **then**

17:         $r_v := r_{dictator}$

18:     **end if**

19: **end for**

20: Return r_v

---

of the pairs proposed by the $t + 1$ nodes in Step 9. This set of nodes will be ranked first. The second set will contain all candidates for which the node has not received any propose message. These candidates will be ranked second and will be dominated by all candidates from the first set. Next, we can rank all candidates in the first set according to the proposed relations, possibly leaving some pairs of the candidates not ranked. In the last step, all candidates which have not been ranked in each of the sets can be ranked by choosing binary relations from the local ranking of the node. This strategy outputs a ranking of candidates in which all proposed binary relations are satisfied.   □

The next lemma summarizes the correctness results of Algorithm 1 and states that the consensus ranking will be valid.

Lemma 3. *At the end of Algorithm 1 all nodes will have agreed on the same ranking which additionally satisfies Pareto-Validity.*

5   Kemeny Median with Byzantine Nodes

Weakly Paretian voting rules are often not sufficient to pick a fair ranking from a set of individual preference rankings. In search of the best possible consensus ranking we have to add restrictions on the voting rules without violating the known impossibility results of Arrow [2]. This leads us to majoritarian SWFs, one of which is the Kemeny rule. In the following we will introduce this rule and use it to derive a better consensus ranking in the presence of Byzantine nodes. Since Byzantine nodes have influence on the final ranking, the corresponding solutions can be qualified with respect to their approximation ratio which we define in Sect. 5.1. In Sect. 5.2, we will derive lower bounds on the approximation ratio of the Kemeny median in the presence of Byzantine nodes and further provide a matching upper bound in Sect. 5.3.

Definition 1 (Kendall's τ distance [24]). *The Kendall's τ distance measures the distance between two rankings r and p on candidates $c_1, \ldots, c_m$ by counting all pairs of candidates on which they disagree:*

$$\tau(r, p) \triangleq |\{(c_i, c_j) \mid c_i \succ_r c_j \text{ and } c_j \succ_p c_i\}|.$$

This metric τ on ballots can be extended to a distance function between a ranking r and a profile $\mathcal{P}$:

$$\tau(r, \mathcal{P}) \triangleq \sum_{p \in \mathcal{P}} \tau(r, p).$$

Definition 2 (Kemeny median). *For a given profile $\mathcal{P}$, the Kemeny median is the ranking r which minimizes $\tau(r, \mathcal{P})$.*

The Kemeny median satisfies many nice properties and to some extent guarantees that the chosen ranking is "fair". The most prominent quality is probably *monotonicity*: if voters increase a candidate's preference level, the ranking result either does not change or the promoted choice increases in overall popularity. This quality makes the median solution more robust to Byzantine behavior. The Kemeny rule is also a Condorcet method, it only depends on the number of voters who prefer one alternative over the other and is reinforcing.

Kendall's τ distance, which is used in the Kemeny rule, essentially captures the nature of multidimensionality in our consensus problem. Although it is not straightforward to properly define dimensions for metric spaces, there exist some widely used definitions such as the equilateral dimension. The equilateral dimension is described by the maximum number of points which lie at equal distance from each other. Using the equilateral dimension makes a lot of sense in many cases, it is for example not difficult to see that the equilateral dimension of a d-dimensional Euclidean space is $d + 1$. Here we also use the equilateral dimension in order to argue that by using the Kemeny rule we are actually solving a multi-dimensional consensus problem. For any m, we can construct rankings $r_i, i = 1, \ldots, \lfloor m/2 \rfloor$ at equal distance as follows: r_i ranks every candidate j as the j-th element in the ranking and only swaps the candidates $2i - 1$ and $2i$. Any pair of rankings in this construction has the same distance 2 to each other and the equilateral dimension of Kendall's τ metric space is therefore at least $\lfloor m/2 \rfloor$.

5.1   Byzantine Setting

The Kemeny median cannot be computed exactly in the presence of Byzantine nodes since they might suggest rankings which have a large distance to the Kemeny median of the correct nodes, thus moving the median ranking away from the actual median. A notion for approximate median rankings is therefore introduced as follows:

Definition 3 (α-approximation of Kemeny median). *Let κ be a Kemeny median of a preference profile $\mathcal{P}$. An α-approximation of κ is a preference ranking κ_α satisfying*

$$\tau(\kappa_\alpha, \mathcal{P}) \leq \alpha \cdot \tau(\kappa, \mathcal{P})$$

As an example consider binary agreement ($m = 2$): Here τ counts the number of correct nodes who disagree with the consensus value. Any binary Byzantine agreement algorithm that satisfies All-Same-Validity will also satisfy $\alpha < n-t-1$.

Unlike binary agreement, it is not straightforward to see what a Byzantine node would choose as its ranking when the Kemeny rule determines the consensus ranking. Since the input vectors of nodes are rankings, each voter has to propose a strict order between candidates and the corresponding preference relation is transitive. A possible strategy for the Byzantine nodes could then be to choose exactly the opposite ranking of the Kemeny median of all correct nodes. While this strategy can be shown to be optimal, such a solution is not unique for most preference profiles. To see this, assume that all correct nodes agree on the preference $c_i \succ c_j$ such that this pair will always belong to the Kemeny median of the correct rankings. Then, the Byzantine nodes can pick either $c_i \succ c_j$ or $c_j \succ c_i$ for their ranking, since this strategy does not have any influence on the Kemeny median of all rankings. It is therefore difficult for the correct nodes to detect which of the rankings might have been Byzantine.

5.2   Lower Bounds on the Approximation Ratio

In this section we discuss preference profiles that are vulnerable to Byzantine nodes. The first case is based on reducing the rankings to binary agreement and gives the highest approximation ratio for $t < n/3$. Binary agreement does however assume that there are two groups of voters who completely disagree in their preferences. This is somewhat unlikely in practical situations when m is sufficiently large. In the second case we therefore exclude such binary instances and provide a lower bound based on Condorcet cycles within a preference profile which converges to the same value for large m. The approximation ratio usually depends on the ratio n/t, which will be denoted k for the sake of simplicity.

For our analysis, we represent the preference profile $\mathcal{P}$ as a weighted *tournament graph*, i.e., a graph where the nodes represent the candidates and weighted edges represent how many voters prefer one candidate to the other. The sum of the forward and the backward edges should be equal to the total number of voters in the corresponding preference profile. The ranking of a node is a directed

Hamiltonian path following the order of the ranking, and all other edges are derived from the transitivity. For any two candidates we denote the edge between these candidates a *majority edge* if its backward edge has a smaller weight. The backward edge we then call a *minority edge*. A Kemeny median of a weighted tournament graph is the ranking that minimizes the sum of the weights of all backward edges of the graph. Note that rankings restrict the power of Byzantine nodes in the sense that Byzantine nodes only can send transitive tournament graphs where every edge has weight 1.

We first consider all possible preference profiles, in which the worst case is the binary case. This case corresponds to a class of tournament graphs where the Byzantine nodes can redirect all edges by adding t rankings to the preference profiles of the correct nodes. Theorem 1 gives a lower bound for the binary case.

Theorem 1. *There exists a tournament graph corresponding to a preference profile for which the Byzantine nodes may change the edge weights such that no deterministic algorithm can output a ranking which is better than a $\frac{k}{k-2}$-approximation of the Kemeny median of all correct nodes, where $k = n/t$. For t close to $n/3$, this gives a 3-approximation.*

Proof. This tournament graph is equivalent to binary agreement. Consider therefore one pair of candidates: t Byzantine nodes are only able to change the median, i.e., the majority edge, between these two candidates if they can swap the majority and minority edge by supporting the minority edge with their ranking. Assume the worst case, where the forward and the backward edge both have the same weight $n/2$ after the Byzantine nodes have added their preferences. In this worst case the tournament graph of correct nodes had the weight $n/2$ for the majority edge. Since the correct nodes will not be able to determine the actual majority edge, they might agree on a minority edge with weight $n/2 - t$ instead. The corresponding approximation ratio is then $\frac{n/2}{n/2-t} = \frac{k}{k-2}$. This result can be easily generalized to m candidates by using opposite rankings.

In the following, we present another lower bound using Condorcet cycles which can result in ambiguous views as well. We start with one directed cycle formed by three nodes on the tournament graph and assume that every majority edge has a weight of more than $(n+t)/2$, thus discarding the possibility to reduce any pair of forward and backward edges in the tournament graph to binary agreement. The main difficulty in finding a good example comes from the fact that not every tournament graph has an underlying preference profile.

Theorem 2. *There exists a preference profile containing directed majority cycles in the corresponding tournament graph, for which the Byzantine nodes can add t rankings such that no deterministic algorithm can output a ranking with a better approximation ratio to the actual median than $k/(k-2)$, for m large.*

Proof. Considering a tournament graph formed by one directed cycle of candidates c_1, c_2, c_3, i.e., a directed cycle formed by majority edges. Assume all

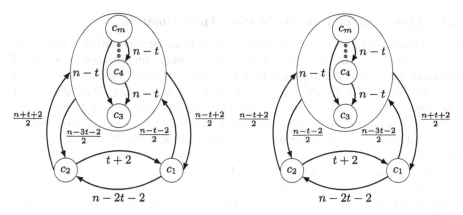

Fig. 1. Two indistinguishable views on m candidates for directed cycles. We have two views which show the profiles of correct nodes only. The left tournament graph results from a profile where $\frac{n-t}{2} - 1$ nodes choose $c_1 \succ c_2 \succ c_m \succ \ldots \succ c_3$, $\frac{n-3t}{2} - 1$ nodes choose $c_m \succ \ldots \succ c_3 \succ c_1 \succ c_2$ and $t + 2$ nodes choose $c_2 \succ c_m \succ \ldots \succ c_3 \succ c_1$. The right tournament graph results from $\frac{n-3t}{2} - 1$ nodes choosing $c_1 \succ c_2 \succ c_m \succ \ldots \succ c_3$, $\frac{n-t}{2} - 1$ nodes choosing $c_m \succ \ldots \succ c_3 \succ c_1 \succ c_2$ and $t + 2$ nodes choosing $c_2 \succ c_m \succ \ldots \succ c_3 \succ c_1$. If the Byzantine nodes add t profiles $c_m \succ \ldots \succ c_3 \succ c_2 \succ c_1$ to the left view, and t profiles $c_2 \succ c_1 \succ c_m \succ \ldots \succ c_3$ to the right view, the resulting profiles become indistinguishable to the correct nodes.

correct nodes receive a view where $n - 2t - 2$ nodes prefer c_1 to c_2, where (c_1, c_2) is a majority edge. Then $(n + t)/2 + 1$ nodes prefer c_2 to c_3 and $(n + t)/2 + 1$ nodes prefer c_3 to c_1. For $n > 3t + 4$, the edge (c_1, c_2) is in the median ranking of all nodes. Since the edges (c_2, c_3) and (c_3, c_1) cannot both be in the median ranking, the nodes have to decide for one of the rankings. In the worst case, one of these two edges was supported by all t Byzantine nodes while the other edge was not supported by any Byzantine node. This leads to two views which are not distinguishable for the correct nodes, as shown in Fig. 1. The approximation ratio for these views is

$$\frac{n + t + 2}{n - t + 2} \approx \frac{k + 1}{k - 1} < \frac{5}{3}$$

An extension to m candidates gives an approximation ratio of

$$\frac{m \cdot n + 2n + t + 2}{m \cdot (n - 2t) + 2n - 3t + 2} \approx \frac{k}{k - 2}$$

for large m. □

The received approximation ratio converges to the same approximation ratio as in the binary case for large m, even though we have excluded the binary case from the tournaments. This lower bound underlines the fact that Byzantine agreement on rankings is more complex than binary Byzantine agreement.

5.3   Algorithm for Kemeny Median Approximation

In this section we present a synchronous algorithm for computing a consensus median which matches the lower bound on the approximation ratio presented in the previous section. A simple idea is to use interactive consistency [11,34]: For $t + 1$ rounds, the nodes exchange all information they have received this far and after the $(t + 1)$-st round they compute the Kemeny median from a set of rankings which they have received often enough. This algorithm guarantees that the set of rankings will be the same for each node and therefore that all nodes will decide on the same ranking. The main drawback of interactive consistency is that it has a large message complexity. The message complexity of this strategy is in $\Theta(mn^t)$ which is exponential for $t \in \Theta(n)$. Also other approaches, such as agreeing on each ranking upfront require the nodes to reliably broadcast their rankings at least once, which results in a message complexity of at least $O(n^3)$ (each node has to forward every received ranking to all other nodes).

Instead of exchanging large amounts of information, we present an approach where we can directly exploit the fact that the Byzantine nodes cannot change a Kemeny median of the preference profile of the correct nodes by more than a transitive tournament graph with edge weights t. This strategy is presented in Algorithm 2.

Algorithm 2. Byzantine agreement for the Kemeny median (for $t < n/3$)

Every node v executes the following algorithm
1: broadcast own ranking r_v
2: compute the Kemeny median of the received preference profile, call it m_v
3: apply Algorithm 1 with m_v as an input value

Algorithm 2 has the same order of round and message complexity as Algorithm 1 as stated in the next theorem.

Theorem 3. *Algorithm 2 terminates within $t + 3$ phases exchanging $O(tn^2m \log m)$ messages. The computed consensus ranking satisfies the lower bounds from Sect. 5.2 and Pareto-Validity.*

6   Discussion and Future Work

In this paper we introduced a new Byzantine agreement problem which extends binary Byzantine agreement to rankings. We showed that rules for choosing a consensus ranking in voting theory fit well with requirements from Byzantine agreement. We further considered a special voting rule, the Kemeny median, for which we provided an optimal Byzantine agreement protocol that can tolerate up to $t < n/3$ Byzantine nodes. We do not claim to have chosen the best voting rule at this point, since such a rule simply does not exist due to impossibility results in voting theory. Instead, we think of our results as an inspiration to consider a larger pool of voting rules, such as approval voting, the Godgson's rule, and many others.

References

1. Ailon, N., Charikar, M., Newman, A.: Aggregating inconsistent information: ranking and clustering. J. ACM **55**(5), 23:1–23:27 (2008)
2. Arrow, K.J.: Social Choice and Individual Values, 1st edn. Cowles Foundation, New Haven (1951)
3. Arrow, K.J.: Social Choice and Individual Values, 2nd edn. Wiley, New York (1963)
4. Bartholdi, J., Tovey, C.A., Trick, M.A.: Voting schemes for which it can be difficult to tell who won the election. Soc. Choice Welfare **6**(2), 157–165 (1989)
5. Bartholdi, J.J., Tovey, C.A., Trick, M.A.: The computational difficulty of manipulating an election. Soc. Choice Welfare **6**(3), 227–241 (1989)
6. Bassett, G.W., Persky, J.: Robust voting. Public Choice **990**(3), 299–310 (1999)
7. Ben-Or, M.: Another advantage of free choice (extended abstract): completely asynchronous agreement protocols. In: Proceedings of the Second Annual ACM Symposium on Principles of Distributed Computing, PODC 1983, pp. 27–30 (1983)
8. Berman, P., Garay, J.A.: Asymptotically optimal distributed consensus. In: Ausiello, G., Dezani-Ciancaglini, M., Della Rocca, S.R. (eds.) ICALP 1989. LNCS, vol. 372, pp. 80–94. Springer, Heidelberg (1989). https://doi.org/10.1007/BFb0035753
9. Berman, P., Garay, J.A., Perry, K.J.: Towards optimal distributed consensus. In: 30th Annual Symposium on Foundations of Computer Science, FOCS, October 1989
10. Betzler, N., Niedermeier, R., Woeginger, G.J.: Unweighted coalitional manipulation under the Borda rule is NP-hard. In: IJCAI, vol. 11, pp. 55–60 (2011)
11. Bracha, G.: Asynchronous Byzantine agreement protocols. Inf. Comput. **75**(2), 130–143 (1987)
12. Brandt, F., Conitzer, V., Endriss, U., Lang, J., Procaccia, A.D.: Handbook of Computational Social Choice, 1st edn. Cambridge University Press, New York (2016)
13. Chauhan, H., Garg, V.K.: Democratic elections in faulty distributed systems. In: Frey, D., Raynal, M., Sarkar, S., Shyamasundar, R.K., Sinha, P. (eds.) ICDCN 2013. LNCS, vol. 7730, pp. 176–191. Springer, Heidelberg (2013). https://doi.org/10.1007/978-3-642-35668-1_13
14. Davies, J., Katsirelos, G., Narodytska, N., Walsh, T.: Complexity of and algorithms for Borda manipulation. In: AAAI, vol. 11, pp. 657–662 (2011)
15. Diaconis, P., Graham, R.L.: Spearman's footrule as a measure of disarray. J. R. Stat. Soc. Ser. B (Methodol.) **39**, 262–268 (1977)
16. Doerr, B., Goldberg, L.A., Minder, L., Sauerwald, T., Scheideler, C.: Stabilizing consensus with the power of two choices. In: Proceedings of the Twenty-third Annual ACM Symposium on Parallelism in Algorithms and Architectures, SPAA (2011)
17. Dolev, D., Lynch, N.A., Pinter, S.S., Stark, E.W., Weihl, W.E.: Reaching approximate agreement in the presence of faults. J. ACM **33**(3), 499–516 (1986)
18. Dwork, C., Kumar, R., Naor, M., Sivakumar, D.: Rank aggregation methods for the web. In: Proceedings of the 10th International Conference on World Wide Web, WWW 2001, pp. 613–622. ACM, New York (2001)
19. Fekete, A.D.: Asymptotically optimal algorithms for approximate agreement. Distrib. Comput. **4**(1), 9–29 (1990)
20. Fischer, M.J., Lynch, N.A.: A lower bound for the time to assure interactive consistency. Inf. Process. Lett. **14**(4), 183–186 (1982)

21. Fischer, M.J., Lynch, N.A., Paterson, M.S.: Impossibility of distributed consensus with one faulty process. J. ACM **32**(2), 374–382 (1985)
22. Kemeny, J.G.: Mathematics without numbers. Daedalus **88**(4), 577–591 (1959)
23. Kemeny, J.G., Snell, J.L.: Mathematical Models in the Social Sciences. Introductions to Higher Mathematics, Blaisdell, Waltham (Mass.) (1962)
24. Kendall, M.G.: A new measure of rank correlation. Biometrika **30**(1/2), 81–93 (1938)
25. King, V., Saia, J.: Byzantine agreement in expected polynomial time. J. ACM **63**(2), 13:1–13:21 (2016)
26. Lamport, L., Shostak, R., Pease, M.: The Byzantine generals problem. ACM Trans. Program. Lang. Syst. **4**(3), 382–401 (1982)
27. May, K.O.: A set of independent necessary and sufficient conditions for simple majority decision. Econometrica **20**(4), 680–684 (1952)
28. Melnyk, D., Wattenhofer, R.: Byzantine agreement with interval validity. In: 37th Annual IEEE International Symposium on Reliable Distributed Systems, SRDS (2018)
29. Mendes, H., Herlihy, M.: Multidimensional approximate agreement in Byzantine asynchronous systems. In: Proceedings of the Forty-fifth Annual ACM Symposium on Theory of Computing, STOC (2013)
30. Mendes, H., Herlihy, M., Vaidya, N., Garg, V.K.: Multidimensional agreement in Byzantine systems. Distrib. Comput. **28**(6), 423–441 (2015)
31. Pareto, V.: Manuale di Economia Politica con una Introduzione alla Scienza Sociale. Società Editrice Libraria (1919)
32. Pease, M., Shostak, R., Lamport, L.: Reaching agreement in the presence of faults. J. ACM **27**(2), 228–234 (1980)
33. Procaccia, A.D., Rosenschein, J.S., Kaminka, G.A.: On the robustness of preference aggregation in noisy environments. In: Proceedings of the 6th International Joint Conference on Autonomous Agents and Multiagent Systems, p. 66. ACM (2007)
34. Srikanth, T., Toueg, S.: Simulating authenticated broadcasts to derive simple fault-tolerant algorithms. Distrib. Comput. **2**(2), 80–94 (1987)
35. Stolz, D., Wattenhofer, R.: Byzantine agreement with median validity. In: 19th International Conference on Principles of Distributed Systems, OPODIS (2015)
36. Tideman, N.: The single transferable vote. J. Econ. Perspect. **9**(1), 27–38 (1995)
37. Tseng, L.: Voting in the presence of Byzantine faults. In: 2017 IEEE 22nd Pacific Rim International Symposium on Dependable Computing (PRDC), January 2017
38. Vaidya, N.H., Garg, V.K.: Byzantine vector consensus in complete graphs. In: Proceedings of the 2013 ACM Symposium on Principles of Distributed Computing, PODC (2013)
39. Wattenhofer, R.: Distributed Ledger Technology: The Science of the Blockchain, 2nd edn. CreateSpace Independent Publishing Platform, Scotts Valley (2017)
40. van Zuylen, A., Williamson, D.P.: Deterministic algorithms for rank aggregation and other ranking and clustering problems. In: Kaklamanis, C., Skutella, M. (eds.) WAOA 2007. LNCS, vol. 4927, pp. 260–273. Springer, Heidelberg (2008). https://doi.org/10.1007/978-3-540-77918-6_21

Robust and Approximately Stable Marriages Under Partial Information

Vijay Menon[✉] and Kate Larson

David R. Cheriton School of Computer Science, University of Waterloo, Waterloo, Canada
{vijay.menon,kate.larson}@uwaterloo.ca

Abstract. We study the stable marriage problem in the partial information setting where the agents, although they have an underlying true strict linear order, are allowed to specify partial orders either because their true orders are unknown to them or they are unwilling to completely disclose the same. Specifically, we focus on the case where the agents are allowed to submit strict weak orders and we try to address the following questions from the perspective of a market-designer: *(i)* How can a designer generate matchings that are robust—in the sense that they are "good" with respect to the underlying unknown true orders? *(ii)* What is the trade-off between the amount of missing information and the "quality" of solution one can get? With the goal of resolving these questions through a simple and prior-free approach, we suggest looking at matchings that minimize the maximum number of blocking pairs with respect to all the possible underlying true orders as a measure of "goodness" or "quality", and subsequently provide results on finding such matchings. In particular, we first restrict our attention to matchings that have to be stable with respect to at least one of the completions (i.e., weakly-stable matchings) and show that in this case arbitrarily filling-in the missing information and computing the resulting stable matching can give a non-trivial approximation factor for our problem in certain cases. We complement this result by showing that, even under severe restrictions on the preferences of the agents, the factor obtained is asymptotically tight in many cases. We then investigate a special case, where only agents on one side provide strict weak orders and all the missing information is at the bottom of their preference orders, and show that in this special case the negative result mentioned above can be circumvented in order to get a much better approximation factor; this result, too, is tight in many cases. Finally, we move away from the restriction on weakly-stable matchings and show a general hardness of approximation result and also discuss one possible approach that can lead us to a near-tight approximation bound.

1 Introduction

Two-sided matching markets have numerous applications, e.g., in matching students to dormitories (i.e., Stable Roommates problem (SR) [11]), residents to

© Springer Nature Switzerland AG 2018
G. Christodoulou and T. Harks (Eds.): WINE 2018, LNCS 11316, pp. 341–355, 2018.
https://doi.org/10.1007/978-3-030-04612-5_23

hospitals (i.e., Hospital-Resident problem (HR) [18]) etc., and hence are ubiquitous in practice. Perhaps unsurprisingly, then, this line of research has received much attention, with plenty of work done on investigating numerous problems like SR and HR, and their many variations (we refer the reader to the excellent books by Gusfield and Irving [9] and Manlove [17] for a survey on two-sided matching problems). The focus of this paper, too, is on one such problem—one that is perhaps the most widely-studied, but yet the simplest—called the *Stable Marriage problem* (SM), first introduced by Gale and Shapley [8]. In SM we are given two disjoint sets (colloquially referred to as the set of men and women) and each agent in one set specifies a strict linear order over the agents in the other set, and the aim is to find a *stable matching*, i.e., a matching where there is no man-woman pair such that each of them prefers the other over their partner in the matching. (Such a pair, if it exists, is called a *blocking pair*).

While the assumption that the agents will be able to specify strict linear orders is not unreasonable in small markets, in general, as the markets get larger, it may not be feasible for an agent to determine a complete ordering over all the alternatives. Furthermore, there may arise situations where agents are simply unwilling to provide strict total orders due to, say, privacy concerns. Thus, it is natural for a designer to allow agents the flexibility to specify partial orders, and so in this paper we assume that the agents submit strict weak orders[1] (i.e., strict partial orders where incomparability is transitive) that are consistent with their underlying true strict linear orders. Although the issue of partially specified preferences has received attention previously, we argue that certain aspects have not been addressed sufficiently. In particular, the common approach to the question of what constitutes a "good" matching in such a setting has been to either work with stable matchings that arise as a result of an arbitrary linear extension of the submitted partial orders (these are known as *weakly-stable matchings*) or to look at something known as *super-stable matchings*, which are matchings that are stable with respect to all the possible linear extensions of the submitted partial orders [12,19]. In the case of the former, one key issue is that we often do not really know how "good" a particular weakly-stable matching is with the respect to the underlying true orders of the agents, and in the case of the latter they often do not exist. Furthermore, we believe that it is in the interest of the market-designer to understand how robust or "good" a matching is with respect to the underlying true orders of the agents, for, if otherwise, issues relating to instability and market unravelling can arise since the matching that is output by a mechanism can be arbitrarily bad with respect to these true orders. Hence, in this paper we propose to move away from the extremes of working with either arbitrary weakly-stable matchings or super-stable matchings, and to find a middle-ground when it comes to working with partial preference information. To this end, we aim to answer two questions from the perspective of a market-designer: *(i)* How should one handle partial information so as to be able to

---

[1] All our negative results naturally hold for the case when the agents are allowed to specify strict partial orders. As for our positive results, most of them can be extended for general partial orders, although the resulting bounds will be worse.

provide some guarantees with respect to the underlying true preference orders? *(ii)* What is the trade-off between the amount of missing information and the quality of a matching that one can achieve? We discuss our proposal in more detail in the following sections.

1.1 How Does One Work with Partial Information?

When agents do not submit full preference orderings, there are several possible ways to cope with the missing information. For instance, one approach that immediately comes to mind is to assume that there exists some underlying distribution from which the agents' true preferences are drawn, and then use this information to find a "good" matching—which is, say, the one with the least number of blocking pairs in expectation. However, the success of such an approach crucially depends on having access to information about the underlying preference distributions which may not always be available. Therefore, in this paper we make no assumptions on the underlying preference distributions and instead adopt a prior-free and absolute-worst-case approach where we assume that any of the linear extensions of the given strict partial orders can be the underlying true order, and we aim to provide solutions that *perform well* with respect to all of them. We note that similar worst-case approaches have been looked at previously, for instance, by Chiesa et al. [5] in the context of auctions.

The objective we concern ourselves with here is that of minimizing the number of blocking pairs, which is well-defined and has been considered previously in the context of matching problems (for instance, see [1,4]). In particular, for a given instance $\mathcal{I}$ our aim is to return a matching $\mathcal{M}_{opt}$ that has the best worst case—i.e., a matching that has the minimum maximum 'regret' after one realises the true underlying preference orders. (We refer to $\mathcal{M}_{opt}$ as the minimax optimal solution.) More precisely, let $\mathcal{I} = (p_U, p_W)$ denote an instance, where $p_U = \{p_{u_1}, \cdots, p_{u_n}\}$, $p_W = \{p_{w_1}, \cdots, p_{w_n}\}$, $U = \{u_i\}_{i \in \{1,2,\cdots,n\}}$ and $W = \{w_i\}_{i \in \{1,2,\cdots,n\}}$ are the set of men and women respectively, and p_i is the strict partial order submitted by agent i. Additionally, let $C(p_i)$ denote the set of linear extensions of p_i, C be the Cartesian product of the $C(p_i)$s, i.e., $C = \times_{i \in U \cup W} C(p_i)$, $bp(\mathcal{M}, c)$ denote the set of blocking pairs that are associated with the matching $\mathcal{M}$ according to some linear extension $c \in C$, and S denote the set of all possible matchings. Then the matching $\mathcal{M}_{opt}$ that we are interested in is defined as $\mathcal{M}_{opt} = \arg\min_{\mathcal{M} \in S} \max_{c \in C} |bp(\mathcal{M}, c)|$.

While we are aware of just one work by Drummond and Boutilier [6] who consider the minimax regret approach in the context of stable matchings (they consider it mainly in the context of preference elicitation; see Sect. 1.4 for more details), the approach, in general, is perhaps reminiscent, for instance, of the works of Hyafil and Boutilier [10] and Lu and Boutilier [15] who looked at the minimax regret solution criterion in the context of mechanism design for games with type uncertainty and preference elicitation in voting protocols, respectively.

Remark: In the usual definition of a minimax regret solution, there is a second term which measures the 'regret' as a result of choosing a particular solution. That is, in the definition above, it would usually be $\mathcal{M}_{opt} =$

arg min$_{\mathcal{M} \in S}$ max$_{c \in C}$ $|bp(\mathcal{M}, c)| - |bp(\mathcal{M}_c, c)|$, where $\mathcal{M}_c$ is the optimal matching (with respect to the objective function $|bp()|$) for the linear extension c. We do not include this in the definition above because $|bp(\mathcal{M}_c, c)| = 0$ as every instance of the marriage problem with linear orders has a stable solution (which by definition has zero blocking pairs). Additionally, the literature on stable matchings uses the term "regret" to denote the maximum cost associated with a stable matching, where the cost of a matching for an agent is the rank of its partner in the matching and the maximum is taken over all the agents (for instance, see [16]). However, here the term regret is used in the context of the minimax regret solution criterion.

1.2   How Does One Measure the Amount of Missing Information?

For the purposes of understanding the trade-off between the amount of missing information and the "quality" of solution one can achieve, we need a way to measure the amount of missing information in a given instance. There are many possible ways to do this, however in this paper we adopt the following. For a given instance $\mathcal{I}$, the amount of missing information, δ, is the fraction of pairwise comparisons one cannot infer from the given strict partial orders. That is, we know that if every agent submits a strict linear order over n alternatives, then we can infer $\binom{n}{2}$ comparisons from it. Now, instead, if an agent i submits a strict partial order p_i, then we denote by δ_i the fraction of these $\binom{n}{2}$ comparisons one cannot infer from p_i (this is the "missing information" in p_i). Our δ here is equal to $\frac{1}{2n} \sum_{i \in U \cup W} \delta_i$. Although, given a strict partial order p_i, it is straightforward to calculate δ_i, we will nevertheless assume throughout that δ is part of the input. Hence, our definition of an instance will be modified the following way to include the parameter for missing information: $\mathcal{I} = (\delta, p_U, p_W)$.

Remark: $\delta = 0$ denotes the case when all the preferences are strict linear orders. Also, for an instance with n agents on each side, the least value of δ when the amount of missing information is non-zero is $\frac{1}{2n} \frac{1}{\binom{n}{2}}$ (this happens in the case where there is only one agent with just one pairwise comparison missing). However, despite this, in the interest of readability, we sometimes just write statements of the form "for all $\delta > 0$". Such statements need to be understood as being true for only realizable or valid values of δ that are greater than zero.

1.3   Our Contributions

The focus of our work is on computing the minimax optimal matching, i.e., a matching that, when given an instance $\mathcal{I}$, minimizes the maximum number of blocking pairs with respect to all the possible linear extensions (see Sect. 2.1 for a formal definition of the problem). Towards this end, we make the following contributions:

– We formally define the problem and show that, interestingly, the problem under consideration is equivalent to the problem of finding a matching that has the minimum number of *super-blocking pairs* (i.e., man-woman pairs where each of them weakly-prefers the other over their current partners).

- While an optimal answer to our question might involve matchings that have man-woman pairs such that each of them strictly prefers the other over their partners, we start by focusing our investigation on matchings that do not have such pairs. Given the fact that any matching with no such pairs are weakly-stable, through this setting we address the question "given an instance, can we find a weakly-stable matching that performs well, in terms of minimizing the number of blocking pairs, with respect to all the linear extensions of the given strict partial orders?" We show that by arbitrarily filling-in the missing information and computing the resulting stable matching, one can obtain a non-trivial approximation factor (i.e., one that is $o(n^2)$) for our problem for many values of δ. We complement this result by showing that, even under severe restrictions on the preferences of the agents, the factor obtained is asymptotically tight in many cases.

- By assuming a special structure on the agents' preferences—one where strict weak orders are specified by just agents on one side and all the missing information is at the bottom of their preference orders—we show that one can obtain a $\mathcal{O}(n)$-approximation algorithm for our problem. The proof of the same is via finding a 2-approximation for another problem (see Problem 3) that might be of independent interest.

- In Sect. 4 we remove the restriction to weakly-stable matchings and show a general hardness of approximation result for our problem. Following this, we discuss one possible approach that can lead to a near-tight approximation guarantee for the same.

1.4   Related Work

There has recently been a number of papers that have looked at problems relating to missing preference information or uncertainty in preferences in the context of matching.

Drummond and Boutilier [6] used the minimax regret solution criterion in order to drive preference elicitation strategies for matching problems. While they discussed computing robust matchings subject to a minimax regret solution criteria, their focus was on providing an NP-completeness result and heuristic preference elicitation strategies for refining the missing information. In contrast, in addition to focusing on understanding the exact trade-offs between the amount of missing information and the solution "quality", we concern ourselves with arriving at approximation algorithms for computing such robust matchings.

Rastegari et al. [19] studied a partial information setting in labour markets. However, again, the focus of this paper was different than ours. They looked at pervasive-employer-optimal matchings, which are matchings that are employer-optimal (see [19] for the definitions) with respect to all the underlying linear extensions. In addition, they also discussed how to identify, in polynomial time, if a matching is employer-optimal with respect to some linear extension.

Recent work by Aziz et al. [2] looked at the stable matching problem in settings where there is uncertainty about the preferences of the agents. They considered three different models of uncertainty and primarily studied the complexity of computing the stability probability of a given matching and the question of finding a matching that will have the highest probability of being stable. In contrast to their work, in this paper we do not make any underlying distributional assumptions about the preferences of the agents and instead take an absolute worst-case approach, which in turn implies that our results hold irrespective of the underlying distribution on the completions.

Finally, we also briefly mention another line of research which deals with partial information settings and goes by the name of *interview minimization* (see, for instance, [7,20]). One of the main goals in this line of work is to come with a matching that is stable (and possibly satisfying some other desirable property) by conducting as few 'interviews' (which in turn helps the agents in refining their preferences) as possible. We view this work as an interesting, orthogonal, direction from the one we pursue in this paper.

2   Preliminaries

Let U and W be two disjoint sets. The sets U and W are colloquially referred to as the set of men and women, respectively, and $|U| = |W| = n$. We assume that each agent in U and W has a true strict linear order (i.e., a ranking without ties) over the agents in the other set, but this strict linear order may be unknown to the agents or they may be unwilling to completely disclose the same. Hence, each agent in U and W specifies a strict partial order over the agents in the other set (which we refer to as their *preference order*) that is consistent with their underlying true orders, and p_U and p_W, respectively, denote the collective preference orders of all the men and women. For a strict partial order p_i associated with agent i, we denote the set of linear extensions associated with p_i by $C(p_i)$ and denote by C the Cartesian product of the $C(p_i)$s, i.e., $C = \times_{i \in U \cup W} C(p_i)$. We refer to the set C as "the set of all completions" where the term *completion* refers to an element in C. Also, throughout, we denote strict preferences by $\succ$ and use $\succeq$ to denote the relation 'weakly-prefers'. So, for instance, we say that an agent c strictly prefers a to b and denote this by $a \succ_c b$ and use $a \succeq_c b$ to denote that either c strictly prefers a to b or finds them incomparable. As mentioned previously, we restrict our attention to the case when the strict partial orders submitted by the agents are strict weak orders over the set of agents in the other set.

Remark: Strict weak orders are defined to be strict partial orders where incomparability is transitive. Hence, although the term *tie* is used to mean indifference, it is convenient to think of strict weak orders as rankings with ties. Therefore, throughout this paper, whenever we say that agent c finds a and b to be tied, we mean that c finds a and b to be incomparable. Additionally, we will use the terms ties and incomparabilities interchangeably.

An instance $\mathcal{I}$ of the stable marriage problem (SM) is defined as $\mathcal{I} = (\delta, p_U, p_w)$, where δ denotes the amount of missing information in that instance and this in turn, as defined in Sect. 1.2, is the average number of pairwise comparisons that are missing from the instance, and p_U and p_W are as defined above. Given an instance $\mathcal{I}$, the aim is usually to come up with a matching $\mathcal{M}$—which in turn is a set of disjoint pairs (m, w), where $m \in U$ and $w \in W$—that is stable. There are different notions of stability that have been proposed and below we define two of them that are relevant to our paper: *(i)* weak-stability and *(ii)* super-stability. However, before we look at their definitions we introduce the following terminology that will be used throughout this paper. (Note that in the definitions below we implicitly assume that in any matching $\mathcal{M}$ all the agents are matched. This is so because of the standard assumption that is made in the literature on SM (i.e., the stable marriage problem where every agent has a strict linear order over all the agents in the other set) that an agent always prefers to be matched to some agent than to remain unmatched).

Definition 1 (blocking pair/obvious blocking pair). *Given an instance $\mathcal{I}$ and a matching $\mathcal{M}$ associated with $\mathcal{I}$, (m, w) is said to be a blocking pair associated with $\mathcal{M}$ if $w \succ_m \mathcal{M}(m)$ and $m \succ_w \mathcal{M}(w)$. The term blocking pair is usually used in situations where the preferences of the agents are strict linear orders, so in cases where the preferences of the agents have missing information, we refer to such a pair as an obvious blocking pair.*

Definition 2 (super-blocking pair). *Given an instance $\mathcal{I}$ where the agents submit partial preference orders and a matching $\mathcal{M}$ associated with $\mathcal{I}$, we say that (m, w) is a super-blocking pair associated with $\mathcal{M}$ if $w \succeq_m \mathcal{M}(m)$ and $m \succeq_w \mathcal{M}(w)$.*

Given the definitions above we can now define weak-stability and super-stability.

Definition 3 (weakly-stable matching). *Given an instance $\mathcal{I}$ and matching $\mathcal{M}$ associated with $\mathcal{I}$, $\mathcal{M}$ is so said to be weakly-stable with respect to $\mathcal{I}$ if it does not have any obvious blocking pairs. When the preferences of the agents are strict linear orders, such a matching is just referred to as a stable matching.*

Definition 4 (super-stable matching). *Given an instance $\mathcal{I}$ and matching $\mathcal{M}$ associated with $\mathcal{I}$, $\mathcal{M}$ is so said to be super-stable with respect to $\mathcal{I}$ if it does not have any super-blocking pairs.*

2.1   What Problems Do We Consider?

As mentioned in the introduction, we are interested in finding the minimax optimal matching where the objective is to minimize the number of blocking pairs, i.e., to find, from the set S of all possible matchings, a matching that has the minimum maximum number of blocking pairs with respect to all the completions. This is formally defined below.

Problem 1 (δ-minimax-matching). Given a $\delta \in [0,1]$ and an instance $\mathcal{I} = (\delta',$ $p_U, p_W)$, where $\delta' \leq \delta$ is the amount of missing information and p_U, p_W are the preferences submitted by men and women respectively, compute $\mathcal{M}_{opt}$ where $\mathcal{M}_{opt} = \arg\min_{\mathcal{M} \in S} \max_{c \in C} |bp(\mathcal{M}, c)|$.

Although the problem defined above is our main focus, for the rest of this paper we will be talking in terms of the following problem which concerns itself with finding an approximately super-stable matching (i.e., a super-stable matching with the minimum number of super-blocking pairs). As we will see below, the reason we do the same is because both the problems are equivalent.

Problem 2 (δ-min-bp-super-stable-matching). Given a $\delta \in [0,1]$ and an instance $\mathcal{I} = (\delta', p_U, p_W)$, where $\delta' \leq \delta$ is the amount of missing information and p_U, p_W are the preferences submitted by men and women respectively, compute $\mathcal{M}_{opt}^{SS}$ where $\mathcal{M}_{opt}^{SS} = \arg\min_{\mathcal{M} \in S} |\text{super-bp}(\mathcal{M})|$ and super-bp($\mathcal{M}$) is the set of super-blocking pairs associated with $\mathcal{M}$ for the instance $\mathcal{I}$.

Below we show that both the problems described above are equivalent. However, before that we prove the following lemma. Unfortunately, due to space constraints, all the proofs are omitted. We refer the reader to the full version of the paper[2] for all the proofs and for more detailed explanations.

Lemma 1. *Let $\mathcal{M}$ be a matching associated with some instance $\mathcal{I} = (\delta, p_U, p_W)$, α denote the maximum number of blocking pairs associated with $\mathcal{M}$ for any completion of $\mathcal{I}$, and β denote the number of super-blocking pairs associated with $\mathcal{M}$ for the instance $\mathcal{I}$. Then, $\alpha = \beta$.*

Given the lemma above, we can now show the following theorem.

Theorem 1. *For any $\delta \in [0,1]$, the δ-minimax-matching and δ-min-bp-super-stable-matching problems are equivalent.*

For the rest of this paper, we assume that we are always dealing with instances which do not have a super-stable matching as this can be checked in polynomial-time [12, Theorem 3.4]. So, now, in the context of the δ-min-bp-super-stable-matching problem, it is easy to show that if the number of super-blocking pairs k in the optimal solution is a constant, then we can solve it in polynomial-time. We state this in the theorem below. Later, in Sect. 4, we will see that the problem is NP-hard, even to approximate.

Theorem 2. *An exact solution to the δ-min-bp-super-stable-matching problem can be computed in $\mathcal{O}(n^{2(k+1)})$ time, where k is the number of super-blocking pairs in the optimal solution.*

3   Investigating Weakly-Stable Matchings

In this section we focus on situations where obvious blocking pairs are not permitted in the final matching. In particular, we explore the space of weakly-stable

---

[2] https://arxiv.org/abs/1804.09156.

matchings and ask whether it is possible to find weakly-stable matchings that also provide good approximations to the δ-min-bp-super-stable-matching problem (and thus the δ-minimax-matching problem).

3.1 Approximating δ-min-bp-super-stable-matching with Weakly-Stable Matchings

It has previously been established that a matching is weakly-stable if and only if it is stable with respect to at least one completion [16, Sect. 1.2]. Therefore, given this result, one immediate question that arises in the context of approximating the δ-min-bp-super-stable-matching problem is "what if we just fill in the missing information arbitrarily and then compute a stable matching associated with such a completion?" This is the question we consider here, and we show that weakly-stable matchings do give a non-trivial (i.e., one that is $o(n^2)$, as any matching has only $\mathcal{O}(n^2)$ super-blocking pairs) approximation bound for our problem for certain values of δ. The proof of the following theorem is through a simple application of the Cauchy-Schwarz inequality and due to space constraints is omitted.

Theorem 3. *For any $\delta > 0$ and an instance $\mathcal{I} = (\delta', p_U, p_W)$ where $\delta' \leq \delta$, any weakly-stable matching with respect to $\mathcal{I}$ gives an $\mathcal{O}\left(\min\left\{n^3\delta, n^2\sqrt{\delta}\right\}\right)$-approximation for the δ-min-bp-super-stable-matching problem.*

3.2 Can We Do Better When Restricted to Weakly-Stable Matchings?

While Theorem 3 established an approximation factor for the δ-min-bp-super-stable-matching problem when considering only weakly-stable matchings, it was simply based on arbitrarily filling-in the missing information. Therefore, there remains the question as to whether one can be clever about handling the missing information and as a result obtain improved approximation bounds. In this section we consider this question and show that for many values of δ the approximation factor obtained in Theorem 3 is asymptotically the best one can achieve when restricted to weakly-stable matchings.

Theorem 4. *For any $\delta \in [\frac{16}{n^2}, \frac{1}{4}]$, if there exists an α-approximation algorithm for δ-min-bp-super-stable-matching that always returns a matching that is weakly-stable, then $\alpha \in \Omega\left(n^2\sqrt{\delta}\right)$. Moreover, this result is true even if we allow only one side to specify ties and also insist that all the ties need to be at the top of the preference order.*

3.3 The Case of One-Sided Top-Truncated Preferences: An $\mathcal{O}(n)$ Approximation Algorithm for δ-min-bp-super-stable-matching

Although Theorem 4 is an inherently negative result, in this section we consider an interesting restriction on the preferences of the agents and show how this negative result can be circumvented. In particular, we consider the case where only agents on one side are allowed to specify ties and all the ties need to be at the bottom. Such a restriction has been looked at previously in the context of matching problems and as noted by Irving and Manlove [13] is one that appears in practise in the Scottish Foundation Allocation Scheme (SFAS). Additionally, restricting ties to only at the bottom models a very well-studied class of preferences known as top-truncated preferences, which has received considerable attention in the context of voting (see, for instance, [3]).

Top-truncated preferences model scenarios where an agent is certain about their most preferred choices, but is indifferent among the remaining ones or is unsure about them. More precisely, in our setting, the preference order submitted by, say, a woman w is said to be a top-truncated order if it is a linear order over a subset of U and the remaining men are all considered to be incomparable by w. In this section we consider one-sided top-truncated preferences, i.e., where only men or women are allowed to specify top-truncated orders, and show an $\mathcal{O}(n)$-approximation algorithm for δ-min-bp-super-stable-matching under this setting. (Without loss of generality we assume throughout that only the women submit strict weak orders.) Although arbitrarily filling-in the missing information and computing the resulting weakly-stable matching can lead to an $\mathcal{O}(n^2\sqrt{\delta})$-approximate matching even for this restricted case (see the full version of the paper for an example), we will see that not all weakly-stable matchings are "bad" and that in fact the $\mathcal{O}(n)$-approximate matching we obtain is weakly-stable.

However, in order to arrive at this result, we first introduce the following problem which might be of independent interest. (To the best of our knowledge, this has not been previously considered in the literature.) Informally, in this problem we are given an instance $\mathcal{I}$ and are asked if we can delete some of the agents to ensure that the instance, when restricted to the remaining agents, will have a perfect super-stable matching.

Problem 3 (min-delete-super-stable-matching). Given an instance $\mathcal{I} = (\delta, p_U, p_W)$, where δ is the amount of missing information and p_U, p_W are the preferences submitted by men and women respectively, compute the set D of minimum cardinality such that the instance $\mathcal{I}_{-D} = (\delta_{-D}, p_{U\setminus D}, p_{W\setminus D})$, where $\delta_{-D} = \frac{1}{|(U\cup W)\setminus D|}\sum_{i\in(U\cup W)\setminus D}\delta_i$, has a perfect super-stable matching (i.e., every agent in $(U \cup W) \setminus D$ is matched in a super-stable matching).

Below we first show that Algorithm 1 gives a 2-approximation for the min-delete-super-stable-matching problem when restricted to the case of one-sided top-truncated preferences. Subsequently, we then use this result in order to get an $\mathcal{O}(n)$-approximation for our problem. Intuitively, the main idea in Algorithm 1, which is inspired by the work of Tan [21], is that some of the entries in each agent's preference list can be deleted by running the proposal-rejection sequence like in Gale-Shapley algorithm and through rotation eliminations, while at the same time maintaining at least one solution of the maximum size. Unfortunately, due to space constraints, we are unable to provide a complete analysis and so we refer the reader to the full version of the paper for detailed explanations and proofs.

Procedure: proposeWith($A, \mathcal{I}$)
1: assign each agent $a \in A$ to be free
2: **while** some $a \in A$ is free **do**
3:     $b \leftarrow$ first agent on a's list
4:     **if** b is already engaged to agent p && b finds p and a incomparable **then**
5:         delete (a, b)
6:     **else**
7:         **if** b is already engaged to agent p **then**
8:             assign p to be free
9:         **end if**
10:         assign a and b to be engaged
11:         **for** each agent c in b's list such that $a \succ_b c$ **do**
12:             delete (c, b)
13:         **end for**
14:     **end if**
15: **end while**
16: **for** each man m **do**
17:     $w \leftarrow$ first woman on m's list
18:     **if** there exists a man m' such that w finds m and m' incomparable **then**
19:         delete (m', w)
20:     **end if**
21: **end for**   ▷ deletions in this loop only happen once and results in the removal of all the remaining ties
22: **return** $\mathcal{I}$ ▷ this returns the updated lists
Main:
Input: a one-sided top-truncated instance $\mathcal{I} = (\delta, p_U, p_W)$
23: $\mathcal{I}' \leftarrow$ proposeWith($U, \mathcal{I}$)
24: $\mathcal{I}' \leftarrow$ proposeWith($W, \mathcal{I}'$)
25: **while** there exists some exposed rotation $(m_1, w_1), (m_2, w_2), \cdots, (m_r, w_r)$ in $\mathcal{I}'$ **do**
26:     delete (m_i, w_i) for all $i \in \{1, \cdots, r\}$
27:     $\mathcal{I}' \leftarrow$ proposeWith($U, \mathcal{I}'$)
28:     $\mathcal{I}' \leftarrow$ proposeWith($W, \mathcal{I}'$)
29: **end while**
30: $\mathcal{M} \leftarrow$ for all men $m \in U$, match m with the only woman in his list
31: construct $G = (V, E)$ where $V = U \cup W$, $(m, w) \in E$ if (m, w) is a super-blocking pair in $\mathcal{M}$
    w.r.t. $\mathcal{I}$
32: $D \leftarrow$ minimum vertex cover of G
33: **for** each $a \in D$ **do**
34:     $D \leftarrow D \cup \mathcal{M}(a)$
35: **end for**
36: **return** $(D, \mathcal{M})$

Algorithm 1. For the case of one-sided top-truncated preferences, the set D returned by the algorithm is a 2-approximation for the min-delete-super-stable-matching problem and the matching $\mathcal{M}$ returned is an $\mathcal{O}(n)$-approximation for δ-min-bp-super-stable-matching

Proposition 1. *Algorithm 1 is a polynomial-time 2-approximation algorithm for the min-delete-super-stable-matching problem when restricted to the case of one-sided top-truncated preferences.*

Given Proposition 1, we can now prove the following theorem.

Theorem 5. *For any $\delta > 0$, Algorithm 1 is a polynomial-time $\mathcal{O}(n)$-approximation algorithm for the δ-min-bp-super-stable-matching problem when restricted to the case of one-sided top-truncated preferences. Moreover, the matching it returns is also weakly-stable.*

Before we end this section, we address one final question as to whether, for the class of one-sided top-truncated preferences, one can obtain a better approximation result if one continues to consider only weakly-stable matchings. In the theorem below we show that for $\delta \in \Omega(\frac{1}{n})$ Algorithm 1 is asymptotically the best one can do under this restriction.

Theorem 6. *For $\delta \leq \frac{1}{2}$, if there exists an α-approximation algorithm for δ-min-bp-super-stable-matching that always returns a matching that is weakly-stable for the case of one-sided top-truncated preferences, then $\alpha \in \Omega\left(\min\left\{n^{\frac{3}{2}}\sqrt{\delta}, n\right\}\right)$.*

4   Beyond Weak-Stability

In the previous section we investigated weakly-stable matchings and we showed several results concerning this situation. Here we move away from this restriction and explore what happens when we do not place any restriction on the matchings. In particular, we begin this section by showing a general hardness of approximation result, and then follow it with a discussion on one possible approach that can lead to a near-tight approximation result.

4.1   Inapproximability Result for δ-min-bp-super-stable-matching

We show a hardness of approximation result for the δ-min-bp-super-stable-matching problem through a gap-producing reduction from the Vertex Cover (VC) problem, which is a well-known NP-complete problem [14]. In the VC problem, we are given a graph $G = (V, E)$, where $V = \{v_1, \cdots, v_k\}$, and a $k_0 \leq k$ and are asked if there exists a subset of the vertices with size less than or equal to k_0 such that it contains at least one endpoint of every edge.

Theorem 7. *For any constant $\epsilon \in (0, 1]$ and $\delta \in (0, 1)$, one cannot obtain a polynomial-time $(n\sqrt{\delta})^{1-\epsilon}$ approx. algorithm for the δ-min-bp-super-stable-matching problem unless $P = NP$.*

4.2   A Possible General Approach for Obtaining a Near-Tight Approximation Factor for δ-min-bp-super-stable-matching

While obtaining a general near-optimal approximation result for the δ-min-bp-super-stable-matching problem is still open, in this section we propose a potentially promising direction for this problem. In particular, we demonstrate how solving even a very relaxed version of the min-delete-stable-matching problem will be enough to get an $\mathcal{O}(n)$-approximation for δ-min-bp-super-stable-matching in general. Below, we first define the relaxation in question, which we refer to as an (α, β)-approximation to the min-delete-super-stable-matching problem.

Definition 5 ((α, β)-**min-delete-super-stable-matching**). *Given an instance $\mathcal{I} = (\delta, p_U, p_W)$, compute a set D' such that $|D'| \leq \alpha \cdot |D_{opt}|$, where $|D_{opt}|$ is the size of the optimal solution to the min-delete-super-stable-matching for the same instance, and the instance $\mathcal{I}_{-D'} = (\delta_{-D'}, p_{U \setminus D'}, p_{W \setminus D'})$, where $\delta_{-D'} = \frac{1}{|(U \cup W) \setminus D'|} \sum_{i \in (U \cup W) \setminus D'} \delta_i$, has a matching with at most β super-blocking pairs.*

Next, we show that an (α, β)-approximation to the min-delete-super-stable-matching problem gives us an $(\alpha n + \beta)$-approximation for δ-min-bp-super-stable-matching. So, in particular, if we have an (α, β)-approximation where α is a constant and $\beta \in \mathcal{O}(n)$, then this in turn gives us an $\mathcal{O}(n)$-approximation for δ-min-bp-super-stable-matching in general.

Proposition 2. *If there exists an (α, β)-approximation algorithm for the min-delete-super-stable-matching problem, then there exists an $(\alpha n + \beta)$-approximation algorithm for the δ-min-bp-super-stable-matching problem.*

5   Conclusion

In this paper we initiated a study on matching with partial information in order to investigate what makes a matching "good" in this context, and to better understand the trade-off between the amount of missing information and the quality of different matchings. Towards this end, we introduced a measure for accounting for missing preference information in an instance, and argued that a natural definition of a "good" matching in this context is one that minimizes the maximum number of blocking pairs with respect to all the possible completions. Subsequently, using an equivalent problem (δ-min-bp-super-stable-matching) we first explored the space of matchings that contained no obvious blocking pairs (i.e., weakly-stable matchings) in order to better understand how missing preference information effected/affected the quality, in terms of approximation with respect to the objective of minimizing the number of super-blocking pairs. Later on, by expanding the space of matchings we considered (i.e., removing the restriction that matches must be weakly-stable), we asked whether it was possible to improve on the approximation factors that were achieved under the restriction to weakly-stable matchings.

There are a number of interesting directions for future work. First, while in Sect. 4.2 we proposed one possible approach that can lead to near-tight approximations, there may be other approaches that can prove fruitful. Second, we believe that the min-delete-super-stable-matching problem, and its relaxation we introduced, are both of independent interest, and so an open question is to see if one can obtain general results on them. In Proposition 1 we saw that a 2-approximation was achievable for the case of one-sided top-truncated preferences and hence it would also be interesting to determine if there are other interesting classes of preferences for which constant-factor approximations are possible. Finally, there are possible extensions, like, for instance, allowing incompleteness—meaning the agents can specify that they are willing to be matched to only a subset of the agents on the other set—that one could consider and ask similar questions like the ones we considered.

References

1. Abraham, D.J., Biró, P., Manlove, D.F.: "Almost stable" matchings in the room-mates problem. In: Erlebach, T., Persinao, G. (eds.) WAOA 2005. LNCS, vol. 3879, pp. 1–14. Springer, Heidelberg (2006). https://doi.org/10.1007/11671411_1
2. Aziz, H., Biró, P., Gaspers, S., de Haan, R., Mattei, N., Rastegari, B.: Stable matching with uncertain linear preferences. In: Gairing, M., Savani, R. (eds.) SAGT 2016. LNCS, vol. 9928, pp. 195–206. Springer, Heidelberg (2016). https://doi.org/10.1007/978-3-662-53354-3_16
3. Baumeister, D., Faliszewski, P., Lang, J., Rothe, J.: Campaigns for lazy voters: truncated ballots. In: Proceedings of the Eleventh International Conference on Autonomous Agents and Multiagent Systems (AAMAS), pp. 577–584 (2012)
4. Biró, P., Manlove, D., Mittal, S.: Size versus stability in the marriage problem. Theor. Comput. Sci. 411(16–18), 1828–1841 (2010)
5. Chiesa, A., Micali, S., Allen Zhu, Z.: Mechanism design with approximate valuations. In: Proceedings of the Third Innovations in Theoretical Computer Science (ITCS), pp. 34–38 (2012)
6. Drummond, J., Boutilier, C.: Elicitation and approximately stable matching with partial preferences. In: Proceedings of the Twenty-Third International Joint Conference on Artificial Intelligence (IJCAI), pp. 97–105 (2013)
7. Drummond, J., Boutilier, C.: Preference elicitation and interview minimization in stable matchings. In: Proceedings of the Twenty-Eighth AAAI Conference on Artificial Intelligence (AAAI), pp. 645–653 (2014)
8. Gale, D., Shapley, L.S.: College admissions and the stability of marriage. Am. Math. Mon. 69(1), 9–15 (1962)
9. Gusfield, D., Irving, R.W.: The Stable Marriage Problem - Structure and Algorithms. Foundations of Computing Series. MIT Press, Cambridge (1989)
10. Hyafil, N., Boutilier, C.: Regret minimizing equilibria and mechanisms for games with strict type uncertainty. In: Proceedings of the Twentieth Conference on Uncertainty in Artificial Intelligence (UAI), pp. 268–277 (2004)
11. Irving, R.W.: An efficient algorithm for the stable roommates problem. J. Algorithms 6(4), 577–595 (1985)
12. Irving, R.W.: Stable marriage and indifference. Discrete Appl. Math. 48(3), 261–272 (1994)
13. Irving, R.W., Manlove, D.: Approximation algorithms for hard variants of the stable marriage and hospitals/residents problems. J. Comb. Optim. 16(3), 279–292 (2008)
14. Karp, R.M.: Reducibility among combinatorial problems. In: Miller, R.E., Thatcher, J.W., Bohlinger, J.D. (eds.) Complexity of Computer Computations. The IBM Research Symposia Series, pp. 85–103. Springer, Boston (1972). https://doi.org/10.1007/978-1-4684-2001-2_9
15. Lu, T., Boutilier, C.: Robust approximation and incremental elicitation in voting protocols. In: Proceedings of the Twenty-Second International Joint Conference on Artificial Intelligence (IJCAI), pp. 287–293 (2011)
16. Manlove, D., Irving, R.W., Iwama, K., Miyazaki, S., Morita, Y.: Hard variants of stable marriage. Theor. Comput. Sci. 276(1–2), 261–279 (2002)
17. Manlove, D.F.: Algorithmics of Matching Under Preferences. Series on Theoretical Computer Science, vol. 2. WorldScientific (2013)
18. Manlove, D.F.: Hospitals/residents problem. In: Kao, M.Y. (ed.) Encyclopedia of Algorithms, pp. 926–930. Springer, Heidelberg (2016). https://doi.org/10.1007/978-3-642-27848-8

19. Rastegari, B., Condon, A., Immorlica, N., Irving, R., Leyton-Brown, K.: Reasoning about optimal stable matchings under partial information. In: Proceedings of the Fifteenth ACM Conference on Economics and Computation (EC), pp. 431–448 (2014)
20. Rastegari, B., Condon, A., Immorlica, N., Leyton-Brown, K.: Two-sided matching with partial information. In: Proceedings of the Fourteenth ACM Conference on Electronic Commerce (EC), pp. 733–750 (2013)
21. Tan, J.J.M.: A maximum stable matching for the roommates problem. BIT Num. Math. **30**(4), 631–640 (1990)

Prophet Inequalities vs. Approximating Optimum Online

Rad Niazadeh[1]([✉])[iD], Amin Saberi[2], and Ali Shameli[2]

[1] Computer Science Department, Stanford University, Stanford, CA 94305, USA
rad@cs.stanford.edu
[2] Management Science and Engineering, Stanford University, Stanford, CA 94305,
USA
{saberi,shameli}@stanford.edu

Abstract. We revisit the classic prophet inequality problem, where the
goal is selling a single item to an arriving sequence of buyers whose values are drawn from independent distributions, to maximize the expected
allocated value. The usual benchmark is the expected value that an omniscient prophet who knows the future can attain. We diverge from this
framework and compare the performance of the best single pricing mechanism with the best optimum online mechanism.

Somewhat surprisingly, we show that the original tight prophet
inequality bounds comparing the single-pricing with the optimum offline
are tight even when we use the optimum online as a benchmark, both for
the identical and non-identical distributions. Moreover, we incorporate
linear programming to characterize this benchmark, and show how this
approach leads to a modular way of designing prophet inequalities, hence
reconstructing the results of [31] and [13] with somewhat simpler proofs.

Keywords: Prophet inequality · Optimal stopping · Online algorithm

1 Introduction

Mechanisms with simple semantics are natural and easy to optimize. The practical advantages of such mechanisms have led to their prevalence in electronic
commerce, where the canonical examples are different forms of pricing mechanisms. One important aspect of such mechanisms is how well they perform versus
an economically justified benchmark. This "simple vs. optimal" comparison [24]
can then yield new insights when designing simple mechanisms.

We revisit a classic problem in the optimal stopping theory, which is inti-
mately related to the design of pricing mechanisms. Consider selling a single
item to a sequence of arriving buyers whose values are drawn from non-identical
and independent distributions, with the goal of maximizing the expected value of
the buyer who eventually receives the item. In such a setting, we can consider the
omniscient prophet benchmark, also termed as the optimum offline, which can
foresee the entire realized value sequence of the buyers and picks the maximum

© Springer Nature Switzerland AG 2018
G. Christodoulou and T. Harks (Eds.): WINE 2018, LNCS 11316, pp. 356–374, 2018.
https://doi.org/10.1007/978-3-030-04612-5_24

value. Another benchmark is the *optimum online* that is the best implementable mechanism in an online fashion for maximizing the expected allocated value, assuming we know the distributions of values of the buyers.

In their seminal work, [27] asked the following question: how does the expected value of the optimum online mechanism compare with the expected value of the omniscient prophet? Garling, Krengel and Sucheston established a tight bound of 2 for this problem, which became the first of many *"prophet inequalities"* (see [29] for a comprehensive survey). The optimum online mechanism in this setting is an adaptive pricing mechanism, where prices change over time based on the observed set of the buyers. Moreover, if the ordering is also known to the mechanism, these prices can be computed by a simple backward induction. In the special case when the distributions are identical, the same backward induction yields a factor strictly better than 2, as first shown by [25]. The tight factor of ≈ 1.342 for identical distributions is established recently by [1] and [13].

Quite surprisingly, [31] showed that one can achieve the same factor 2 guarantee as the result of Garling et al. using a simple and elegant "single pricing" mechanism. This bound is tight due to a simple example; there are two buyers, where the first buyer value is deterministically equal to 1 and the second buyer value is equal to $1/\epsilon$ with probability ϵ and 0 otherwise for an arbitrarily small ϵ. The expectation of the maximum value is $2 - \epsilon$, while no online policy can achieve a value higher than 1 in expectation.

What if we use the optimum online as the benchmark? Clearly, in the above example, while there is a gap between the prophet and any online policy, there is no gap between the single threshold and the optimum online. This discrepancy motivates us to seek the answer to the following question.

In a single-item prophet inequality problem (with identical or non-identical distributions), does single pricing achieve an improved approximation guarantee when compared to the optimum online solution as the benchmark, instead of the optimum offline solution?

Our Results. Our main result is to answer the above question in the negative, and hence proving that by switching the benchmark from the optimum offline to the optimum online, no improved approximation guarantees are possible. In a nutshell, we proceed in a step-by-step fashion by investigating simpler problems and special cases first, e.g. when the distributions are identical or non-atomic. We then build up on top of the solutions for these special cases to get the final answer. In our study, we identify hard instances for approximating the optimum online and offline benchmarks, discover simpler proofs for the performance guarantees of single pricing mechanisms, and find linear programming characterizations for the optimum online benchmark. While these technical pieces play key roles in proving our result, some of them are also of independent interest.

We start by considering the special case of identical distributions; We identify an i.i.d. sequence of discrete distributions under which no single pricing mechanism obtains an approximation factor smaller than 2 with respect to the optimum online, which is clearly tight thanks to the [31]'s original prophet inequality. However, this approximation factor is simply an artifact of *tie-breaking*: we re-derive similar results of [13,17] by a new simpler proof, and show for any non-atomic instance, posting the price that is accepted by a single buyer with probability $\frac{1}{n}$ is an $\frac{e}{e-1}$-approximation to the optimum online (and also offline), where n is the number of the buyers in the sequence.

For general distributions, we arm the single pricing with a *randomized tie-breaking*, i.e. flipping an independent coin with a fixed given probability to break the tie and overcome the above issue. As a simple corollary of the non-atomic i.i.d. result we show under any i.i.d. sequence of distributions (no matter atomic or non-atomic) there is a single pricing with randomized tie-breaking that is $\frac{e}{e-1}$-approximation to the optimum online (and also offline). Essentially, the role of randomized tie-breaking can be seen as perturbing the discrete distribution to make it non-atomic. We further prove this factor is tight by identifying another i.i.d. sequence of discrete distributions under which the best single pricing with the best randomized tie-breaking probability is no better than a $\frac{e}{e-1}$-approximation to the optimum online.

We then consider the case of non-identical distributions. The above investigation still leaves the following question open in this case: what is the tight approximation factor of the best single pricing with randomized tie-breaking with respect to the optimum online? or equivalently, can single pricing with randomized tie-breaking beat the approximation factor of 2 when compared to the optimum online and when the distributions are non-atomic? (note that the factor 2 is tight if we allow atomic distributions as discussed earlier.) Somewhat surprisingly, we also answer this question in the negative: there exists a sequence of non-identical, yet non-atomic, distributions under which no single pricing mechanism obtains more than half of the expected value of the optimum online mechanism. We conclude that replacing the prophet benchmark with the optimum online does not allow a worst-case approximation factor smaller than 2 for single pricing with randomized tie-breaking mechanisms, if the distributions are allowed to be non-identical.

Finally, we switch gears to a slightly stronger, yet computable, benchmark than the optimum online and seek to understand this new benchmark through the lens of linear programming. The considered benchmark is basically the optimum online mechanism when the ordering of the buyers is also known to the mechanism. While the previous negative results trivially apply to this benchmark as well, as a good news this benchmark can be characterized completely through a *linear program*. This LP is then used as the main tool to design competitive algorithms with respect to the optimum offline benchmark: we show how to start from a relaxation of the optimum offline benchmark (termed as the *ex-ante relaxation* [3,15,28]) and then modify its solution to make it feasible for this LP, which can then be rounded exactly by a sequential pricing mechanism (with randomized tie-breakings in case of discrete distributions). For the case of non-identical distributions, our proposed approach loses a factor of 2, and hence re-deriving the celebrated result of [27] with a simpler proof. Furthermore, for identical distributions, we re-derive the approximation factor of $\frac{e}{e-1}$ of [25] using this LP and with a much simpler proof. We believe the improved approximation factor in [1] and the tight approximation factor in [13] can also be extracted using this LP and the same techniques, which we leave as two interesting open problems.

Studying new benchmarks can offer new insights. In some cases they are approximable with a better ratio and lead to the design of new algorithms (e.g. see [4]). In our case they lead to simpler proofs; while trying to understand the optimum online through linear programming, all of a sudden a two-line proof for the well-studied prophet inequalities magically pops out.

Further Related Work. Besides the classic prophet inequalities, many variations such as prophet inequalities with limited samples form the distributions [5], i.i.d. and random order [1,18], and finally ordered prophets (a.k.a. the free-order sequential posted pricing problem) [6,8,14,32] have been explored, and discovering connections to the price of anarchy [15], online contention resolution schemes [20], and online combinatorial optimization [21] have been of particular interest in this literature. Generalizations of the simple prophet inequality problem to combinatorial settings have also been studied, where the examples are matroids [26], knapsack [20], k-uniform matroids (for better bounds) [2,22], or even general downward-closed [30]. Finally, techniques and results in this literature had an immense impact on mechanism design [7,9–13,16,19]. For a full list of recent and old related results, refer to [29].

Notations. We are considering a *single item prophet inequality* problem, in which a seller is interested in selling an item to a sequence of n arriving buyers. Each buyer i has a value v_i for the item. This value is independently drawn from a distribution $\mathcal{F}_i$. Buyers arrive one by one and reveal their values.[1] Upon the arrival of a buyer, the seller decides whether to sell the item or move on to the next buyer. The goal is to maximize the expected value of the selected buyer. We consider the setting where the sequence of distributions $\mathcal{F}_1, \mathcal{F}_2, \ldots, \mathcal{F}_n$ is picked by an oblivious adversary up front. We assume the seller knows the distributions in advance but does not know the order in which the buyers arrive.

In this paper we consider the following types of mechanisms/benchmarks for the seller:

- *Optimum offline benchmark*: Assume the seller is assisted by an omniscient prophet who knows the values of all the buyers in advance and helps the seller to pick the buyer with the maximum value.
- *Optimum online benchmark*: It is the mechanism that achieves the maximum expected value possible by looking at the values of the buyers sequentially and adaptively deciding whether to pick/reject the current buyer by inferring from the past observed values as well as the distributions $\mathcal{F}_1, \mathcal{F}_2, \ldots, \mathcal{F}_n$.

---

[1] Although the focus of this paper is not on truthful mechanism design, all of our mechanisms are pricing and hence truthful.

- *Sequential pricing mechanism*: the seller sets a price T_i for each buyer i, and picks the first buyer i whose value is at least T_i, i.e. $v_i \geq T_i$.
- *Single pricing mechanism*: the seller sets a price T, and picks the first buyer i (if any) whose value is at least T, i.e. $v_i \geq T$.

Throughout this paper, we compare the performance of single pricing mechanisms to the optimum online mechanism as well as the optimum offline mechanism. The expected value obtained by the sequential pricing mechanism is always upper bounded by that of the optimum online and can assist us in obtaining some of our bounds.

2 Optimum Online and Beating Prophet Inequalities

In this section, we investigate a new benchmark, namely the optimum online mechanism, for the single item prophet inequality problem. We ask how it is compared with different variations of the single pricing mechanism, with and without (randomized) tie-breaking, in terms of expected value, and whether optimum online is amenable to improved approximation factors by these single pricing mechanisms. We start our journey by considering an important special case: independent and identical distributions. It turns out, as we will see in Sect. 2.1, this important special case paves the path to crack the problem for general non-identical distributions in Sect. 2.2.

2.1 Independent and Identical Distributions

Suppose we allow the adversarial instance to have distributions with point-masses. Even in this case, although ties can happen, the single pricing mechanism of [31] or [26] can still achieve at least half of the value obtained by the optimum online in expectation (as it achieves half of the expected value obtained by the optimum offline). Interestingly, if the single pricing is not armed with any tie-breaking, we can show this bound is tight, even for i.i.d. distributions.

Proposition 1. *For every $\epsilon > 0$, there exists a large enough n and an identical sequence of (discrete) distributions $\mathcal{F}_1, \ldots, \mathcal{F}_n$, where the optimal single pricing obtains at most $(\frac{1}{2} + \epsilon)$ fraction of the expected value of the optimum online for this instance.*

Proof. Consider the simple example below:

Example 1. There are n i.i.d. buyers, where the value of buyer i, denoted by v_i, is 1 with probability $(1 - \frac{1}{n})^{\frac{1}{n}}$ and n otherwise.

First, we calculate the expected value of the optimum offline. We have:

$$\mathbb{E}\left[\max_{i \in [n]}(v_i)\right] = ((1 - \frac{1}{n})^{\frac{1}{n}})^n + n(1 - ((1 - \frac{1}{n})^{\frac{1}{n}})^n) = 1 - \frac{1}{n} + n\frac{1}{n} = 2 - \frac{1}{n}$$

Moreover, the optimum online mechanism is a sequential pricing mechanism in this example; one can see that to always pick the maximum value we can post the price n for the buyers arriving at times $t = 1, \ldots, n - 1$, and then post the price 1 for the last buyer. Therefore, to compare the single pricing with the optimum online, it suffices to compare the single pricing with $\mathbb{E}\left[\max_{i \in [n]} v_i\right]$.

Without loss of the generality, suppose the single pricing mechanism posts a price $T \in [1, n]$. By construction, with probability $1 - \left((1 - \frac{1}{n})^{\frac{1}{n}}\right)^n = \frac{1}{n}$, the random variable $\max_{i \in [n]} v_i$ is equal to n. Therefore, If $T > 1$, the expected obtained value by posting T is equal to 1. On the other hand, if $T \leq 1$, the expected obtained value is equal to:

$$(1 - \frac{1}{n})^{\frac{1}{n}} + n(1 - (1 - \frac{1}{n})^{\frac{1}{n}}) = 1 + O(\frac{1}{n^2}) + n\left(1 - 1 + O(\frac{1}{n^2})\right) = 1 + O(\frac{1}{n})$$

Finally, letting n to be large enough finishes the proof.

Now, here is one concrete question: is the tightness of the factor $\frac{1}{2}$ in Proposition 1 a consequence of having discrete distributions, or can we obtain the same upper-bound for non-atomic distributions as well? Not much surprisingly, we prove that for such distributions, the optimum single pricing mechanism can always achieve $1 - \frac{1}{e}$ fraction of the value obtained by the optimum offline mechanism, and hence that of the optimum online. Similar results have been reported in the literature, e.g. in [13] by incorporating the Bernoulli selection lemma or in [18] by reducing to the prophet secretary problem. Yet, we provide a simpler and more direct proof here, which might also be of independent interest.

Inspired by [2, 20], we start by defining the *value curve*. This curve is defined in the probability space (a.k.a. the *quantile space*) and captures the expected value obtained by posting a price to a single buyer when the *ex-ante* probability of sale is some fixed quantity.

Definition 1. *Consider a single buyer with value $v \sim \mathcal{F}$, and suppose $\mathcal{F}$ is a non-atomic distribution. For every sale probability $q \in [0, 1]$, referred to as the quantile q, the value curve $V(q)$ is defined as $V(q) \triangleq \mathbb{E}_{v \sim \mathcal{F}}[v \cdot \mathbb{1}\{v \geq T(q)\}]$, where $T(q)$ is the minimum price at which the item is sold to the buyer with probability q, i.e. $T(q) \triangleq \underset{p \in \mathbb{R}}{\operatorname{argmin}} \, \mathbb{P}_{v \sim \mathcal{F}}[v \geq p] = q$. We refer to $T(q)$ as the pricing curve.*

The following lemma is immediate, which describes some basic properties of the value curve.

Lemma 1. *For a given value curve $V(.)$ as in Definition 1, we have $\frac{\partial V}{\partial q}(q) = T(q)$. Moreover, $V(q)$ is non-decreasing and $T(q)$ is non-increasing (and therefore $V(q)$ is concave).*

Proof. The proof is straightforward. First of all, $T(q)$ is non-increasing as by increasing the price the sale probability can only decrease. Therefore $V(q)$ is also non-decreasing as by increasing q the price can only decrease and hence a larger set of buyer values can get accepted. Now consider moving from quantile q to $q + \epsilon$ for infinitesimal $\epsilon > 0$. This changes the price from $T(q)$ to $T(q + \epsilon)$. Remember that $V(q)$ is the expected obtained value, and hence new values in $[T(q + \epsilon), T(q)]$ will be added to this expectation by the mentioned change. Moreover, $T(q)$ is a continuous function (as $\mathcal{F}$ is non-atomic), and therefore the marginal change in $V(.)$ is equal to $\epsilon T(q)$. Hence, $\frac{\partial V}{\partial q} = T(q)$.

Proposition 2. *If all the distributions $\mathcal{F}_1, \mathcal{F}_2, \ldots, \mathcal{F}_n$ are independent, identical, and non-atomic, then the optimal single pricing mechanism obtains at least $(1 - \frac{1}{e})$ fraction of the value procured by both the optimum offline and the optimum online mechanism in expectation.*

Proof. We use the optimum offline as a surrogate to approximate the optimum online. Let $V(q)$ and $T(q)$ be the value curve and the pricing curve for the distribution $\mathcal{F}_1$ respectively, as in Definition 1. Now consider posting a single price $T(\frac{1}{n})$. The expected value obtained by the seller is equal to:

$$V(\tfrac{1}{n}) + (1 - \tfrac{1}{n})V(\tfrac{1}{n}) + \ldots + (1 - \tfrac{1}{n})^{n-1}V(\tfrac{1}{n}) = n\left(1 - (1 - \tfrac{1}{n})^n\right)V(\tfrac{1}{n})$$
$$\geq n(1 - \tfrac{1}{e})V(\tfrac{1}{n}).$$

To complete the proof, it suffices to show $\mathbb{E}\left[\max_{i \in [n]} v_i\right] \leq nV(\frac{1}{n})$. We have

$$\mathbb{E}\left[\max_{i \in [n]} v_i\right] = \sum_{i=1}^{n} \mathbb{E}\left[v_i \cdot \mathbf{1}\{v_i \geq \max v_{-i}\}\right] = n\mathbb{E}_{v_{-1}}\left[\mathbb{E}_{v_1}\left[v_1 \cdot \mathbf{1}\{v_1 \geq \max v_{-1}\}|v_{-1}\right]\right] = n\mathbb{E}_{v_{-1}}\left[V(\bar{q})\right],$$

where $\bar{q}$ is the quantile corresponding to $\bar{v} = \max_{i \neq 1} v_i$, or equivalently $\bar{q} = 1 - \mathcal{F}_1(\bar{v})$. Note that $\bar{q}$ is a random variable with $\mathbb{E}[\bar{q}] = \frac{1}{n}$. This is because for each v_i, $2 \leq i \leq n$, the quantile q_i corresponding to v_i is a random variable drawn from the uniform distribution on $[0, 1]$ (by using the standard connections between the value space and the quantile space for a given distribution, e.g. see [23]). Since $\bar{v} = \max_{i \neq 1} v_i$, we have $\bar{q} = \min_{i \neq 1} q_i$, where each $q_i \sim \text{unif}[0, 1]$, and hence $\mathbb{E}[\bar{q}] = \frac{1}{n}$. Finally, by the concavity of $V(.)$ (Lemma 1) and using Jensen inequality:

$$nV(\tfrac{1}{n}) = nV(\mathbb{E}[\bar{q}]) \geq n\mathbb{E}[V(\bar{q})] = \mathbb{E}\left[\max_{i \in [n]} v_i\right].$$

Even though we stated Proposition 2 for the case of non-atomic distributions, we can obtain the same result for any distribution (atomic or non-atomic) by allowing the single pricing to do *randomized tie breaking*: in case of a tie, the mechanism flips a random independent coin and break the tie based on the outcome of this coin flip. We first prove this lower-bound of $1 - \frac{1}{e}$ through the following proposition and combining it with Proposition 2. Later we will show that this bound is in fact tight for the case of identical distributions and when single pricing is allowed to randomly break the ties, even when we consider optimum online as a benchmark.

Proposition 3. *Given a single pricing mechanism M_1 that always achieves in expectation, fraction $0 \leq \alpha \leq 1$ of the value obtained by the optimum offline* *mechanism over i.i.d. and non-atomic distributions, we can devise a mechanism M_2 that achieves the same performance guarantee for i.i.d. and general (maybe atomic) distributions.*

Proof. Assume each buyer i has a valuation $v_i \sim \mathcal{F}$, where $\mathcal{F}$ can (potentially) have point-masses. We create a non-atomic distribution $\mathcal{F}'$ as follows. For any $v \in \text{supp}(\mathcal{F})$ with point mass probability p, we add a continuous and uniform distribution with probability mass p to $\mathcal{F}'$. This distribution will be defined on interval $[v, v + \epsilon]$ and therefore we must have $\frac{\partial}{\partial x}\mathcal{F}'(x) = \frac{p}{\epsilon}$ for $x \in [v, v + \epsilon]$. Let $M_1(\mathcal{F}')$ denote the price picked by M_1 given the distribution $\mathcal{F}'$. Without loss of generality, we can assume that $M_1(\mathcal{F}') \in [v, v + \epsilon]$ for some $v \in \text{supp}(\mathcal{F})$ (potentially the end points). We devise a new mechanism $M_2(\mathcal{F})$ for the original distribution $\mathcal{F}$, as follows. Given value $v' \sim \mathcal{F}$ of a buyer, If v' is bigger/smaller than v, we pick/reject the buyer with probability 1. On the other hand, if $v' = v$, we pick the buyer with probability $\frac{(v+\epsilon-M_1(\mathcal{F}'))}{\epsilon}$. Note that the probability of $M_2(\mathcal{F})$ picking buyer i, is the same as the probability of $M_1(\mathcal{F}')$ picking buyer i. When we move from $M_1(\mathcal{F}')$ to $M_2(\mathcal{F})$, there is at most an ϵ difference in the expected value obtained from buyer i conditioned on the item is not yet allocated upon her arrival. Also, the probability of the item is not yet being allocated when buyer i arrives is the same. So their expected obtained values from buyer i differ by at most ϵ. Also note that in the limit as $\epsilon \to 0$, the expected value obtained by the optimum offline remains unchanged. Therefore by taking the limit of ϵ to 0, M_2 will achieve an approximation factor of at least α on the original problem instance when compared to the optimum offline.

To obtain a matching upper-bound for the case of identical distributions and when the single pricing mechanism is armed with randomized tie-breaking, consider Example 2. This example is inspired by [17], and shows the existence of an i.i.d. instance (obviously with atomic distributions) under which we cannot achieve more than $1 - \frac{1}{e}$ fraction of the value obtained by the optimum offline (or the optimum online since they are equal), even with the best possible single price and the best possible randomized tie-breaking.

Example 2. There are n i.i.d. buyers, where the value of buyer i, denoted by v_i, is 1 with probability $(1 - \frac{1}{n})^{\frac{1}{n}}$ and $\frac{n}{e-2}$ otherwise.

Proposition 4. *For any $\epsilon > 0$, there exists a large enough n such that the optimum single pricing mechanism with tie breaking cannot obtain any better than $(1 - \frac{1}{e} + \epsilon)$ fraction of both the optimum online and the optimum offline in Example 2.*

Proof. Note that the optimal sequential pricing mechanism (and therefore the optimum online) can always achieve the maximum value by picking price 1 on the last day and price $\frac{n}{e-2}$ on all other days. Now we show that the optimal single pricing mechanism (with the optimal tie breaking probability) cannot achieve, in expectation, anything better than $(1 - \frac{1}{e})\mathbb{E}\left[\max_{i\in[n]} v_i\right]$. Note that $\mathbb{E}\left[\max_{i\in[n]} v_i\right] = 1 + \frac{1}{e-2} = \frac{e-1}{e-2}$. Moreover, the mechanism is allowed to pick any price T and probability q. At each time, if the value of the buyer is exactly equal to T, the item gets allocated with probability q. If the value was lower (or higher) than T, the mechanism rejects (or allocates to) the buyer with probability 1. Now consider two possible price choices: If $1 < T < \frac{n}{e-2}$, then in this case the expected obtained value is equal to $\left(1 - \left((1 - \frac{1}{n})^{\frac{1}{n}}\right)^n\right) \cdot \frac{n}{e-2} = \frac{1}{e-2}$, hence the ratio is strictly smaller than $1 - \frac{1}{e}$. Suppose the mechanism posts $T = 1$. Denote the expected obtained value of this mechanism by $U(n)$. We have:

$$U(n) = \sum_{i=0}^{n-1}((1-q)(1-\frac{1}{n})^{\frac{1}{n}})^i\left(q(1-\frac{1}{n})^{\frac{1}{n}} + \frac{n}{e-2}(1-(1-\frac{1}{n})^{\frac{1}{n}})\right)$$

$$= \frac{1-(1-q)^n(1-\frac{1}{n})}{1-(1-q)(1-\frac{1}{n})^{\frac{1}{n}}}\left(q(1-\frac{1}{n})^{\frac{1}{n}} + \frac{n}{e-2}(1-(1-\frac{1}{n})^{\frac{1}{n}})\right)$$

$$= \frac{1-(1-q)^n(1-\frac{1}{n})}{1-(1-q)(1-\frac{1}{n})^{\frac{1}{n}}}\left(q(1-\frac{1}{n})^{\frac{1}{n}} + \frac{1}{n(e-2)} + o(\frac{1}{n})\right)$$

We first claim that $\lim_{n\to+\infty}\frac{U(n)}{\mathbb{E}[\max_{i\in[n]}v_i]}$ is again strictly less than $1 - \frac{1}{e}$ if $q = o(\frac{1}{n})$ or $q = \Omega(\frac{1}{n})$. In the former case:[2]

$$\frac{1-(1-q)^n(1-\frac{1}{n})}{1-(1-q)(1-\frac{1}{n})^{\frac{1}{n}}} \leq \frac{\frac{1}{n}}{\frac{1}{n^2}} + o(1), \quad q(1-\frac{1}{n})^{\frac{1}{n}} + \frac{1}{n(e-2)} + o(\frac{1}{n}) \approx \frac{1}{n(e-2)},$$

and hence the ratio converges to $\lim_{n\to+\infty}\frac{U(n)}{\mathbb{E}[\max_{i\in[n]}v_i]} = \frac{1}{e-1} < 1 - \frac{1}{e}$.

If $q = \Omega(\frac{1}{n})$, we can again drop the lower-order terms and we have:

$$\frac{1-(1-q)^n(1-\frac{1}{n})}{1-(1-q)(1-\frac{1}{n})^{\frac{1}{n}}} \approx \frac{1}{q}, \quad q(1-\frac{1}{n})^{\frac{1}{n}} + \frac{1}{n(e-2)} + o(\frac{1}{n}) \approx q,$$

and hence the ratio converges to $\frac{e-2}{e-1} < 1 - \frac{1}{e}$. Next, suppose $q = \frac{c}{n}$ for some constant $c > 0$. We have:

$$U(n) = \frac{1-(1-\frac{c}{n})^n(1-\frac{1}{n})}{1-(1-\frac{c}{n})(1-\frac{1}{n})^{\frac{1}{n}}}(\frac{c}{n}(1-\frac{1}{n})^{\frac{1}{n}} + \frac{n}{e-2}(1-(1-\frac{1}{n})^{\frac{1}{n}})).$$

By taking the limit as $n \to \infty$, we end up with $\lim_{n\to+\infty} U(n) = \frac{1-e^{-c}}{c}\frac{(e-2)c+1}{(e-2)}$. So by looking at the ratio of $U(n)$ to the expected maximum value and taking the limit, we have:

$$\lim_{n\to+\infty}\frac{U(n)}{\mathbb{E}[\max_{i\in[n]}v_i]} = \frac{1-e^{-c}}{c}\frac{(e-2)c+1}{(e-1)}. \tag{1}$$

---

[2] We drop lower order terms by using $\approx$.

By taking the derivative from the right-hand side of the Eq. 1 and setting it to zero, and also by checking the direction of concavity of this right-hand side term, we can easily verify that the maximum value of the above function is attained at $c = 1$. With $c = 1$, the right-hand side of Eq. 1 evaluates to $1 - \frac{1}{e}$, which finishes the proof.

We further show that we can turn the distribution in Example 2 into a non-atomic distribution such that the expected values of the optimum online and optimum offline mechanisms do not change. This along with Propositions 4 and 3, indirectly proves that the result of Proposition 2 is tight. More importantly, it shows that we do not benefit by moving the benchmark from the optimum offline to optimum online for the case of i.i.d. distributions (or equivalently for general distributions when randomized tie-breaking is allowed).

Lemma 2. *Consider a perturbed version of Example 2 where each buyer has a value uniformly distributed on the $[1, 1 + \epsilon]$ with probability $(1 - \frac{1}{n})^{\frac{1}{n}}$ and on $[\frac{n}{e-2}, \frac{n}{e-2} + \epsilon]$ otherwise. Then the expected value obtained by the optimal online (or offline) approaches the expected value obtained by the original optimal online (or offline) in Example 2 as $\epsilon \to 0$.*

Proof. We first show that the expected value collected by the optimum online and optimum offline are at most ϵ apart. To see this, note that we can design a sequential pricing mechanism that has price $\frac{n}{e-2}$ for buyers $1, 2, \ldots, n-1$ and price 1 for buyer n. This mechanism always attains a value that is at least the maximum value minus ϵ, and therefore, the expected value obtained by the optimum online is no less than that of the optimum offline minus ϵ. Next we show that the expected value obtained by the optimum offline is at most ϵ apart from the expected value obtained by the optimum offline in Example 2. By a simple argument, we can map each realized perturbed value profile (i.e. values of all the buyers) to a realized value profile in the discrete instance, so that the maximum value in the first profile is in the ϵ neighborhood of the second profile. By linearity of expectation, the expected maximum value of the first profile is also in the ϵ neighborhood of the maximum value of the second profile, as desired. Taking the limit as $\epsilon \to 0$ finishes the proof.

2.2 Independent and Non-identical Distributions

As we have shown so far, the worst case ratio between the expected value of the single pricing mechanism and the expected value of the optimum online mechanism is different when the distributions are non-atomic as opposed to when they are discrete. More precisely, for the case of non-atomic distributions this ratio is $1 - \frac{1}{e}$ whereas for discrete distributions (without tie breaking) this ratio is $\frac{1}{2}$. This gives rise to a natural question. Is this difference an artifact of having non-continuous distributions? And more importantly, does this gap between these ratios still exists for the more general case of non-identical distributions. More formally, we show the following proposition.

Proposition 5. *For the case of non-identical and non-atomic distributions, there exists an example of $\mathcal{F}_1, \mathcal{F}_2, \ldots, \mathcal{F}_n$ such that the expected value obtained by the optimum single pricing mechanism is at most half of the expected value obtained by the optimum online.*

Proof. Now we describe how to transform the distributions in Example 1 into continuous and non-identical distributions in a way that the $\frac{1}{2}$ ratio remains tight. For any $\epsilon > 0$ and $\delta = \frac{\epsilon}{n}$, let $\mathcal{F}_i$ be uniformly distributed on $[1+\epsilon-i\delta, 1+\epsilon-(i-1)\delta]$ with probability $(1-\frac{1}{n})^{\frac{1}{n}}$ and uniformly distributed on $[n-\epsilon, n+\epsilon)$ otherwise. Note that similar to the argument in Example 1, by picking price $n-\epsilon$ for buyers $1, 2, \ldots, n-1$ and price 1 for buyer n, the optimal sequential pricing mechanism achieves value $2+O(\epsilon)$ in expectation. On the other hand, one can see that for any choice of price $T \leq 1+\epsilon-\delta$, the first buyer will always be picked by the mechanism and therefore the expected value obtained will be at most $1+\epsilon$. On the other hand, if $T > 1+\epsilon$ the mechanism does not pick buyer 1, since only accepts a buyer if his value is bigger than or equal to $n-\epsilon$ and therefore the expected value obtained cannot be more than $(n+\epsilon)(1-((1-\frac{1}{n})^{\frac{1}{n}})^n) = 1+\frac{\epsilon}{n}$. Finally, we show that for large n, picking $T \in (1+\epsilon-\delta, 1+\epsilon]$ does not yield to an expected allocated value more than $1+O(\epsilon)$. To show this, suppose $p = \frac{1+\epsilon-T}{\delta}$. Then the expected obtained value is at most:

$$\left(p(1+\epsilon)(1-\frac{1}{n})^{\frac{1}{n}} + (n+\epsilon)(1-(1-\frac{1}{n})^{\frac{1}{n}}) \right) + \left((1-p)(n+\epsilon)(1-((1-\frac{1}{n})^{\frac{1}{n}})^{n-1}) + O(\epsilon) \right)$$

where the first parenthesis is the contribution of the first buyer, and the second parenthesis is the contribution of the remaining buyers. Straightforward calculations shows this is at most $1 + O(\epsilon)$ in the limit as $n \to \infty$ for any value of $0 < p < 1$. This in turn shows that the optimal single pricing mechanism obtains, in expectation, at most half of the expected value obtained by the mentioned sequential pricing, and hence at most half of the optimum online.

Corollary 1. *Proposition 5 implies that for non-identical distributions, no improved approximation guarantees can be obtained for the single pricing mechanism (even with randomized tie-breaking), when compared against the optimum online benchmark instead of the optimum offline.*

3   Optimum Online and Linear Programming

In this section, we use linear programming to study (a variant of) the optimum online benchmark for the single item prophet inequality problem. The benefits of this linear programming approach are two-fold. On the one hand, the LP gives us a systematic way of describing the optimum online benchmark, which can then be easily generalized to other more combinatorial domains (e.g. matroids). On the other hand, we show how to use this LP to design approximate pricing mechanisms with respect to the optimal offline in a *modular way*, and therefore re-deriving simpler proofs for a couple of already existing prophet inequalities.

We believe this approach can be useful for other settings as well, which we leave as future research directions. For the ease of exposition, we focus on non-atomic distributions in this section. The case of general distributions can be easily handled by adding randomized tie-breaking to our mechanisms in a straightforward fashion.

3.1 LP Characterization of the Optimum Online Benchmark

We consider a slightly stronger version of the optimum online benchmark used in Sect. 2, which is the optimum online mechanism that knows the ordering of the buyers in advance. We say that buyers arrive with the ordering $\pi : [n] \to [n]$ if at each time $t = 1, \ldots, n$ buyer $\pi(t)$ arrives. Now, for an ordering π over the buyers, let OPT-ONLINE(π) be the optimum online mechanism knowing π. We seek to find a linear programming characterization for OPT-ONLINE(π) for a fixed π. Note that using a simple backward induction one can find such an optimum policy; However, the introduced LP sheds more insight on the structure of this policy and helps us with designing approximate policies with respect to the optimum offline (i.e. prophet inequalities).

For a given online policy for the single item prophet inequality problem, let $\mathcal{X}_t(v)$ denote the probability that the policy allocates the item to the buyer arriving at time t conditioned on the event that the value of this buyer is equal to v. We term $\{\mathcal{X}_t(v)\}$ as the *allocation probabilities* associated with a given online policy. Our linear programming has variables $\mathcal{X}_t(v)$ for every $t \in [n]$ and every $v \in \text{supp}(\mathcal{F}_{\pi(t)})$, where supp(.) denotes the support of its input distribution.[3] We then try to impose constraints on these variables to guarantee that the solution of the LP is *implementable* by an online policy, without losing anything in the expected allocated value. Formally speaking, consider the following linear program, which we denote by LP-ONLINE(π):

$$(\text{LP-ONLINE}(\pi))$$

$$\text{maximize} \sum_{t=1}^{n} \mathbb{E}_{v_{\pi(t)} \sim \mathcal{F}_{\pi(t)}} \left[v_{\pi(t)} \cdot \mathcal{X}_t(v_{\pi(t)}) \right]$$

$$\text{subject to } \mathcal{X}_t(v) \leq 1 - \sum_{t' < t} \mathbb{E}_{v_{\pi(t')} \sim \mathcal{F}_{\pi(t')}} \left[\mathcal{X}_{t'}(v_{\pi(t')}) \right], \quad \forall v \in \text{supp}(\mathcal{F}_{\pi(t)}),\ t = 2, \ldots, n$$

$$\mathcal{X}_1(v) \leq 1, \quad \forall v \in \text{supp}(\mathcal{F}_{\pi(1)})$$

$$\mathcal{X}_t(v) \geq 0, \quad \forall v \in \text{supp}(\mathcal{F}_{\pi(t)}),\ t = 1, \ldots, n$$

It is not hard to see that every feasible online policy induces a feasible solution for LP-ONLINE(π), by setting $\mathcal{X}_t(v)$ to be the allocation probabilities of this policy. In fact, no allocation happens at time t if the item has been allocated at some time $t' < t$. Therefore, by taking an expectation with respect to the buyer values arriving at times $t' = 1, \ldots, t - 1$, $\mathcal{X}_t(v)$ will be at most equal to $1 - \sum_{t' < t} \mathbb{E}_{v_{\pi(t')} \sim \mathcal{F}_{\pi(t')}} \left[\mathcal{X}_{t'}(v_{\pi(t')}) \right]$. More interestingly, the converse is also true.

---

[3] In the case of non-atomic distributions, this LP is essentially a continuous program with uncountably many variables. In the case of discrete distributions, the LP has finitely many variables.

Proposition 6. *Given any feasible assignment $\{\mathcal{X}_t(v)\}$ for LP-ONLINE(π), there exists a feasible online policy with an expected allocated value equal to the objective value of the LP under $\{\mathcal{X}_t(v)\}$.*

Proof. Define $q_t \triangleq 1 - \sum_{t'<t} \mathbb{E}_{v_{\pi(t')} \sim \mathcal{F}_{\pi(t')}} \left[\mathcal{X}_{t'}(v_{\pi(t')}) \right]$ for $t \geq 2$, and $q_1 \triangleq 1$. Consider the following randomized rounding policy: at time $t \geq 1$, if the item has already been allocated do nothing. If it has not yet been allocated, upon realizing the value $v_{\pi(t)}$ flip an independent coin with heads probability of $\frac{\mathcal{X}_t(v_{\pi(t)})}{q_t}$. Now, if the coin flips heads allocate the item and terminate. Otherwise, continue to the next buyer.

Clearly the above policy is online and feasible, i.e. it sells the item to only one buyer. To compare its expected allocated value with the objective value of the LP under the assignment $\{\mathcal{X}_t(v)\}$, we first claim that q_t is equal to the probability that this randomized policy reaches time t, i.e. with probability q_t the policy does not sell the item to any buyer arriving before time t. We prove this claim by induction. Clearly $q_1 = 1$ satisfies this property. As the induction hypothesis, suppose the policy reaches time $t \geq 2$ with probability q_t. To prove the inductive step, we have:

$$
\begin{aligned}
\mathbb{P}\left[\text{reaching time } t+1\right] &= \mathbb{P}\left[(\text{reaching time } t) \ \& \ (\text{no allocation at time } t)\right] \\
&= \mathbb{P}\left[\text{no allocation at time } t | \text{reaching time } t\right] \cdot q_t \\
&= \mathbb{E}_{v_{\pi(t)} \sim \mathcal{F}_t}\left[\mathbb{P}\left[\text{no allocation at time } t | (\text{reaching time } t) \ \& \ v_{\pi(t)}\right]\right] \cdot q_t \\
&= \mathbb{E}_{v_{\pi(t)} \sim \mathcal{F}_t}\left[1 - \frac{\mathcal{X}_t(v_{\pi(t)})}{q_t}\right] \cdot q_t = q_t - \mathbb{E}_{v_{\pi(t)} \sim \mathcal{F}_t}\left[\mathcal{X}_t(v_{\pi(t)})\right] = q_{t+1}
\end{aligned}
$$

Next we claim that by conditioning the realized value at time t to be v, the policy allocates the item with probability $\mathcal{X}_t(v)$. This is simply true because the policy reaches time t with probability q_t, and then conditioned on reaching time t and realizing value v allocates the item with probability $\frac{\mathcal{X}_t(v)}{q_t}$. Finally, as $\{\mathcal{X}_t(v)\}$ are the allocation probabilities of the policy (as we just proved), the expected allocated value at time t is equal to $\mathbb{E}_{v_{\pi(t)} \sim \mathcal{F}_{\pi(t)}} \left[v_{\pi(t)} \cdot \mathcal{X}_t(v_{\pi(t)}) \right]$. The proof of the proposition is then finished by summing over all t.

3.2   Ex-ante Relaxation and Rounding

The goal of this section is to propose two sequential pricing policies, one for the case of non-identical distributions and one for the case of identical distributions, so that they obtain $\frac{1}{2}$ and $1 - \frac{1}{e}$ fractions of the expected value of the omniscient prophet benchmark, respectively. To this end, we use LP-ONLINE(π) and the rounding algorithm proposed in Proposition 6, and in a modular fashion design new algorithms satisfying the classic prophet inequality of [27] and the semi-optimal prophet inequality of [13] and [18].

Our approach is based on a relaxation of the omniscient prophet benchmark which we term as the *ex-ante relaxation*. Suppose the seller intends to sell the item, but rather than selling the item to only one buyer for every profile of buyer values, it has a relaxed constraint of selling the item to one person in expectation over buyer values. Without loss of generality, assume $\forall t : \pi(t) = t$ (this will be cleared shortly). In the ex-ante relaxation benchmark the seller only needs to sell the item to each buyer t with probability q_t, where $\sum_t q_t \leq 1$. Clearly, the maximum expected value of the ex-ante relaxation is an upper-bound on the expected value of the optimum offline mechanisms, as the omniscient prophet allocates the item to only one buyer point-wise. Moreover, the following linear program capture the ex-ante relaxation:

$$\text{(Ex-ante-LP)}$$

$$\text{maximize } \sum_{t=1}^{n} \mathbb{E}_{v_t \sim \mathcal{F}_t} [v_t \cdot \mathcal{X}_t(v_t)]$$

$$\text{subject to } \sum_{t \in [n]} \mathbb{E}_{v_t \sim \mathcal{F}_t} [\mathcal{X}_t(v_t)] \leq 1,$$

$$\mathcal{X}_t(v) \geq 0, \qquad\qquad \forall v \in \text{supp}(\mathcal{F}_t), \ t = 1, \ldots, n$$

Fix a feasible assignment $\{\mathcal{X}_t(v)\}$ for the above LP, and let $q_t = \mathbb{E}_{v_t \sim \mathcal{F}_t} [\mathcal{X}_t(v_t)]$. Note that $\sum_t q_t \leq 1$. Now, one can replace $\{\mathcal{X}_t(v)\}$ with the following assignment, which obtains at least as much expected value as before and is a feasible assignment for Ex-ante-LP:

$$\hat{\mathcal{X}}_t(v) = \begin{cases} 1 & v \geq T_t(q_t), \\ 0 & \text{o.w.} \end{cases}$$

where $T_t(q_t)$ is the price corresponding to the quantile q_t of the distribution $\mathcal{F}_t$, as in Definition 1. Under the pricing allocations $\{\hat{\mathcal{X}}_t(v)\}$, the objective value of the ex-ante LP is equal to $\sum_{t=1}^{n} V_t(q_t)$, where $V_t(q_t)$ is the concave value-curve of the distribution $\mathcal{F}_t$, as in Definition 1. By putting all the pieces together, the optimal solution to Ex-ante-LP is $\mathcal{X}_t^*(v) = \mathbb{1}\{v \geq T_t(q_t^*)\}$, where $\mathbf{q}^*$ is the optimal solution of the following convex program:

$$\text{maximize } \sum_{t=1}^{n} V_t(q_t)$$

$$\text{(Ex-ante-Conv)}$$

$$\text{subject to } \sum_{t \in [n]} q_t \leq 1, \ q_t \geq 0 \qquad t = 1, \ldots, n$$

Remark 1. Note that the optimal solution of the ex-ante relaxation (Ex-ante-LP) can be computed by only knowing the set of distributions $\{\mathcal{F}_t\}_{t \in [n]}$, and without the need to know the ordering π. In other words, this benchmark, similar to optimum offline, is *order oblivious*; no matter what the ordering π is, the ex-ante relaxation yields the same solution.

Rounding for the Non-Identical Distributions. We now start with the optimal solution of the ex-ante relaxation described above, i.e. $\{\mathcal{X}_t^*(v)\}$, and modify it so that it becomes online implementable. In a nutshell, consider $\mathcal{X}_t(v) = \frac{1}{2}\mathcal{X}_t^*(v)$. We show this solution is feasible for LP-ONLINE(π), for any π as the true ordering of the buyers. Below, without loss of generality, assume $\forall t : \pi(t) = t$.

Proposition 7. *The expected value obtained by Algorithm 1 is at least $\frac{1}{2}$ of optimum offline.*

Proof. Suppose $\mathcal{X}_t^*(v) = \mathbb{1}\{v \geq T_t(q_t^*)\}$ is the optimal assignment of (Ex-ante-LP), where $\mathbf{q}^*$ is the optimal solution of the convex program Ex-ante-Conv. Consider $\mathcal{X}_t(v) = \frac{1}{2}\mathcal{X}_t^*(v)$. We have:

$$\mathcal{X}_t(v) = \frac{1}{2}\mathcal{X}_t^*(v) \overset{(1)}{\leq} \frac{1}{2} \overset{(2)}{\leq} 1 - \frac{1}{2}\sum_{t'<t} q_{t'}^* \overset{(3)}{=} 1 - \mathbb{E}_{v_{t'} \sim \mathcal{F}_{t'}}\left[\mathcal{X}_{t'}(v_{t'})\right]$$

where inequality (1) holds as $\mathcal{X}_t^*(v) \leq 1$, inequality (2) holds as $\sum_{t'<t} q_{t'}^* \leq \sum_{t'} q_{t'}^* \leq 1$, and equality (3) holds as $\mathbb{E}_{v_{t'} \sim \mathcal{F}_{t'}}\left[\mathcal{X}_{t'}(v_{t'})\right] = \frac{1}{2}\mathbb{E}_{v_{t'} \sim \mathcal{F}_{t'}}\left[\mathcal{X}_{t'}^*(v_{t'})\right] = \frac{1}{2}q_{t'}^*$. Therefore, $\{\mathcal{X}_t(v)\}$ forms a feasible assignment for the LP-ONLINE(π).[4] By applying Proposition 6, there exists a feasible randomized policy that implements $\{\mathcal{X}_t(v)\}$; this policy obtains exactly the same allocation probabilities as $\{\mathcal{X}_t(v)\}$ and obtains an expected value equal to the objective value of LP-ONLINE(π) for this assignment. Clearly, this objective value is at least $\frac{1}{2}$ of the expected value of the optimum offline, as the optimal value of the ex-ante relaxation program is an upper-bound on the expected maximum value. Finally, note that the randomized policy implementing $\mathcal{X}_t(v)$, described in the proof of Proposition 6, is exactly equivalent to Algorithm 1.

Algorithm 1: Online policy for non-identical distributions

---

input: Distributions $\{\mathcal{F}_1, \ldots, \mathcal{F}_n\}$
Compute the optimal solution of (Ex-ante-Conv). Let $\mathbf{q}^*$ be this optimal solution.
$t \leftarrow 0$
while [*item is not allocated*] & [$t \leq n$] **do**
    $t \leftarrow t + 1$
    Post a price $p_t \triangleq T_t(q_t^*)$.
    if *price p_t gets accepted ($v_t \geq p_t$)* **then**
        Allocate the item with probability $\dfrac{1}{2 - \sum_{t'<t} q_{t'}^*}$.

---
[4] Note that we assume the true π is the unitary ordering, but that is without loss of generality.

Rounding for the identical distributions. Can we round the ex-ante optimal solution for the case of identical distributions, and obtain the improved bound of $1 - \frac{1}{e}$, or even the optimal bound in [13]? Interestingly, by incorporating a careful rounding of the ex-ante and using the LP of optimum online, we can obtain a mechanism which is posting the single price of $T(\frac{1}{n})$ and show that it achieves at least $1 - \frac{1}{e}$ fraction of the optimum offline (hence an alternative proof for Proposition 2 and a similar result in [13] and [18]).

Proof (Proof of Proposition 2 using LP). Due to the symmetry, the optimal solution of Ex-ante-Conv is attained at $q_i^* = \frac{1}{n}$, and hence $\mathcal{X}_t^*(v) = \mathbb{1}\{v \geq T(\frac{1}{n})\}$. Let $\gamma \triangleq 1 - \frac{1}{n}$, and consider the solution $\mathcal{X}_t(v) = \gamma^t \cdot \mathcal{X}_t^*(v)$. Note that $\mathbb{E}_{v \sim \mathcal{F}}[\mathcal{X}_t^*(v)] = \frac{1}{n} = 1 - \gamma$ for all t. Moreover, we have $\mathcal{X}_t(v) = \gamma^t \cdot \mathcal{X}_t^*(v) \leq \gamma^t$, simply because $\mathcal{X}_t^*(v) \leq 1$. Therefore,

$$1 - \sum_{t' < t} \mathbb{E}_{v_{t'} \sim \mathcal{F}}[\mathcal{X}_{t'}(v_{t'})] = 1 - \sum_{t' < t} \gamma^{t'} \cdot \mathbb{E}_{v_{t'} \sim \mathcal{F}}[\mathcal{X}_{t'}^*(v_{t'})] = 1 - \frac{1}{n}\frac{1 - \gamma^t}{1 - \gamma} = \gamma^t$$

where in the last equality we used $\gamma = 1 - \frac{1}{n}$. So, $\mathcal{X}_t(v) \leq 1 - \sum_{t' < t} \mathbb{E}_{v_{t'} \sim \mathcal{F}}[\mathcal{X}_{t'}(v_{t'})]$, and hence forms a feasible solution to LP-ONLINE(π) for any π. Proposition 6 suggests that there exists a randomized policy that implements this feasible assignment. In fact, similar to the proof of Proposition 6, the final policy should post the price $T(\frac{1}{n})$, and if $v \geq T(\frac{1}{n})$ should accept it with probability:

$$\frac{\gamma^t}{1 - \sum_{t' < t} \mathbb{E}_{v_{t'} \sim \mathcal{F}}[\mathcal{X}_{t'}(v_{t'})]} = \frac{\gamma^t}{1 - \frac{1}{n}\sum_{t' < t}\gamma^{t'}} = \frac{\gamma^t}{1 - \frac{1}{n}\frac{1 - \gamma^t}{1 - \gamma}} = 1,$$

where the last equality again holds because $\gamma = 1 - \frac{1}{n}$. So, the exact rounding policy simply suggests posting the single price $T(\frac{1}{n})$. To compare the expected value of this policy with that of the optimum offline, we only need to compare the objective value of LP-ONLINE(π) at this solution with the ex-ante objective value, thanks to Proposition 6. We have:

$$\sum_{t=1}^{n} \mathbb{E}_{v_t \sim \mathcal{F}}[v_t \cdot \mathcal{X}_t(v_t)] = \sum_{t=1}^{n} \gamma^t \cdot \mathbb{E}_{v_t \sim \mathcal{F}}\left[v_t \cdot \mathbb{1}\{v_t \geq T(\frac{1}{n})\}\right] = V(\frac{1}{n})\sum_{t=1}^{n}\gamma^t = V(\frac{1}{n})\frac{1 - \gamma^{n+1}}{1 - \gamma},$$

where the right-hand-side is equal to $n \cdot V(\frac{1}{n}) \cdot (1 - (1 - \frac{1}{n})^{n+1})$ as $\gamma = 1 - \frac{1}{n}$. Finally, the ex-ante optimal objective is equal to $n \cdot V(\frac{1}{n})$ and $(1 - (1 - \frac{1}{n})^{n+1}) \geq 1 - \frac{1}{e}$, which completes the proof.

4   Discussion

When designing a simple mechanism, one key question to ask is how do these simple mechanisms perform compared to a *benchmark optimal mechanism*. Finding the answer to this question occasionally sheds insights on the structure of the simple mechanisms and potentially leads to the design of better ones. But, which benchmark should be selected? Quite often the designer has multiple economically justified choices for the benchmark, and the answer to the above "simple vs. optimal" question relies heavily on this choice. It is therefore valuable to characterize and compare the different existing benchmarks, and understand how well each one is amenable to approximations by simple mechanisms.

In this paper we take the first stab to consider approximations with respect to the optimum online benchmark for the single item prophet inequality setting. We believe this approach can be used for other settings as well and can lead to more insights on designing more meaningful approximation mechanisms. We leave these directions as future research.

References

1. Abolhassani, M., Ehsani, S., Esfandiari, H., HajiAghayi, M., Kleinberg, R., Lucier, B.: Beating $1 - 1/e$ for ordered prophets. In: Proceedings of the 49th Annual ACM SIGACT Symposium on Theory of Computing, pp. 61–71. ACM (2017)
2. Alaei, S.: Bayesian combinatorial auctions: expanding single buyer mechanisms to many buyers. SIAM J. Comput. **43**(2), 930–972 (2014)
3. Alaei, S., Hartline, J., Niazadeh, R., Pountourakis, E., Yuan, Y.: Optimal auctions vs. anonymous pricing. In: 2015 IEEE 56th Annual Symposium on Foundations of Computer Science (FOCS), pp. 1446–1463. IEEE (2015)
4. Anari, N., Niazadeh, R., Saberi, A., Shameli, A.: Nearly optimal pricing algorithms for production constrained and laminar bayesian selection. arXiv preprint arXiv:1807.05477 (2018)
5. Azar, P.D., Kleinberg, R., Weinberg, S.M.: Prophet inequalities with limited information. In: Proceedings of the Twenty-Fifth Annual ACM-SIAM Symposium on Discrete Algorithms, pp. 1358–1377. Society for Industrial and Applied Mathematics (2014)
6. Azar, Y., Chiplunkar, A., Kaplan, H.: Prophet secretary: Surpassing the $1 - 1/e$ barrier. In: Proceedings of the 2018 ACM Conference on Economics and Computation, pp. 303–318. ACM (2018)
7. Babaioff, M., Immorlica, N., Lucier, B., Weinberg, S.M.: A simple and approximately optimal mechanism for an additive buyer. ACM SIGecom Exch. **13**(2), 31–35 (2015)
8. Beyhaghi, H., Golrezaei, N., Leme, R.P., Pal, M., Siva, B.: Improved approximations for free-order prophets and second-price auctions. arXiv preprint arXiv:1807.03435 (2018)
9. Cai, Y., Daskalakis, C., Weinberg, S.M.: Optimal multi-dimensional mechanism design: reducing revenue to welfare maximization. In: 2012 IEEE 53rd Annual Symposium on Foundations of Computer Science (FOCS), pp. 130–139. IEEE (2012)
10. Cai, Y., Devanur, N.R., Weinberg, S.M.: A duality based unified approach to Bayesian mechanism design. In: Proceedings of the Forty-Eighth Annual ACM Symposium on Theory of Computing, pp. 926–939. ACM (2016)
11. Chawla, S., Hartline, J.D., Malec, D.L., Sivan, B.: Multi-parameter mechanism design and sequential posted pricing. In: Proceedings of the Forty-Second ACM Symposium on Theory of Computing, pp. 311–320. ACM (2010)
12. Chawla, S., Miller, J.B.: Mechanism design for subadditive agents via an ex-ante relaxation. In: Proceedings of the 2016 ACM Conference on Economics and Computation, pp. 579–596. ACM (2016)

13. Correa, J., Foncea, P., Hoeksma, R., Oosterwijk, T., Vredeveld, T.: Posted price mechanisms for a random stream of customers. In: Proceedings of the 2017 ACM Conference on Economics and Computation, pp. 169–186. ACM (2017)
14. Correa, J., Saona, R., Ziliotto, B.: Prophet secretary through blind strategies. arXiv preprint arXiv:1807.07483 (2018)
15. Düetting, P., Feldman, M., Kesselheim, T., Lucier, B.: Prophet inequalities made easy: stochastic optimization by pricing non-stochastic inputs. In: 2017 IEEE 58th Annual Symposium on Foundations of Computer Science (FOCS), pp. 540–551. IEEE (2017)
16. Dütting, P., Fischer, F., Klimm, M.: Revenue gaps for discriminatory and anonymous sequential posted pricing. arXiv preprint arXiv:1607.07105 (2016)
17. Ehsani, S., Hajiaghayi, M., Kesselheim, T., Singla, S.: Prophet secretary for combinatorial auctions and matroids. In: Proceedings of the Twenty-Ninth Annual ACM-SIAM Symposium on Discrete Algorithms, pp. 700–714. SIAM (2018)
18. Esfandiari, H., Hajiaghayi, M., Liaghat, V., Monemizadeh, M.: Prophet secretary. SIAM J. Discrete Math. 31(3), 1685–1701 (2017)
19. Feldman, M., Fu, H., Gravin, N., Lucier, B.: Simultaneous auctions are (almost) efficient. In: Proceedings of the Forty-Fifth Annual ACM Symposium on Theory of Computing, pp. 201–210. ACM (2013)
20. Feldman, M., Svensson, O., Zenklusen, R.: Online contention resolution schemes. In: Proceedings of the Twenty-Seventh Annual ACM-SIAM Symposium on Discrete Algorithms, pp. 1014–1033. Society for Industrial and Applied Mathematics (2016)
21. Göbel, O., Hoefer, M., Kesselheim, T., Schleiden, T., Vöcking, B.: Online independent set beyond the worst-case: secretaries, prophets, and periods. In: Esparza, J., Fraigniaud, P., Husfeldt, T., Koutsoupias, E. (eds.) ICALP 2014. LNCS, vol. 8573, pp. 508–519. Springer, Heidelberg (2014). https://doi.org/10.1007/978-3-662-43951-7_43
22. Hajiaghayi, M.T., Kleinberg, R., Sandholm, T.: Automated online mechanism design and prophet inequalities. AAAI. vol. 7, pp. 58–65 (2007)
23. Hartline, J.D.: Approximation in mechanism design. Am. Econ. Rev. 102(3), 330–336 (2012)
24. Hartline, J.D., Roughgarden, T.: Simple versus optimal mechanisms. In: Proceedings of the 10th ACM Conference on Electronic Commerce, pp. 225–234. ACM (2009)
25. Hill, T.P., Kertz, R.P.: Comparisons of stop rule and supremum expectations of IID random variables. Ann. Probab. 10(2), 336–345 (1982)
26. Kleinberg, R., Weinberg, S.M.: Matroid prophet inequalities. In: Proceedings of the Forty-Fourth Annual ACM Symposium on Theory of Computing, pp. 123–136. ACM (2012)
27. Krengel, U., Sucheston, L.: On semiamarts, amarts, and processes with finite value. Probab. Banach Spaces 4, 197–266 (1978)
28. Lee, E., Singla, S.: Optimal online contention resolution schemes via ex-ante prophet inequalities. In: LIPIcs-Leibniz International Proceedings in Informatics, vol. 112. Schloss Dagstuhl-Leibniz-Zentrum fuer Informatik (2018)
29. Lucier, B.: An economic view of prophet inequalities. ACM SIGecom Exch. 16(1), 24–47 (2017)
30. Rubinstein, A.: Beyond matroids: secretary problem and prophet inequality with general constraints. In: Proceedings of the Forty-eighth Annual ACM Symposium on Theory of Computing, pp. 324–332. ACM (2016)

31. Samuel-Cahn, E.: Comparison of threshold stop rules and maximum for independent nonnegative random variables. Ann. Probab. **12**(4), 1213–1216 (1984)
32. Yan, Q.: Mechanism design via correlation gap. In: Proceedings of the Twenty-Second Annual ACM-SIAM Symposium on Discrete Algorithms, pp. 710–719. Society for Industrial and Applied Mathematics (2011)

Optimal Mechanism Design with Risk-Loving Agents

Evdokia Nikolova, Emmanouil Pountourakis[✉], and Ger Yang

The University of Texas at Austin, Austin, USA
`manolis@utexas.edu`

Abstract. One of the most celebrated results in mechanism design is Myerson's characterization of the revenue optimal auction for selling a single item. However, this result relies heavily on the assumption that buyers are indifferent to risk. In this paper we investigate the case where the buyers are risk-loving, i.e. they prefer gambling to being rewarded deterministically. We use the standard model for risk from expected utility theory, where risk-loving behavior is represented by a convex utility function.

We focus our attention on the special case of exponential utility functions. We characterize the optimal auction and show that randomization can be used to extract more revenue than when buyers are risk-neutral. Most importantly, we show that the optimal auction is simple: the optimal revenue can be extracted using a randomized take-it-or-leave-it price for a single buyer and using a loser-pay auction, a variant of the all-pay auction, for multiple buyers. Finally, we show that these results no longer hold for convex utility functions beyond exponential.

1 Introduction

The classic mechanism design problem, pioneered by Myerson's seminal work (Myerson 1981), considers designing an auction that maximizes the auctioneer's revenue. There is rich literature on this mechanism design problem under different settings. However, most prior work assumes the buyers are utility maximizers with quasilinear utility functions, where the utility function is linear in either the payment or the buyer's value. These assumptions often make the problem simple and easy to analyze. In the real world, agents need not follow such assumptions. In fact, under different behavioral models for the buyers, the auctioneer is able to draw more revenue than under the standard setting, where the buyers are maximizing their linear utility functions. One particular example is when the buyers are risk-averse (Chawla et al. 2018; Maskin and Riley 1984). In this case, the seller/auctioneer can design an "insurance"-based auction to extract more revenue from risk-averse buyers. In this paper, we study the setting where the buyers are *risk-loving*. We ask whether the auctioneer can take advantage of such risk-loving behavior, and if so, what can be achieved?

Recently, experiments in electricity markets and transportation networks have demonstrated the importance of designing a mechanism for risk-loving

© Springer Nature Switzerland AG 2018
G. Christodoulou and T. Harks (Eds.): WINE 2018, LNCS 11316, pp. 375–392, 2018.
https://doi.org/10.1007/978-3-030-04612-5_25

agents. Electric utility companies are considering how to incentivize customers to reduce their electricity consumption in peak load times so as to alleviate the strain on the grid and to prevent expensive line capacity and transformer upgrades. Some of the more successful attempts to achieve a desired "demand response" have included offering lottery coupons to consumers for scaling back demand (Li et al. 2015). In transportation networks, similar lottery schemes have been applied to reduce congestion in the rush hour (Lu 2015; Merugu et al. 2009; Pluntke and Prabhakar 2013). In both cases, more consumer response was elicited from lotteries, where a consumer was offered a small chance to win a big reward, than from small fixed payments. Hence, there is a need for a theoretical foundation and analysis of the optimal lottery schemes to improve the consumer response and experience in these nation critical infrastructure applications.

In economics, von Neumann-Morgenstern's expected utility theory (Von Neumann and Morgenstern 1945) has been a standard model to describe people's preferences. According to this theory, an agent evaluates the payoff of an event by applying a utility function on the wealth it generates, and takes the expectation over all possible events to evaluate the payoff of a given action. As such, expected utility theory provides a simple way to describe how people behave when facing *risk*—a risk-averse player has a concave utility function, whereas a risk-loving player has a convex utility function. Consider a payment scheme where a buyer can choose one of two payment options. In the first option, the buyer either pays $100 or $0, each with probability 50%. In the second option, she has to pay $50 with certainty. These two options have the same expected gains. A risk-neutral buyer, who has a linear utility function, is indifferent between these two options. A risk-averse buyer is going to choose the second option because she prefers the less risky payment scheme. A risk-loving buyer will choose the first option because she is more willing to take risks.

In the above example, the expected payment the seller receives is $50. If the buyer is risk-loving, we can extract more revenue by replacing the first option with an even more risky payment option. For example, we can offer another payment option in which the buyer pays $110 or 0, each with probability 50%. From the risk-loving buyer's perspective, this new payment option is still preferable to the second option. Therefore, the expected payment the seller receives will increase to $55. In fact, it has been shown by Hinnosaar (2017) that in the absence of any regulation, the seller is able to extract infinite expected revenue from a risk-loving buyer by simply taking advantage of this trick—offering a menu option that asks the buyer to pay a very high amount with a very small probability. Therefore, in this paper, we will mainly focus on the *bounded transfer* setting, i.e. where we upper bound the *ex-post* payment that the seller may ask the buyer to pay. In other words, the amount of payment by the buyer is upper-bounded by some specific value under all circumstances. Particularly, the *bounded transfer* requirement can be shown to be equivalent to the buyers having a publicly known really high yet still bounded budget.

1.1 Our Results and Techniques

In this paper, we focus on a special case of risk-loving agents, that use an exponential utility function of the form $u(x) = \beta(e^{\alpha x} - 1)$. We seek to design individually rational and incentive compatible mechanisms that maximize the revenue. We assume bounded transfers, that is the maximum payment of the mechanism is bounded, and characterize the optimal mechanism. Surprisingly, we show that if the value distribution of the agents is well behaved, then the optimal revenue can be extracted using a randomized take-it-or-leave-it price for a single buyer and a loser-pay auction, a variant of the all-pay auction where the winner gets a refund, for multiple buyers.

Our analysis combines a generalized virtual value function similarly to Myerson's analysis (Myerson 1981) and the duality framework developed by Cai et al. (2016). In particular, we upper bound the revenue of the optimal mechanism by defining a dual solution that can be interpreted as a generalization of the virtual value function. Then we show that this solution matches the revenue obtained by a randomized-take-it-or-leave-it price and the loser-pay auction, for a single buyer and for multiple buyers, respectively. To our surprise, the virtual value function that captures the marginal revenue is different in the single buyer and multiple buyer settings, which may be explained due to the additional uncertainty introduced by the extra buyers.

These results are in stark contrast with the risk-averse setting where the seller can improve the revenue by offering a plethora of lotteries each with a deterministic price but different allocation probabilities (Maskin and Riley 1984). The risk-averse buyer opts to pay for lotteries that are priced close to her value and the risk is used as a deterrent for under-bidding. On the other hand, we can extract more revenue from a risk-loving buyer by randomizing the payment. This is because the buyer gains more utility from gambles so that we can increase the probability that the price is accepted. This difference in how risk behavior is exploited explains the conceptually different nature of revenue maximizing mechanisms in the two settings.

2   Related Work

Most work on optimal mechanism design beyond the risk-neutral setting has focused on risk-averse preferences. The classic results of Maskin and Riley (1984) and Matthews (1983) provide a characterization of the optimal mechanism with concave utility functions. A recent result in this area by Dughmi and Peres (2012) is that any mechanism designed for risk-neutral buyers can be adjusted to also align the incentives of risk-averse buyers and obtain similar guarantees. Fu et al. (2013) consider the design of prior-independent mechanisms (that have no access to the buyers' private value distributions) for risk-averse buyers. Finally, Chawla et al. (2018) study the design of robust mechanisms under the cumulative prospect theory model.

To the best our knowledge, the only work on mechanism design under risk-loving behavior is by Hinnosaar (2017), who shows that in the absence of regulations, the seller can extract infinite revenue from the buyer with asymptotically risk-loving behavior under both the expected utility theory and prospect theory models.

Recently, the duality theory framework has drawn attention in the mechanism design community for understanding optimal mechanisms for selling multiple items. For example, Daskalakis et al. (2017, 2013) and Giannakopoulos and Koutsoupias (2014, 2015) discovered the connection between the dual problem and the optimal transport (bipartite matching) problem. Cai et al. (2016) consider a duality framework via linear programming, and identify a connection between the virtual valuations and the dual variables. In our setting, the problem results in a different form of dual problem than in the multi-item setting, hence we seek to establish a new duality framework that diverges from the multi-item setting to different behavior models.

3   Problem Statement

We study revenue maximization for a single seller and n symmetric buyers. The seller has a single item to sell and each buyer i has a private value t_i for the item. We use $t = (t_1, \ldots, t_n)$ to denote the values of all buyers. We let $V = \{v_1, v_2, \ldots, v_K\}$ denote the set of all possible values, which is shared by all buyers. For simplicity, we assume $v_1 = 0$ and $v_1 < v_2 < \cdots < v_K$. Additionally, we assume each buyer's private value is drawn independently from a known identical distribution with probability mass function f. Without loss of generality, we assume $f(v) > 0$ for all $v \in V$. Further, we let $\mathcal{P} = \{z_1, z_2, \ldots, z_M\}$ denote the set of allowed payments, where $z_1 = 0$ (no positive transfers) and $z_1 < z_2 < \cdots < z_M$. Here we implicitly assume that $\mathcal{P}$ is upper-bounded by z_M[1]. This implies that our setting becomes equivalent with the case where the payments are unconstrained but the buyer has a publicly known budget of z_M as we show in Subsect. 3.2. We additionally require that $z_M > v_K$, that is, the upper bound of the payment is larger than the largest possible buyer's value.

Each buyer seeks to maximize her utility given by a function $u : \mathbb{R} \rightarrow \mathbb{R}$, which we assume is strictly increasing and $u(0) = 0$. If u is linear, then we say the buyers are *risk-neutral* and if u is convex, then we say that the buyers are *risk-loving*. For the rest of the paper, we focus on a special case of convex utility, specifically the exponential utility function given by $u(x) = \beta(e^{\alpha x} - 1)$ for some $\alpha > 0$ and $\beta > 0$. Unless otherwise noted, we will assume such an exponential utility function for the buyers.

Notation. Let $[R]$ denote the set $\{1, 2, \ldots, R\}$, for any positive integer R. For any vector v, we use v_{-i} to denote the vector generated by removing the i-th coordinate from v. Also, we use (v, v_{-i}) to denote the vector generated by replacing the i-th coordinate of v with v.

---

[1] Without this assumption, it can be shown that there exists a mechanism that attains infinite revenue from risk-loving buyers Hinnosaar (2017).

3.1 Direct Mechanisms and Bayesian Incentive Compatibility

In a direct mechanism the auctioneer elicits bids from each buyer and then decides on their allocation probabilities and payments. We represent such a mechanism by $\mathcal{M}_d = (X, P)$, where $X : V^n \to \{0, 1\}^n$ is a random allocation function and $P : V^n \to \mathcal{P}^n$ is a payment function which can also be randomized. Given all buyers' values $t = (t_1, \ldots, t_n)$, we refer to the random variable $(X(t), P(t))$ as the *outcome* of the mechanism at t.

We require that our mechanism $\mathcal{M}_d$ is *Bayesian incentive compatible* (BIC), that is, for each buyer, it is in her best interest to truthfully report her value in expectation. Note that this expectation takes into account the randomness of the mechanism as well as the uncertainty about the other buyers' values. Formally, $\mathcal{M}_d = (X, P)$ is BIC if for any $i \in [n]$ and for any $v \in V$ and $v' \in V$, it holds that

$$\mathbb{E}[u(vX(v, t_{-i}) - P(v, t_{-i}))] \geq \mathbb{E}[u(vX(v', t_{-i}) - P(v', t_{-i}))], \tag{1}$$

where the expectation is taken over X, P, and t_{-i}. A mechanism is *individually rational* (IR) if it guarantees a non-negative expected utility for every buyer that truthfully reveals her value, i.e., for any $i \in [n]$ and for any $v \in V$, it holds that

$$\mathbb{E}[u(vX(v, t_{-i}) - P(v, t_{-i}))] \geq 0. \tag{2}$$

Note that if we only allow non-negative payments, then we must have $P(0, t_{-i}) \leq 0$ almost surely.

3.2 Bounded Transfers and Budgeted Buyers

As we stated in the beginning of Sect. 3 we require the mechanism to charge ex-post payments from a finite pool of $\mathcal{P}$, where z_M is the largest ex-post price that also satisfies that $z_M > v_K$, i.e., the upper-bound on the payment is larger than the highest value of the buyer. The finiteness of z_M can be thought of as buyers having a finite budget equal to z_M. Particulalry, for the case where $\mathcal{P}$ was unbounded but the buyers had a budget of z_M then no revenue maximizing IR mechanism would ever charge an ex-post price larger than z_M. Similarly, in any feasible mechanism under the bounded-transfer setting, the buyers with budget greater than the upper-bound on the ex-post price behave as if they had no budget at all. As a result, in both of those cases the revenue-maximizing BIC and IR mechanism are the same.

3.3 Myerson's Mechanism and Virtual Values

One of the fundamental results of auction theory is Myerson's characterization of revenue optimal mechanisms for risk-neutral buyers (Myerson 1981). This is achieved by an amortized analysis that expresses the revenue of any mechanism via the *virtual value function* $\phi(v)$, which captures the marginal revenue of allocating to a buyer with value v. The virtual value function is defined for a continuous distribution of values (and can be similarly defined for a discrete one), with cumulative distribution function F and probability density function f, as

$$\phi(v) = v - \frac{1 - F(v)}{f(v)}. \tag{3}$$

The revenue of the mechanism equals the expected virtual surplus, i.e., the expected virtual value of the winner. As a result, if the value distributions satisfy certain properties, the optimal mechanism turns out to be quite simple: for a single buyer it is just a take-it-or-leave-it price and for multiple symmetric buyers it is the second price auction with a common reserve. However, this definition of virtual values heavily relies on the risk-neutrality assumption. Our analysis generalizes this definition for risk-loving buyers in Definition 2 in order to derive our results.

3.4   Revenue Maximization as an Optimization Problem

Our goal is to characterize the optimal mechanism for revenue maximization. To that end, we model the mechanism design question as an optimization problem. We define the decision variables $\{y_{i,j}^0, y_{i,j}^1\}_{i \in [n], j \in [M]}$, where $y_{i,j}^0 : V^n \to [0, 1]$ and $y_{i,j}^1 : V^n \to [0, 1]$, that encode the mechanism $\mathcal{M}_d$ as follows: $y_{i,j}^0(t)$ represents the probability that buyer i does not get the item and pays z_j when the buyers' values are t. Similarly, $y_{i,j}^1(t)$ represents the probability that buyer i gets the item and pays z_j, given the buyers' values are t.

Those decision variables capture both the allocation and the payment of the mechanism given any reported values. To see this, the allocation probability that buyer i gets the item given values t is $\sum_j y_{i,j}^1(t)$ and the expectation of her randomized payment is $\sum_j z_j y_{i,j}^1(t) + \sum_j z_j y_{i,j}^0(t)$ where the first and second summand correspond to her expected payment if she wins or loses the item respectively.

For the sake of succinctness of our optimization problem formulation, we further define the *interim* version of the decision variables $y_{i,j}^1(t), y_{i,j}^0(t)$, denoted by $y_{i,j}^1(v_k), y_{i,j}^0(v_k)$. Namely, given that the buyer has value v_k, what is the expected probability of winning/losing the item and paying value z_j in expectation over the values of the other buyers v_{-i}? These interim variables are given by:

$$y_{i,j}^1(v_k) = \sum_{v_{-i} \in V^{n-1}} y_{i,j}^1(v_k, v_{-i}) f(v_{-i}), \quad y_{i,j}^0(v_k) = \sum_{v_{-i} \in V^{n-1}} y_{i,j}^0(v_k, v_{-i}) f(v_{-i}).$$

$$\tag{4}$$

We can express the interim allocation $x_i(k)$ of buyer i at value v_k as $x_i(k) = \sum_j y_{i,j}^1(v_k)$ and her interim payments in case of win $p_i(k)$ and loss $q_i(k)$ as:

$$p_i(k) = \frac{\sum_j z_j y_{i,j}^1(v_k)}{x_i(k)}, \qquad q_i(k) = \frac{\sum_j z_j y_{i,j}^0(v_k)}{1 - x_i(k)}.$$

The above follow from the definition of conditional probability. With this notation, we can rewrite the BIC constraint as:

$$\sum_j \left[y_{i,j}^1(v_k)u(v_k - z_j) + y_{i,j}^0(v_k)u(-z_j) \right] \geq \sum_j \left[y_{i,j}^1(v_{k'})u(v_k - z_j) + y_{i,j}^0(v_{k'})u(-z_j) \right], \forall k, k' \in [K],$$

(5)

where the first and second summand on the left hand side correspond to the expected utility if buyer i wins and loses, respectively, after truthfully reporting v_k. Similarly, the first and second summand on the right hand side correspond to the expected utility if buyer i wins and loses, respectively, after misreporting $v_{k'}$.

In addition, we can write the IR constraint as

$$\sum_j \left[y_{i,j}^1(v_k)u(v_k - z_j) + y_{i,j}^0(v_k)u(-z_j) \right] \geq 0, \qquad \forall k \in [K].$$

(6)

Finally, we need to satisfy the feasibility constraints

$$\sum_j \left(y_{i,j}^0(v) + y_{i,j}^1(v) \right) = 1, \qquad \sum_i \sum_j y_{i,j}^1(v) \leq 1, \qquad \forall v \in V^n.$$

(7)

Therefore, we can find the optimal mechanism by solving the following linear program:

$$\text{Maximize} \quad \sum_i \sum_{v \in V} f(v) \sum_j z_j \left[y_{i,j}^0(v) + y_{i,j}^1(v) \right]$$

$$\text{Subject to} \quad \text{Constraints (4), (5), (6), and (7).}$$

$$y_{i,j}^0(v) \geq 0, \quad y_{i,j}^1(v) \geq 0, \qquad \qquad \forall v \in V^n. \quad (8)$$

3.5   Overview of Main Theorems and Results

The main result of this paper is that there is no need to actually solve the linear program (8) in order to compute the optimal mechanism. Instead, we take advantage of the linear program formulation (8) of the problem to help us derive simple mechanisms that are optimal. In particular, when there is a single risk-loving buyer with an exponential utility function, we show that the optimal mechanism is a *randomized "take-it-or-leave-it" price*, which offers the buyer a single randomized price irrespectively of her value. We present this result in the following theorem:

Theorem 1 (Restatement of Theorem 5). *Consider a single risk-loving buyer with exponential utility. The optimal mechanism is the revenue maximizing randomized take-it-or-leave-it price.*

When there are multiple symmetric risk-loving buyers, we show that the optimal mechanism is a *loser-pay* auction with a reserve price. In a loser-pay auction, the item is awarded to the buyer with the highest value but only the buyers who do not get the item are paying. Similarly to the single buyer case, all payments are randomized between the minimum and the maximum price. We will show the following theorem:

Theorem 2 (Restatement of Theorem 6). *Consider $n \geq 2$ risk-loving buyers with exponential utility $u(x) = \beta(e^{\alpha x} - 1)$. Assume $z_M \gg \alpha$. Then, the optimal mechanism is a loser-pay auction.*

4   Optimal Mechanism Design for a Single Buyer

In this section, we characterize the revenue-maximizing mechanism for selling an item to a single risk-loving agent. Specifically, we show that the optimal mechanism is a *randomized "take-it-or-leave-it" price*, that offers the buyer a single randomized price irrespectively of her value. To prove that, we first characterize the revenue generated by the optimal randomized take-it-or-leave-it price. Then, we prove that this mechanism remains optimal if we allow an arbitrary BIC mechanism, by utilizing the optimization problem formulation and the duality framework to find a matching upper bound. Note that this result is quite similar to what the Myerson characterization implies for the single risk-neutral buyer setting with the exception that the optimal "take-it-or-leave-it" price is always deterministic. However, as Example 1 demonstrates, this is no longer true for the risk-loving setting, and randomizing the price is required even in the case where the allocation probability is deterministic.

For the rest of this section, we are going to use the following *menu of options* interpretation that provides an equivalent description of BIC mechanisms in the single buyer case: A menu consists of a tuple (X_i, P_i) where X_i and P_i represent the allocation and payment random variables. Instead of the buyer revealing her value, the seller offers the buyer a collection of menu options and the buyer chooses the menu option that maximizes her utility. For example, a deterministic take-it-or-leave-it price can be described with the menu options $(0,0), (1, P)$ where P is a point-mass at some price p.

4.1   Sub-optimality of Deterministic "Take-It-or-Leave-It" Prices

We illustrate that randomizing a take-it-or-leave-it price can result in increased revenue as the buyer shifts from a risk-neutral to a risk-loving utility function. For the sake of the presentation we use a continuous distribution but similar results can be derived using a discrete value distribution.

Example 1. Assume the buyer's value t for the item is distributed according to the Uniform distribution $U(0,1)$. Then, the optimal mechanism with a risk-neutral buyer is to offer the buyer a take-it-or-leave it price of $1/2$, producing a revenue of $1/4$. Now, consider the case of a risk-loving buyer with utility u. Her utility of accepting this price is $u(t-1/2)$. Now, consider a different scheme using a randomized take-it-or-leave-it price: with probability $1/2$ pay nothing and with probability $1/2$ pay 1. Note that this scheme has the same expected payment as the first one. However, the expected utility of the buyer for this option is $\frac{1}{2}u(t) + \frac{1}{2}u(t-1)$ and by Jensen's inequality, we get that $\frac{1}{2}u(t) + \frac{1}{2}u(t-1) > u\left(t - \frac{1}{2}\right)$, which indicates that the expected utility function for the randomized menu option is always above the utility function of the menu option $(1,1/2)$. This means that the probability that a buyer accepts the randomized menu option is greater than $1/2$, i.e. $\Pr[\frac{1}{2}u(t) + \frac{1}{2}u(t-1) \geq 0] \geq \Pr[u(t-1/2) \geq 0] = 1/2$. Therefore, offering the randomized menu option earns more revenue.

4.2   Optimal Take-It-or-Leave-It Randomized Price

In Example 1, we saw that randomizing the take-it-or-leave-it price increased the revenue extracted by a risk-loving buyer. This gives rise to the question: What is the revenue maximizing randomized take-it-or-leave-it price?

Definition 1. *A randomized take-it-or-leave-it price with allocation and price P is a mechanism that contains only two menu options $(0,0)$ and $(1, P)$.*

In a randomized take-it-or-leave-it price scheme, the seller posts a (possibly randomized) price P of the item, and asks the buyer to accept it or not. If the buyer rejects the price, then her allocation is zero and she pays nothing. According to Myerson (1981), we know that with a risk-neutral buyer, it is optimal to post P with the expectation of P equal to $\arg\max_v(v\Pr[t \geq v])$. In contrast, for a risk-loving buyer we show that P must be randomized and specifically have positive support only for the maximum and minimum allowable price. Formally, we state this in the following theorem:

Theorem 3. *With a risk-loving buyer, the randomized take-it-or-leave-it price with following randomized payment rule*

$$P = \begin{cases} z_M, & w.p. \ \frac{u(v^*)}{u(v^*)-u(v^*-z_M)} \\ 0, & otherwise \end{cases}$$

where $v^* = \arg\max_{v\in V} \frac{z_M \Pr[t\geq v]u(v)}{u(v)-u(v-z_M)}$, is the optimal randomized take-it-or-leave-it price. The optimal take-it-or-leave-it price has revenue $\max_{v\in V} \frac{z_M \Pr[t\geq v]u(v)}{u(v)-u(v-z_M)}$.

Note that for a linear utility function u this implements the optimal deterministic take-it-or-leave-it price for a risk neutral agent and produces the exact same revenue. We call the optimal take-it-or-leave-it mechanism the *revenue maximizing randomized take-it-or-leave-it price*.

Theorem 3 can be proved in two steps. In the first step, we apply Jensen's inequality to show that given any pricing rule P, we can construct another pricing rule P' that randomizes between 0 and z_M, and achieves a larger utility than P for any value $v \in V$. In the second step, we use the individual rationality constraint to derive the optimal pricing. Due to space constraints, we defer the full proof to the full version of this paper (Nikolova et al. 2018).

4.3   Duality Theory for Optimal Mechanisms

In this section, we prove our main theorem (Theorem 4), namely that the revenue-maximizing take-it-or-leave-it randomized price is optimal among all possible mechanisms for revenue extraction. This is similar in nature to the risk-neutral setting where Myerson (1981) showed that a take-it-or-leave-it payment is optimal. It is important to note that the same is not true for the case of risk-averse agents where multiple menu options can be used to extract revenue that approaches the social welfare given sufficient aversion to risk.

We prove this theorem by upper bounding the revenue of the optimal mechanism using the optimization problem formulation (8) and employing the duality framework. Specifically, we identify dual variables of the Lagrangian dual program and show that it matches the maximum revenue obtained by a take-it-or-leave-it randomized price. The core idea is to define a *virtual value function* that captures the marginal revenue and assume a regularity condition to close the duality gap. The virtual values used in interpreting Myerson's result heavily rely on the assumption of risk neutrality, therefore we need a new definition of virtual value.

Definition 2 (Virtual value function for a single buyer). *In the single buyer setting, the virtual value function* $\phi_u : [K] \to \mathbb{R}$ *with respect to utility function u is defined as*

$$\phi_u(k) = \frac{1}{f(v_k)} \left(\sum_{k' \geq k} f(v_{k'}) \cdot \frac{u(v_k)}{u(v_k) - u(v_k - z_M)} - \sum_{k' \geq k+1} f(v_{k'}) \cdot \frac{u(v_{k+1})}{u(v_{k+1}) - u(v_{k+1} - z_M)} \right).$$

Note that this definition of the virtual value function reduces to Myerson's virtual value function for linear u. To see how this new definition of virtual value function is related to the marginal revenue, consider a take-it-or-leave-it mechanism with a pricing rule that asks the buyer to pay z_M with probability p, and pay zero otherwise. If we would like to guarantee that this pricing rule is accepted by a buyer with value greater than v, then by individual rationality, we can find that the largest p, the probability of paying z_M, we can set is $\frac{u(v)}{u(v)-u(v-z_M)}$. Therefore, the expected revenue we have from this mechanism is $\sum_{v' \geq v} f(v') z_M \frac{u(v)}{u(v)-u(v-z_M)}$. As a result, we can find that $f(v_k)\phi_u(k)z_M$ is indeed the marginal increase on the revenue by moving the threshold value from v_{k+1} to v_k. Given the definition of the virtual value function above, we need to define the following regularity condition:

Definition 3 (Regular distribution for a single buyer). *In the single buyer setting, a distribution f is regular if the corresponding virtual value function is monotone increasing, i.e. for any $k, k' \in [K]$ and $k > k'$, it holds that $\phi_u(k) > \phi_u(k')$.*

Note that with this regularity condition, in the optimal take-it-or-leave-it mechanism considered in Theorem 3, the optimal quantile can also be found by looking at the smallest $k \in \{0, 1, \ldots, K\}$ such that $\phi_u(k) > 0$. We are now ready to state our first main result in Theorem 4, which says that the revenue maximizing randomized take-it-or-leave-it price that we derived in Theorem 3 is the optimal mechanism for a single buyer, assuming the regularity condition. Due to space constraints, we will only give a proof sketch that demonstrates our duality framework, and we defer the full proof to the full version of this paper (Nikolova et al. 2018).

Theorem 4. *Consider a single risk-loving buyer with exponential utility whose value is drawn from a regular distribution f. Then, the optimal mechanism is the revenue maximizing randomized take-it-or-leave-it price.*

In order to prove this theorem, first note that the revenue of the optimal mechanism is upper-bounded by the Lagrangian dual program of the linear program (8). For the single buyer case the dual program of (8) can be simplified as

$$\text{Minimize} \quad \sum_{k \in [K]} \nu_k \tag{9}$$

Subject to the following constraints:

$$f(v_k)z_j + \sum_{k'} (\lambda_{kk'} u(v_k - z_j) - \lambda_{k'k} u(v_{k'} - z_j)) + \mu_k u(v_k - z_j) \leq \nu_k, \quad \forall k \in [K], j \in [M]$$

$$f(v_k)z_j + \sum_{k'} (\lambda_{kk'} - \lambda_{k'k}) u(-z_j) + \mu_k u(-z_j) \leq \nu_k, \qquad \forall k \in [K], j \in [M]$$

$$\mu_k \geq 0, \lambda_{kk'} \geq 0, \qquad \forall k, k' \in [K],$$

where $\lambda_{kk'}$ corresponds to the BIC constraints (5), μ_k corresponds to the IR constraints (6), and ν_k corresponds to the feasibility constraints (7). For the first two sets of the constraints in program (9), we define

$$\Gamma_k(z; \lambda, \mu) = f(v_k)z + \sum_{k'} (\lambda_{kk'} u(v_k - z) - \lambda_{k'k} u(v_{k'} - z)) + \mu_k u(v_k - z)$$

$$\Pi_k(z; \lambda, \mu) = f(v_k)z + \sum_{k'} (\lambda_{kk'} - \lambda_{k'k}) u(-z) + \mu_k u(-z).$$

In the dual program (9), the constraint $\Gamma_k(z; \lambda, \mu) \leq \nu_k$ corresponds to the variable $y^1_{1,j}(v_k)$ in the primal (8) and the constraint $\Pi_k(z; \lambda, \mu) \leq \nu_k$ corresponds to the variable $y^0_{1,j}(v_k)$. In fact, for any $k \in [K]$, we can show that among the sets of constraints $\{\Gamma_k(z_j; \lambda, \mu) \leq \nu_k\}_{j \in [M]}$ and $\{\Pi_k(z_j; \lambda, \mu) \leq \nu_k\}_{j \in [M]}$, only the following four can be binding: $\Gamma_k(0; \lambda, \mu) \leq \nu_k$, $\Gamma_k(z_M; \lambda, \mu) \leq \nu_k$, $\Pi_k(0; \lambda, \mu) \leq \nu_k$, and $\Pi_k(z_M; \lambda, \mu) \leq \nu_k$. This can be observed by the following lemma:

Lemma 1. *In the dual program (9), both $\Gamma_k(z; \lambda, \mu)$ and $\Pi_k(z; \lambda, \mu)$ are either increasing or strongly convex in z for $z \geq 0$.*

Now, we are ready to demonstrate how we can construct a set of dual variables that helps prove our main theorem in the following proof sketch. Due to space constraints, the proof of Lemma 1 and the detailed discussion of Theorem 4 can be found in the full version of this paper (Nikolova et al. 2018).

Proof (Proof sketch of Theorem 4). To prove the theorem, we construct a set of feasible dual variables that upper bound the value of the dual program (9) by the revenue of the optimal take-it-or-leave-it randomized price. For any $k, k' \in [K]$, we set $\mu_k = 0$ and

$$
\lambda_{kk'} = \begin{cases} \sum_{\ell \geq k} f(v_\ell) z_M \frac{1}{u(v_k) - u(v_k - z_M)}, & \text{if } k' = k - 1 \\ 0, & \text{otherwise.} \end{cases} \tag{10}
$$

For simplicity, we define $k^* = \min\{k : \phi_u(k) > 0\}$. For any $k \in [K]$, we set

$$
\nu_k = \max\left\{0, f(v_k)\phi_u(k)z_M\right\}.
$$

Under this choice of ν_k, given the regularity condition, we can write the objective of the dual program (9) as

$$
\sum_k \nu_k = \sum_{k : \phi_u(k) > 0} f(v_k)\phi_u(k)z_M = \sum_{k \geq k^*} f(v_k)\phi_u(k)z_M.
$$

Then, by the definition of the virtual value function, we have

$$
\sum_k \nu_k = \sum_{k \geq k^*} \left(\frac{\sum_{k' \geq k} f(v_{k'})u(v_k)}{u(v_k) - u(v_k - z_M)} - \frac{\sum_{k' \geq k+1} f(v_{k'})u(v_{k+1})}{u(v_{k+1}) - u(v_{k+1} - z_M)} \right) z_M
$$

$$
= \sum_{k \geq k^*} f(v_k)z_M \cdot \frac{u(v_{k^*})}{u(v_{k^*}) - u(v_{k^*} - z_M)}, \tag{11}
$$

where the last equality results from taking a telescopic sum. We can find that the expression in the last line (11) is equal to the revenue of the optimal take-it-or-leave-it mechanism.

It remains to verify that this choice of the dual variables is feasible for the dual program (9). Since we have argued that the sets of constraints $\{\Gamma_k(z_j; \lambda, \mu) \leq \nu_k\}_{j \in [M]}$ and $\{\Pi_k(z; \lambda, \mu) \leq \nu_k\}_{j \in [M]}$ can only be binding at $j = 0$ and $j = M$, it suffices to check whether $\Gamma_k(0; \lambda, \nu) \leq \nu_k$, $\Gamma_k(z_M; \lambda, \nu) \leq \nu_k$, $\Pi_k(0; \lambda, \nu) \leq \nu_k$, and $\Pi_k(z_M; \lambda, \nu) \leq \nu_k$. We defer the detailed derivation to the full proof in the full version of the paper. Here, we only show that by bringing our choice of the dual variables into these constraints, we have

$$
\Gamma_k(0; \lambda, \mu) = f(v_k)\phi_u(k)z_M, \qquad \Gamma_k(z_M; \lambda, \mu) = f(v_k)\phi_u(k)z_M,
$$

$$
\Pi_k(0; \lambda, \mu) = 0, \qquad \Pi_k(z_M; \lambda, \mu) = (1 - e^{-\alpha z_M})f(v_k)\phi_u(k)z_M.
$$

By definition of ν_k, we can find that this assignment of dual variables is feasible. Therefore, weak duality implies that the objective value of program (8) is upper bounded by the revenue maximizing take-it-or-leave-it randomized price, which shows that this mechanism is optimal.

In addition, we can find that given $k \in [K]$, as long as $\phi_u(k) > 0$, the only binding dual constraints are $\Gamma_k(0; \lambda, \mu) \leq \nu_k$ and $\Gamma_k(z_M; \lambda, \mu) \leq \nu_k$. Therefore, by complementary slackness, in the optimal mechanism, the pricing scheme must be a randomization of 0 and z_M, and must ask the buyer to pay only if she is given the item. This coincides with the revenue-maximizing randomized take-it-or-leave-it price, which is our claimed optimal primal solution.

4.4   Optimal Mechanism Beyond Regularity Condition

In the next step, we would like to extend our duality framework described in Theorem 4 to the case without a regularity condition. Namely, we would like to prove that the revenue-maximizing take-it-or-leave-it randomized price is still optimal even though the virtual value function is not an increasing function, using the similar argument that we have made in Sect. 4.3. Formally, we would like to prove the following theorem:

Theorem 5. *Consider a single risk-loving buyer with exponential utility. The optimal mechanism is the revenue-maximizing randomized take-it-or-leave-it price.*

The technique that we use to prove this theorem consists of two steps. First, we construct an *ironed virtual value* $\widetilde{\phi}_u$. The ironed virtual value function is an increasing function constructed based on the original virtual value function, which is similar to the risk-neutral case as in Myerson's work. After that, we can slightly modify the dual variables that we specified in Sect. 4.3 to match the ironed virtual value and then claim the optimality of the revenue-maximizing randomized take-it-or-leave-it price by strong duality.

Surprisingly, having a non-linear utility function does not complexify the ironing process. We can directly apply Myerson's ironing for a risk-loving buyer by *convexifying* the cumulative virtual value function. Formally, we can define the "ironed virtual value" function in the following way:

Definition 4 (Ironed virtual value for a single buyer). *Given a virtual value function ϕ_u. Let $\{[a_1, b_1], [a_2, b_2], \ldots, [a_m, b_m]\}$ denote the intervals that are not convex on the cumulative virtual value function $F_\phi(k) = \sum_{\ell \leq k} \phi_u(\ell)$. Then, the ironed virtual value function is defined as*

$$
\widetilde{\phi}_u(k) = \begin{cases} \dfrac{\sum_{\ell=a_i}^{b_i} f(v_\ell)\phi_u(\ell)}{\sum_{\ell=a_i}^{b_i} f(v_\ell)}, & \text{if } k \in [a_i, b_i], \text{ for any } i \in [m] \\ \phi_u(\ell), & \text{otherwise.} \end{cases}
$$

Although the ironing process is straightforward, how to modify the dual variables to match the ironed virtual value is more tricky. Recall that under the choice of λ as specified in (10), the left hand side of the first dual constraint is upperbounded by the virtual value, i.e. $\Gamma_k(0; \lambda, \mu) = \Gamma_k(z_M; \lambda, \mu) = f(v_k)\phi_u(k)z_M$. Motivated by Cai et al. (2016), we show that we can add *loops* to λ to alter the value of Γ_k.[2] More precisely, consider some $k' < k$. If we add $Ae^{-\alpha v_k}$ to $\lambda_{k,k'}$ and add $Ae^{-\alpha v_{k'}}$ to $\lambda_{k',k}$ for some $A > 0$, then $\Gamma_k(0; \lambda, \mu)$ is increased by $A[e^{-\alpha v_{k'}} - e^{-\alpha v_k}]$ and $\Gamma_{k'}(0; \lambda, \mu)$ is decreased by $A[e^{-\alpha v_{k'}} - e^{-\alpha v_k}]$. This distorted modification guarantees that the binding structure does not change, i.e. $\Gamma_k(0; \lambda, \mu) = \Gamma_k(z_M; \lambda, \mu)$ and $\Gamma_{k'}(0; \lambda, \mu) = \Gamma_{k'}(z_M; \lambda, \mu)$. With this idea,

---

[2] In Cai et al. (2016), the dual variables can be interpreted as *flows*. However, in our setting, this interpretation no longer holds. We need to handle the distortion caused by the non-linear utility function.

we are able to apply Myerson's ironing by iterative adding loops to λ in the way such that $\Gamma_k(0; \lambda, \mu) = f(v_k)\tilde{\phi}_u(k)z_M$ holds after the modification. After that, we are able to follow the steps in the proof of Theorem 1 to prove the theorem. Due to the space constraint, we defer the detailed proof of Theorem 5 to the full version of the paper (Nikolova et al. 2018).

5   Optimal Mechanism Design with Multiple Symmetric Buyers

In this section, we extend our analysis of optimal mechanism design to multiple symmetric buyers ($n \geq 2$). Since the buyers are symmetric, their values come from the same distribution and they have the same utility function. We show that the *loser-pay auction* achieves the maximum revenue in the multiple-buyer case. In a loser-pay auction, the buyer with the highest value wins the item and only the buyers that do not obtain the item pay. In addition, similarly to the randomized take-it-or-leave-it pricing, all payments are made using a mixing of the minimum and the maximum price. This auction could be thought of as implementing an incentive-compatible version of the all-pay auction but adjusting it to achieve maximum discrepancy between the two outcomes.

When characterizing the revenue maximizing mechanisms, our analysis is similar to the analysis in Sect. 4.3 in that it uses the virtual value formulation and the duality framework to upper bound the revenue obtained by the optimal mechanism. In what follows, we first give an example of the loser-pay auction and show how it improves the revenue compared to the second price auction, which is optimal for risk-neutral buyers.

5.1   An Example

Example 2. Consider two buyers. Assume the private values of both buyers are distributed independently according to the uniform distribution $U(\{0, 1\})$, and $u(x) = e^{\alpha x} - 1$. Also assume $3(e^{\alpha x} - 1) < 1 - e^{-\alpha z_M}$. Consider the following mechanism:

1. The item is allocated to the buyer who reports the higher value. If both buyers report 1, the item is allocated uniformly at random. If both buyers report 0, the item is not allocated to anyone.
2. If a buyer reports 1, she gets the item and does not pay anything. However, if she reports 1 and she does not get the item, then she pays z_M with probability $\frac{3(e^\alpha - 1)}{1 - e^{-\alpha z_M}}$.

To verify this mechanism is BIC and IR, we check the utility curve if a buyer reports 1 to the seller. Consider buyer 1. Given her true value is t_1, her expected utility is

$$x_1(2)u(t_1) + (1 - x_1(2))\frac{3(e^\alpha - 1)}{1 - e^{-\alpha z_M}}u(-z_M) = \frac{3}{4}\left(e^{\alpha t_1} - 1\right) - \frac{3}{4}(e^\alpha - 1),$$

which is 0 if $t_1 = 1$ and $-\frac{3}{4}(e^\alpha - 1)$ if $t_1 = 0$. This verifies that the mechanism is BIC and IR. Next, we can find that the revenue of this mechanism is

$$\text{Rev} = \sum_{i \in \{1,2\}} \Pr[t_i = 1](1 - x_i(2)) \frac{3(e^\alpha - 1)}{1 - e^{-\alpha z_M}} z_M = \frac{3}{4} \frac{e^\alpha - 1}{1 - e^{-\alpha z_M}} z_M.$$

From the above example, we make two observations. First, we find that compared with the case of a single buyer, the revenue is increased by a factor of $\frac{3}{2} e^\alpha$. Note that in the case of risk-neutral buyers, this factor is only $\frac{3}{2}$. The additional factor of e^α comes from the second observation—the buyer *pays if she does not get the item*. In the rest of this section, we show that these two properties hold in the optimal mechanism.

5.2   The Loser-Pay Auction

Recall that in the setting with risk-neutral buyers, the second-price auction with reserve price is optimal. A natural question here is, when the buyers are risk-loving, does the optimal mechanism take a similar form? We show that, given the assumption that $z_M \gg \alpha$, i.e., the maximum allowed price is far greater than the level of risk-loving, the optimal mechanism corresponds to the revenue maximizing loser-pay auction. From Example 2, we already know that the optimal payment rule is to ask the buyer who loses to pay a randomized price. The reserve price in our risk-loving setting, similarly to the risk-neutral setting, is going to be computed via a *virtual value function*.

We have already defined a virtual value function for the convex utility function in Sect. 4.3. However, recall that the virtual value function is the marginal revenue in the quantile space. According to Example 2, there is an additional e^α factor in the revenue of the multi-buyer setting, hence now we need a different definition of the virtual value function than in the single buyer case. Specifically, in the multi-buyer setting, we need to consider the following new virtual value function that takes this e^α factor into account:

Definition 5 (Virtual value function for multiple buyers). *In the multi-buyer case, the virtual value function $\Phi_u : [K] \to \mathbb{R}$ with respect to an exponential utility function u is defined as*

$$\Phi_u(k) = \frac{1}{f(v_k)} \left(\sum_{k' \geq k} f(v_{k'}) \cdot \frac{e^{\alpha v_k} u(v_k)}{u(v_k) - u(v_k - z_M)} - \sum_{k' \geq k+1} f(v_{k'}) \cdot \frac{e^{\alpha v_{k+1}} u(v_{k+1})}{u(v_{k+1}) - u(v_{k+1} - z_M)} \right).$$

In the full version of the paper we show that this additional e^α factor comes from the *competition* that only happens when there are multiple buyers. Similar to the single buyer case, we call a distribution f a *regular distribution* if its corresponding virtual value function is a monotone increasing function. If f is not regular, then we can apply Myerson's ironing to the new virtual value function as what we have described in Sect. 4.4. We let $\widetilde{\Phi}_u$ denote the ironed virtual value of Φ_u. The formal definition of $\widetilde{\Phi}_u$ can be found in the full version of this paper (Nikolova et al. 2018).

Now, we are ready to describe the loser-pay auction—the mechanism that we claim to be optimal when the buyers are risk-loving:

Definition 6 (The loser-pay auction). *A loser-pay auction is a direct mechanism with the following allocation and payment rule:*

1. *Suppose each buyer i bids v_{k_i}. Then, the auctioneer allocates the item to buyer i if $\widetilde{\Phi}_u(k_i) > \widetilde{\Phi}_u(k_{i'})$ for every other buyer $i' \neq i$ provided $\widetilde{\Phi}_u(k_i) > 0$.*
2. *Suppose each buyer i bids v_{k_i}. If buyer i submits the bid with the largest ironed virtual value and ties with $n_t - 1$ other buyers, then she gets the item with probability $1/n_t$ provided $\widetilde{\Phi}_u(k_i) > 0$.*
3. *If buyer i bids v_k and gets the item, she pays nothing.*
4. *If buyer i bids v_k with $\Phi_u(k) \leq 0$, she pays nothing, i.e. $q_i(k) = 0$.*
5. *If buyer i bids v_k with $\widetilde{\Phi}_u(k) > 0$ and does not get the item, then she pays z_M with probability*

$$q_i(k) = \frac{1}{1 - x_i(k)} \sum_{k'=k^*}^{k} \frac{[x_i(k') - x_i(k' - 1)]u(v_{k'})}{-u(-z_M)},$$

where $k^* = \min\{k \in [K] : \Phi_u(k) > 0\}$ is the index of the reserve price.

For simplicity, in this section, we use $k = (k_1, \ldots, k_n)$ to denote the indices of bids from all buyers. Also, we use $f(k) = \prod_{i \in [n]} f(v_{k_i})$ to denote the probability that $t = (v_{k_1}, v_{k_2}, \ldots, v_{k_n})$. If buyer i submits her bid v_k that is no less than the reserve price v_{k^*}, then the interim probability that she gets the item is

$$x_i(k) = \sum_{k_{-i} \in [K]^{n-1}} f(k_{-i}) \left[\frac{\mathbb{1}\left\{\widetilde{\Phi}(v_{k_i}) \geq \widetilde{\Phi}(v_{k_{i'}}), \forall i' \neq i\right\}}{\sum_{i' \in [n]} \mathbb{1}\left\{\widetilde{\Phi}(v_{k_i}) = \widetilde{\Phi}(v_{k_{i'}})\right\}} \right],$$

where $\mathbb{1}\{A\}$ is the indicator function that equals 1 if event A is true, and 0 otherwise. In order to guarantee that the payment rule of the loser-pay auction is feasible, we need to make the following assumption, which plays an important role in guaranteeing $q_i(k) < 1$:

(A1) $z_M \gg \alpha$ so that for each $v \in V$, it holds that $\frac{1 - \frac{1}{n}f(v)}{\frac{1}{n}f(v)} \cdot (e^{\alpha v} - 1) < 1 - e^{-\alpha z_M}$.

In the full version of this paper (Nikolova et al. 2018), we show that the loser-pay auction is feasible, individually rational, and Bayesian incentive compatible under Assumption (A1). In the following theorem, we formally state that the loser-pay auction is an optimal mechanism. The full proof can also be found in the full version of this paper.

Theorem 6. *Consider $n \geq 2$ buyers with exponential utility. With Assumption (A1), the loser-pay auction is the optimal mechanism.*

6 Conclusion

In this paper we studied the revenue-c mechanism for the special case of risk-loving agents with exponential utility functions. We demonstrate that for both a single and multiple symmetric buyers the optimal auction is simple. A natural question is whether or not the same results extend beyond the expoential utility function. Unfortunately, it can be shown that optimal auction for the case of a single agent with quadratic utility function is more complicated than the simple randomized take-it-or-leave it offer and requires us to utilize at least one more menu option[3].

References

Cai, Y., Devanur, N.R., Weinberg, S.M.: A duality based unified approach to bayesian mechanism design. In: Proceedings of the Forty-eighth Annual ACM Symposium on Theory of Computing, pp. 926–939. ACM (2016)

Chawla, S., Goldner, K., Miller, J.B., Pountourakis, E.: Revenue maximization with an uncertainty-averse buyer. In: Proceedings of the Twenty-Ninth Annual ACM-SIAM Symposium on Discrete Algorithms, pp. 2050–2068. SIAM (2018)

Daskalakis, C., Deckelbaum, A., Tzamos, C.: Mechanism design via optimal transport. In: Proceedings of the Fourteenth ACM Conference on Electronic Commerce, pp. 269–286. ACM (2013)

Daskalakis, C., Deckelbaum, A., Tzamos, C.: Strong duality for a multiple-good monopolist. Econometrica **85**(3), 735–767 (2017)

Dughmi, S., Peres, Y.: Mechanisms for risk averse agents, without loss. arXiv preprint arXiv:1206.2957 (2012)

Fu, H., Hartline, J., Hoy, D.: Prior-independent auctions for risk-averse agents. In: Proceedings of the Fourteenth ACM Conference on Electronic Commerce, pp. 471–488. ACM (2013)

Giannakopoulos, Y., Koutsoupias, E.: Duality and optimality of auctions for uniform distributions. In: Proceedings of the Fifteenth ACM Conference on Economics and Computation, pp. 259–276. ACM (2014)

Giannakopoulos, Y., Koutsoupias, E.: Selling two goods optimally. In: Halldórsson, M.M., Iwama, K., Kobayashi, N., Speckmann, B. (eds.) ICALP 2015. LNCS, vol. 9135, pp. 650–662. Springer, Heidelberg (2015). https://doi.org/10.1007/978-3-662-47666-6_52

Hinnosaar, T.: On the impossibility of protecting risk-takers. Econ. J. (2017). https://doi.org/10.1111/ecoj.12446. ISSN 1468-0297

Li, J., et al.: Energy coupon: a mean field game perspective on demand response in smart grids. ACM SIGMETRICS Perform. Eval. Rev. **43**(1), 455–456 (2015)

Lu, F.: Framework for a lottery-based incentive scheme and its influence on commuting behaviors: an MIT case study. PhD thesis, Massachusetts Institute of Technology (2015)

Maskin, E., Riley, J.: Optimal auctions with risk averse buyers. Econ.: J. Econ. Soc. **52**, 1473–1518 (1984)

---

[3] The detailed discussion can be found in the full version of this paper (Nikolova et al. 2018).

Matthews, S.A.: Selling to risk averse buyers with unobservable tastes. J. Econ. Theory **30**(2), 370–400 (1983)

Merugu, D., Prabhakar, B.S., Rama, N.S.: An incentive mechanism for decongesting the roads: a pilot program in Bangalore. In: Proceedings of ACM NetEcon Workshop. ACM (2009)

Myerson, R.B.: Optimal auction design. Math. Oper. Res. **6**(1), 58–73 (1981)

Nikolova, E., Pountourakis, E., Yang, G.: Optimal mechanism design with risk-loving agents. CoRR, abs/1810.02758 (2018). http://arxiv.org/abs/1810.02758

Pluntke, C., Prabhakar, B.: INSINC: a platform for managing peak demand in public transit. JOURNEYS Land Transp. Auth. Acad. Singap., 31–39 (2013)

Von Neumann, J., Morgenstern, O.: Theory of games and economic behavior. Bull. Amer. Math. Soc **51**(7), 498–504 (1945)

Robust Bounds on Choosing from Large Tournaments

Christian Saile[1] and Warut Suksompong[2(✉)]

[1] Department of Informatics, Technical University of Munich, Munich, Germany
saile@in.tum.de
[2] Department of Computer Science, Stanford University, Stanford, USA
warut@cs.stanford.edu

Abstract. Tournament solutions provide methods for selecting the "best" alternatives from a tournament and have found applications in a wide range of areas. Previous work has shown that several well-known tournament solutions almost never rule out any alternative in large random tournaments. Nevertheless, all analytical results thus far have assumed a rigid probabilistic model, in which either a tournament is chosen uniformly at random, or there is a linear order of alternatives and the orientation of all edges in the tournament is chosen with the same probabilities according to the linear order. In this work, we consider a significantly more general model where the orientation of different edges can be chosen with different probabilities. We show that a number of common tournament solutions, including the top cycle and the uncovered set, are still unlikely to rule out any alternative under this model. This corresponds to natural graph-theoretic conditions such as irreducibility of the tournament. In addition, we provide tight asymptotic bounds on the boundary of the probability range for which the tournament solutions select all alternatives with high probability.

1 Introduction

Tournaments play an important role in numerous situations as a means of representing entities and a dominance relationship between them. For instance, both the outcome of a round-robin sports competition and the majority relation of voters in an election can be represented by a tournament. A question that occurs frequently is therefore the following: Given a tournament, how can we choose the "best" alternatives in a consistent manner? This question has been addressed by a rich and beautiful literature on tournament solutions, which have found applications in areas ranging from sports competitions [31] to multi-criteria decision analysis [2,4] to biology [1,15,27,30]. Over the past half century several tournament solutions have been proposed, two of the oldest and best-known of which are the *top cycle* [12,20,28] and the *uncovered set* [21].[1]

---

[1] For a thorough treatment of tournament solutions, we refer the reader to excellent surveys by Laslier [16] and Brandt et al. [5].

© Springer Nature Switzerland AG 2018
G. Christodoulou and T. Harks (Eds.): WINE 2018, LNCS 11316, pp. 393–407, 2018.
https://doi.org/10.1007/978-3-030-04612-5_26

Given that the purpose of tournament solutions is to discriminate the "best" alternatives from the remaining ones, it perhaps comes as a surprise that many common tournament solutions—including the top cycle, the uncovered set, the Banks set, and the minimal covering set—select all alternatives with high probability in a large random tournament [9,29]. Put differently, the aforementioned tournament solutions almost never exclude any alternative in a tournament chosen at random. Nevertheless, these results are based on the *uniform random model*, in which all tournaments are drawn with equal probability, or equivalently each edge is oriented in one direction or the other with equal probability independently of other edges. For a large majority of applications of tournaments, one would not expect that this assumption holds. Indeed, stronger teams are likely to beat weaker teams in a sports competition, and candidates with a large base of support have a higher chance of winning an election. Moreover, real-world tournaments often exhibit a certain degree of transitivity: If alternatives a, b, and c are such that a dominates b and b dominates c, then it is more likely that a dominates c than the other way around.

A more general model of random tournaments is the *Condorcet random model*, previously considered by Frank [11], Łuczak et al. [18], Vassilevska Williams [32] and Kim et al. [14]. In this model, there is a linear order of alternatives, which can be interpreted as an ordering of the alternatives from strongest to weakest. For each pair of alternatives, the probability that the edge is oriented from the alternative that occurs later in the linear order to the alternative that occurs earlier in the linear order is p, independently of other pairs of alternatives.[2] Crucially, the value of p is the same for all pairs of alternatives. The Condorcet random model generalizes the uniform random model, since the latter can be obtained from the former by taking $p = 1/2$. Łuczak et al. [18] showed that under the Condorcet random model, the top cycle selects all alternatives as long as $p \in \omega(1/n)$. The same authors show furthermore that this bound is tight, that is, the statement no longer holds if $p \in O(1/n)$.[3]

Although the Condorcet random model addresses the issues raised above with regard to the uniform random model, it is still rather unrealistic for two important reasons. Firstly, in tournaments in the real world, the orientation of different edges are typically determined by different probabilities. For instance, in a sports tournament the probability that a very strong team beats a very weak team is usually higher than the probability that a moderately strong team beats a moderately weak team; a similar phenomenon can be observed in elections. Secondly, even though one can roughly order the alternatives in a tournament according to their strength, it is often the case that not all probabilities of the orientation of the edges respect the ordering. Indeed, this precisely corresponds to the notion of "bogey teams"—weak teams that nevertheless frequently beat certain supposedly stronger teams. Given the limitations of the uniform random model and the Condorcet random model, it is natural to ask whether previous results continue to hold under more general and realistic models of random

---

[2] By symmetry, we may assume without loss of generality that $p \leq 1/2$.

[3] See, e.g., Cormen et al. [8] for the definitions of asymptotic notations.

tournaments, or whether they break down as soon as we move beyond these restricted models.

In this paper, we show that a number of tournament solutions, including the top cycle and the uncovered set, still choose all alternatives with high probability under a significantly more general model of random tournaments. Unlike the Condorcet random model, our model does not rely on an ordering of the alternatives. Instead, the orientation of each edge is determined by probabilities within the range $[p, 1 - p]$ for some parameter p, and these probabilities are allowed to vary across edges. The only substantive assumption that we make is that the orientations of different edges are chosen independently from one another. Under this model, which is more general than both the uniform random model and the Condorcet random model, we establish in Sect. 3 that the top cycle almost never rules out any alternative as long as $p \in \omega(1/n)$, thus generalizing the result by Łuczak et al. [18]. We also show that our bound is asymptotically tight, and that analogous results hold for two other tournament solutions based on the set of Condorcet winners and losers as well. Moreover, we prove in Sect. 4 that the uncovered set is likely to include the whole set of alternatives when $p \in \omega(\sqrt{\log n/n})$. This bound is again asymptotically tight, and the same holds for another tournament solution based on the uncovered set. Since the condition that the top cycle or the uncovered set chooses all alternatives have meaningful graph-theoretic interpretations—the top cycle is the whole set of alternatives if and only if the tournament is strongly connected,[4] and the uncovered set fails to exclude any alternative exactly when all alternatives are *kings*[5]—we believe that our results are of independent interest in graph theory and discrete mathematics. Furthermore, the generality of our model allows us to derive consequences in Sect. 5 for a different model in which tournaments are generated from random voter preferences, and we complement our theoretical results with experimental data in Sect. 6.

1.1   Related Work

The study of the behavior of tournament solutions in large random tournaments goes back to Moon and Moser [23], who showed that the top cycle almost never rules out any alternative in a large tournament chosen uniformly at random. In fact, they proved a stronger statement that the probability that the top cycle excludes at least one alternative is inverse exponential in the number of alternatives; the estimate was later made more precise by Moon [22] in his seminal book on tournaments. Bell [3] also considered the top cycle but assumed that tournaments are generated from the preferences of a large number of voters, each with

---

[4] A strongly connected tournament is also said to be *strong*. Strong connectedness is equivalent to *irreducibility* and to the property of having a Hamiltonian cycle [22].

[5] A *king* is an alternative that can reach any other alternative via a directed path of length at most two [19]. Therefore, all alternatives of a tournament are kings if and only if every pair of alternatives can reach each other via a directed path of length at most two. Such a tournament has been studied in graph theory and called an *all-kings tournament* [25].

a uniform random ranking over the alternatives; he likewise found that the top cycle selects all alternatives with high probability under this assumption. Fey [9] and later Scott and Fey [29] established results on several tournament solutions including the uncovered set, the Banks set, the Copeland set, the minimal covering set, and the bipartisan set using the uniform random model. While the uncovered set, the Banks set, and the minimal covering set are likely to include all alternatives in a large random tournament, the same event is unlikely to occur for the Copeland set. On the other hand, the bipartisan set chooses on average half of the alternatives in a random tournament of any fixed size [10]; it is the unique most discriminating tournament solution satisfying standard properties proposed in the literature [6].

The discriminative power of tournament solutions has also been investigated empirically by Brandt and Seedig [7]. Building on the observation that the distributions of real-world tournaments are typically far from uniform, these authors examined the behavior of eleven common tournament solutions on tournaments generated according to stochastic preference models and empirical data. The stochastic models that they used include the impartial culture model, the Mallows mixtures model, and the Pólya-Eggenberger urn model. They reported that under these more realistic models, most tournament solutions are in fact much more discriminating than the analytical results for uniform random tournaments suggest.

2   Preliminaries

A tournament T consists of a set $A = \{a_1, a_2, \ldots, a_n\}$ of alternatives and a dominance relation. The dominance relation is an asymmetric and connex binary relation on A represented by a directed edge between each unordered pair of distinct alternatives in A. We say that alternative a_i *dominates* another alternative a_j if there is an edge from a_i to a_j. An alternative is said to be a *Condorcet winner* if it dominates all of the remaining alternatives, and a *Condorcet loser* if it is dominated by all of the remaining alternatives. We extend the dominance relation to sets and say that a set $A' \subseteq A$ of alternatives dominates another set $A'' \subseteq A$ of alternatives disjoint from A' if for all $a' \in A'$ and $a'' \in A''$, a' dominates a''. A tournament is commonly interpreted as the outcome of a round-robin sports competition and as the majority relation of an odd number of voters with linear preferences. In the former interpretation, alternative a_i dominating alternative a_j means that the player or team represented by a_i beats the player or team represented by a_j in the competition. In the latter interpretation, the same dominance relation signifies that more than half of the voters prefer a_i to a_j.

We are interested in tournament solutions, which are functions that map each tournament to a nonempty subset of its alternatives, usually referred to as the *choice set*. Two simple tournament solutions are *COND*, which chooses a

Condorcet winner if one exists and chooses all alternatives otherwise,[6] and the set of *Condorcet non-losers* (*CNL*), which consists of all alternatives that are not Condorcet losers. Other tournament solutions considered in this paper are the following:

- The *top cycle* (*TC*) is the (unique) smallest set of alternatives such that all alternatives in the set dominate all alternatives not in the set;
- The *uncovered set* (*UC*) consists of all alternatives that can reach all other alternatives via a domination path of length at most two;[7]
- The *iterated uncovered set* (UC^∞) is the result of iteratively computing the uncovered set until there is no further reduction.

The inclusions $UC^\infty(T) \subseteq UC(T) \subseteq TC(T) \subseteq CNL(T)$ and $TC(T) \subseteq COND(T)$ hold for any tournament T.

Next, we describe the random models for generating tournaments that we consider in this paper. We will work with the first model in Sects. 3 and 4 and the second model in Sect. 5.

- Model 1: For each pair of distinct alternatives a_i, a_j, there is an edge from a_i to a_j with probability $p_{i,j}$ and an edge from a_j to a_i with probability $p_{j,i} = 1 - p_{i,j}$, independently of other pairs of alternatives.
- Model 2: There is a constant number k of voters, where k is odd. For each voter v and each pair of distinct alternatives a_i, a_j, the voter prefers a_i to a_j with probability $q_{v,i,j}$ and prefers a_j to a_i with probability $q_{v,j,i} = 1 - q_{v,i,j}$, independently of other voters and other pairs of alternatives.[8] The majority relation, in which alternative a_i dominates another alternative a_j if and only if more than half of the voters prefer a_i to a_j, forms a tournament with A as its set of alternatives.

Several models for generating random tournaments considered in previous work are special cases of our models. For example, the *uniform random model* [9, 29] corresponds to taking $p_{i,j} = 1/2$ for all i, j in Model 1 or taking $q_{v,i,j} = 1/2$ for all v, i, j in Model 2 with any k. The *Condorcet random model* [11, 14, 18, 32] corresponds to taking $p_{i,j} = p$ for all $i < j$ in Model 1, for some fixed value of p. The *Condorcet random model for voters* [7] corresponds to taking $q_{v,i,j} = p$ for all v and all $i < j$ in Model 2, for some fixed value of p. Following standard

---

[6] Note that the set of Condorcet winners is not a tournament solution because it can be empty.

[7] This is known in graph theory as the set of *kings* (cf. Footnote 5). An alternative definition, which is also the origin of the name "uncovered set", is based on the *covering relation*. An alternative a_i is said to *cover* another alternative a_j if (i) a_i dominates a_j, and (ii) any alternative that dominates a_i also dominates a_j. The uncovered set corresponds to the set of alternatives that are not covered by any other alternative.

[8] One way to interpret the possible intransitivity of the preferences is as a result of noise in the voters' true preferences. Laslier [17] introduced the term *Rousseauist cultures* for this kind of models.

terminology, we say that an event occurs "with high probability" or "almost surely" if the probability that the event occurs converges to 1 as n, the number of alternatives, goes to infinity.

Due to space constraints, all omitted proofs can be found in the full version of this paper [26].

3 Top Cycle

In this section, we consider the top cycle. We show that when each probability $p_{i,j}$ is between $f(n)$ and $1 - f(n)$ for some function $f(n) \in \omega(1/n)$, TC chooses all alternatives with high probability (Theorem 3.1). By using the inclusion relationships between TC, $COND$, and CNL, we obtain analogous statements for $COND$ and CNL. We also show that our results are asymptotically tight—for all three tournament solutions, the statement ceases to hold if $f(n) \in O(1/n)$ (Theorem 3.2).

We begin with our main result of the section.

Theorem 3.1. *Let $f : \mathbb{Z}^+ \to \mathbb{R}_{\geq 0}$ be a function such that $f(n) \leq 1/2$ for all n and $f(n) \in \omega(1/n)$. Assume that a tournament T is generated according to Model 1, and that*

$$p_{i,j} \in [f(n), 1 - f(n)]$$

for all $i \neq j$. Then with high probability, $TC(T) = A$.

Theorem 3.1 generalizes a result by Luczak et al. [18] that establishes the claim for the case where $p_{i,j} = f(n)$ for all $i < j$ (or, by symmetry, the case where $p_{i,j} = 1 - f(n)$ for all $i < j$). We remark that their proof relies crucially on the assumption that there is a linear order of alternatives and all edges are more likely to be oriented in one direction than in the other direction according to the order. Indeed, this assumption allows the authors to show that with high probability, any alternative can be reached by the strongest alternative and can reach the weakest alternative via a domination path of length at most two each. Moreover, with the assumption $f(n) \in \omega(1/n)$ one can show that the weakest alternative can almost surely reach the strongest alternative via a domination path of length four, thus establishing the strong connectivity of the tournament. In contrast, we do not assume that the edges in the tournament are likely to be oriented in one direction or the other. As such, we will need a completely different approach for our proof.

We give here a high-level overview of the proof of Theorem 3.1; the full proof can be found in the full version of this paper [26]. We observe that $TC(T) \neq A$ exactly when there exists a proper, nontrivial subset of alternatives B that dominates the complement set of alternatives $A \backslash B$. Using the union bound, we then upper bound the probability that $TC(T) \neq A$ by the sum over all sets B of the probabilities that B dominates $A \backslash B$. This sum can be written entirely in terms of the variables $p_{i,j}$ for $i < j$ and is moreover linear in all of these variables, implying that its maximum is attained when all variables take on a value at one of the two boundaries of their domain. Using a number of helper lemmas, which include an interesting combinatorial extension of Karamata's inequality, we show that the sum is in fact maximized when all variables take on a value at the same boundary. This allows us to bound the sum directly by plugging in the value at a boundary and complete the proof.

Since $TC(T) \subseteq COND(T)$ and $TC(T) \subseteq CNL(T)$, we immediately obtain the following corollary.

Corollary 3.1. *Let $f : \mathbb{Z}^+ \to \mathbb{R}_{\geq 0}$ be a function such that $f(n) \leq 1/2$ for all n and $f(n) \in \omega(1/n)$. Assume that a tournament T is generated according to Model 1, and that $p_{i,j} \in [f(n), 1 - f(n)]$ for all $i \neq j$. Then with high probability, $COND(T) = CNL(T) = A$.*

Next, we show that Theorem 3.1 and Corollary 3.1 are tight in the sense that if $f(n) \in O(1/n)$, the results no longer hold.

Theorem 3.2. *Let $c \geq 0$ be a constant. Assume that a tournament T is generated according to Model 1, and that $p_{i,j} \leq \frac{c}{n}$ for all $i > j$. Then for large enough n, with at least constant probability both $TC(T)$ and $COND(T)$ contain a single alternative. Moreover, for large enough n, with at least constant probability $CNL(T)$ does not contain all alternatives.*

Theorems 3.1 and 3.2 and Corollary 3.1 allow us to obtain the following corollary on the Condorcet random model.

Corollary 3.2. *Let $f : \mathbb{Z}^+ \to \mathbb{R}_{\geq 0}$ be a function such that $f(n) \leq 1/2$ for all n. Assume that a tournament T is generated according to Model 1, and that $p_{i,j} = f(n)$ for all $i > j$.*

- *If $f(n) \in \omega(1/n)$, then with high probability, $TC(T) = COND(T) = CNL(T) = A$.*
- *If $f(n) \in o(1/n)$, then with high probability, $TC(T)$ and $COND(T)$ contain a single alternative, and $CNL(T)$ does not contain all alternatives.*
- *If $f(n) \leq c/n$ for some constant $c \geq 0$, then for large enough n, with at least constant probability $TC(T)$ and $COND(T)$ contain a single alternative. Moreover, for large enough n, with at least constant probability $CNL(T)$ does not contain all alternatives.*

Luczak et al. [18] also considered the case where $p_{i,j} = c/n$ for all $i > j$ and showed that the probability that TC selects all alternatives converges to $(1 - e^{-c})^2$ in this special case. Our next theorem establishes an analogous result for $COND$ and CNL.

Theorem 3.3. *Let $c \geq 0$ be a constant. Assume that a tournament T is generated according to Model 1, and that $p_{i,j} = \frac{c}{n}$ for all $i > j$. Then the probability that $COND(T) = A$ converges to $1 - e^{-c}$ as $n \to \infty$. The same statement holds for CNL.*

Proof. We show the result for *COND*; a similar argument holds for *CNL*. We have

$$\Pr[COND(T) \neq A] = \sum_{i=1}^{n} \Pr[a_i \text{ is a Condorcet winner}]$$

$$= \sum_{i=1}^{n} \left(1 - \frac{c}{n}\right)^{n-i} \left(\frac{c}{n}\right)^{i-1}$$

$$= \left(1 - \frac{c}{n}\right)^{n-1} \cdot \sum_{i=0}^{n-1} \left(\frac{c}{n-c}\right)^{i}.$$

The first term converges to e^{-c} as $n \to \infty$. For the second term, notice that it is always at least 1. Moreover, when $n \geq (k+1)c$ for some positive $k > 1$, the term is at most

$$1 + \frac{1}{k} + \frac{1}{k^2} + \cdots = \frac{k}{k-1},$$

which approaches 1 for large n. Hence the second term converges to 1, and therefore the probability that $COND(T) \neq A$ converges to e^{-c}, yielding the desired result. $\square$

4   Uncovered Set

In this section, we turn our focus to the uncovered set. We show that when each probability $p_{i,j}$ is between $f(n)$ and $1 - f(n)$ for some function $f(n) \geq c\sqrt{\log n/n}$ with $c > \sqrt{2}$ a constant, UC chooses all alternatives with high probability (Theorem 4.1). As with TC, we also show that our result is asymptotically tight—if $f(n) \leq 0.6\sqrt{\log n/n}$, the statement no longer holds (Theorem 4.2). It follows that similar results hold for UC^∞, implying that $\Theta(\sqrt{\log n/n})$ is the threshold where the two tournament solutions go from almost always choosing all alternatives to excluding at least one alternative with high probability.

Our first result of the section shows that UC chooses the whole set of alternatives for a wide range of distributions over tournaments.

Theorem 4.1. *Let $c > \sqrt{2}$ be a constant. Assume that a tournament T is generated according to Model 1, and that $p_{i,j} \in \left[c\sqrt{\frac{\log n}{n}}, 1 - c\sqrt{\frac{\log n}{n}}\right]$ for all $i \neq j$. Then with high probability, $UC(T) = A$.*

We give here a high-level overview of the proof of Theorem 4.1; the full proof can be found in the full version of this paper [26]. Let $A_1 = \{a_1, a_2, \ldots, a_{\lfloor n^{0.49} \rfloor}\}$, and let A_2 be the set of alternatives that a_n dominates. Our key claim is that, with high probability, the following two events occur simultaneously: (i) a_n does not dominate any of the alternatives in A_1, and (ii) $|A_2| \leq 0.61\sqrt{n \log n}$. When these two events occur, a_n can reach all of the alternatives in A_1 via a domination path of length at most two if and only if each alternative in A_1 is dominated by some alternative in A_2. Moreover, the event that this holds for a particular alternative in A_1 is independent of the corresponding events for other alternatives in A_1. This allows us to show that the probability that a_n belongs to the uncovered set goes to 0 for large n.

Since the uncovered set is the finest tournament solution satisfying the axioms of Condorcet consistency, neutrality, and expansion [24], Theorem 4.1 implies that any tournament solution that satisfies these three axioms also selects all alternatives with high probability when the tournament is generated according to the assumptions of the theorem.

Next, we show that the statement of Theorem 4.1 breaks down if $f(n) \leq 0.6\sqrt{\log n/n}$, thus confirming that the assumption of the theorem cannot be relaxed asymptotically.

Theorem 4.2. *Assume that a tournament T is generated according to Model 1, and that*

$$p_{i,j} \leq 0.6\sqrt{\frac{\log n}{n}}$$

for all $i > j$. Then with high probability, $UC(T) \neq A$.

Since $UC(T) = A$ exactly when $UC^\infty(T) = A$, we immediately have the following corollary.

Corollary 4.1. *Assume that a tournament T is generated according to Model 1.*

- *Let $c > \sqrt{2}$ be a constant. If $p_{i,j} \in \left[c\sqrt{\frac{\log n}{n}}, 1 - c\sqrt{\frac{\log n}{n}} \right]$ for all $i \neq j$, then with high probability, $UC^\infty(T) = A$.*
- *If $p_{i,j} \leq 0.6\sqrt{\frac{\log n}{n}}$ for all $i > j$, then with high probability, $UC^\infty(T) \neq A$.*

Theorems 4.1 and 4.2 and Corollary 4.1 allow us to obtain the following corollary on the Condorcet random model.

Corollary 4.2. *Let $f : \mathbb{Z}^+ \to \mathbb{R}_{\geq 0}$ be a function such that $f(n) \leq 1/2$ for all n. Assume that a tournament T is generated according to Model 1, and that $p_{i,j} = f(n)$ for all $i > j$.*

- *If $f(n) \in \omega\left(\sqrt{\log n/n}\right)$ or $f(n) \geq c\sqrt{\log n/n}$ for some constant $c > \sqrt{2}$, then with high probability, $UC(T) = UC^\infty(T) = A$.*
- *If $f(n) \in o\left(\sqrt{\log n/n}\right)$ or $f(n) \leq 0.6\sqrt{\log n/n}$, then with high probability, $UC(T) \neq A$ and $UC^\infty(T) \neq A$.*

5   Majority Tournaments

Thus far, we have established probabilistic results for a general model in which the distribution over tournaments is defined by the probabilities that an alternative dominates another alternative in the tournament (Model 1). As we mentioned in Sect. 2, a common interpretation of tournaments is as the majority relation of an odd number of voters who are endowed with linear preferences over a set of alternatives. In this section, we investigate a more specific model in which the distribution over tournaments is determined by the probability that a voter prefers an alternative to another alternative (Model 2). It turns out that the generality of our results for Model 1 allows us to derive similar results for Model 2 as consequences.

Theorem 5.1. *Let* $f : \mathbb{Z}^+ \to \mathbb{R}_{\geq 0}$ *be a function such that* $f(n) \leq 1/2$ *for all* n, *and* $f(n) \in \omega\left(1/n^{2/(k+1)}\right)$. *Assume that a tournament* T *is generated according to Model 2, and that* $q_{v,i,j} \in [f(n), 1 - f(n)]$ *for all voters* v *and all* $i \neq j$. *Then with high probability,* $TC(T) = COND(T) = CNL(T) = A$.

Theorem 5.2. *Let* $c > \sqrt{2}$ *be a constant. Assume that a tournament* T *is generated according to Model 2, and that* $q_{v,i,j} \in \left[c\left(\frac{\log n}{n}\right)^{\frac{1}{k+1}}, 1 - c\left(\frac{\log n}{n}\right)^{\frac{1}{k+1}}\right]$ *for all voters* v *and all* $i \neq j$. *Then with high probability,* $UC(T) = UC^{\infty}(T) = A$.

6   Experiments

To complement our theoretical results, we investigate the asymptotic behavior of random tournaments according to the Condorcet random model. Starting from a set of alternatives $\{a_1, a_2, \ldots, a_n\}$, we generated random tournaments by inserting for each pair of alternatives a_i, a_j with $i < j$ an edge from a_i to a_j with probability p and an edge in the reverse direction with probability $1 - p$. The tournament solutions that we consider can all be computed efficiently: A simple counting algorithm suffices to compute $COND$, a depth-first search algorithm computes TC in linear time, and the asymptotic running time for computing UC equals that of matrix multiplication [13]. In our experimental setup, we drew 10000 random tournaments of each size $n \in \{5, 10, 20, 30, \ldots, 100\}$ for each $p \in \{0.5, 0.3, 1/n, 1/n^2, \sqrt{2 \log n/n}, 0.6\sqrt{\log n/n}\}$ and checked for each tournament solution $S \in \{COND, UC, TC\}$ whether it selects all alternatives.[9,10] Out of that, we computed the percentage of tournaments in which all alternatives are selected. The resulting graphs are displayed in Fig. 1.

---

[9] Our setting is slightly different for the last two values of p, as we explain later in this section.

[10] Since the probability that CNL selects all alternatives is equal to the corresponding probability for $COND$ for any fixed n by symmetry, and UC^{∞} selects all alternatives exactly when UC does, the results for CNL and UC^{∞} are captured by those for $COND$ and UC, respectively.

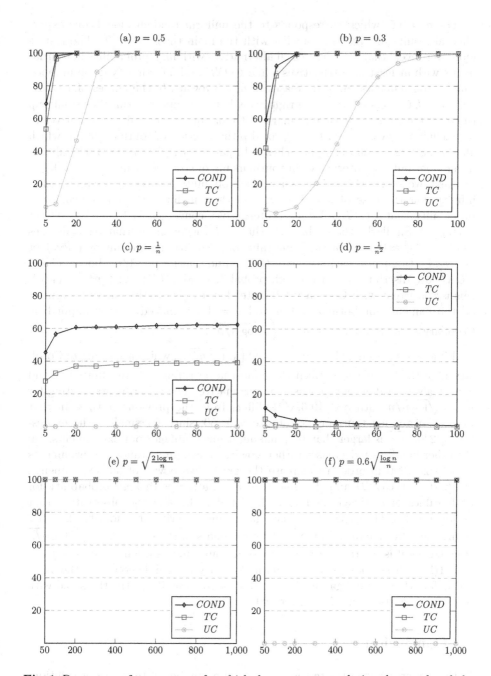

Fig. 1. Percentage of tournaments for which the tournament solution chooses the whole set of alternatives. The horizontal and vertical axes correspond to the number of alternatives in the tournament and the percentage, respectively. Averages are taken over 10000 runs.

For $p = 0.5$, which corresponds to the uniform random model, our experimental results in Fig. 1(a) coincide with the main theorem of [9]. The results moreover reveal that UC chooses all alternatives with high probability in tournaments with at least 50 alternatives while $COND$ and TC already do so in much smaller tournaments. As p decreases from 0.5 toward 0, the curves of $COND$, TC, and UC are shifted to the right; this is to be expected since for smaller p the tournament is more skewed, making it more likely for weaker alternatives to be excluded. Nevertheless, for any fixed p the fraction of tournaments in which all alternatives are chosen approaches 1. In particular, when $p = 0.3$, UC almost never rules out any alternative in tournaments of size 100 or more (Fig. 1(b)).

Next, we look at the regimes where the probability p goes to 0 as n approaches infinity. For the case of $p = 1/n$ we find that, in line with Theorem 3.3, the probability that $COND$ selects all alternatives converges to $1 - e^{-1} \approx 0.6321$ (Fig. 1(c)). Similarly, the probability that TC selects all alternatives converges to $(1 - e^{-1})^2 \approx 0.3996$ for the same value of p, confirming a result by Łuczak et al. [18]. Letting p approach 0 even faster, we find that for $p = 1/n^2$, both TC and $COND$ are discriminative with high probability (Fig. 1(d)). As $1/n^2 \in o(1/n)$, this is consistent with Corollary 3.2. Note that UC is discriminative for almost all tournaments for both $p = 1/n$ and $p = 1/n^2$; indeed, this is implied by Corollary 4.2 since already $1/n \in o\left(\sqrt{\log n/n}\right)$.

Finally, we consider the regime $p = \Theta\left(\sqrt{\log n/n}\right)$, which according to Corollary 4.2 is the boundary between UC almost never ruling out any alternative and almost always ruling out at least one alternative. The experimental setting for $p = c\sqrt{\log n/n}$ with $c \in \{0.6, \sqrt{2}\}$ differs from the previous settings in that we only examined tournaments of size $n \geq 50$, since for small n the expression $\sqrt{2\log n/n}$ is larger than 0.5, making it unsuitable for our experiments. On the other hand, as p decreases rather slowly, we examined random tournaments up to size 1000 in order to increase the expressive power of our experiments. We find that $COND$ and TC select all alternatives with high probability for both values of c; this is in line with Corollary 3.2 and the observation that $c\sqrt{\log n/n} \in \omega(1/n)$. On the other hand, our experiments indicate that UC returns all alternatives in almost all tournaments in the case of $p = \sqrt{2\log n/n}$ (Fig. 1(e)) but is discriminative in almost all tournaments when $p = 0.6\sqrt{\log n/n}$ (Fig. 1(f)). These findings coincide with Corollary 4.2 and demonstrate the interesting fact that a small gap in the constant factor constitutes the threshold with regard to the discriminative power of UC.

7   Conclusion

In this paper, we investigate the behavior of a number of tournament solutions in large random tournaments under a general probabilistic model. We establish tight asymptotic bounds on the boundary of the probability range for which each tournament solution is unlikely to exclude any alternative. In particular, we illustrate a difference between the discriminative power of the top cycle and the uncovered set; this difference is not evident in previous studies that focused on more restricted models. Indeed, while both tournament solutions include all alternatives with high probability in the uniform random model, our results suggest that the uncovered set is in fact considerably more discriminative than the top cycle.

Our work leaves many interesting open questions for future study. A natural next step would be to investigate the asymptotic behavior of other tournament solutions that have been previously studied in the uniform random model—including the Banks set [9], the minimal covering set [29], and the bipartisan set [10]—using our general probabilistic model. For instance, it is conceivable that the approach used by Fey [9] to show that the Banks set almost never rules out any alternative in the uniform random model can be extended to establish an analogous statement when each edge probability is drawn from some constant range. It is not clear, however, whether the approach would still work if we allow the range to depend on the number of alternatives in the tournament like we do in the current work.

From a broader point of view, we believe that an important direction is to apply our model to other tournament problems beyond those concerning tournament solutions, for example the problem of finding a dominating set of minimum size. It is well-known that a dominating set of size at most $\log_2(n + 1)$ always exists and can be found using a simple greedy algorithm. While a dominating set can be as small as a singleton in tournaments that admit a Condorcet winner, Scott and Fey [29] showed that for uniform random tournaments, a dominating set of logarithmic size is the best that one can hope for. More precisely, these authors showed that given any constant $0 < c < 1$, the smallest dominating set of a tournament chosen uniformly at random contains at least $c \log_2 n$ alternatives with high probability. Establishing a similar result in our general probabilistic model is an intriguing technical challenge that would allow us to better understand the behavior of such structures in the real world.

Acknowledgments. This material is based upon work supported by the Deutsche Forschungsgemeinschaft under grant BR 2312/11-1 and by a Stanford Graduate Fellowship. The authors thank Felix Brandt, Pasin Manurangsi, and Fedor Petrov for helpful discussions and the anonymous reviewers for helpful comments.

References

1. Allesina, S., Levine, J.M.: A competitive network theory of species diversity. Proc. Natl. Acad. Sci. (PNAS) **108**(14), 5638–5642 (2011)
2. Arrow, K.J., Raynaud, H.: Social Choice and Multicriterion Decision-Making. MIT Press, Cambridge (1986)
3. Bell, C.E.: A random voting graph almost surely has a Hamiltonian cycle when the number of alternatives is large. Econometrica **49**(6), 1597–1603 (1981)
4. Bouyssou, D.: Monotonicity of 'ranking by choosing': a progress report. Soc. Choice Welf. **23**(2), 249–273 (2004)

5. Brandt, F., Brill, M., Harrenstein, P.: Tournament solutions. In: Brandt, F., Conitzer, V., Endriss, U., Lang, J., Procaccia, A.D. (eds.), Handbook of Computational Social Choice. Cambridge University Press, Cambridge (2016). Chapter 3
6. Brandt, F., Brill, M., Seedig, H.G., Suksompong, W.: On the structure of stable tournament solutions. Econ. Theory **65**(2), 483–507 (2018)
7. Brandt, F., Seedig, H.G.: On the Discriminative power of tournament solutions. In: Lübbecke, M., Koster, A., Letmathe, P., Madlener, R., Peis, B., Walther, G. (eds.) Operations Research Proceedings 2014. ORP, pp. 53–58. Springer, Cham (2016). https://doi.org/10.1007/978-3-319-28697-6_8
8. Cormen, T.H., Leiserson, C.E., Rivest, R.L., Stein, C.: Introduction to Algorithms, 3rd edn. MIT Press, Cambridge (2009)
9. Fey, M.: Choosing from a large tournament. Soc. Choice Welf. **31**(2), 301–309 (2008)
10. Fisher, D.C., Ryan, J.: Tournament games and positive tournaments. J. Graph Theory **19**(2), 217–236 (1995)
11. Frank, O.: Stochastic competition graphs. Rev. Int. Stat. Inst. **36**(3), 319–326 (1968)
12. Good, I.J.: A note on condorcet sets. Public Choice **10**(1), 97–101 (1971)
13. Hudry, O.: A survey on the complexity of tournament solutions. Math. Soc. Sci. **57**(3), 292–303 (2009)
14. Kim, M.P., Suksompong, W., Vassilevska Williams, V.: Who can win a single-elimination tournament? SIAM J. Discret. Math. **31**(3), 1751–1764 (2017)
15. Landau, H.G.: On dominance relations and the structure of animal societies: III. The condition for a score structure. Bull. Math. Biophys. **15**(2), 143–148 (1953)
16. Laslier, J.-F.: Tournament Solutions and Majority Voting. Springer, Heidelberg (1997)
17. Laslier, J.-F.: In silico voting experiments. In: Laslier, J.-F., Sanver, M.R. (eds.) Handbook on Approval Voting. Studies in Choice and Welfare, pp. 311–335. Springer, Heidelberg (2010). https://doi.org/10.1007/978-3-642-02839-7_13. Chapter 13
18. Łuczak, T., Ruciński, A., Gruszka, J.: On the evolution of a random tournament. Discret. Math. **148**(1–3), 311–316 (1996)
19. Maurer, S.B.: The king chicken theorems. Math. Mag. **53**, 67–80 (1980)
20. Miller, N.R.: Graph-theoretic approaches to the theory of voting. Am. J. Polit. Sci. **21**(4), 769–803 (1977)
21. Miller, N.R.: A new solution set for tournaments and majority voting: further graph-theoretical approaches to the theory of voting. Am. J. Polit. Sci. **24**(1), 68–96 (1980)
22. Moon, J.W.: Topics on Tournaments. Holt, Reinhard and Winston, New York (1968)
23. Moon, J.W., Moser, L.: Almost all tournaments are irreducible. Can. Math. Bull. **5**, 61–65 (1962)
24. Moulin, H.: Choosing from a tournament. Soc. Choice Welf. **3**(4), 271–291 (1986)
25. Reid, K.B.: Every vertex a king. Discret. Math. **38**(1), 93–98 (1982)
26. Saile, C., Suksompong, W.: Robust bounds on choosing from large tournaments. CoRR, abs/1804.02743 (2018)
27. Schjelderup-Ebbe, T.: Beiträge zur Sozialpsychologie des Haushuhns. Z. für Psychol. **88**, 225–252 (1922)
28. Schwartz, T.: Rationality and the myth of the maximum. Noûs **6**(2), 97–117 (1972)
29. Scott, A., Fey, M.: The minimal covering set in large tournaments. Soc. Choice Welf. **38**(1), 1–9 (2012)

30. Slater, P.: Inconsistencies in a schedule of paired comparisons. Biometrika **48**(3–4), 303–312 (1961)
31. Ushakov, I.A.: The problem of choosing the preferred element: an application to sport games. In: Machol, R.E., Ladany, S.P., Morrison, D.G. (eds.) Management Science in Sports, pp. 153–161. North-Holland, Amsterdam (1976)
32. Vassilevska Williams, V.: Fixing a tournament. In: Proceedings of the 24th AAAI Conference on Artificial Intelligence (AAAI), pp. 895–900. AAAI Press (2010)

Equilibria in Routing Games
with Edge Priorities

Robert Scheffler[1](✉), Martin Strehler[1], and Laura Vargas Koch[2]

[1] BTU Cottbus-Senftenberg, 03046 Cottbus, Germany
{schefrob,strehler}@b-tu.de
[2] RWTH Aachen, 52056 Aachen, Germany
laura.vargas@oms.rwth-aachen.de

Abstract. In this paper, we present a new routing model with edge priorities. We consider network users that route packages selfishly through a network over time and try to reach their destinations as fast as possible. If the number of packages that want to enter an edge at the same time exceeds the inflow capacity of this edge, edge priorities with respect to the preceding edge solve these conflicts. For this class of games, we show the existence of equilibrium solutions for single-source-single-sink games and we analyze structural properties of these solutions. We present an algorithm that computes Nash equilibria and we prove bounds both on the Price of Stability and on the Price of Anarchy. Moreover, we introduce the new concept of the Price of Mistrust and analyze the connection to earliest arrival flows.

Keywords: Routing game · Algorithmic game theory
Pure Nash equilibria · Edge priorities · Price of Mistrust

1 Motivation

Routing games are a common way to model network congestion in a variety of settings. Network users, consequently called players, behave selfishly and try to independently improve their travel times. These routing games over time are widely studied [6,12,13] due to various applications, e.g., internet routing and other communication networks, road and air traffic control, transportation, logistic in production systems, and financial flows. Depending on applications, there is a huge range of models (see, e.g., [11–13]).

The constrained inflow capacity of edges is an essential ingredient of these models. If more players than the capacity allows want to enter an edge at the same time, one needs to decide which players can go first. Some models maintain the overall arrival order with a first-in first-out rule [6,12,13], i.e., only players with the exact same arrival time are re-arranged. Other models allow overtaking by higher prioritized players with global priority lists [11,12] or local priority lists [11] of players, that is, priority is a property that is inherent to the players and players are linearly ordered. A very natural way to motivate priority has origin in road

© Springer Nature Switzerland AG 2018
G. Christodoulou and T. Harks (Eds.): WINE 2018, LNCS 11316, pp. 408–422, 2018.
https://doi.org/10.1007/978-3-030-04612-5_27

traffic. At intersections of road networks, priority is usually determined by road signs. One road has the way of right, whereas the traffic on the crossing road is prompted to give way by yield or stop signs. This motivates a priority that depends on the chosen edges and not only on the road users.

In this paper, we introduce a routing game over time on a network $G = (V, E)$ where priority on every edge is assigned to players depending on their preceding edge. Every edge $e = (v, w)$ has an ordering of the v entering edges. That is, if two players arrive at a node $v \in V$ at the same time via two different edges $e_1 \in E$ and $e_2 \in E$, and both players are going to enter the same outgoing edge e, then the player from the edge that is first in the order of edge e, say e_1, always moves first and the player on the other edge, here e_2, always has to wait. Moreover, if a third player arrives in the subsequent time step at v via edge e_1, then this player also gets priority and the player on edge e_2 has to wait another round. In other words, for this edge-based priority the first-in first-out (FIFO) property does not hold in nodes. Nevertheless, we assume that the first-in-first-out property holds on edges, i.e., players may queue up before traversing an intersection.

Our Contribution. In this paper, we focus on the symmetric game, i.e., there is only one source and one sink for all players. We show that Nash equilibria are guaranteed to exist in symmetric games and we present a tight bound on the Price of Anarchy as well as a lower bound on the Price of Stability in the same magnitude. Surprisingly, even in the best Nash equilibrium, strategies may include the use of cycles. Beyond that, we show that some players may have to visit a node up to $\mathcal{O}(k)$-times in a network with k players and $|E| \in \mathcal{O}(k)$ edges. Furthermore, we introduce the new concept of the Price of Mistrust. Here, a player mistrusts the other players and chooses a best response such that she can be delayed by as few as possible other players. We show that this concept yields equilibria with values strictly in between the Price of Stability and Price of Anarchy. Furthermore, we present an algorithm to compute mistrustful equilibria for symmetric routing games.

The paper is organized as follows. In Sect. 2, we discuss related results from literature and categorize our problem. Section 3 gives a formal problem definition and fixes the terminology. Afterwards, we present basic results on Nash Equilibria in Sect. 4. In Sect. 5, we present an algorithm that computes such equilibria and introduce the Price of Mistrust. Finally, in Sect. 6 we show the connection to earliest arrival flows. Due to space constraints, some of the proofs are omitted. They can be found in a preliminary version of this paper on arXiv [19].

2   Pointers to the Literature

Routing games are a widely used approach to model traffic situations and there are a lot of different variants. Rosenthal [17] introduced atomic selfish routing games in 1973. Much work on the existence and efficiency of equilibria in this kind of games has been done [1,2,5,18]. For a profound survey by Roughgarden see Chap. 18 in [16].

However, these routing games have the drawback of being static. Hoefer et al. [12] introduced temporal network congestion games, where players block each other only when they enter or use an edge at the same moment in time. The authors analyze the existence of equilibria as well as best response dynamics for several priority rules. Harks et al. [11] proved bounds on the Price of Anarchy and Stability for competitive packet routing games where players have a global or local ranking. In contrast to the player-dependent forwarding/priority rules in the work just mentioned, Cao et al. [4] use a FIFO policy with an edge dependent tie-breaking. They focus on the comparison between Nash equilibria and equilibria in a game with full control (subgame perfect equilibria) on acyclic graphs. In the paper at hand, we use the subordinated tie-breaking of Cao et al. [4] as our main criterion in a non-FIFO setting.

More generally, Ford and Fulkerson introduced a dynamic model for *flows over time*, also called *dynamic network flows*, in their seminal works [8,9]. The authors used time-expanded networks to keep track of the dynamic movement of flow particles and compute system optimal solutions. An excellent introduction to dynamic flows is given by Skutella [20]. A very fascinating concept in this context is the *earliest arrival flow* (EAF). Given a network, an earliest arrival flow simultaneously maximizes the amount of flow that has already reached the sink for all points in time [10]. A game theoretical extension of this model was given by Koch and Skutella [13,14], who presented a characterization of Nash equilibria for flows over time. However, the results were obtained for non-atomic players, i.e., a flow particle is arbitrarily small, while we consider atomic players.

3   A Model for Routing with Edge Priorities

The object under consideration in this paper is a *routing game with edge priorities* which we define as follows.

Playing Field and Rules. The routing game is played on a directed network $G = (V, E)$, where V is the node set with $n = |V|$ nodes and E is the edge multiset with $m = |E|$ edges. We allow multiple edges, i.e., more than one edge starting at the same node and ending at the same node, as well as loops, i.e., an edge starting and ending at the same node. Further, $u : E \to \mathbb{N}$ are *integral capacities* on the edges. This capacity limits the inflow rate, that is, the amount of flow entering an edge $e \in E$ per time unit. This means the total number of players per edge is not restricted. Furthermore, the edges of G are equipped with *integral transit times* $\tau : E \to \mathbb{N}_0$. The transit time or cost $\tau(e)$ denotes the time a player needs to traverse edge $e \in E$. We use constant transit times here, that is, players use edges independently of other players and there is no delay due to congestion. We allow edges with transit time zero, since they are useful to model capacities on nodes. The throughput capacity of a node v can be limited by replacing this node by an edge (v', v'') with the desired capacity and cost zero. Similar to the work of Harks et al. [11], we restrict to integral transit times, since each player blocks exactly one capacity unit for one time unit on each edge of her path.

Additionally, each routing game fixes two distinguished nodes, a source s and a sink t. We assume w.l.o.g. that the source has no incoming edges, i.e., $\deg^-(s) = 0$ with $\deg^-(v) := |\delta^-(v)|$. Every *player* in the set $N = \{1, \ldots, k\}$ of players chooses a path P_i from the set of s-t-paths $\mathcal{P}_{st}$ and travels over time through the network. To be more specific, a player can leave an edge e at the earliest $\tau(e)$ time units after entering e. Moreover, we only consider discrete time steps, since we have integral transit times. Please note that we do not restrict to simple paths, since it may be advantageous for a player to visit a node more than once, as we will see in the upcoming analysis. Since all the players have the same set of strategies, we call it a *symmetric* or *single-commodity* game.

It might happen that more than $u(e)$ players try to enter an edge $e = (v, w)$ at the same time. To decide which players are allowed to proceed directly and which players need to wait at least one time unit, we define a *priority order* $\pi(e) = (e_1, e_2, \ldots, e_{\deg^-(v)})$, i.e., for every $v \in V$ we assign an ordered list $\pi(e)$ of all incoming edges $e_i = (u_i, v) \in \delta^-(v)$ of v to each outgoing edge $e = (v, w) \in \delta^+(v)$ of v. If edge e has remaining capacity at time T, a player seeking to enter edge e at time T may do so, if the incoming edge of this player has the lowest possible index in the ordered list $\pi(e)$ among all players who want to enter the link e. This applies iteratively. Thus, after the first player has entered the link, we choose the next player with the lowest possible incoming edge in $\pi(e)$ from the remaining players, if e still has capacity left.

Among the players waiting on an edge e' the *first-in first-out* rule (FIFO) applies. This means, if Player i and Player j both try to enter edge e from the same edge e', the player who arrived on edge e' first will be preferred. If several players have entered e' at the same time, but the desired edge e does not provide enough capacity for all of them, we use the number of the player as a global tie-breaker. That is, in case of a tie, Player i moves before Player j if $i < j$. In particular, this rule applies at the source, i.e., Player 1 is always the first player to leave source s.

If for every node $v \in V$ it holds that $\pi(e) = \pi(e')$ for all priority lists of outgoing edges $e, e' \in \delta^+(v)$, then we call a game *global* and else *local*. In case of a global game, we may simply define a total order on all edges, since such an order canonically defines the priority list of each edge. This total order is of course not unique. For the sake of simplicity in a global game, we always relabel the edges of G to $e_1, e_2, \ldots, e_m$ such that e_i has higher priority than e_j whenever $i < j$. Summarizing, we determine priority lexicographically according to the order: *edge list > FIFO > player ID*. A *routing game with edge priorities* is defined as a tuple (G, u, τ, N, π).

Goal of the Game. In the game, a *strategy* of a Player i is an s-t path P_i. Let P be the *profile* or *state* of the game with the strategies of all the players, that is, P consists of k paths P_1 to P_k. Now, we denote the *arrival time* of Player i as $C_i(P)$, which is the sum of the transit times $\tau(e)$ of all the edges of the chosen path P_i and the *waiting time* on those edges. Obviously, the former is independent of the strategies of the other players due to constant transit times, but the latter significantly depends on P.

A profile P is *socially optimal* if it minimizes the total cost given by $C(P) = \sum_{i \in N} C_i(P)$. However, we assume players to behave *selfishly*, i.e., each player aims to minimize her own arrival time. We call a state a (pure) Nash equilibrium (PNE) if the chosen strategies separately minimize the costs for each player. Let P_{-i} be the state P without the strategy of Player i. Furthermore, with P_i', P_{-i} we denote replacing the strategy of Player i in P by P_i'. More formally, a routing game with state P and strategy P_i for Player i is in a PNE if $C_i(P_i, P_{-i}) \leq C_i(P_i', P_{-i})$ $\forall P_i' \in \mathcal{P}_{st}$, for all players $i \in N$. In other words, no player can reduce her cost by switching from P_i to another path P_i'.

Making the Game Well-Defined. For a well-defined game, it is necessary to have a unique mapping of a strategy profile P to costs $C(P)$ of the players. Unfortunately, the current model can still lead to some paradoxical situations in connection with zero cost cycles. An example with two players is given in Fig. 1. Assume, Player 1 chooses the path $(s, v_1, v_3, v_2, v_4, t)$ and Player 2 chooses $(s, v_2, v_4, v_1, v_3, t)$. Note that the paths intersect twice. Player i hits node v_i before node v_{3-i} on the respective path.

Now, assume the red wavy edges (v_4, v_1) and (v_3, v_2) in the cycle have priority over the black straight entering edges (s_1, v_1) and (s_2, v_2) to proceed on (v_1, v_3) and (v_2, v_4), respectively. Furthermore, all edges have zero transit time. On the one hand, Player 1 could reach v_2 in zero time and block Player 2 there. If Player 2 is blocked at v_2, she cannot block Player 1 at v_1. Thus, Player 1 reaches t at time 0, and Player 2 reaches t at time 1. On the other hand, the same argument is valid vice versa and, thus, there is no unique mapping from the strategy P to arrival times of players $C(P)$.

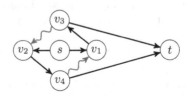

Fig. 1. Non-unique embedding due to zero cost cycles. Red wavy edges have higher priority. (Color figure online)

Since we do not want to forbid zero transit times in general, we exclude all networks with directed zero cost cycles from our consideration. Moreover, we will compute ratios of various solutions, e.g., the Price of Anarchy. To avoid division by zero, we additionally exclude all games where source s and sink t have distance zero.

Given this analog to [11] we can show that there is a unique embedding of a given set of paths. The basic idea is to use a Dijkstra-like algorithm and always forward players which are in nodes without entering 0-edges of the paths of other players.

4   Basic Results on Equilibria

In this section, we highlight some important properties of equilibria in routing games with edge priorities. Before we can claim the existence of a PNE in every symmetric game, we have to study structural properties of equilibria, which will be exploited in a constructive proof in the next section.

We start by recognizing that not only is the equilibrium not unique with regard to the profile, but even the cost $C(P)$ of equilibria can vary for a given game.

Lemma 1. *The value of PNEs in a routing game with edge priorities is not unique.*

Proof. We use a graph based on the famous example of Braess [3]. We refer to the graph with b parallel paths from s to t as b-*Braess graph*. Each path consists of three edges. Furthermore, edges connect the third node of the i-th path with the second node of the $(i + 1)$-th path for all $1 \leq i < b$. A 4-Braess graph is depicted in Fig. 2. Consider the b-Braess graph where all s-leaving edges have transit time one and all the other edges have cost zero, while all edges have unit capacity. In this network, priority follows the scheme depicted in Fig. 2, that is, the red wavy edges connecting the parallel paths are always prioritized over the edges in the direct paths.

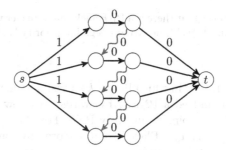

Fig. 2. The 4-Braess graph, where the value of a PNE is not unique. Red wavy edges have higher priority (Color figure online)

We study the game with $k = b$ players. Obviously, using the b parallel paths in this network is a PNE, which is also socially optimal with total cost $C(P) = k$.

However, it is also a PNE for all players to go along the zigzag-path in order of the player IDs, i.e., every player uses all the red wavy edges. In this case, Player i has cost i and no improving move. In total $C(P) = \frac{k(k+1)}{2}$.

Surprisingly, in routing games with edge priorities, a player may visit a node arbitrarily often. Furthermore, there are networks where this can happen in every PNE. We start with the following lemma.

Lemma 2. *In a PNE in a symmetric routing game with edge priorities, players arrive at the sink t in unchanged order. Further, for every $1 \leq i \leq k - 1$, Player $i + 1$ arrives at t at most one time unit after the preceding Player i.*

Note that this statement does not hold for intermediate nodes. There, players may arrive much earlier on a subordinate edge and wait until they can proceed.

Theorem 3. *For every $k \in \mathbb{N}$, $k \geq 3$, there exists a routing game with edge priorities with k players on a network with $2k - 1$ edges and unit capacities $u \equiv 1$ such that there is a player who visits a node $k - 1$ times in every PNE.*

Proof. We construct an instance with a unique PNE. The network consists of three nodes s, v, and t. The source s is connected to v with k parallel edges $e_k, \ldots, e_{2k-1}$. Furthermore, there are $k - 2$ loops at v, namely $e_2, \ldots, e_{k-1}$, and a single edge e_1 from v to the sink t. All edges have capacity $u \equiv 1$ and transit time $\tau \equiv 1$. An example of the network is given in Fig. 3. We use global edge priorities, i.e., the index of the edge is equivalent to its priority.

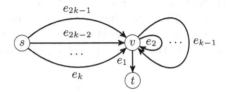

Fig. 3. This example shows that there is a PNE such that a player visits a node multiple times. All edges have unit travel time. We use a global priority list where e_i is preferred over e_j if $i < j$.

First, we define a state P in which Player $k-1$ visits node v exactly $k-1$ times and prove that this is indeed a PNE. Afterwards, we show that the strategies of the Players 1 to $k - 1$ coincide in every PNE. For Player $i \in \{1, \ldots, k - 1\}$ define $P_i = (e_{k+i-1}, e_i, \ldots, e_1)$. Player k can choose any path arriving at time $k + 1$. By construction, Player i visits node v exactly i times and arrives at time $i + 1$ for $i \in \{1, \ldots, k - 1\}$. Assume for the sake of contradiction that the state $P = (P_1, \ldots, P_k)$ is not a PNE. Then there is a player that can improve her strategy. Since the players arrive one after another, improving is only possible by delaying a preceding player. This is not possible, since every player goes along edges with the highest possible priorities.

Assume there is a PNE P' which differs from P by another path than P_k. Let i be the first player deviating from the strategy in P. Player i cannot arrive earlier than in P as argued above. Moreover, Player i cannot arrive later than in profile P in a PNE, since strategy P_i is always a valid choice for arriving at time $i + 1$. Thus, P' realizes the same arrival time as P_i. In this case, Player $i + 1$ can now improve by playing $(e_x, e_i, \ldots, e_1)$ where e_x is an unused edge among e_{k+i-1} and e_{2k-1}. With this strategy, Player $i + 1$ overtakes Player i and arrives at time $i + 1$ which is an improving move. Thus, P' was not a PNE.

Note that we used multi-edges for simplicity, here. In a network without multi-edges, one can still achieve $\mathcal{O}(k)$ visits of a node by a single player by subdividing each edge.

Contrariwise, there is no such result on the edges, as we show in Corollary 6 in Sect. 5. Since edges and FIFO determine priority, there is no advantage to use an edge twice.

5   Computing Equilibria and the Price of Mistrust

Computing an Equilibrium. In this section, we present an efficient algorithm for computing PNEs in symmetric routing games with edge priorities. Intuitively, the players choose their strategies one by one in order of the tie breaking rule. Each player chooses a strategy minimizing her cost, and among all those strategies she takes the one, where she cannot be delayed by any of the following players. We start with an outline of the main algorithmic ideas, before we give a detailed description of the algorithm in pseudo-code in Algorithm 1.

The algorithm consists of three steps. The first step initializes a kind of shortest path network and this step is executed only once. In step 2, a path for the next player is found within this shortest path network, and in step 3, the network is updated to renew the earliest arrival property for the upcoming player. Step 2 and 3 are executed once for each player. In detail:

1. A modified Dijkstra search [7] is executed starting in s. We determine two functions $d : V \to \mathbb{N}_0$ and $\varepsilon : E \to \mathbb{N}_0$. Here, $d(v)$ describes, at which time step node v can be reached at the earliest, i.e., at this initialization step $d(v)$ is the standard label set by Dijkstra's algorithm. The function value $\varepsilon(e)$ defines the earliest point in time at which edge e can be left. Hence, $d(v) \leq \varepsilon(e)$ where $e = (u, v)$. In the initialization, it holds $\varepsilon(e) = d(u) + \tau(e)$ for $e = (u, v)$, since there are no waiting times. Please note that the sub network $G' = (V, E')$ where $E' = \{e = (u, v) \in E : \varepsilon(e) = d(v)\}$ contains all shortest paths from s to t.
2. We now perform a backward search in this sub network $G' = (V, E')$ to find a path p that reaches t at time $d(t)$ for every Player i, starting with Player 1. The algorithm iteratively adds edges to p until reaching s. Here, $\Delta : V \to \mathbb{N}_0$ describes the latest time when $v \in V$ must be reached in order to arrive at the following node of the path in time. It is updated to $\Delta(v) := \varepsilon(e) - \tau(e)$ always when adding edge e to p. Among all possibilities to arrive at v in the time interval $[d(v), \Delta(v)]$, we choose the edge with the highest available priority with respect to the subsequent edge, which was already determined. Moreover, we use this edge at the earliest possible moment in time.
3. After assigning Player i to the path p constructed in step 2, we have to update the values of d and ε. If Player i exhausted the capacity of an edge $e = (u, v)$, we increment $\varepsilon(e)$ by 1. We now perform a modified Dijkstra search to check whether we have to increment other labels $d(v)$ and $\varepsilon(e)$, too. Now, we can go back to step 2 to compute the path for the next Player.

Algorithm 1. PATHFINDER

Input: $G = (V, E)$ with priorities π, $s, t \in V$ and set of players $N = \{1, \ldots, k\}$

Output: A walk for every player $i \in N$

1 calculate $d(v)$ and $\varepsilon(e)$ $\forall v \in V, \forall e \in E$;

2 **for** $i := 1$ *to* k **do**

3 $\quad$ $e :=$ arbitrary edge of $\delta_{\mathrm{in}}(t)$ with $\varepsilon(e) = d(t)$;

4 $\quad$ $v := \mathrm{tail}(e)$;

5 $\quad$ $\Delta(v) := d(t) - \tau(e)$;

6 $\quad$ $p := \{e\}$;

7 $\quad$ **while** $v \neq s$ **do**

8 $\quad\quad$ let $e' \in \delta_{in}(v)$, with $\varepsilon(e') \leq \Delta(v)$ and maximal priority for e;

9 $\quad\quad$ $v := \mathrm{tail}(e')$;

10 $\quad\quad$ $\Delta(v) := \varepsilon(e') - \tau(e')$;

11 $\quad\quad$ $p := p \cup \{e'\}$;

12 $\quad\quad$ $e := e'$;

13 $\quad$ print "Path of player i is p";

14 $\quad$ **foreach** $e \in p$ **do**

15 $\quad\quad$ **if** *capacity of e at entry time $\Delta(\mathrm{tail}(e))$ is exhausted* **then**

16 $\quad\quad\quad$ $\varepsilon(e) := \varepsilon(e) + 1$;

17 $\quad$ $d(s) := 0$ and $d(v) := \infty$ $\forall v \in V, v \neq s$;

18 $\quad$ $\Pi := \mathrm{heap}(V, d)$;

19 $\quad$ **while** $\Pi \neq \varnothing$ **do**

20 $\quad\quad$ $v := \mathrm{getMin}(\Pi)$ and remove v from Π;

21 $\quad\quad$ **foreach** $e \in \delta_{\mathrm{out}}(v)$ **do**

22 $\quad\quad\quad$ $\varepsilon(e) := \max\{\varepsilon(e), d(v) + \tau(e)\}$;

23 $\quad\quad\quad$ $d(\mathrm{head}(e)) := \min\{d(\mathrm{head}(e)), \varepsilon(e)\}$;

Let us remark some observations. Firstly, for the first player, the situation is quite simple with $\Delta(v) = d(v)$ for all nodes of the constructed path in step 2, since the network is not yet congested and no waiting times occur. However, this is not true for subsequent players. Here, some edge $e = (v, w)$ could be blocked at time $d(v)$ by preceding players. Secondly, $d(v)$ denotes the earliest arrival time at each node. In particular, $d(t)$ is the earliest time to reach the sink and we always construct a path achieving this time bound. Thirdly, we always use the earliest option for each edge under consideration. Therefore, no subsequent player can use an earlier option and nobody can displace Player i. And fourthly, by choosing the edge $e' = (u, v)$ with the highest priority in $\pi(e)$, we can guarantee that we can go from e' to e at time $\Delta(v)$. No subsequent player can use edge e' at an earlier time, i.e., there is no additional delay on e' to be expected. Moreover, this path has the way of right over all subsequent players who arrive at v via other edges.

Theorem 4. *Let (G, u, τ, N, π) be a symmetric routing game with edge priorities and $k := |N| \geq 1$. Then Algorithm 1 has running time $\mathcal{O}(k(m^2 + n \log n))$ and computes a PNE.*

Note that the run-time bound is given with respect to the network size. However, it is essentially determined by looping through the edge priority lists. Even when we visit a node twice or more as in Fig. 3, priority increases locally, i.e., we only have to run through all priority lists once. Since the total length of all edge priority lists is in $\mathcal{O}(m)$ for global games, we can improve the run-time analysis for this case to $\mathcal{O}(k(m + n \log n))$. However, these lists are part of the input, too. Hence, run-time is even linear regarding this part of the input. Since we have to return a path for every player, the run-time is obviously only pseudo-polynomial in the number of players k, but it is polynomial, when we use a unary encoding of players. The previous result also implies the existence of equilibria.

Theorem 5. *In every symmetric routing game with edge priorities there exists a PNE.*

So in contrast to Theorem 3, there is always a PNE where each player uses each edge at most once. Intuitively this follows from the definition of the priority rules, a formal proof is given in Theorem 4 [19] since there we show that every edge is processed only once while executing the algorithm.

Corollary 6. *In every symmetric routing game with edge priorities, there exists a PNE such that every edge is used at most once by each player.*

Yet, this does not imply that an optimal strategy does not contain an edge multiple times. Furthermore, please note that the presented algorithm does not necessarily compute the best PNE. In particular, the cost of the computed PNE significantly depends on the way in which we choose the incoming edge to the sink t. Since we do not have any priority here, one may choose this edge randomly or in any fixed order among all those edges $e = (v, t)$ with $d(t) = \varepsilon(e)$. As an example, consider again the graph in Fig. 2. If we choose the bottom edge to t first, the algorithm computes the zigzag-path for the first player and every following player. Thus, the algorithm computes the worst possible PNE in this scenario. Contrary, if the algorithm treats the incoming edges of t in a fixed order from top to bottom, it computes the best PNE.

Bounding the PoS and the PoA. On the one hand, we have seen in Lemma 1 that there may exist several user equilibria with different values for a routing game with edge priorities. Consequently, we ask for the highest quotient of the cost of an equilibrium compared to the cost of a social optimum. This ratio is known as the *Price of Anarchy* (PoA) (cf. [15]). On the other hand, if a social optimal state is not a user equilibrium, then it is not stable, i.e., some players have occasion to change routes. If every social optimal solution is not stable, then even the best equilibrium has additional cost compared to the social optimum. This ratio is called the *Price of Stability* (PoS) (cf. [1]).

In this subsection, we give a tight bound of the PoA for routing games with edge priorities. Furthermore, we show that there are instances of a routing game with edge priorities where every social optimum is not stable and we present a lower bound example of the PoS.

Theorem 7. *The PoA in a symmetric routing game with edge priorities is at most $\frac{k+1}{2}$, where k is the number of players. This upper bound is tight.*

The idea of the proof is to use Lemma 2 to bound the cost of Player i in the PNE from above by $\ell + i - 1$, where ℓ is the length of a shortest path. This is similar to the proof in [11].

For similar routing games, e.g. [11], it turns out that there is a stable socially optimal state, i.e., the PoS is equal to 1. Surprisingly, this is not true for the routing game with edge priorities and the PoS can be of the same order of magnitude as the PoA.

Theorem 8. *The PoS in a symmetric routing game with edge priorities can be $\geq \frac{k+1}{4}$, where k is the number of players.*

Price of Mistrust. Algorithm 1 calculates PNEs that have a special property, which we formalize in the following.

Definition 9. *For Player i, given a strategy profile P, a mistrustful best response is a best response to P_{-i} that minimizes the number of players that can delay Player i through a strategy change. A PNE is mistrustful, if every player plays a mistrustful best response.*

From Theorem 4 (see [19]), we get that Player i can always find a mistrustful best response, such that she can maximally be delayed by $i - 1$ players (the preceding players who can delay her by coping the strategy).

Observation 10. *Every PNE that is computed by Algorithm 1 is mistrustful.*

It is an interesting question how much social cost is caused by this kind of mistrust. For the best PNE, it is often necessary that players co-operate and trust each other. Similar to the PoS, we define the *Price of Mistrust (PoM)* as the quotient of the mistrustful PNE with minimal cost and the social optimum. As an example, consider the graph in Fig. 2. If the first player chooses the lowest horizontal path, every following player can find a path that has optimal cost and does not block the first player. Unfortunately, any of these players may also use a path blocking the first player, maybe because of ignorance or because of malignity. Thus, mistrust forces the first player to choose another path.

In the following, we present some routing games which show that the PoM can reach exactly the same extremal value of $\frac{k+1}{2}$ of the PoA. Nevertheless, we also show that there is an instance where the PoM is strictly smaller than the pessimistic PoA, but strictly larger than the optimistic PoS. For this purpose, we use the construction of the b-Braess graph in the proof of Lemma 1 and extend it by a twisted second copy. Roughly speaking, we add a second zigzag path after the first one, which connects the parallel paths in the opposite directions. We call these graphs *b-double-Braess graphs* and Fig. 4 shows two 4-double-Braess graphs. In these graphs, each outgoing edge of s has cost one, each other edge has cost zero and all edges have unit capacities. For $k = b$ players the cost of the social optimum is k, since the players can use parallel paths with cost one. Since there is no cheaper path, this is also a PNE.

Now, we consider two different choices of priorities, depicted by red wavy edges in Fig. 4. Firstly, on the left hand side, both zigzag paths have higher priority than any other path. No matter which edge to t is chosen by the first player, Player 1 always has to visit node v and she has to use the first zigzag path completely as otherwise any subsequent player would be able to block her. For the subsequent players the same argumentation holds and the players follow each other. The total cost is $\frac{k(k+1)}{2}$, so the PoM is equal to $\frac{k+1}{2}$, which is also the worst case PoA. In consequence, we obtain the following corollary.

Corollary 11. *There are symmetric routing games with edge priorities, where Algorithm 1 is not able to compute a PNE with minimal cost.*

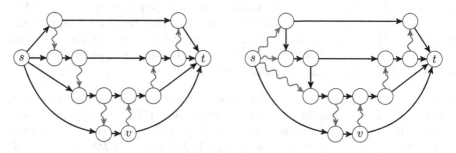

Fig. 4. On the left hand side the PoM is equal to the PoA. On the right hand side the PoM lies between PoS and PoA. The red wavy edges have higher priority. (Color figure online)

Secondly, consider the priorities on the right hand side of Fig. 4. Again, the social optimum, which is also a PNE (with cost k), is not mistrustful, but the PoS is equal to 1. Moreover, the mistrustful first player has to visit node v or otherwise, her strategy would be vulnerable on the second zigzag path by any other player who visits v. To reach v on a prioritized path, Player 1 has to choose the second lowest outgoing edge of s. In the best case, Player 1 will not use the second zigzag path, but she will use edge (v, t). Now, each subsequent Player i has to choose the $(i − 1)$th lowest horizontal path if this player mistrusts all upcoming players. The last Player k can choose any path, but she will definitely not reach t before time step 2. Thus, the *PoM* is $\frac{k+1}{k}$. In the worst case, Player 1 uses the second zigzag path completely, which again yields a PoA of $\frac{k+1}{2}$ as in the previous example. Thus, we finally conclude with the following observation.

Observation 12. *There are symmetric routing games with edge priorities, where the PoM is greater than the PoS and less than the PoA.*

6   Connection to Earliest Arrival Flows

A socially optimal state of a routing game in a network with a single source and a single sink and *without* priority lists is an earliest arrival flow (EAF). Furthermore, if we restrict to unit capacities and follow the same argumentation as in Harks et al. [11], we can show that the socially optimal state of a given symmetric game *with* edge priorities is an EAF, too. Similar to the model with player priorities, there is always a socially optimal solution with edge priorities, where players only wait at the source node in an EAF in a network with unit capacities due to the fact that an EAF fulfills strong flow conservation. But at the source, the global tie-breaking order of the players applies and there is no priority rules of any entering edges in our model. Thus, both models, our model and the model with player priorities, coincide in this case.

However, in [11], an EAF is also always an equilibrium, i.e., the PoS is equal to one. This is not the case in our model with edge priorities as we have seen in Theorem 8. Yet, given a network initially without edge priorities, it may still be possible to find a priority rule on the edges, such that there is an equilibrium that is an EAF. In other words, can a network operator always establish a system optimal flow in a network with edge priorities by choosing an appropriate priority list? Unfortunately, this is not possible in general.

Theorem 13. *There exist networks where an earliest arrival flow cannot be established by any priority order of the edges as a PNE with edge priorities. Further, for arbitrarily many players and arbitrarily high edge costs, the additive gap of the social costs of an optimal flow (EAF) and the best PNE is arbitrarily large.*

Nevertheless, there are graph classes which always allow the construction of appropriate priority lists.

Proposition 14. *On series-parallel networks with unit capacities, there exist edge priorities such that the best mistrustful PNE is an EAF.*

7   Discussion

Motivated by right of way rules, routing over time with edge priorities brings some surprises like optimal Nash equilibrium solutions with many cycles. Yet, there are also many open questions. Among them, what are efficient network topologies, i.e., for which graph classes can the system optimum be established by a (mistrustful) equilibrium? How can one compute the best equilibrium? Moreover, the gap between social optimum (EAF) and PoS makes routing with edge priorities seem unattractive in terms of 'protocol design'. On the other hand, it is a natural rule for self-organizing flows like road traffic. Hence, it is quite natural to ask for a generalization of the results to multi-commodity games. Yet, it is unlikely that equilibria always exist in this case. Under what conditions can we guarantee the existence of equilibria?

References

1. Anshelevich, E., Dasgupta, A., Kleinberg, J., Tardos, É., Wexler, T., Roughgarden, T.: The price of stability for network design with fair cost allocation. SIAM J. Comput. **38**(4), 1602–1623 (2008)
2. Awerbuch, B., Azar, Y., Epstein, A.: The price of routing unsplittable flow. In: Proceedings of the Thirty-seventh Annual ACM Symposium on Theory of Computing, pp. 57–66. ACM (2005)
3. Braess, D.: Über ein Paradoxon aus der Verkehrsplanung. Unternehmensforschung **12**, 258–268 (1968)
4. Cao, Z., Chen, B., Chen, X., Wang, C.: A network game of dynamic traffic. arXiv preprint arXiv:1705.01784 (2017)
5. Christodoulou, G., Koutsoupias, E.: The price of anarchy of finite congestion games. In: Proceedings of the Thirty-Seventh Annual ACM Symposium on Theory of Computing, pp. 67–73. ACM (2005)
6. Cominetti, R., Correa, J.R., Larré, O.: Existence and uniqueness of equilibria for flows over time. In: Aceto, L., Henzinger, M., Sgall, J. (eds.) ICALP 2011. LNCS, vol. 6756, pp. 552–563. Springer, Heidelberg (2011). https://doi.org/10.1007/978-3-642-22012-8_44
7. Dijkstra, E.W.: A note on two problems in connexion with graphs. Numer. Math. **1**(1), 269–271 (1959). https://doi.org/10.1007/BF01386390
8. Ford, L.R., Fulkerson, D.R.: Constructing maximal dynamic flows from static flows. Oper. Res. **6**, 419–433 (1958)
9. Ford, L.R., Fulkerson, D.R.: Flow in Networks. Princeton University Press, Princeton (1962)
10. Gale, D.: Transient flows in networks. Mich. Math. J. **6**, 59–63 (1959)
11. Harks, T., Peis, B., Schmand, D., Koch, L.V.: Competitive packet routing with priority lists. In: Faliszewski, P., Muscholl, A., Niedermeier, R. (eds.) 41st MFCS 2016. LIPIcs, vol. 58, pp. 49:1–49:14. Schloss Dagstuhl-Leibniz-Zentrum fuer Informatik, Dagstuhl, Germany (2016)
12. Hoefer, M., Mirrokni, V.S., Röglin, H., Teng, S.-H.: Competitive routing over time. In: Leonardi, S. (ed.) WINE 2009. LNCS, vol. 5929, pp. 18–29. Springer, Heidelberg (2009). https://doi.org/10.1007/978-3-642-10841-9_4
13. Koch, R., Skutella, M.: Nash equilibria and the price of anarchy for flows over time. In: Mavronicolas, M., Papadopoulou, V.G. (eds.) SAGT 2009. LNCS, vol. 5814, pp. 323–334. Springer, Heidelberg (2009). https://doi.org/10.1007/978-3-642-04645-2_29
14. Koch, R., Skutella, M.: Nash equilibria and the price of anarchy for flows over time. Theory Comput. Syst. **49**(1), 71–97 (2011)
15. Koutsoupias, E., Papadimitriou, C.: Worst-case equilibria. In: Meinel, C., Tison, S. (eds.) STACS 1999. LNCS, vol. 1563, pp. 404–413. Springer, Heidelberg (1999). https://doi.org/10.1007/3-540-49116-3_38
16. Nisan, N., Roughgarden, T., Tardos, É., Vazirani, V.V.: Algorithmic Game Theory, vol. 1. Cambridge University Press, Cambridge (2007)

17. Rosenthal, R.W.: The network equilibrium problem in integers. Networks **3**(1), 53–59 (1973)
18. Roughgarden, T., Tardos, É.: How bad is selfish routing? J. ACM (JACM) **49**(2), 236–259 (2002)
19. Scheffler, R., Strehler, M., Koch, L.V.: Nash equilibria in routing games with edge priorities. arXiv preprint arXiv:1803.00865 (2018)
20. Skutella, M.: An introduction to network flows over time. In: Cook, W., Lovász, L., Vygen, J. (eds.) Research Trends in Combinatorial Optimization, pp. 451–482. Springer, Berlin Heidelberg (2009). https://doi.org/10.1007/978-3-540-76796-1_21

Cost-Sharing Games in Real-Time Scheduling Systems

Tami Tamir[✉]

School of Computer Science, The Interdisciplinary Center (IDC), Herzliya, Israel
tami@idc.ac.il

Abstract. We apply non-cooperative game theory to analyze the server's activation cost in *real-time scheduling* systems. An instance of the game consists of a single server and a set of unit-length jobs. Every job needs to be processed along a specified time interval, defined by its release-time and due-date. Jobs may also have variable weights, which specify the amount of resource they require. We assume that jobs are controlled by selfish agents who act to minimize their own cost, rather than to optimize any global objective.

The jobs processed in a specific time-slot cover the server's activation cost in this slot, with the cost being shared proportionally to the jobs' weights. Known result on cost-sharing games do not exploit the special interval-structure of the strategy space in our game, and are therefore not tight. We present a complete analysis of equilibrium existence, computation, and inefficiency in real-time scheduling cost-sharing games. Our tight analysis covers various classes of instances, and distinguishes between unilateral and coordinated deviations.

1   Introduction

The emergence of cloud systems as a common computation resource gives rise to plenty of optimization problems whose input is a *real-time scheduling* instance, consisting of time-sensitive jobs which are often business-critical. Each job needs to be processed along a specified time interval, defined by its release-time and due-date. Jobs may also have variable lengths and weights, corresponding to their resource demand [10,11,26,30].

Traditional research interest in cluster systems has been high performance, such as high throughput, low response time, or load balancing [10,30]. In this paper we apply non-cooperative game theory to study the problem of minimizing the server's activation cost, a recent trend in cluster computing which aims at *reducing power consumption* (see, e.g., [4,14,28]).

The jobs should be processed by a server available during the whole schedule. We assume that time is slotted, and a job needs to be processed along one time-slot in order to be completed. For every time-slot, we are given the server's cost for this slot. The server has unlimited capacity, and the cost is independent of the load (as long as it is non-zero). In other words, the cost is associated

© Springer Nature Switzerland AG 2018
G. Christodoulou and T. Harks (Eds.): WINE 2018, LNCS 11316, pp. 423–437, 2018.
https://doi.org/10.1007/978-3-030-04612-5_28

with *activating* the server. This scenario arises in cloud computing management, as well as in several applications in media-on-demand systems, optical-network design, and shuttle services.

As a real-life toy example, consider a big carousel (merry-go-round) in an amusement park. Rides may start in every quarter of an hour (defining the time slots), and every operation of the carousel costs a predefined amount, which is independent of the number of riders. The park attracts many groups of kids along the day. All the groups would like to have a ride on the carousel during their visit in the park. The carousel owner would like to schedule the group rides in a way that minimizes the total activation cost. The carousel's capacity is not a problem, as the number of simultaneous visitors in the park never exceeds the carousel capacity.

In this paper, we analyze the game corresponding to this job-scheduling scenario. We assume that jobs are controlled by selfish agents who act to minimize their own cost, rather than to optimize any global objective. Thus, each agent *chooses* the slot in which its job is processed instead of being assigned to one by a central authority. Back to our merry-go-round example, in the corresponding game, every group selects its riding time, with the understanding that groups riding together share the carousel's activation cost with the share being proportional to the groups' sizes.

While game theory has become an essential tool in their study, many real-world applications do not necessarily fit the basic framework assumed in their common analysis. In particular, the setting of real-time scheduling induces a game in which the resources form a line and the strategy space of each player is defined by an interval in this line. Thus, our paper belongs to the rich literature on congestion and cost-sharing games with limited strategy space (e.g., matroids, paths in graphs, etc.). As we show, the strategies' interval structure induces a game which is more stable than general singleton cost-sharing games. While some of our results are simple adaptations of previously studied games, most of them require different techniques and new tools, that exploit the unique interval-strategy structure.

1.1   Preliminaries

An instance G of our game consists of a set $\mathcal{J}$ of n unit-length jobs, and a single server. Every job $j \in \mathcal{J}$ is associated with a time interval, $I(j) = [r_j, d_j)$, where r_j and d_j denote its release-time and due-date. In addition, every $j \in \mathcal{J}$ has a *weight* $w_j > 0$.

Let $T = max_{j \in \mathcal{J}} d_j$ be the maximal deadline of a job. We assume that the server is available along the interval $[0, T)$. Time is slotted, and a job can start its processing only at integral time points. For $t = 1, \ldots, T$ we refer to $[t - 1, t)$ as the t-th *slot*. Let c_t denote the *activation cost* of the server in slot t.

A schedule S determines for every job j the time slot s_j in which it is processed, such that $[s_j - 1, s_j) \subseteq I(j)$. We say that the server is *busy* at time-slot t if it processes at least one job in slot t. Otherwise, the server is *idle* at time t. Every feasible schedule S corresponds to a profile of the game. For a profile S, the load on slot t is denoted $\ell_t(S)$ and is given by the total weight of jobs processed in slot t. That is, $\ell_t(S) = \sum_{j|s_j=t} w_j$. The jobs processed in slot t share the server's activation cost c_t in a way proportional to the load they generate, given by their weights. Formally, the cost of job j in a profile S is $cost_j(S) = w_j \cdot c_{s_j}/\ell_{s_j}(S)$. The total cost of a schedule is $cost(S) = \sum_{j \in \mathcal{J}} cost_j(S)$. Note that $cost(S)$ also equals the total activation cost of non-idle slots, that is, $cost(S) = \sum_{t|\ell_t(S)>0} c_t$. This cost-sharing scheme fits the commonly used proportional cost-sharing rule for weighted players (e.g., [6,9,22,32]), when the cost of a resource splits among its users proportional to their demand.

For a profile S, a job $j \in \mathcal{J}$, and a slot $s'_j \subseteq I(j)$, let (S_{-j}, s'_j) denote the profile obtained from S by replacing the strategy of job j by s'_j. That is, the profile resulting from a migration of job j from slot s_j to slot s'_j. A profile S is a *pure Nash equilibrium* (NE) if no job can benefit from unilaterally deviating from his strategy in S to another strategy; i.e., for every job j and every slot $s'_j \subseteq I(j)$ it holds that $cost_j(S_{-j}, s'_j) \geq cost_j(S)$.[1]

Best-Response Dynamics (BRD) is a local-search method where in each step some player is chosen and plays its best improving deviation (if one exists), given the strategies of the other players. Since BRD corresponds to actual dynamics in real-life applications, the question of BRD convergence and the quality of possible BRD outcomes are major issues in the study of resource allocation games in applied systems.

It is well known that decentralized decision-making may lead to sub-optimal solutions from the point of view of the society as a whole. For a game G, let $P(G)$ be the set of feasible profiles of G. We denote by $OPT(G)$ the cost of a social optimal (SO) solution; i.e., $OPT = \min_{S \in P(G)} cost(S)$. We quantify the inefficiency incurred due to self-interested behavior according to the *price of anarchy* (PoA) [29] and *price of stability* (PoS) [6] measures. The PoA is the worst-case inefficiency of a pure Nash equilibrium, while the PoS measures the best-case inefficiency of a pure Nash equilibrium. Formally,

Definition 1. Let $\mathcal{G}$ be a family of games, and let G be a game in $\mathcal{G}$. Let $\Upsilon(G)$ be the set of pure Nash equilibria of the game G. Assume that $\Upsilon(G) \neq \emptyset$.

- The *price of anarchy* of G is the ratio between the *maximal* cost of a NE and the social optimum of G. That is, $PoA(G) = \max_{S \in \Upsilon(G)} cost(S)/OPT(G)$. The *price of anarchy* of the family of games $\mathcal{G}$ is $PoA(\mathcal{G}) = sup_{G \in \mathcal{G}} PoA(G)$.
- The *price of stability* of G is the ratio between the *minimal* cost of a NE and the social optimum of G. That is, $PoS(G) = \min_{S \in \Upsilon(G)} cost(S)/OPT(G)$. The *price of stability* of the family of games $\mathcal{G}$ is $PoS(\mathcal{G}) = sup_{G \in \mathcal{G}} PoS(G)$.

A firmer notion of stability requires that a profile is stable against *coordinated deviations*. A set of players $\Gamma \subseteq \mathcal{J}$ forms a *coalition* if there exists a joint move where each job $j \in \Gamma$ strictly reduces its cost. When BRD is applied with

---

[1] Throughout this paper, we consider pure strategies, as is the case for the vast literature on cost-sharing games. Unlike mixed strategies, pure strategies may not be random, or drawn from a distribution.

coordinated deviations, in every step some coalition performs a joint beneficial move. A profile S is a *Strong Equilibrium* (SE) if there is no coalition $\Gamma \subseteq \mathcal{J}$ that has a beneficial joint move from S [7]. The *strong price of anarchy* (SPoA) and the *strong price of stability* (SPoS), introduced in [5], are defined similarly to the PoA and PoS in Definition 1, where $\Upsilon(G)$ refers to the set of strong equilibria.

1.2   Related Work

This paper links two well-studied areas (i) cost-sharing games, and in particular cost-sharing games with singleton strategies, and (ii) real-time scheduling, and in particular efficient energy allocation. Each of these areas has been widely studied. We survey below the papers we find most relevant to our work.

Game-theoretic analysis became an important tool for analyzing systems that are controlled by users with strategic consideration. In particular, systems in which a set of resources is shared by selfish users. *Congestion games* [12, 29,32] consist of a set of resources and a set of players who need to use these resources. Players' strategies are subsets of resources. In *cost-sharing games*, such as network formation games, each resource has an activation cost that is shared by the players using it according to some sharing mechanism. With unit-weight players and uniform cost-sharing, this is a potential game, a NE exists and the PoS is logarithmic in the number of players [6]. On the other hand, *Weighted* cost-sharing games, with proportional cost-sharing need not have a pure NE and the PoS may be as high as the number of players [6,16].

The paper [33] studies the complexity of equilibria in a wide range of cost-sharing games. The results on singleton cost-sharing games correspond to our model with unit-weight jobs. Other related work studies the impact of the strategies' combinatorial structure [1,17,25]. In a more general setting, players' strategies are multisets of resources. Thus, a player may need multiple uses of the same resource and his cost for using the resource depends on the number of times he uses the resource [8]. Job scheduling on unrelated machines is a special case of this class [9].

Variants of cost-sharing games have been the subject of extensive research. It is well-known that games with player-specific costs [31] as well as other sharing variants need not have a NE. Another line of research study the effect of different cost-sharing mechanisms on the equilibrium inefficiency [16,20,23,24]. A lot of attention has been given to scheduling congestion games (e.g., [18,34]), which can be thought of as a special case of weighted congestion games with singleton strategies. The paper [21] provides bounds on the PoS for singleton congestion games, with weighted and unweighted players.

The SPoA and SPoS measures were introduced by [5], which study a scheduling game, with the goal of minimizing the cost of the highest paying player. The SPoA and SPoS were studied also for job scheduling on unrelated machines [9], and for network formation games [3,5].

To the best of our knowledge, none of the above rich literature on job-scheduling games consider the special structure of players' strategies in the setting of *real-time* scheduling.

There is a wide literature also on *real-time scheduling*, either on a single or on parallel machines (see surveys in [14,26]). All previous work on real-time scheduling consider systems controlled by a centralized authority determining the jobs' assignment. We are not aware of any results in which this setting is analyzed as a non-cooperative game. When the server has a limited capacity, and jobs have variable weights, many problems such as minimizing the number of late jobs, or minimizing the servers' busy time are NP-hard, even with unit-length jobs [4,13]. On the other hand, with unit-weight unit-length jobs, these problems are polynomially solvable [10,14]. The papers [19,28] provide constant approximation algorithms for the minimum busy-time problem with variable-length, variable-weight jobs.

1.3 Our Results

We provide a complete analysis of equilibrium existence, computation, and inefficiency in real-time scheduling cost-sharing games. Our analysis distinguishes between instances with *unit slot-activation costs*, in which $c_t = 1$ for all $1 \leq t \leq T$, and instances with *unit job-weights*, in which $w_j = 1$ for all $j \in \mathcal{J}$. Specifically, we analyze the following four classes of games:

$\mathcal{G}_{1,1} = \{$games with unit slot-activation costs and unit job-weights$\}$.
$\mathcal{G}_{1,v} = \{$games with unit slot-activation costs and variable job-weights$\}$.
$\mathcal{G}_{v,1} = \{$games with variable slot-activation costs and unit job-weights$\}$.
$\mathcal{G}_{v,v} = \{$games with variable slot-activation costs and variable job-weights$\}$.

We first show that, independent of the instance class, any application of best-response dynamics, of unilateral or coordinated deviations, converges to a NE or a SE, respectively. Also, a SE can be computed efficiently. In addition, $\text{PoS}(\mathcal{G}_{1,v}) = 1$ and for this class we present an $O(n^2)$-time algorithm for computing, for any $G \in \mathcal{G}_{1,v}$, a NE profile $S^\star$ such that $cost(S^\star) = OPT(G)$. This result heavily exploits the interval-structure of the players' strategy space, and is in contrast to other singleton cost-sharing games, in which computing an optimal stable solution in NP-hard, even with unit-weight players [15]. The guaranteed existence of a SE is in contrast to other singleton cost-sharing games in which a SE may not exist [9]. Finally, we present an $O(n^2 + T)$-time algorithm for computing a social optimum profile for general instances.

In Sect. 3 we consider instances with unit slot-activation costs. While in many singleton cost-sharing games, $\text{PoA} = n$ even with unit-weight players, unit-cost resources, and a restricted strategy space [6,9], the PoA in our game is only $\Theta(\sqrt{n})$ with unit job-weights, and $n/2 + 1$ with variable job-weights, and its unique analysis relies on the interval-structure of the strategies.

In Sect. 4 we study instances with variable slot-activation costs. The bad news is that the limited strategy-structure does not help in reducing the PoA. That is, the PoA may be as high as the number of players, n, even if $\max_t c_t / \min_t c_t$ is arbitrarily close to 1. On the other hand, while in other singleton unweighted cost-sharing games $\text{PoS} = \Omega(\log n)$ [6], we show that $\text{PoS}(\mathcal{G}_{v,1})$ is the constant $\frac{8}{3}$.

Moreover, when combined with our algorithm for computing a social optimum, the PoS upper-bound proof is constructive.

Our results for the equilibrium inefficiency with respect to unilateral deviations are summarized in Table 1. All the bounds specified in the table are tight, and all PoS upper bounds are constructive, that is, for each of the four classes we present an algorithm for computing a NE whose cost is at most $PoS(\mathcal{G})\cdot OPT(G)$.

Table 1. Our results for equilibrium inefficiency with respect to unilateral deviations.

Slot-activation costs	Job-weights	Pure Nash Equilibrium	
		PoS	PoA
Unit	Unit	1	$\sqrt{4n+1}-1$
	Variable	1	$n/2+1$
Variable	Unit	8/3	n
	Variable	n	n

In Sect. 5 we study the equilibrium inefficiency with respect to coordinated deviations. By definition, for every game G, PoA(G) $\geq$ SPoA(G) $\geq$ SPoS(G) $\geq$ PoS(G). For instances with variable slot-activation costs, and variable job-weights, our analysis for unilateral deviations implies that all four measures are as high as the number of jobs.

For instances with unit slot-activation costs, our analysis of coordinated deviations is more positive and a bit surprising – showing no difference between unit and variable job-weights, and no difference between the worst and best strong equilibrium. Specifically, we show that $SPOA(\mathcal{G}_{1,v}) = SPOS(\mathcal{G}_{1,1})$ and both measures are a constant – arbitrarily close to 2. Combined with our convergence proof, we conclude that natural dynamics, even with coordinated deviations allowed, are guaranteed to converge to a solution whose cost is less than $2OPT$. This result distinguishes our game from other games in which the strong price of anarchy was analyzed and shown to be either equal to the PoS ($O(\log n)$ in network formation games, and $O(n)$ in scheduling on unrelated machines) or to 1 (single-source connection games) [2,5].

In general, our results show that games in which the players' strategies have an interval structure, are more stable than general singleton cost-sharing games, the loss due to selfish behavior is smaller, and it is possible to compute efficiently a stable and optimal or close to optimal solutions. We conclude in Sect. 6 with some open problems and directions for future work. Due to space constraints, some of the proofs are omitted.

2   Equilibrium Existence and Computation

In this section we study the stability of real-time scheduling games. We first show that any application of best-response dynamics, with unilateral or coordinated deviations, converges to a NE or a SE, respectively. We then present an

$O(n^2)$ algorithm for calculating a strong equilibrium. Both results are valid for general instances – with variable slot-activation costs and variable job-weights. The algorithm generalizes an algorithm from [33] for finding a NE in unweighted singleton games.

Theorem 1. *For every $G \in \mathcal{G}_{v,v}$, any application of BRD, with unilateral or coordinated deviations, converges to a NE or a SE, respectively.*

Theorem 2. *For every $G \in \mathcal{G}_{v,v}$, a strong equilibrium exists, and can be computed efficiently.*

We turn to consider the class $\mathcal{G}_{1,v}$. We show that for any $G \in \mathcal{G}_{v,1}$, a NE assignment whose cost equals the social optimum exists, and can be computed in time $O(n^2)$.

Theorem 3. $PoS(\mathcal{G}_{1,v}) = 1$, *and for every $G \in \mathcal{G}_{1,v}$, a NE whose cost is $OPT(G)$ can be computed efficiently.*

Proof. We present an optimal algorithm that computes a NE solution whose cost is $OPT(G)$. It consists of two phases: In the first phase, a social optimum solution, $S^\star$, is computed. This solution is not necessarily a NE. In the second phase, the jobs are assigned in the busy slots of $S^\star$, such that the resulting schedule is stable.

Algorithm 1. Computes a NE schedule of cost $OPT(G)$ for $G \in \mathcal{G}_{1,v}$

1: Sort the jobs such that $d_1 \leq d_2 \leq \cdots \leq d_k$.
2: **while** there are unassigned jobs **do**
3:     Let j be the next unassigned job. Activate slot d_j and assign every job k such that $d_j \subseteq I_k$ in slot d_j.
4:     Remove the assigned jobs from the instance.
5: **end while**
6: Let $b_1, \ldots, b_m$ be the set of slots in which the server is busy.
7: Remove all the jobs from the server and reassign them as follows:
8: **while** there are unassigned jobs **do**
9:     For every slot b_i, let $A(b_i)$ be the total weight of jobs for which $b_i \subseteq I_j$.
10:     Let $i^\star = \arg\min_i c_{b_i}/A(b_i)$.
11:     Assign all jobs j for which $b_{i^\star} \subseteq I_j$ in slot $b_{i^\star}$.
12:     Remove the assigned jobs from the instance.
13: **end while**

The proof of the algorithm combines two claims. The first claim, whose proof is based on an exchange argument, shows that the number of slots open during the first phase is minimal. The second claim refers to the stability of the schedule produced in the second phase.

The first phase can be implemented in linear time after the jobs are sorted by due-dates and release-times, and it therefore takes $O(n \log n)$. The calculation and the updates of $A(b_i)$ take $O(n^2)$. Thus, the time complexity of Algorithm 1 is $O(n^2)$ and is independent of T.

Note that the resulting schedule does not produce a strong equilibrium. In Sect. 5 we show that $SPOS = 2$. Specifically, Algorithm 1 fails when coordinated deviations are allowed, since moving to an idle slot may be beneficial for a coalition, but never for a single job.

Finally, we consider the problem of computing a (not necessarily stable) social optimum profile for instances with variable activation costs.

Theorem 4. *For every* $G \in \mathcal{G}_{v,v}$, *a profile whose cost is* $OPT(G)$ *can be computed efficiently.*

Proof. Let $G \in \mathcal{G}_{v,v}$. Since the server's capacity on each slot is not limited, the social optimum is independent of the jobs' weights. Also, we assume that for every two jobs j_1, j_2 it holds that if $r_{j_1} < r_{j_2}$ then $d_{j_1} \leq d_{j_2}$. In other words, no interval is contained in another (such an instance is commonly denoted *proper*). This assumption is w.l.o.g, since if $I(j_2) \subseteq I(j_1)$, then j_1 can be removed from the instance, and be assigned in $j_2's$ slot once the assignment is done.

By the above, the jobs in $\mathcal{J}$ can be sorted such that $r_1 \leq \cdots \leq r_n$ and $d_1 \leq \cdots \leq d_n$. Our algorithm is based on *dynamic programming*. For every $j_1 \leq j_2$, let

$$\alpha(j_1, j_2) = \begin{cases} min_{t \in \{r_{j_2}+1,\ldots,d_{j_1}\}} c_t & \text{if } r_{j_2} < d_{j_1} \\ \infty & \text{otherwise} \end{cases}$$

In words, $\alpha(j_1, j_2)$ is the cost of a cheapest slot in $I(j_1) \cap I(j_2)$. After the table α is computed, the algorithm advances by computing for every $1 \leq j \leq n$ the minimal cost $C(j)$ of an assignment of jobs $1, \ldots, j$. The base case is $C(0) = 0$. Then, for $j = 1, \ldots, n$, let

$$C(j) = \min_{k < j} C(k) + \alpha(k+1, j).$$

That is, for every $k < j$, we consider the cheapest assignment in which the rightmost busy slot processes the jobs $\{k+1, \ldots, j\}$, and select the cheapest among these candidates. In particular, $C(n)$, is the social optimum.

Standard DP backtracking can be used to retrieve the busy slots (rather than their costs). The calculation of $\alpha(j_1, j_2)$ for all $1 \leq j_1 \leq j_2 \leq n$ takes time $O(T + n^2)$. Calculating $C(j)$ takes $O(j)$, for a total of $O(n^2)$ for the whole table C. Thus, the total time complexity of the algorithm is $O(T + n^2)$.

3   Instances with Unit Slot-Activation Costs

3.1   Unit Job-Weights, Unit Slot-Activation Costs

This section discusses the equilibrium inefficiency of the class $\mathcal{G}_{1,1}$. Being a subclass of $\mathcal{G}_{1,v}$, Theorem 3 implies that $\mathrm{PoS}(\mathcal{G}_{1,1}) = 1$. We show that the interval-structure of the players' strategies, limits the PoA to $\Theta(\sqrt{n})$.

Theorem 5. $PoA(\mathcal{G}_{1,1}) = \sqrt{4n+1} - 1$.

Proof. We begin with the lower bound. Let $n = h^2 + h$ for some integer h. We present a game $G \in \mathcal{G}_{1,1}$ for which $OPT(G) = 1$ and some NE profile has cost $2h = \sqrt{4n+1} - 1$. An example for $n = 20$ and $h = 4$ is presented in Fig. 1. The game is played over n unit-weight jobs. For $i = 0 \ldots, h-1$, the set $\mathcal{J}$ includes $h - i$ jobs for which $I_j = [i, h+1)$. For $i = 1, \ldots, h$, the set $\mathcal{J}$ includes i jobs for which $I_j = [h, h+1+i)$.

In our example, the interval of each of the 4 jobs assigned in slot 1 is $[0, 5)$. Symmetrically, the interval of each of the 4 jobs assigned in slot 9 is $[4, 9)$, and so on. Note that for all $j \in \mathcal{J}$ it holds that $[h, h+1) \subseteq I(j)$, thus, an optimal solution assigns all the jobs in slot $h+1$ (slot 5 in our example). A possible NE profile S assigns i jobs for $i = 1, \ldots h$, in each of the slots $h - i + 1$ and $h + i + 1$. The profile S is a NE, since jobs can only migrate towards slot $h+1$, that is, to slots with a lower load. Since the server is busy in $[0, h)$ and $[h+1, 2h)$, $cost(S) = 2h$ and the PoA bound follows.

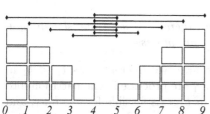

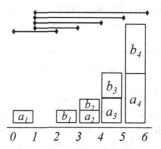

Fig. 1.   A   NE   achieving PoA $= \sqrt{4n+1} - 1 = 8$ for $n = 20$ unit-weight jobs. The jobs' intervals are shown above the schedule.

Fig. 2. A NE achieving PoA $= \frac{n}{2} + 1 = 5$ for $n = 8$ variable-weight jobs.

For the upper bound, let $S^\star$ be a social optimum schedule and assume that $cost(S^\star) = m$. Let $b_1 < b_2 < \cdots < b_m$ be the sequence of slots in which the server is busy in $S^\star$. Let S be a NE schedule. Partition the jobs into at most $2m$ sets $L_1, R_1, \ldots, L_m, R_m$, in the following way: For every job j, let $s_j^\star \in \{b_1, \ldots, b_m\}$ be the slot in which Job j is processed in $S^\star$, and let s_j be the slot in which j is processed in S. If $s_j \leq s_j^\star$, then let i be the minimal index such that $s_j \leq b_i$, and let $j \in L_i$. In other words, j belongs to the L-set of the earliest busy slot in $S^\star$ that can process it. Symmetrically, if $s_j > s_j^\star$, then let i be the maximal index such that $t_j > b_i$, and let $j \in R_i$. In other words, j belongs to the R-set of the latest busy slot in $S^\star$ that can process it. The partition into sets implies that if $j \in L_i \cup R_i$ then Job j can be processed in slot b_i, that is, $b_i \subseteq I(j)$.

The following observations will be used in our analysis. The structure of S is sketched in Fig. 3. We first show that the loads on non-idle slots accommodating jobs from the same set form a strictly decreasing or increasing sequence.

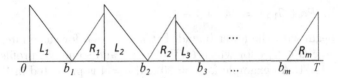

Fig. 3. The structure of a NE given that in an optimal solution the server is busy in slots $b_1, b_2, \ldots, b_m$.

Observation 6. *In S, if two slots $t_1 < t_2$ both accommodate jobs from L_i, then $\ell_{t_1}(S) > \ell_{t_2}(S)$. If two slots $t_1 < t_2$ both accommodate jobs from R_i, then $\ell_{t_1}(S) < \ell_{t_2}(S)$.*

Proof. Assume by contradiction that in some NE, S, two jobs $\{j_1, j_2\} \subseteq L_i$ are processed in different slots, $t_1 < t_2$ such that $\ell_{t_1}(S) \leq \ell_{t_2}(S)$. Since $t_1 < t_2$, we have that t_2 is closer to b_i. Since $b_i \subseteq I_{j_1}$ and $t_1 < t_2 \leq b_i$, it must be that $t_2 \subseteq I_{j_1}$, thus, j_1 can migrate to t_2 and reduce its cost to $\frac{1}{\ell_{t_2}(S)+1} < \frac{1}{\ell_{t_1}(S)}$, contradicting the stability of S. The analysis for R_i is symmetric (note that the word 'symmetric' is accurate here).

Observation 7. *In S, for every $1 \leq i \leq m$, there is at most one slot in $[b_i + 1, b_{i+1})$ in which jobs from both R_i and L_{i+1} are processed.*

Proof. Assume by contradiction that there are two different slots $t_1 < t_2$ in $[b_i, b_{i+1})$, in which jobs from both R_i and L_{i+1} are processed. The partition into sets implies that moving to the right, towards b_{i+1}, is feasible for $j \in L_{i+1}$, and moving to the left, towards b_i, is feasible for every $j \in R_i$. In particular, some job currently assigned in t_2 can migrate to t_1 and some job, currently assigned in t_1 can migrate to t_2. This implies that S cannot be a NE - as a job from a least loaded slot among t_1 and t_2 can perform a beneficial move.

We conclude that S has the following structure: during $[0, b_1)$, jobs from L_1 are processed in some slots with decreasing loads. During $[b_1 + 1, b_2)$, jobs from R_1 are processed in some slots with increasing loads, then a single slot may process jobs from $R_1 \cup L_2$, and then jobs from L_2 are processed in some slots with decreasing loads. This middle slot with the jobs from $R_1 \cup L_2$ has the maximal load. The same structure continues until, during $[b_m + 1, T)$ jobs from R_m are processed in some slots with increasing loads.

In the sequel, we assume that no slot accommodates jobs from both R_i and L_{i+1}. It can be shown that an instance with such a slot, t, can be replaced by an instance in which all the jobs processed in t are from R_i (specifically, their interval is $(b_i, t]$), with the same social optimum and the same worst NE.

Observation 8. *If k jobs are assigned on h slots with distinct loads then $h \leq \frac{1}{2}(\sqrt{8k + 1} - 1)$.*

Proof. The number of slots is maximized if the loads are $1, 2, \ldots, h$. Thus, in order to utilize h slots, at least $\sum_{i=1}^{h} i = \frac{1}{2}(h^2 + h)$ jobs are required, implying $h \leq \frac{1}{2}(\sqrt{8k+1} - 1)$.

Let $f(k) = \frac{1}{2}(\sqrt{8k+1} - 1)$. By Observation 8, and the structure of S, at most $f(L_1)$ slots are busy in $[0, b_1)$, at most $f(R_m)$ slots are busy in $[b_m + 1, T)$, and for every $1 \leq i < m$, at most $f(|R_i|) + f(|L_{i+1}|)$ slots are busy in $[b_i + 1, b_{i+1})$.

Given that $OPT = cost(S^\star) = m$, the PoA is at most $\frac{1}{m} \sum_{i=1}^{m}(f(|L_i|) + f(|R_i|))$. Since $f(k)$ is convex and $\sum_{i=1}^{m}(|L_i| + |R_i|) = n$, by Jensen's inequality [27], the PoA gets its maximal value when $m = 1$ and $L_1 = R_1 = n/2$. Specifically, for every $G \in \mathcal{G}_{1,1}$, we have $PoA(G) \leq 2f(n/2) = 2 \cdot \frac{1}{2}(\sqrt{4n+1} - 1) = \sqrt{4n+1} - 1$.

3.2   Variable Job-Weights, Unit Slot-Activation Costs

We turn to analyze instances with variable job-weights. Here again, the PoA is lower than n - the PoA in general cost-sharing games with singleton strategies, however, it is still $\Theta(n)$.

Theorem 9. $PoA(\mathcal{G}_{1,v}) = \frac{n}{2} + 1$.

Proof. We begin with the upper bound and show that $PoA(G) \leq \frac{n}{2} + 1$ for every $G \in \mathcal{G}_{v,1}$. First note that if the social optimum assigns the jobs on two or more slots, then $PoA(G) \leq n/2$ follows form the fact that the maximal cost of a solution is n. Assume that $OPT(G) = 1$, and let t be a slot such that $t \subseteq I(j)$ for every $j \in \mathcal{J}$. Assume by contradiction that in some NE profile S, the jobs are assigned on at least $\frac{n}{2} + 2$ slots. This implies that for at least three slots, a single job is assigned in each of these slots. Moreover, at least two of these three slots are either in $[t-1, T)$, or in $[0, t)$. Assume w.l.o.g., that two jobs j_1, j_2, are assigned alone on two different slots $t_1 < t_2$ in $[0, t)$. Since slot t is feasible for both jobs, the job assigned on t_1 can join the job on t_2. This migration reduces its cost from 1 to $w(j_1)/(w(j_1) + w(j_2))$, contradicting the assumption that S is a NE.

We proceed to prove the lower bound. for every even integer n, we describe a game $G \in \mathcal{G}_{v,1}$ over n jobs, such that $PoA(G) = \frac{n}{2} + 1$. An example for $n = 8$ is given in Fig. 2. Let $n = 2z$. The set of jobs consists of z pairs, $a_1, b_1, \ldots, a_z, b_z$. Each of the four jobs a_1, b_1, a_2, b_2 has weight 1. For $3 \leq j \leq z$, $w(a_j) = w(b_j) = 2^{j-2}$. The intervals of the jobs are $I(a_1) = [0, 2)$ and $I(b_1) = [1, 2)$. For $2 \leq j \leq z$, Jobs $I(a_j) = I(b_j) = [1, j+2)$. Note that for all jobs $j \in \mathcal{J}, [1, 2) \subseteq I(\mathcal{J})$. Thus, $OPT(G) = 1$ is achieved by assigning all the jobs in the single slot $[1, 2)$. A possible NE leaves slot 2 idle and assigns a_1 in slot 1, b_1 in slot 3 and for $2 \leq j \leq z$, Jobs a_j and b_j are assigned in slot $j + 2$. We show that S is a NE: the cost for each of a_0 and b_0 is 1, however, these jobs cannot join any other job, as they can only move towards slot 2 which is idle. The other jobs are paired with an equal-weight job, so each has cost $1/2$. These jobs can move towards slot 2, but each of the busy slots they can move to has load not larger than their current pair's load. Thus, no migration is beneficial, and S is a NE. The social cost of S is $\frac{n}{2} + 1$, implying the lower bound of the PoA.

4   Instances with Variable Slot-Activation Costs

In this section we discuss the equilibrium inefficiency of the classes $\mathcal{G}_{v,1}$ and $\mathcal{G}_{v,v}$. As we show, allowing variable slot-activation costs, may increase significantly the equilibrium inefficiency, even if $\max_t c_t / \min_t c_t$ is arbitrarily close to 1. On the other hand, while the PoS equals $O(\log n)$ in other singleton unweighted cost-sharing games [6], the interval strategy structure of real-time scheduling game guarantees that with unit-weight players, the PoS is $O(1)$. Moreover, our proof is constructive. First, a social optimum profile is computed (as shown in Theorem 4), and then the SO is converted to a stable profile whose cost is at most $\frac{8}{3} \cdot OPT(G)$.

Theorem 10. $PoA(\mathcal{G}_{v,1}) = n$ and $PoS(\mathcal{G}_{v,1}) = \frac{8}{3}$.

Theorem 11. $PoA(\mathcal{G}_{v,v}) = n$ and for every $\varepsilon > 0$, there exists a game $G \in \mathcal{G}_{v,v}$ for which $PoS(G) = n - \varepsilon$.

5   Coordinated Deviations

In this section we study the equilibrium inefficiency with respect to coordinated deviations. By definition, for every game G, $PoA(G) \geq SPoA(G) \geq SPoS(G) \geq PoS(G)$. For general instances, the following upper bound follows from simple standard arguments, and the lower bound follows from Theorem 11.

Theorem 12. $SPoA(\mathcal{G}_{v,v}) < n$ and for every $\varepsilon > 0$, there exists a game $G \in \mathcal{G}_{v,v}$ for which $SPoS(G) \geq n - \varepsilon$.

For instances with unit slot-activation costs, we showed in Sect. 3 that PoS $= 1$ and the PoA is $\Theta(n)$ or $\Theta(\sqrt{n})$ depending on the uniformity of job-weights. Our analysis of the SPOA and SPOS is therefore a bit surprising - showing no difference between unit- and variable-weight jobs, and no difference between the worst and best strong equilibrium. All measures turned out to be the same constant – arbitrarily close to 2. Formally,

Theorem 13. $SPoA(\mathcal{G}_{1,v}) < 2$, and for every $\varepsilon > 0$, there exists a game $G \in \mathcal{G}_{1,1}$ for which $SPOS(G) \geq 2 - \varepsilon$.

6   Conclusions and Open Problems

In this paper we analyzed, using game theoretic tools, the server's activation cost in real-time job-scheduling systems. We showed that the limited interval-structure of players' strategies induces a game which is more stable than general singleton cost-sharing games. Specifically, a strong equilibrium exists even in the most general setting, and the equilibrium inefficiency bounds are significantly lower than in other singleton cost-sharing games with uniform-cost resources or unweighted players. Our results imply that if the system is controlled by rational selfish users, then the increase in its activation cost is limited. This is valid especially if the server's activation cost does not vary over time, or if clients have uniform resource demand, and even if users can form coalitions and coordinate their assignment.

This is the first work that studies real-time scheduling games, and it can be extended in various directions:

1. Consider games with negative congestion effect. In our setting, the slot-activation cost is shared by the jobs assigned in it, thus, jobs have an incentive to join other jobs. Games in which jobs' costs increases with the congestion require different analysis.
2. Study games with variable-length jobs, in which every job is associated with a processing time p_j, and should select its processing interval $[t_{j,1}, t_{j,2}) \subseteq [r_j, d_j)$ such that $t_{j,2} - t_{j,1} = p_j$. The cost of processing a job is the total cost of its process. With variable-length jobs, preemptions may be allowed, inducing a different game, in which the strategy space of job j consists of all subsets of size p_j of $\{r_j + 1, r_j + 2, \ldots, d_j\}$.
3. Another interesting direction is to consider systems with limited server's capacity. Formally, for a given parameter B, at most B jobs may be processed in every slot. In this setting, the cost-sharing mechanism should also handle the challenge of convergence to a feasible solution.

References

1. Ackermann, H., Röglin, H., Vöcking, B.: On the impact of combinatorial structure on congestion games. J. ACM **55**(6), 25:1–25:22 (2008)
2. Adany, R., Tamir, T.: Algorithms for battery utilization in electric vehicles. Appl. Artif. Intell. **28**(3), 272–291 (2014)
3. Albers, S.: On the value of coordination in network design. SIAM J. Comput. **38**(6), 2273–2302 (2009)
4. Albers, S.: Energy-efficient algorithms. Commun. ACM **53**(5), 86–96 (2010)
5. Andelman, N., Feldman, M., Mansour, Y.: Strong price of anarchy. Games Econ. Behav. **65**(2), 289–317 (2009)
6. Anshelevich, E., Dasgupta, A., Kleinberg, J., Tardos, E., Wexler, T., Roughgarden, T.: The price of stability for network design with fair cost allocation. SIAM J. Comput. **38**(4), 1602–1623 (2008)
7. Aumann, R.: Acceptable points in general cooperative n-person games. In: Contributions to the Theory of Games IV, vol. 4 (1959)
8. Avni, G., Kupferman, O., Tamir, T.: Network-formation games with regular objectives. J. Inf. Comput. **251**, 165–178 (2016)
9. Avni, G., Tamir, T.: Cost-sharing scheduling games on restricted unrelated machines. Theor. Comput. Sci. **646**, 26–39 (2016)

10. Baptiste, P.: Batching identical jobs. Math. Methods Oper. Res. **52**(3), 355–367 (2000)
11. Bar-Noy, A., Guha, S., Naor, J., Schieber, B.: Approximating the throughput of multiple machines in real-time scheduling. SIAM J. Comput. **31**(2), 331–352 (2001)
12. Caragiannis, I., Flammini, M., Kaklamanis, C., Kanellopoulos, P., Moscardelli, L.: Tight bounds for selfish and greedy load balancing. Algorithmica **61**(3), 606–637 (2011)
13. Chang, J., Erlebach, T., Gailis, R., Khuller, S.: Broadcast scheduling: algorithms and complexity. ACM Trans. Algorithms **7**(4), 47:1–47:14 (2011). https://doi.org/10.1145/2000807.2000815. Article No. 47
14. Chang, J., Gabow, H.N., Khuller, S.: A model for minimizing active processor time. Algorithmica **70**(3), 368–405 (2014)
15. Chekuri, C., Chuzhoy, J., Lewin-Eytan, L., Naor, J., Orda, A.: Non-cooperative multicast and facility location games. IEEE J. Sel. Areas Commun. **25**(6), 1193–1206 (2007)
16. Chen, H., Roughgarden, T.: Network design with weighted players. Theory Comput. Syst. **45**(2), 302–324 (2009)
17. de Jong, J., Klimm, M., Uetz, M.: Efficiency of equilibria in uniform matroid congestion games. In: Gairing, M., Savani, R. (eds.) SAGT 2016. LNCS, vol. 9928, pp. 105–116. Springer, Heidelberg (2016). https://doi.org/10.1007/978-3-662-53354-3_9
18. von Falkenhausen, P., Harks, T.: Optimal cost sharing for resource selection games. Math. Oper. Res. **38**(1), 184–208 (2013)
19. Flammini, M., Monaco, G., Moscardelli, L., Shachnai, H., Shalom, M., Tamir, T., Zaks, S.: Minimizing total busy time in parallel scheduling with application to optical networks. Theor. Comput. Sci. **411**(40–42), 3553–3562 (2010)
20. Fotakis, D., Kontogiannis, S., Spirakis, P.: Selfish unsplittable flows. Theor. Comput. Sci. **348**(2–3), 226–239 (2005)
21. Gairing, M., Schoppmann, F.: Total latency in singleton congestion games. In: Deng, X., Graham, F.C. (eds.) WINE 2007. LNCS, vol. 4858, pp. 381–387. Springer, Heidelberg (2007). https://doi.org/10.1007/978-3-540-77105-0_42
22. Gkatzelis, V., Kollias, K., Roughgarden, T.: Optimal cost-sharing in general resource selection games. J. Oper. Res. **64**(6), 1230–1238 (2016)
23. Harks, T., Klimm, M.: On the existence of pure nash equilibria in weighted congestion games. Math. Oper. Res. **37**(3), 419–436 (2012)
24. Harks, T., Miller, K.: The worst-case efficiency of cost sharing methods in resource allocation games. Oper. Res. **59**(6), 1491–1503 (2011)
25. Ieong, S., McGrew, R., Nudelman, E., Shoham, Y., Sun, Q.: Fast and compact: a simple class of congestion games. In: Proceedings of the 20th AAAI, pp. 489–494 (2005)
26. Irani, S., Pruhs, K.R.: Algorithmic problems in power management. SIGACT News **36**(2), 63–76 (2005)
27. Jensen, J.L.W.V.: Sur les fonctions convexes et les ingalits entre les valeurs moyennes. Acta Math. **30**, 175–193 (1906)
28. Khandekar, R., Schieber, B., Shachnai, H., Tamir, T.: Real-time scheduling to minimize machine busy time. J. Sched. **18**(6), 561–573 (2015)
29. Koutsoupias, E., Papadimitriou, C.: Worst-case equilibria. Comput. Sci. Rev. **3**(2), 65–69 (2009)
30. Leung, J., Kelly, L., Anderson, J.H.: Handbook of Scheduling: Algorithms, Models, and Performance Analysis. CRC Press Inc., Boca Raton (2004)

31. Milchtaich, I.: Congestion games with player-specific payoff functions. Games Econ. Behav. **13**(1), 111–124 (1996)
32. Rosenthal, R.W.: A class of games possessing pure-strategy nash equilibria. Int. J. Game Theory **2**, 65–67 (1973)
33. Syrgkanis, V.: The complexity of equilibria in cost sharing games. In: Saberi, A. (ed.) WINE 2010. LNCS, vol. 6484, pp. 366–377. Springer, Heidelberg (2010). https://doi.org/10.1007/978-3-642-17572-5_30
34. Vöcking, B.: Selfish load balancing. In: Algorithmic Game Theory. Cambridge University Press (2007)

Abstracts

How to Hide in a Network?

Francis Bloch[1], Bhaskar Dutta[2], and Marcin Dziubiński[3]([⊠])

[1] Université Paris 1 and Paris School of Economics,
48 Boulevard Jourdan, 75014 Paris, France
`francis.bloch@univ-paris1.fr`
[2] University of Warwick and Ashoka University, Coventry CV4 7AL, UK
`b.dutta@warwick.ac.uk`
[3] Institute of Informatics, University of Warsaw, Banacha 2, 02-097 Warsaw, Poland
`m.dziubinski@mimuw.edu.pl`

We propose a model of strategic hiding in a network in face of a hostile authority. Given a set of nodes, the hider chooses a network over these nodes together with a node. The network chosen by the hider is observed by the seeker (the hostile authority) but the location choice is not observed. The seeker chooses one of the nodes in the network to inspect. The inspected node is removed from the network. If the hider hides in the inspected node or one of its neighbours, he is caught by the seeker and suffers a penalty. Otherwise, he enjoys the benefits from the network that are a convex and increasing function of the number of nodes (including himself) that the hider can access (directly or not) in the network. This form of network benefits, first proposed by [3], is in line with the celebrated Metcalfe's law, where the function is identity. The objectives of the seeker are to minimize the payoff of the hider and the proposed model takes the form of a two-stage zero-sum game.

The hide and seek stage in our model is similar to the hide and seek games on graphs of [2], with the difference that in their case the penalty from being caught is 0 and benefits from not being caught are fixed and independent of the graph. Unlike in the model of [1], in our model the authorities choose their seeking strategy knowing the network and only one node chooses the network topology to hide himself. This is similar to the model of [4]. However, unlike in their model, the authorities are strategic and they take into account the incentives and strategic behaviour of the hider when choosing the seeking strategy. Although very stylised and simple, the model allows us to capture the trade-off between secrecy and network benefits.

We provide optimal networks for the hider and characterize optimal strategies of the two players on these networks. In general, the optimal networks consists of a number of singleton nodes and a connected component which is either a cycle or a core-periphery network. If the component is a cycle, in equilibrium the hider mixes uniformly across its nodes. If the component is a core-periphery network, the hider mixes uniformly across the periphery nodes. This provides theoretical support to the claim that the hider chooses networks where his centrality is small and indistinguishable from the centralities of the other nodes.

This work was supported by Polish National Science Centre through Grant 2014/13/B/ST6/01807.

G. Christodoulou and T. Harks (Eds.): WINE 2018, LNCS 11316, pp. 441–442, 2018.
https://doi.org/10.1007/978-3-030-04612-5

Full version of the paper is available at: http://www.mimuw.edu.pl/
~amosild/hnsnet-full.pdf.

References

1. Baccara, M., Bar-Isaac, H.: How to organize crime. Rev. Econ. Stud. **75**(4), 1039–1067 (2008)
2. Fisher, D.: Two person zero-sum games and fractional graph parameters. Congressus Numerantium **85**, 9–14 (1991)
3. Goyal, S., Vigier, A.: Attack, defence, and contagion in networks. Rev. Econ. Stud. **81**(4), 1518–1542 (2014)
4. Waniek, M., Michalak, T., Wooldridge, M., Rahwan, T.: Hiding individuals and communities in a social network. Nat. Hum. Behav. **2**, 139–147 (2018)

Multiplicative Pacing Equilibria in Auction Markets

Vincent Conitzer[1], Christian Kroer[2]([✉]), Eric Sodomka[2],
and Nicolas E. Stier-Moses[2]

[1] Econorithms LLC, and Duke University, Durham, USA
`vincent.conitzer@duke.edu`
[2] Core Data Science, Facebook Inc., Cambridge, USA
{`chrkroer,sodomka,nstier`}`@fb.com`

Abstract. Budgets play a significant role in real-world sequential auction markets such as those implemented by Internet companies. To maximize the value provided to auction participants, spending is smoothed across auctions so budgets are used for the best opportunities. Motivated by a mechanism used in practice by several companies, this paper considers a smoothing procedure that relies on *pacing multipliers*: on behalf of each bidder, the auction market applies a factor between 0 and 1 that uniformly scales the bids across all auctions. Reinterpreting this process as a game between bidders, we introduce the notion of *pacing equilibrium*, and prove that they are always guaranteed to exist. We demonstrate through examples that a market can have multiple pacing equilibria with large variations in several natural objectives. We show that pacing equilibria refine another popular solution concept, competitive equilibria, and show further connections between the two solution concepts. Although we show that computing either a social-welfare-maximizing or a revenue-maximizing pacing equilibrium is NP-hard, we present a mixed-integer program (MIP) that can be used to find equilibria optimizing several relevant objectives. We use the MIP to provide evidence that: (1) equilibrium multiplicity occurs very rarely across several families of random instances, (2) static MIP solutions can be used to improve the outcomes achieved by a dynamic pacing algorithm with instances based on a real-world auction market, and (3) for our instances, bidders do not have an incentive to misreport bids or budgets provided there are enough participants in the auction.

Keywords: ad auctions · Budget constraints · Internet advertising Pacing · Repeated auctions

This work was done while the first author was visiting Facebook Core Data Science. A full version of this paper can be found at https://arxiv.org/abs/1706.07151.

G. Christodoulou and T. Harks (Eds.): WINE 2018, LNCS 11316, p. 443, 2018.
https://doi.org/10.1007/978-3-030-04612-5

Trading Networks with Bilateral Contracts

Tamás Fleiner[1], Zsuzsanna Jankó[2], Akihisa Tamura[3],
and Alexander Teytelboym[4(✉)]

[1] Budapest University of Technology and Economics,
Műegyetem rkp. 3, Budapest 1111, Hungary
fleiner@cs.bme.hu
[2] Corvinus University, Fővám tér 8, Budapest 1093, Hungary
jzsuzsy@cs.elte.hu
[3] Department of Mathematics, Keio University,
3-14-1 Hiyoshi, Kohoku-ku, Yokohama 223-8522, Japan
aki-tamura@math.keio.ac.jp
[4] Department of Economics, University of Oxford,
10 Manor Road, Oxford OX1 3UQ, UK
alexander.teytelboym@economics.ox.ac.uk

Abstract. We consider a model of matching in trading networks in which firms can enter into bilateral contracts. In trading networks, *stable* outcomes, which are immune to deviations of arbitrary sets of firms, may not exist. We define a new solution concept called *trail stability*. Trail-stable outcomes are immune to consecutive, pairwise deviations between linked firms. We show that any trading network with bilateral contracts has a trail-stable outcome whenever firms' choice functions satisfy the full substitutability condition. For trail-stable outcomes, we prove results on the lattice structure, the rural hospitals theorem, strategy-proofness, and comparative statics of firm entry and exit. We also introduce *weak* trail stability which is implied by trail stability under full substitutability. We describe relationships between the solution concepts.

Keywords: Matching markets · Market design · Trading networks
Supply chains · Trail stability · Weak trail stability · Chain stability
Stability · Contracts

Fleiner: Research was supported by the OTKA K108383 research project and the MTA-ELTE Egerváry Research Group; part of the research was carried out during two working visits at Keio University. Jankó: Research was supported by the OTKA K109240 research project and the MTA-ELTE Egerváry Research Group. Tamura: Research was supported by Grants-in-Aid for Scientific Research (B) from JSPS. Teytelboym: This work was supported by the Economic and Social Research Council grant number ES/R007470/1.

© Springer Nature Switzerland AG 2018
G. Christodoulou and T. Harks (Eds.): WINE 2018, LNCS 11316, p. 444, 2018.
https://doi.org/10.1007/978-3-030-04612-5

Markets for Public Decision-Making

Nikhil Garg, Ashish Goel, and Benjamin Plaut[✉]

Stanford University, Stanford, CA, USA
{nkgarg,ashishg,bplaut}@stanford.edu

A public decision-making problem consists of a set of issues, each with multiple possible alternatives, and a set of competing agents, each with a preferred alternative for each issue. We study adaptations of market economies to this setting, focusing on binary issues. Issues have prices, and each agent is endowed with artificial currency that she can use to purchase probability for her preferred alternatives (we allow randomized outcomes). We first show that when each issue has a single price that is common to all agents, market equilibria can be arbitrarily bad.

This negative result motivates a different approach. We present a novel technique called *pairwise issue expansion*, which transforms any public decision-making instance into an equivalent Fisher market, the simplest type of private goods market. This is done by expanding each issue into many goods: one for each pair of agents who disagree on that issue. We show that the equilibrium prices in the constructed Fisher market yield a *pairwise pricing equilibrium* in the original public decision-making problem which maximizes Nash welfare. More broadly, pairwise issue expansion uncovers a powerful connection between the public decision-making and private goods settings; this immediately yields several interesting results about public decisions markets, and furthers the hope that we will be able to find a simple iterative voting protocol that leads to near-optimum decisions.

The full version of the paper can be found at https://arxiv.org/pdf/1807.10836.pdf.

Acknowledgements. This research was supported in part by NSF grant CCF-1637418, ONR grant N00014-15-1-2786, and the NSF Graduate Research Fellowship under grants DGE-114747 and DGE-1656518.

G. Christodoulou and T. Harks (Eds.): WINE 2018, LNCS 11316, p. 445, 2018.
https://doi.org/10.1007/978-3-030-04612-5

Information Signal Design for Incentivizing Team Formation (Extended Abstract)

Chamsi Hssaine[(✉)] and Siddhartha Banerjee

School of Operations Research and Information Engineering,
Cornell University, Ithaca, NY 14853, USA
{ch822,sbanerjee}@cornell.edu

Abstract. We study the use of Bayesian persuasion (i.e., strategic use of information disclosure/signaling) in endogenous team formation. This is an important consideration in settings such as crowdsourcing competitions, open science challenges and group-based assignments, where a large number of agents organize themselves into small teams which then compete against each other. A central tension here is between the strategic interests of agents who want to have the highest-performing team, and that of the principal who wants teams to be balanced. Moreover, although the principal cannot choose the teams or modify rewards, she often has additional knowledge of agents' abilities, and can leverage this information asymmetry to provide signals that influence team formation. Our work uncovers the critical role of self-awareness (i.e., knowledge of one's own abilities) for the design of such mechanisms. For settings with two-member teams and binary-valued agents, we provide signaling mechanisms which are asymptotically optimal when agents are agnostic of their own abilities. On the other hand, when agents are self-aware, then we show that there is no signaling mechanism that can do better than not releasing information, while satisfying agent participation constraints.

Our work focuses on the use of strategic signaling for incentivizing team formation. The main idea is that many strategic settings have an inherent information asymmetry, where the principal has more information than the participating agents. We seek to understand if there is any way of leveraging this information asymmetry to *influence endogenous team formation*, with the objective of creating balanced teams.

We consider a setting with n agents who form teams of two, leading to some utility for both the agents and the principal. The teams are chosen endogenously by the agents, in the form of a *stable matching*; the principal however can influence agents' preferences via strategic release of information. Each agent has an intrinsic (numerical) type, drawn from some publicly-known prior. Crucially, we assume that each agent's type is *known to the principal*, but *unknown to other agents*. Moreover, an agent's utility is an increasing function of her and her teammates' types, while the principal's utility function depends on the set of resulting teams, and favors having more 'balanced' teams; thus, the principal's and agents'

© Springer Nature Switzerland AG 2018
G. Christodoulou and T. Harks (Eds.): WINE 2018, LNCS 11316, pp. 446–447, 2018.
https://doi.org/10.1007/978-3-030-04612-5

incentives are misaligned. We focus on settings with a binary type-space $\{0, 1\}$, and a constant number (K) of prior distributions.

Any signaling policy designed by the principal must induce a *stable matching* of agents, as well as obey *individual rationality* constraints, which enforce that each agent be weakly better off by agreeing to receive the signal. We show that it is enough to restrict to signals that are rank-orderings of agents according to expected posterior types. For $K \geq 1$ prior distributions, we propose the *Cluster First Best* signaling policy, in which agents with types drawn from the same prior distribution are always matched to agents of opposing realized type (i.e., high-type agents are always matched to low-type agents, and vice versa). Our main results are the following:

Theorem 1. *When agents do not know their own types, Cluster First Best is asymptotically optimal in n.*

Theorem 2. *When agents do know their own types, no signaling policy can do better than random matching.*

Our results indicate the importance of self-awareness in determining the success of signaling mechanisms. Showing this strategy is asymptotically optimal requires a novel dual-certification argument, which may be useful in related settings. Moreover, our work provides important insights and techniques for the design of Bayesian persuasion schemes for general team formation settings, as well as more general bipartite matching settings. For details, refer to our full version: https://arxiv.org/abs/1809.00751.

Acknowledgment. We acknowledge support from the National Science Foundation (grant DMS 1839346), and the Army Research Office (grant W911NF-17-1-0094).

Revenue Management on an On-Demand Service Platform

Vijay Kamble[(⊠)]

Department of Information and Decision Sciences College of Business Administration,
University of Illinois at Chicago, Chicago, USA
kamble@uic.edu

Abstract. I consider the optimal hourly (or per-unit-time in general) pricing problem faced by a worker (or a service provider) on an on-demand service platform. Service requests arriving while the worker is busy are lost forever. Thus, the optimal hourly prices need to capture the average hourly opportunity costs incurred by accepting jobs. Due to potential asymmetries in these costs, price discrimination across jobs based on duration, characteristics of the arrival process, etc., may be necessary for optimality, even if the customers' hourly willingness to pay is believed to be identically distributed. I first establish that such price discrimination is not necessary if the customer arrival process is Poisson: in this case, the optimal policy charges an identical hourly rate for all jobs. This result holds even if the earnings are discounted over time. I then consider the case where the customers belong to different classes that are differentiated in their willingness to pay. I present a simple and practical iterative procedure to compute the optimal prices in this case under standard regularity assumptions on the distributions of customer valuations.

Keywords: Optimal pricing · On-demand services

A full draft of the paper is available at https://arxiv.org/abs/1803.06797.

G. Christodoulou and T. Harks (Eds.): WINE 2018, LNCS 11316, p. 448, 2018.
https://doi.org/10.1007/978-3-030-04612-5

Implementing the Lexicographic Maxmin Bargaining Solution

Ashish Goel and Anilesh K. Krishnaswamy[✉]

Stanford University, Stanford, USA
{ashishg,anilesh}@stanford.edu

Abstract. A major question which bargaining theory deals with is that of implementation – designing a mechanism for which a desired bargaining solution is the unique subgame perfect outcome, with each player having complete information, and the social planner/designer having no knowledge of the players' preferences. There has been much work on exhibiting mechanisms that implement various bargaining solutions, in particular the Kalai-Smorodinsky solution and the Nash Bargaining solution. However, to the best of our knowledge, there is no known (subgame perfect) implementation of the lexicographic maxmin solution.

The lexicographic maxmin solution is obtained by a repeated application of the maxmin criterion: first, selecting feasible outcomes that maximize the utility of the worst-off player, then, among these outcomes, selecting those that maximize the utility of the next worst-off player, and so on. The utility gains are measured with respect to the disagreement point. The lexicographic maxmin solution has also had a long history outside of the literature on bargaining. It corresponds directly to the notion of maxmin fairness which has been extensively studied in network routing, bandwidth allocation and other resource allocation problems.

This paper is devoted to designing a mechanism for the (subgame perfect) implementation of the lexicographic maxmin solution. We do so by first defining the *Knockout* mechanism on any two given outcomes. This construction is based on a novel notion, namely *disagreement dominance* (a relation defined on pairs of vectors), which we believe is interesting in its own right. We then use the Knockout mechanism as a subroutine in constructing our full mechanism: a binary tree of games, where each node corresponds to a Knockout mechanism with two outcomes. The workings of our overall mechanism rely crucially on an original combinatorial result we establish, that the lexicographic maxmin solution disagreement dominates any other outcome.

A full draft of the paper is available at https://arxiv.org/pdf/1810.01042v1.pdf.

Ashish Goel—Supported in part by NSF grant CCF-1637418, ONR grant N00014-15-1-2786, and ARO grant W911NF-14-1-0526.

Anilesh K. Krishnaswamy—Also supported by the grants mentioned above.

G. Christodoulou and T. Harks (Eds.): WINE 2018, LNCS 11316, pp. 449–450, 2018.
https://doi.org/10.1007/978-3-030-04612-5

Our mechanism uses the standard assumption that the space of outcomes is such that in any player's best outcome, all the surplus goes to her, and every one else gets no utility. This assumption is commonplace in most of the literature on implementation of bargaining solutions.

Keywords: Bargaining · Implementation · Maxmin fairness
Mechanism design

Exploration vs. Exploitation in Team Formation

Ramesh Johari[1], Vijay Kamble[3(✉)], Anilesh K. Krishnaswamy[2],
and Hannah Li[1]

[1] Department of Management Science and Engineering,
Stanford University, Stanford, USA
{rjohari,hannahli}@stanford.edu

[2] Department of Electrical Engineering, Stanford University, Stanford, USA
anilesh@stanford.edu

[3] Department of Information and Decision Sciences,
University of Illinois at Chicago, Chicago, USA
kamble@uic.edu

Abstract. Modern labor platforms face the online learning problem of optimizing matches between jobs and workers of unknown abilities. This problem is complicated by the rise of complex jobs on these platforms that require teamwork, such as web development and product design. Successful completion of such a job depends on the abilities of all workers involved, which can only be indirectly inferred by observing the aggregate performance of the team. Observations of the performance of various overlapping teams induce correlations between the unknown abilities of different workers at any given time. Tracking the evolution of this correlation structure across a large number of workers on the platform as new observations become available, and using this information to adaptively optimize future matches, is a challenging problem.

To study this problem, we develop a stylized model in which teams are of size 2 and each worker is drawn i.i.d. from a binary (good or bad) type distribution. Under this model, we analyze two natural settings: when the performance of a team is dictated by its strongest member and when it is dictated by its weakest member. We find that these two settings exhibit stark differences in the trade-offs between exploration (i.e., learning the performance of untested teams) and exploitation (i.e., repeating previously tested teams that resulted in a good performance). We establish fundamental regret bounds and design near-optimal algorithms that uncover several insights into these tradeoffs.

Keywords: Team formation · Online learning · Online labor platforms

A full draft of the paper is available at https://arxiv.org/abs/1809.06937.

R. Johari—Supported by NSF Grant No. CNS-1544548 and the Stanford Cyber Initiative.

V. Kamble—Supported in part by ARO grant W911NF-14-1-0526.

A. K. Krishnaswamy—Supported in part by NSF grant CCF-1637418, ONR grant N00014-15- 1-2786, and ARO grant W911NF-14-1-0526.

H. Li—Supported by NSF Grant No. CNS-1544548, the Stanford Cyber Initiative, and the Stanford Management Science & Engineering Graduate Fellowship.

© Springer Nature Switzerland AG 2018
G. Christodoulou and T. Harks (Eds.): WINE 2018, LNCS 11316, p. 451, 2018.
https://doi.org/10.1007/978-3-030-04612-5

Author Index

Printed in the United States
By Bookmasters